Atomistic Simulation of Materials
Beyond Pair Potentials

Atomistic Simulation of Materials
Beyond Pair Potentials

Edited by

Vaclav Vitek
University of Pennsylvania
Philadelphia, Pennsylvania

and

David J. Srolovitz
The University of Michigan
Ann Arbor, Michigan

Plenum Press • New York and London

Library of Congress Cataloging in Publication Data

Atomistic simulation of materials: beyond pair potentials / edited by Vaclav Vitek and David J. Srolovitz.
p.   cm.
"Proceedings of an international symposium on atomistic simulation of materials: beyond pair potentials, a satellite conference of the ASM World Materials Congress, held September 25–30, 1988, in Chicago, Illinois" — T.p. verso.
Includes bibliographical references.
ISBN-13: 978-1-4684-5705-6   e-ISBN-13: 978-1-4684-5703-2
DOI: 10.1007/ 978-1-4684-5703-2
1. Metallography — Congresses. 2. Electronic structure — Congresses. I. Vitek, V. II. Srolovitz, David J.
TN689.2.A85   1989   89-16268
669′.95 — dc20   CIP

Proceedings of an international symposium on Atomistic Simulation of Materials: Beyond Pair Potentials, a satellite conference of the ASM World Materials Congress, held September 25–30, 1988, in Chicago, Illinois

© 1989 Plenum Press, New York
Softcover reprint of the hardcover 1st edition 1989
A Division of Plenum Publishing Corporation
233 Spring Street, New York, N.Y. 10013

PREFACE

This book contains proceedings of an international symposium on *Atomistic Simulation of Materials: Beyond Pair Potentials* which was held in Chicago from the 25th to 30th of September 1988, in conjunction with the ASM World Materials Congress. This symposium was financially supported by the Energy Conversion and Utilization Technology Program of the U. S Department of Energy and by the Air Force Office of Scientific Research. A total of fifty four talks were presented of which twenty one were invited.

Atomistic simulations are now common in materials research. Such simulations are currently used to determine the structural and thermodynamic properties of crystalline solids, glasses and liquids. They are of particular importance in studies of crystal defects, interfaces and surfaces since their structures and behavior play a dominant role in most materials properties. The utility of atomistic simulations lies in their ability to provide information on those length scales where continuum theory breaks down and instead complex many body problems have to be solved to understand atomic level structures and processes.

The success atomistic simulations have in mimicking nature is most often limited by the accuracy of the description of the interaction between atoms. While advances in the state of electronic structure calculations now make it possible to calculate total energies with tremendous accuracy, these methods (whether real space cluster methods or reciprocal space supercell methods) are still largely inapplicable to atomistic simulations needed in materials science owing to their computer resource imposed limitations on the number of independent atoms which may be accounted for.

Historically, pair-wise interactions (i.e. pair potentials) were first employed in atomistic simulations. In most cases, these have been obtained by fitting empirical potential forms to a variety of experimental data while in some cases such potentials have also been derived from electronic theory (e.g. from pseudo-potential methods for simple metals). In situations where the bonding is predominantly covalent, three-body (or bond bending) potentials have been used. Calculations employing pair potentials have revealed a number of important structural characteristics many of which now became generally accepted concepts. However, pair potentials possess severe limitations, the most prominent being that they are not applicable to the situations where the density of the material differs significantly from that of a chosen reference state. This is, of course, the case for many crystal defects, interfaces and in particular at surfaces. In the latter case, for example, pair potentials yield atomic relaxations in complete contradiction to experimental findings.

Recently, new descriptions of atomic interactions have come to use the common denominator of which is that they include the many body nature of bonding in the condensed matter. On the empirical side the most popular are the Embedded Atom Method and the Finnis-Sinclair type N-body potentials, two approaches which are closely related. They are principally applicable to metals but analogous developments are now taking place for semiconductors and ionic crystals. An approach more directly based on the theoretical quantum mechanical description of atomic interactions is the tight binding method which is being applied with increasing regularity. There are many different realizations of this approach and the method itself is under active, widespread investigation. The fully self-consistent total energy calculations are, of course, also utilized with increasing frequency in atomistic simulations.

The purpose of this symposium was to bring together a wide spectrum of researchers who share an interest in descriptions of atomic interactions. Participants range from those involved in highly accurate electronic structure methods to those working on empirical descriptions of atomic interactions and researchers who carry out atomistic simulations as part of their research. The materials of interest to the participants included metals, ceramics, and semiconductors. The symposium thus brought together those whose research leads to the development of new, more accurate descriptions of atomic interactions with those who apply these developments in studies of materials properties. This interaction is most important for further development of this field and we believe these proceedings will serve the same purpose.

We wish to extend our sincere gratitude to Dr. J. Eberhart of the US Department of Energy and Dr. A. Rosenstein of the US Air Force Office of Scientific Research whose encouragement and help was invaluable for the success of the symposium. We also wish to thank most sincerely Ms. Denice Gilbert whose typing skill and patience made the preparation of these proceedings possible.

Vaclav Vitek, Philadelphia
David J. Srolovitz, Ann Arbor
April, 1989

CONTENTS

TOTAL ENERGY AND FORCE CALCULATIONS WITH

THE LMTO METHOD

O.K. Andersen, M. Methfessel, C.O. Rodriguez,
P. Blöchl, and H.M. Polatoglou

Max-Planck-Institut für Festkörperforschung
D-7000 Stuttgart 80
Federal Republic of Germany

INTRODUCTION AND OVERVIEW

During the past 15 years it has become possible to perform quantum-mechanical calculations of many properties of simple materials with good accuracy *using as input merely the positions of the atoms and the atomic numbers and masses.* In particular, low-temperature structural properties of pure crystals and their surfaces have been obtained with astonishing accuracy through calculation of the total energy as a function of the atomic positions. Also, the Fermi surfaces of metals, the magnitude and order of magnetic moments in transition metals and many of their alloys, as well as important aspects of the electronic and atomic structures of impurities in metals and semiconductors have been accurately reproduced.[1,2,3]

This advance has largely been based on the *density-functional formalism*[4,5] which reduces the problem of finding properties of the many-electron ground state to that of solving Schrödingers equation for one electron moving in the electrostatic potential from the nuclei and a potential from the electrons. The latter potential is local, i.e. it is the same $V_e(v)$ for all electrons, and it must be determined self consistently with the electron density that it generates through solution of the Schrödinger-equation and subsequent filling of the one-electron states according to Fermi-Dirac statistics. The exact form of this potential is unknown, but the so-called local density approximation (LDA) provides an educated - and, in particular for structural properties, highly successful - guess for it.

The second prerequisite for this advance has, needless to say, been electronic *computers*, and the third has been the development of powerful *methods for solving the self-consistent one-electron problem.* Here, essentially two lines of thought have been followed. In one, originating from Fermi,[6] the core-electron degrees of freedom are neglected in the solid and the potential acting on a valence electron is substituted by a *pseudopotential* whose wavefunctions are nodeless inside the atoms. As a result, wave (LAPW) *method*,[9] which is presumably the most accurate, generally applicable method presently available.[14] It is, however, much more cumbersome to work with sets have been enormously successful in reproducing and predicting structural properties of materials with broad sp-like valence bands.[1] The pseudopotential, however, becomes deep for materials containing less broad bands, such as oxides, halides, transition-metal, rare-earth and actinide compounds; so deep, that the use of unsophisticated basis sets becomes very costly, and the use of PW's often impossible. A material like SiO_2, for instance, is presently at the border-line and requires about 400 PWs per atom.[7]

The other line of thought, going back to Slater,[8] came from the desire to be able to compute the electronic structures also for those materials. Based on the observation that the potential acting the electrons is nearly spherically symmetric near each atom, the idea was to approximate this potential by its so-called *muffin-tin* (MT) average which is spherically symmetric inside non-overlapping MT-spheres and flat in between. For such a potential the solutions of Schrödingers equation for the solid can then be constructed from the solutions for each MT-sphere of the radial Schrödinger equations, which are trivial to integrate numerically. There is no need to eliminate the atomic oscillations of the wavefunctions through pseudizing the potential, and there is no need to exclude (semi-) core electrons from the calculation.

In the so-called *linear* band-structure methods[2,9] this MT-potential is used to construct a relatively small, *dedicated basis set* which is then used to solve Schrödingers equation. The *linear muffin-tin-orbital* (LMTO) *method*, together with its descendant the augmented spherical wave (ASW) method,[10] use typically 9 orbitals per MT-well (i.e., a 1s1p1d basis) and they have been very successful in calculations of electronic, cohesive, and magnetic properties for a large number of s-, p-, d-, or f-band materials.[2] The LMTO method has been used for very large supercells, the largest being a 212 atom model for amorphous Si,[11] as well as for Greens-function calculations for localized and extended defects. The LMTO set may also be transformed exactly into short ranged, so-called first-principles tight-binding basis,[2] and this has been used together with the recursion method for topologically disordered systems,[12] and together with the coherent-potential approximation (CPA) for substitutionally disordered alloys.[13]

In all these applications the LMTO method has, however, always been used with the so-called *atomic-spheres approximation* (ASA) in which the Coulomb energies are calculated after the charge density has been spheridized inside "space-filling" - and hence slightly overlapping - MT-spheres. (In the present paper we define the ASA to have the so-called combined correction[9] included). For open structures, space-filling with acceptable overlaps can only be obtained by including spheres also at interstitial sites. It is obvious that the ASA makes it impossible to calculate structural energy differences associated with symmetry-lowering displacements of the atoms. Such energies are the center of interest in the present symposium, and their calculation seems to require treating the full, non-spheridized charge density and potential.

Until now this has only been done successfully with the *linear augmented-plane-*

Schrödingers equation can often be solved using *unsophisticated basis functions* such as plane waves (PWs) or Gaussians. First-principles pseudopotentials and plane-wave basis LAPWs than with PWs, and the full-potential LAPW method is therefore mostly used in those cases where the number of PWs are inhibiting. For closely packed materials the number of LAPWs needed per atom is between 30 (sp-electrons) and 100 (f-electrons). When going to open structures, the increase in the number of basis functions is, of course, the same for LAPWs as for PWs. An example of a calculation which could only be performed with the LAPW method is the recent first-principles calculation[15] of a few zone-center and zone-boundary phonons for La_2CuO_4.

Current interest is moving towards *modeling*, rather than reproducing, the structures and properties of materials. The density-functional calculations of the past 15 years have let us experience that the physical properties depend crucially on where we put the atoms, so that a first goal must be to *calculate structures*. Fortunately, this is where the LDA has proven most reliable. In order to model the structures of amorphous systems and liquids, surfaces and interfaces, the structures and dynamics of localized and extended defects, melting and crystal growth it is necessary to treat at the order of at least 100 atoms per cell and to let them move through a large number of configurations, possibly with the techniques of molecular dynamics or simulated annealing. This is too demanding for conventional, self-consistent density-functional calculations.

Biswas and Hamann[16] tried to solve this problem for Si by first designing a *classical, inter-atomic three-body potential* which reproduced LDA calculations of the total energy as a function of volume (see Fig. 2) for eight different 4- to 12-fold coordinated crystal structures,[17] plus a surface structure, and then to use this interatomic potential for molecular dynamics studies. It turned out that this classical potential was unable to account properly for the stability of the covalent bond.

Quantum-mechanical calculations are most simply performed with the *empirical tight-binding method* in which the total energy is the sum of a quantum-mechanical electronic hopping term, taken for instance from Harrisons parameterization,[18] plus an inter-atomic two-body repulsion fitted to experiments or to LDA calculations.[19] With this form of the total energy it is, however, not possible to fit to the LDA energy-volume curves for the eight crystalline Si phases particularly well. The empirical tight-binding method was recently used for extensive studies of Si grain boundaries using the technique of molecular statics.[20]

Four years ago Car and Parinello succeeded in performing the first, *unified molecular-dynamics density-functional* (MD-DF) *calculation*.[21] Their method consisted in a number of clever techniques by the help of which they, for instance, were able to avoid matrix diagonalizations (a process which scales with the number of basis functions cubed) as well as electronic self-consistency iterations at each step of the ions. Unfortunately, many of these techniques were only applicable (and needed?) for calculations using PWs and pseudopotentials local in momentum space. The restriction to local pseudopotentials has now been relaxed to separable pseudopotentials,[7] and MD-DF calculations have been performed for ≈ 60-atom supercell models for amorphous and liquid Si,[22] for similar sized models for the Ge (001) Σ = 5 twist boundary[23] and for the Ge (100) surface,[24] for small (<20 atom) Si,[25] S, and Se clusters,[26] and for a 12-atom SiO_2 supercell.[7] Still, there seems to be a long way to go before MD-DF calculations can be performed for anything but broad-band materials.

In this paper we take a few steps in this direction by first showing that, for self-consistent LDA calculations, the LMTO method *without the ASA*, can yield total energies to 10meV accuracy for low symmetry situations using an atom centered LMTO basis with about 20 orbitals per atom (e.g., a 3s3p2d basis) and *no orbitals at interstitial sites*.[27] The power of this LMTO technique is demonstrated by entering the home-ground of the first-principles pseudopotentials[1,17] and calculating the energy-volume curves for the crystalline Si phases,[28] as well as, in the diamond phase, the elastic constants, the frequencies of the phonons at Γ and X, including the splitting of the degeneracy of the LTO(Γ) phonon under tetragonal shear, and the mode Grüneisen parameters.[29] Secondly, we discuss two ways in which self-consistency iterations may be omitted in structural energy calculations. One[30] is a procedure in which atom-centered, spherical, overlapping potentials are rigidly moved ("frozen-potential approach"), and the other[31] is based on the approximate energy-functional of Harris[32] with the electron density obtained by superposition of atomic densities.

These new total energy calculations are admittedly for very small systems, but they may be nearly as fast as conventional LMTO-ASA calculations.[33] With the real-space techniques developed for the tight-binding LMTO representation it should soon be possible to treat large supercells. Seen on the background of the general applicability and increasing popularity of the LMTO-ASA method, we believe that MD-DF calculations for general systems are slowly coming within reach.

THE SILICON BENCHMARK

A linear muffin-tin-orbital (LMTO) is a Hankel function of kinetic energy κ^2 times a spherical Harmonic in the interstitial region, and linear combinations of radial Schrödinger-or rather, scalar relativistic Dirac-solutions inside the MT-spheres. The matching at the sphere surfaces is continuous and once differentiable.

Conventional solid-state LMTOs form a single-kappa basis with $\kappa^2 \equiv 0$. Due to the restricted variational freedom in the interstitial region, the MT-spheres are chosen to be space-filling Wigner-Seitz spheres and, hence, slightly overlapping. It can be shown[34] that the one-electron potential actually solved for in an LMTO-ASA calculation is the *superposition* of the potentials $v_i(\underline{r})$ from the individual MT-wells, i.e.: $V(\underline{r}) = \Sigma_i [v_i(|\underline{r} - \underline{R}_i|) - V_{mtz}] + V_{mtz}$, where V_{mtz} is the MT-zero and $\underline{R}_i$ the sphere positions. This holds to leading (2nd) order in the sphere overlap, $\eta \equiv (s_1 + s_2 - d_{12})/d_{12}$. Here, s is a sphere-radius and d is the distance between two spheres. In practice, the errors are negligible when the overlap is below ≈ 15 percent. Larger overlaps are avoided by using, in addition to the atom-centered spheres and their LMTOs, interstitial spheres and LMTOs centered at interstitial sites. A MT-potential for Si in the cubic diamond structure thus has spheres at the Si- *and* at the tetrahedral interstitial sites. With equal-sized spheres this packing is body-centered cubic and the overlap is only 14 percent. In a conventional, self-consistent LMTO-ASA calculation the overlapping MT-potential is determined self-consistently to minimize the total energy of an ASA energy functional. The self-consistent overlapping MT-potential for Si in the cubic diamond structure is shown in Fig. 1 along the line Si-Si-E-E- in the [111]-direction. Here, E denotes a tetrahedral interstitial site. The figure starts at the left-hand side midway between two Si atoms, and it stops at the right-hand side midway between two interstitial sites. The dashed line is the MT-zero and the dashed curves are the spherical potentials in the Si and E MT-wells. The MT-potential, which is the superposition of the two, is given by the dot-dashed curve. Although discontinuous, this potential is very close to the full, self-consistent potential (FP)[35,36] shown by the full curve. Also shown in the figure are the extent of the Si valence band (B) and the lowest part of the conduction band (CB).

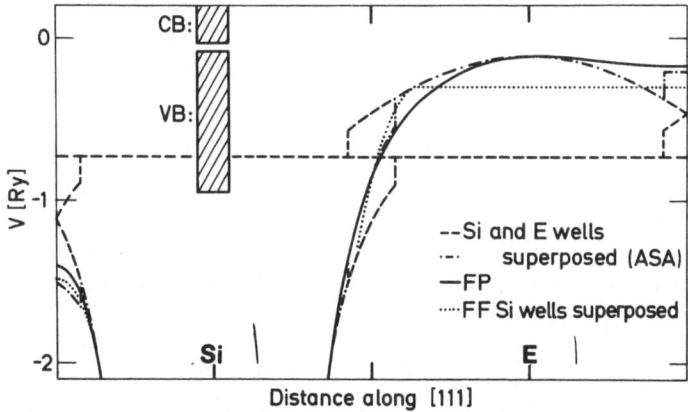

Fig. 1   One-electron potentials and bands in Si as explained in the text.

In Fig. 2 we show calculated total energy vs. volume curves for Si in the cubic diamond (dia), β-tin, simple cubic (sc), body-centered cubic (bcc), and face-centered cubic (fcc) crystal structures. The volumes are relative to V_0, the experimental zero-pressure volume of the diamond structure, and the total energies are per atom with the calculated energy minimum for diamond Si subtracted. At the bottom to the right are the first-principles pseudopotential results of Yin and Cohen.[17] Experimentally,[37] Si goes through the following pressure-induced phase transitions: *dia* → *β-tin* at ≈ 110 kbar with V_{dia}/V_0 ≈ at 0.91, *β-tin* → *simple hexagonal* at ≈ 140 kbar and V_{tin}/V_0 ≈ 0.71, *simple hexagonal* → ? at 340 kbar, ? → *hexagonal close-packed* at ≈ 400kbar, and *hcp* → *fcc* at 780 kbar and V_{hcp}/V_0 ≈ 0.48. These transitions are in good agreement with the common tangent of the energy-volume curves of Yin and Cohen, who for instance calculate a critical pressure of

90 kbar for the *dia* → β-*tin* transition, and in equally good agreement with the LMTO-ASA calculations of McMahan who used the so-called Ewald-correction to the total energy in the ASA.[38] In Fig. 2 we have not shown all the physically relevant phases because the physical situation has been discussed extensively in the literature.

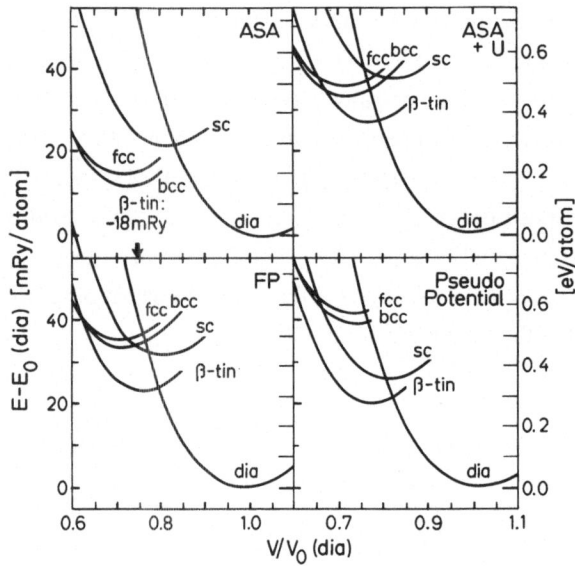

Fig. 2   Calculated energy versus volume curves for crystalline phases of Si.

The results of our conventional LMTO-ASA total-energy calculations are shown at the top left-hand part of Fig. 2. For the fcc, bcc, and β-tin structures we used no interstitial spheres, whereby the overlaps are 10, 14, and 24 percent, respectively. The sc and diamond structures were packed with equal-sized Si and interstitial spheres so that, in both cases, the packing is bcc and the overlap 14 percent. This conventional LMTO-ASAmethod gives good equilibrium volumes and bulk moduli, but some of the structural energy differences are highly inaccurate. One reason is that the ASA, i.e., the spheridization of the charges inside the spheres, and hence the approximation of the long-ranged part of the electrostatic energy by that of point charges, is too crude for the differences in Coulomb energies between different structures. As a result, the energy-volume curves for the closely packed fcc and bcc structures lie 0.3 eV - and that for the sc structure lie 0.1 eV - too low with respect to that for the diamond structure. This is the problem that McMahan[38] solved for the close-packed structures by use of the Ewald correction which substitutes the point-charge electrostatics by that of point charges on a homogeneous background. The second problem, dramatically demonstrated by the fact that the β-tin curve lies 0.5 eV too low with respect to the diamond curve, is that for some structures there is no natural way of filling space with spheres. For β-tin we chose to have no empty spheres and the resulting large overlap caused large kinetic-energy errors, in addition to the errors caused by the spheridization of the charge density. A similar situation arises if the calculation for the sc structure is performed without interstitial spheres. In that case the overlap increases from 14 to 24 percent and the total-energy (not shown in the figure) decreases by as much as 0.7 eV so that the minimum lies 0.1 eV below the minimum for the β-tin structure. The second reason for the inaccuracy of the structural energies is therefore that the conventional LMTO basis treats different structures with different accuracy depending on the sphere-packing. Now, the overlap may be reduced through the use of spheres which are not space-filling, but then the conventional single-kappa basis provides to little variational freedom in the interstitial region.

It thus seems obvious that for future molecular-dynamics calculations we must, first of all, evaluate the *Coulomb energy for the full charge density*, without spheridizing the latter. The schemes, which have been developed to do this, employ either plane-wave[35,36] or Hankel-function[33,39] fits to the smooth part of the charge-density. Secondly, we must remove the condition that the MT-spheres, used in the definition of the LMTO basis set, be space-filling. Thirdly, the use of empty spheres should be avoided. Hence, we must employ a set of LMTOs which is reasonably complete, also in situations where the interstitial region occupies a large fraction of the cell. Consequently, we shall use *atom-centered, double- or triple-kappa LMTO sets.*[28,39]

At the top-right part of Fig. 2 we show results of an approximation named ASA+U. This approximation for the total energy is, first of all, based on the observation that the non-spheridized charge density coming out of a conventional, self-consistent LMTO-ASA calculation is quite accurate.[40] The ASA+U thus primarily consists of evaluating the total-energy functional for this density, and it amounts to calculating the proper Coulomb energy (U) on top of a conventional, self-consistent LMTO-ASA calculation. Secondly, the ASA+U uses for the last LMTO calculation, which provides the kinetic energy, a double-kappa (2s2p2d) LMTO set constructed with non-overlapping spheres. This means that the potential-spheres overlap, but the orbital-spheres do not. The ASA potentials used were those from the ASA calculation in the top-left part of the figure, except for the sc structure where no interstitial spheres were used, i.e. interstitial spheres were used only for the diamond structure. The energy-volume curves resulting from the ASA+U are seen to be reasonably accurate. With this approximation we have also calculated phonon frequencies and Grüneisen parameters for the diamond structure and they turn out to have fair accuracy.[29] Such ASA+U frozen-phonon calculations are extremely fast because the potential wells (see Fig. 1) can be kept frozen, independent of the phonon displacements, so that no self-consistency iterations are needed.

The schemes[33,35,36] developed for fitting the smooth part of the charge density can also be used to evaluate all the matrix elements of the Hamiltonian for a general potential and, hence, to perform *full-potential* (FP) self-consistent calculations. In the following we shall refer to the results obtained with the scheme using Hankel-function fitting and a triple-kappa (3s3p2d) basis set of atom-centered LMTOs with no sphere overlap.[27] This set is seen to employ 22 orbitals per atom which is just slightly more than the 18 single-kappa LMTOs (Si 1s1p1d and E 1s1p1d) traditionally used for the diamond structure. The energy-volume curves shown at the bottom to the left in Fig. 2 are highly accurate; the *dia* $\rightarrow$ *β-tin* transition is calculated to occur at 115 kbar which is in close agreement with experiment. In the first three columns of Table 1 we compare for the diamond structure the experimental, the full-potential LMTO, and the Yin and Cohen[17] results for the equilibrium lattice constant, the cohesive energy, the bulk modulus, the elastic constants, a third-order elastic constant, the phonon frequencies at the Γ and X points, and the pressure derivatives of the latter, the so-called mode Grüneisen parameters. Except for the cohesive energy, our results are seen to be highly accurate. In Fig. 3 we show how the degenerate LTO (Γ) phonon splits by the application of a tetragonal strain. The measured strain derivative, indicated in the figure by the weak lines, is nearly twice the derivative calculated by Nielsen and Martin using first-principles pseudopotentials and about 270 PWs per atoms.[41] The LMTO result is seen to be in perfect agreement with the experiment.

The discrepancy between the experimental and the calculated cohesive energy seen in Table 1 is now known to be a failure of the LDA. We used the Hedin-Lundquist parameterization which is a more accurate parametrization of the exchange-correlation energy of the homogeneous electron gas than the Wigner parameterization used in the pseudopotential calculation.[17] Nevertheless, the Hedin-Lundquist parameterization is known to give a cohesive energy of Si which is 0.2 eV larger than the one obtained with the Wigner paramaterization. This is one reason for the difference between the cohesive energies calculated by us and by Yin and Cohen. The remaining discrepancy is due to incomplete basis-set convergence: Yin and Cohen found that their cohesive energy increased by 0.17 eV when increasing the basis from about 90 to about 200 PWs per atom. With the 22 LMTOs per atom we believe to be converged to better than 10 meV. This is demonstrated in Table 2 where we give the decrements in total energy for diamond Si as we add atom-centered orbitals to the basis. The decrements turn out to be fairly independent of the order in which the orbitals are added. The second and third κ^2 - values were uncritical, we used -1.0 and -1.3 Ry.

Our results shown so far were obtained by calculations of the total energy as a function of some coordinate. It is, however, often of great advantage if forces or stresses can be evaluated directly. The results of Nielsen and Martin were obtained by stress evaluation. In Fig. 4 we show a result of a force calculation for the LTO (Γ) phonon.[36] The force is evaluated as the Hellmann-Feynman force on the frozen core, and a problem is that the evaluation is only accurate if full-potential self-consistency has been reached. The result shown in Fig. 4 was obtained from a FP calculation using the single-kappa Si- and E-centered non-overlapping basis set and it is seen to be inferior to that obtained by the total-energy calculation and shown in Table 1 and Fig. 3. Our result "th2" contained a so-called Pulay correction[42] whereas "th1" did not.

In molecular dynamics calculations it is inconvenient to perform self-consistency iterations for the electrons at each step of the ions. We believe that *frozen-potential calculations*, like those described for the ASA+U phonon calculation[29] and somewhat similar to those often used in LMTO-ASA calculations for evaluation of energy differences

Table 1. Static and Dynamic Properties of Silicon in the Diamond Structure

	Exp	FP LMTO 22/Si	Pseu Pot ≈90/Si	FF LMTO 22/Si	H LMTO 36/Si
Lattice Constant (a_0)	10.26	10.23	10.30	10.53	10.20
Cohesive Energy (eV/Si)	4.63	5.16	4.67		
Bulk Modulus (Mbar)	0.98	0.99	0.98	0.89	1.01
Elastic Constants (Mbar):					
C_{11}- C_{12}	1.02	1.02	1.07	0.86	0.90
C_{44}	0.80	0.83		0.75	0.60
K_{xyz} (eV/a_0^3)	-35.1	-39.1	-32.8	-45.0	
Phonon Frequencies (THz):					
LTO (Γ)	15.53	15.52	15.16	15.29	15.11
TO (X)	13.90	13.75	13.48	13.44	12.70
LOA (X)	12.32	11.82	12.16	12.01	11.46
TA (X)	4.49	4.50	4.45	3.90	4.04
Grüneisen Parameters:					
LTO (Γ)	0.98	0.99	0.92		
TO (X)	1.5	1.51	1.34		
LOA (X)	0.9	1.03	0.92		
TA (X)	-1.4	-1.42	-1.50		

Pseu Pot refers to the DF calculations of Yin and Cohen.[17]
Pseu Pot/FP LMTO used Wigner/Hedin-Lundquist exchange-correlation.
Our calculated cohesive energy included contributions of -0.70 eV from spin-polarization of the atom and -0.07 eV from zero-point vibration in the solid.
FF and H refer to non-self-consistent calculations employing respectively, the Fat & Frozen potential and the Harris functional with superposed atomic densities.

Table 2. Convergence of Atom-Centered, Multiple-Kappa LMTO Basis Sets.

Lowering of total energy of Si as obtained by addition of further LMTOs to the 1s1p-set:

	+1d	+1f	+2s	+2p	+2d	+2f	+3s	+3p	+3d	+3f
meV	360	15	330	1150	50	5	170	70	2	0

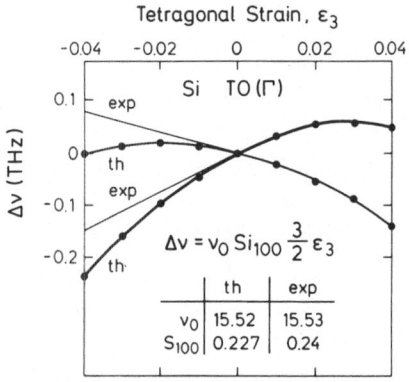

Fig. 3   Calculated[27] and experimental strain dependence of frequency of the
TO (Γ) phonon.

between close packed structures,[2,38] will be useful. In those applications a self-consistent
ASA potential is calculated for one structure and then exported to another structure with the
same volume. So going to full-potential self-consistency is not very important for the total
energy. This is entirely consistent with our finding that the ASA+U and the FP results in
Fig. 2 are rather similar. The remaining problem is that in a molecular dynamics calculation
close-packing and volume are not conserved locally. We therefore would like to have *one,*
frozen, atom-centered potential which can be used for *all* structures and volumes.[30] Such
a potential can not, of course, have lower point symmetry than the highest one encountered
and, hence, it must be spherical. In order that it can reproduce the actual low symmetry it
must be overlapping with a fair range. Now, as long as the potential-sphere only has a
slight overlap (< 15-20 percent) with the neighboring LMTO sphere, the Hamiltonian
matrix has the usual, simple ASA form and the errors are negligible. One may therefore
increase the range of the potential and still keep the simple form of the matrix elements,
provided that the LMTO-spheres are shrunk. For Si in its most open crystal structure,
cubic diamond, it turns out that with the 22 orbital basis set one may decrease the radius of
the LMTO-sphere to 90 percent of the touching sphere radius without notable loss of
accuracy; and this then allows one to increase the size of the potential-sphere to 130
percent. From the plot in Fig. 1 of the ASA potential along the bond and along the back-
bond it is now easy to construct a "fat and frozen" (FF) Si potential whose superposition
(dotted curve) on the diamond structure at the equilibrium volume reproduces the self-
consistent ASA (or full-potential reasonably well. The question is finally, how well this fat
and frozen potential does, say for the phase diagram and for the phonons. The answer,
given on the left hand side of Fig. 5 and in the 4th column of Table 1, is "not too bad",
considering the crude way in which this potential was constructed. It is conceivable that a
bit more of fitting would do the rest. Total energy calculations with this potential are
extremely fast.

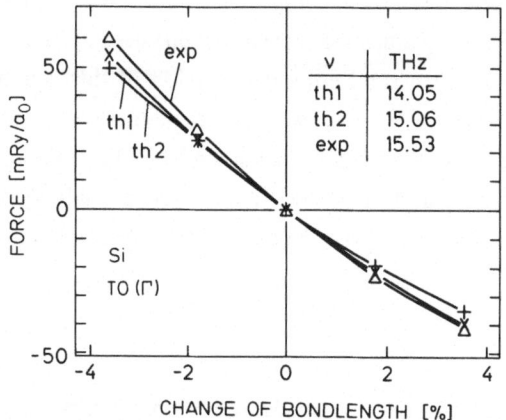

Fig. 4   Hellmann-Feynman forces calculated with (th1) and without (th2) correcting for Pulay forces.[36]

There is another approach to fast non-self-consistent calculations which we[31] have tried. Whereas in the frozen-potential approach one evaluates the energy functional for the output charge density of a one-electron calculation for a guessed potential, the approach invented by Harris[32] uses a guessed input charge density for the Coulomb and the exchange-correlation energies. The potential obtained from this density is used in the one-electron calculation, which then delivers the kinetic energy. If the charge density for the Harris approach is chosen to be the superposition of neutral atom densities, the Coulomb contribution to the energy is simply a repulsive pair-potential. The exchange correlation

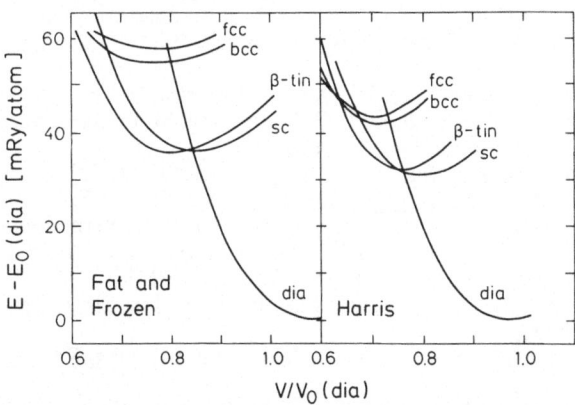

Fig. 5   Non-self-consistent calculations of the energy vs. volume for Si as explained in the text.

contribution unfortunately, has a more complicated appearance. Otherwise, the expression for the total energy looks very much the same as the one used in the empirical tight-binding scheme. In fact, the modern semi-empirical version of this scheme takes the Harris functional as its starting point.[19] Now, we have found that with the Harris approach it is exceedingly important to use the full potential in the LMTO calculation. When this is done we obtain the results shown at the right hand side of Fig. 5 and in the last column of Table 1. We believe that they have as good or better quality as those used in present day MD-DF calculations.

REFERENCES

1.   M.L. Cohen in: "Highlights of Condensed-Matter Theory", edited by F. Bassani, F. Fumi, and M.P. Tosi (North Holland, NY 1985).
2.   O.K. Andersen, O. Jepsen, and D. Glötzel in: "Highlights of Condensed-Matter Theory", edited by F. Bassani, F. Fumi, and M.P. Tosi (North Holland, NY 1985). O.K. Andersen, O. Jepsen, and M. Sob in: "Electronic Band Structure and Its Applications", ed. M. Yussouff (Springer Lecture Notes, 1987).
3.   M. Schlüter in: "Highlights of Condensed-Matter Theory", edited by F. Bassani, F. Fumi, and M.P. Tosi (North Holland, NY 1985).
4.   P. Hohenberg and W. Kohn, Phys. Rev. B 136:864 (1964); W. Kohn and L.J. Sham, Phys. Rev. A 140:1133 (1965).
5.   W. Kohn in: "Highlights of Condensed-Matter Theory", edited by F. Bassani, F. Fumi, and M.P. Tosi (North Holland, NY 1985).
6.   E. Fermi, Nuovo Cimento, 11:157 (1934).
7.   D.C. Allan and M.P. Teter, Phys. Rev. Lett. 59:1136 (1987).
8.   J.C. Slater, Phys. Rev. 51:151 (1937).
9.   O.K. Andersen, Phys. Rev. B 12:3060 (1975); H.L. Skriver in "The LMTO Method" (Springer, NY 1984).
10.   A.R. Williams, J. Kübler and C.D. Gelatt, Phys. Rev. B 19:6094 (1979).
11.   D. Sokolovski, O.K. Andersen and M. Methfessel, to be published.
12.   See for instance, S. Bose, S.S. Jaswal, O.K. Andersen, and J. Hafner, Phys. Rev. B 37:9955 (1988).
13.   J. Kudrnovsky, V. Drchal, and J. Masek, Phys. Rev. B 35:2487 (1987).
14.   E. Wimmer, H. Krakauer, M. Weinert, and A.J. Freeman, Phys. Rev. B, 24:864 (1981); L.F. Mattheiss and D.R. Hamann, Phys. Rev. B, 32:823 (1986).
15.   R.E. Cohen, W.E. Pickett, H. Krakauer, and L.L. Boyer, Physica C 153-155:202 (1988) and Phys. Rev. Lett. (1989) to be published.
16.   R. Biswas and D.R. Hamann, Phys. Rev. Lett. 55:2001 (1985).
17.   M.T. Yin and M.L. Cohen, Phys. Rev. B 25:7403 (1982); Phys. Rev. B 26:3259 (1982).
18.   W.A. Harrison, "Electronic Structure" (Freemann, San Francisco, 1980).
19.   See articles by D.J. Chadi, by D.G. Pettifor, by M. Foulkes, and by A.T. Paxton, in these Proceedings.
20.   A.T. Paxton and A.P. Sutton, J. Phys. C, 21:L481 (1988).
21.   R. Car and M. Parinello, Phys. Rev. Lett. 55:2471 (1985).
22.   R. Car and M. Parinello in: "Proceedings of the 18th International Conference on the Physics of Semiconductors, Stockholm 1986", edited by O. Engström (World Scientific, Singapore, 1987) Vol. 2, 1165.
23.   M.C. Payne, P.D. Bristowe, and J.D. Joannopoulos, Phys. Rev. B 58:1348 (1987).
24.   M. Needels, M.C. Payne, and J.D. Joannopoulos, Phys. Rev. Lett. 58:1765 (1987).
25.   P. Ballone, W. Andreoni, R. Car, and M. Parinello, Phys. Rev. Lett. 60:271 (1988).
26.   D. Hohl, R.O. Jones, R. Car, and M. Parinello, Chem. Phys. Lett. (1988).

27. M. Methfessel, C.O. Rodriguez, and O.K. Andersen, to be published.
28. C.O. Rodriguez, M. Methfessel, P. Blöchl, and O.K. Andersen, to be published.
29. C.O. Rodriguez, M. Methfessel, and O.K. Andersen, to be published.
30. M. Methfessel and O.K. Andersen, to be published.
31. H.M. Polatoglou and M. Methfessel, Phys. Rev. B 37:10403 (1988) and to be published.
32. J. Harris, Phys. Rev. B 31:1770 (1985); and M. Foulkes, PhD Thesis, Cambridge University, 1987 (unpublished).
33. M. Methfessel, Phys. Rev. B 38:1537 (1987); M. Methfessel and M. van Schilfgaarde, to be published.
34. O.K. Andersen, to be published.
35. K.H. Weyrich, Phys. Rev. B 37:10269 (1988).
36. P. Blöchl, PhD Thesis, Stuttgart University, 1988 (unpublished), and P. Blöchl and O.K. Andersen, to be published.
37. S.J. Duclos, Y.K. Vohra, and A. Ruoff, Phys. Rev. Lett. 58:775 (1987) and references therein.
38. A.K. McMahan and J.A. Moriarty, Phys. Rev. B 27:3235 (1983); A.K. McMahan, Phys. Rev. B 30:5835 (1984).
39. M. Springborg and O.K. Andersen, J. Chem. Phys. 87:7125 (1987).
40. O.K. Andersen, Z. Pawlowska, and O. Jepsen, Phys. Rev. B 34:5253 (1986).
41. O.H. Nielsen and R.M. Martin, Phys. Rev. B 32:3792 (1985).
42. P. Pulay, in: "Modern Theoretical Chemistry" (Plenum, NY, 1977), Vol. 4, p. 153.

CONCENTRATION DEPENDENT EFFECTIVE CLUSTER INTERACTIONS IN SUBSTITUTIONAL ALLOYS

A. Gonis and P.E.A. Turchi

Department of Chemistry and
Materials Science
Lawrence Livermore National Lab
Livermore, CA 94550

X.-G. Zhang

Department of Physics
and Astronomy
Northwestern University
Evanston, IL 60201

G.M. Stocks, D.M. Nicholson and W.H. Butler

Metals and Ceramics Division
Oak Ridge National Laboratory
Oak Ridge, TN 37831

INTRODUCTION

In attempting to understand the thermodynamic properties of substitutional alloys, and specifically to construct alloy phase diagrams, it is natural to seek recourse in the vast body of knowledge acquired with respect to magnetic (spin) systems, in particular the 3D Ising model.[1] As is well known, all thermodynamic properties of the Ising model can be calculated from the partition function

$$Z = e^{-\beta H}/\text{Tr}\,[e^{-\beta H}] \tag{1.1}$$

where $\beta = 1/k_B\,T$, with k_B being Boltzmann's constant and T denoting the absolute temperature, and where Tr denotes the trace with respect to configurations. In the Ising model the Hamiltonian, H, is taken to be of the familiar and "simple" form

$$H = -\sum_{i,j} J_{ij}\,\sigma_i\,\sigma_j \tag{1.2}$$

where J_{ij} is an exchange integral between spins on sites i and j (spin-spin interaction) and the occupation numbers σ_i denote the ζ component of the spin, $\sigma^i = 1(-1)$ for a spin pointing along the positive (negative) ζ axis. It is to be noted that in this model the magnetic interaction, J_{ij}, is independent of both temperature and of the spin configuration of the system.

At first glance, it might appear that the Hamiltonian of Eq. (1.1) could be used to describe a class of systems entirely different from the Ising model, namely, random, substitutional binary alloys, $A_c B_{1-c}$, where c denotes the concentration of atoms of kind A. In such a description, the occupation numbers σ^i could be interpreted as denoting the species of atom occupying site i, $\sigma^i = 1(-1)$ for an $A(B)$ atom at site i, and J_{ij} would be some effective interaction between atoms of different species on sites i and j. In fact, with J_{ij}^{AA}, J_{ij}^{AB} and J_{ij}^{BB} denoting the interaction between two A, an A and a B, and two B atoms at sites i and j, from Eq. (1.2) the configuration-dependent part of the energy can be written in the form,

$$H^{alloy} = \sum_{i,j} V_{ij} p_i p_j ,\qquad\qquad(1.3)$$

where $V_{ij} = (J_{ij}^{AA} + J_{ij}^{BB} - 2J_{ij}^{AB})/2$ is an effective pair interaction and the occupation numbers p_i take the values 1(0) for an $A(B)$ atom at site i. In principle, one could now attempt to treat Eq. (1.3) in the same manner as Eq. (1.2), and thus obtain all relevant thermodynamic information, e.g., phase stability, phase diagram for the alloy under study. Unfortunately, the treatment of realistic alloys is quite a bit more complicated than that.

There are two major discrepancies that quickly become evident when one attempts to use the Ising model-like model Hamiltonian, Eq. (1.2), to describe an alloy system. First, the magnetic interaction, J_{ij}, in that model is postulated to be independent of both the temperature and the number of spins in either direction (concentration). On physical grounds, there is no compelling reason to assume that interatomic interactions in alloys possess these properties; in fact, quite the contrary must be the case. Second, the Ising Hamiltonian is restricted to interactions between pairs of spins, whereas multisite interactions may play a crucial role in determining alloy properties. The dependence of atomic interactions on concentration and temperature is the more fundamental of these two effects and we should attempt to understand it before proceeding further.

A simple analogy with the way electrons interact both outside and inside a material might help illustrate this point. In vacuum, the interaction between two electrons a distance r apart is simply e^2/r (e = electronic charge). In a material the electron-electron interaction can be significantly modified by the presence of the ions and the other electrons, a modification manifested in the dielectric constant of the material. In a sense, asking for temperature and concentration independent atomic interactions in alloys is equivalent to asking for a material-independent dielectric constant! Thus, a rigorous treatment of binary (and multicomponent) alloys consists of two rather major tasks: The determination of unique, concentration and temperature dependent atomic interactions, and the development of methods for obtaining thermodynamic quantities in terms of such general parameters.

In the remainder of this lecture, we will be concerned primarily with part of the first, and somewhat easier, half of the problem, the accurate determination of concentration dependent multisite atomic interactions in alloys. We shall concentrate our discussion on phase transformations based on the parent lattice (coherent phase transformations), but many of our results can be readily extended to incoherent transitions. Furthermore, our discussion will be presented almost exclusively in terms of binary alloys, although a number of the formal results to be cited can be generalized to multicomponent systems in a straightforward way.

MULTISITE EFFECTIVE INTERATIONS

As mentioned above, the basic tools for studying ordering and phase stability in substitutional alloys, and ultimately their phase diagrams are based on the 3-D Ising model. Implementation of this model requires the use of effective interactions in alloys, and in this section we review some of the methods that have been devised for the determination of such interactions. We shall restrict our discussion to methods that can be employed within a first-principles approach and which can be extended in a conceptually and computationally straightforward way to the calculation of multisite (triplet, quadruplet) effective interactions as well as effective pair interactions.

The basis idea behind the determination of effective interactions is that the internal energy can be written as a rapidly convergent sum of multisite interactions. This statement has been demonstrated to be true in the case of normal metals and their alloys[2] through the use of pseudopotential theory. In the case of transition metal alloys, where strong disorder effects can be important, such an expansion of the total energy has not been justified. The success of phenomenological theories for the study of ordering processes has initiated extensive work in order to provide a direct link between electronic structure calculations and statistical models. We shall discuss specifically three methods aiming at establishing this link, namely, the Connolly-Williams method (CWM)[3], the generalized perturbation method (GPM)[4], and the related embedded cluster method (ECM).[5] These three methods have the common characteristic of attempting a direct determination of multisite interactions in real space. An alternative approach, based on the concentration functional theory of ordering,[6] also called the theory of concentration waves (CW's), can also be used to yield effective pair interactions in reciprocal space. Numerical calculations[5] have illustrated that Fourier transformations into real space of the pair-correlation functions obtained in the CW's method are essentially indistinguishable from the effective pair-interactions obtained directly within the ECM (or the GPM). In a practical sense, the theory of CW's can be expected to be more useful than real-space methods in cases in which effective interactions extend over long distances, e.g., Cu Pd alloys.[7] On the other hand, the determination of higher-order interactions, known to be necessary for the determination of accurate alloy phase diagrams in many cases, becomes increasingly cumbersome within concentration functional theory as the order of the interaction increases. As we are particularly interested in the determination of higher-order effective interactions, and because CW's will be addressed in another lecture in these proceedings, we will not discuss the method of concentration waves any further.

Having obtained a set of multisite interactions, it remains to specify the statistical model to be used to calculate thermodynamic properties of interest. In this regard, the cluster variation method (CVM)[8] and Monte Carlo simulations[9] can be mentioned. These techniques have evolved to the stage in which they can incorporate both concentration dependent interactions, to the determination of which we now turn.

The Method of Connolly and Williams

In the approach proposed by Connolly and Williams,[3] and used in a number of applications,[10] it is assumed that the total energy for a given configuration, c, of a binary alloy can be written in the form,

$$E^{(c)}(r) = \sum_{\gamma} V_{\gamma}(r) \, \xi_{\gamma}^{c} \, . \tag{2.1}$$

Here, $V_{\gamma}(r)$ and ζ_{γ}^{c} are the *configuration-independent* multisite interactions and the multisite correlation functions, respectively, associated with a cluster γ, r denotes the lattice parameter, and the sum runs over all cluster types (the "empty lattice" included) on a fixed lattice. The correlation functions are given by

$$\xi_{\gamma}^{c} = \frac{1}{N_{\gamma}} \sum_{\{n_i\}} \sigma_{n_1} \sigma_{n_2} \cdots \sigma_{n_{\gamma}} \, , \tag{2.2}$$

where σ_n takes the values +1 or -1 depending on the occupation of site n, N_{γ} is the number of sites in cluster γ, and the sum runs over all γ^{th} order clusters of a given type on the lattice.

It is possible to ascribe a concentration dependence to the effective interactions, $V_{\gamma}(r)$, because the lattice constant is determined self-consistently at each concentration. However, in spite of this desirable feature and of recent applications of the CWM, this method is beset by a number of conceptual and computational difficulties. The most important among these regards the uniqueness of the interactions, which in the CWM are determined through a fitting to the total energies corresponding to selected ordered structures. In other words, the question arises whether or not the cluster interactions depend on the specific ordered structures contrary to the assumption of configuration independence underlying Eq. (2.1). An equally relevant question can be raised regarding the rate of convergence of Eq. (2.1). Finally, what effect does the configuration dependence of the atomic interactions have on the calculated alloy phase diagrams, known to depend rather critically on those interactions?

The results of calculations[11] to be discussed in the following sections indicate that in fact Eq. (2.1) is poorly justified, if at all. In particular, one can conclude that the condition of rapid convergence of the expansion in Eq. (2.1) is not completely fulfilled, and (ii) that the V_{γ}, within the same cluster truncation, are not unambiguously defined. Furthermore, the calculation of total energies within the CWM suffers from a lack of uniqueness, an effect which can adversely affect the determination of alloy phase diagrams. By contrast, the effective cluster interactions determined within the GPM and the ECM do not suffer from such ambiguities, as is discussed in the following subsection.

Methods Based on the Coherent Potential Approximation - the GPM and the ECM

The difficulties associated with the uniqueness of the effective interactions determined within the CWM can be overcome through the use of methods which allow a proper and direct treatment of the random part of the disorder characterizing a substitutional alloy. Here, the central idea is to begin by determining an appropriate reference medium which properly interpolates between the various alloy configurations, and to study statistical fluctuations by means of either perturbation theory, as is done in the GPM, or through the exact treatment of cluster configurations in the alloy, the technique underlying the ECM. Both of these methods rely on the coherent potential approximation (CPA)[12,13,14] for the determination of a proper effective medium to describe the completely disordered state of an alloy. Because the CPA has been reviewed[14] rather extensively in the literature, we shall present only a condensed discussion of the method emphasizing those aspects that are most convenient for the introduction of the GPM and the ECM. Although we shall base our discussion on a first-principles, multiple scattering formalism, on occasion we shall also indicate the forms taken by the various formulae in the case of a tight-binding (TB) description of the Hamiltonian of an alloy. Hopefully, this correspondence will help illustrate the generality of the underlying formalism and clarify the ensuing discussion.

The Coherent Potential Approximation

It is now generally acknowledged that an application of the CPA, within the Green-function formalism introduced by Korringa[15], Kohn and Rostoker[16] in the study of ordered materials, (KKR-CPA), provides a reliable first-principles description of electronic structure and related properties of substitutionally disordered alloys. In the KKR-CPA method, the one electron random alloy potentials, which for simplicity are taken to be of the non-overlapping muffin tin form, are characterized at energy ε by the partial wave scattering amplitudes $f_{i,L}(\varepsilon)$ (i refers to the species and L stands for the angular momentum indices l,m), which are supposed to be configurationally independent. The reference medium is described within the CPA by an ordered array of effective potentials which are self-consistently defined by the effective wave scattering amplitudes $f_{c,LL'}(\varepsilon)$ (the index c refers to quantities calculated in the framework of the CPA), obtained as the solutions of the KKR-CPA equation:

$$m_{c;n,L} = c_A m_{A;n,L} + c_B m_{B;n,L} + (m_{A;n,L} - m_{c;n,L}) \, \tau_{LL'}^{c,nn} \, (m_{B;n,L'} - m_{c;n,L'}) \qquad (2.3)$$

where m_α ($\alpha = c, A$ or B) is the inverse of the "on-the energy shell" t matrix ($f = - \sqrt{\varepsilon} \, t$), and is diagonal in the angular momentum space for spherical muffin-tin potentials; and $\tau_{LL'}^{c,nn}$ are the site diagonal "on the energy shell" matrix elements of the scattering path operator introduced by Gyorffy and Stott.[17] These last elements are given by the inverse of the real-space matrix M,

$$\tau_{LL'}^{c,nm} = [M_c^{-1}]_{LL'}^{nm} \qquad (2.4)$$

where

$$M_{LL'}^{nm} = t_{c;n,L}^{-1} \, \delta_{nm} \, \delta_{LL'} - G_{LL'}^{nm}$$

with the usual definition for the free electron structure constants $G_{LL'}^{nm}$.[14]

The above self consistency equation can be formally rewritten as [18,19]

$$\sum_{i=A,B} c_i \, \chi_{n,L}^j \, \delta_{nm} \, \delta_{LL'} = 0 \qquad (2.5)$$

where

$$\chi_{n,L}^i = \frac{\Delta \, m_{i;n,L}}{1 - \Delta \, m_{i;n,L} \, \tau_{LL}^{c,nn}} \qquad (i = A,B) \ \text{ and } \Delta \, m_{i;n,L} = m_{c;n,L} - m_{i;n,L}$$

As a guideline we note the formal analogy which exists between the KKR-multiple scattering formalism and the more widely used tight binding method.[19]

In fact, Eqs. (2.4) and (2.5) can be used intact within a TB-CPA under the following redefinition of variables: $t_c^{-1} \rightarrow z - \sigma$, $t_i^{-1} \rightarrow z - \varepsilon_i$, $(i = A \text{ or } B)$, $G_{n,m} \rightarrow W_{nm}$, and $\tau^{c,nn} \rightarrow \overline{G}^{nn}$, where bold fact characters denote matrices in orbital (L) space, and σ, ε_i, W_{nm} and $\overline{G}^{nm}$ are, respectively, the site-diagonal CPA self-energy, the on-site energies associated with atoms of kind A or B, the hopping terms between sites n and m, (assumed for the purposes of this discussion to be species independent), and the site-diagonal element of the CPA-medium Green function.

In the following, we consider only the band structure contribution to the thermodynamic potential of an alloy,

$$\Omega_e (\mu) = - \int_{-\infty}^{+\infty} d\varepsilon \, \theta \, (\varepsilon - \mu) \, N \, (\varepsilon,\mu) \qquad (2.6)$$

where $N(\varepsilon,\mu)$ is the integrated density of states (total number of particles), and $\theta(x)$ is the familiar step function. Within the local density approximation to the total energy of an alloy,[20,21] it can be shown[22] that $\Omega_e(\mu)$ can be written as a sum of two terms: (i) the concentration dependent but configuration independent energy $\overline{\Omega}_e(c_i,\mu)$ of the CPA reference medium, and (ii) a configuration dependent energy $\Omega_e'(\{p_n^i\},\mu)$. This configuration-dependent term is expressed as

$$\Omega_e' \left(\left\{ p_n^i \right\}, \mu \right) = -\frac{Im}{\pi N} \int_{-\infty}^{\mu} d\varepsilon \, \theta \, (\varepsilon - \mu) \, \mathrm{Tr} \ln (1 - X \, \hat{t}^c) \, , \qquad (2.7)$$

where the trace is taken over both site and angular momentum indices. We note that X is a site-diagonal operator[22] (matrix),

$$X = \sum_{n,L,i} p_n^i X_{n,L} \mid n \, L > < nL \mid \qquad (l \leq 2) \qquad (2.8)$$

and $\hat{t}^c$ denotes the strictly off-diagonal part of t^c. In Eq. (2.8), the occupation numbers p_n are defined as equal to 1 if site n is occupied by an atom of species i, and as equal to zero otherwise. For $l > 2$, the matrices X and t_c acquire off-diagonal elements in the L representation, a mild complication that can be easily accommodated within a straightforward generalization of the present formalism.

Now expanding the logarithm in Eq. (2.7), we obtain the result,

$$\Omega_e' \left(\left\{ p_n^i \right\}, \mu \right) = -\frac{Im}{\pi N} \sum_{k=2}^{\infty} \int_{-\infty}^{\mu} d\varepsilon \, \theta \, (\varepsilon - \mu) \, \mathrm{Tr} \frac{(X \, \hat{t}^c)^k}{k} \, . \qquad (2.9)$$

By making use of the CPA condition, and considering explicitly the case of a binary alloy $A_c B_{1-c}$, we can write X in the form

$$X = \sum_{n,L} (p_n - c) \, \Delta X_{n,L} \mid n,L > < n,L \mid \, , \qquad (2.10)$$

where p_n, c and $\Delta X_{n,L}$ refer to p_n^A, c_A and $X_{n,L}^A - X_{n,L}^B$.

The GPM

Using Eq. (2.10), the *configuration* and *concentration* dependent part of the ground state energy, Eq. (2.7), can be expressed in the generalized perturbation method (GPM) in the form,

$$\Omega_e' \left(\left\{ p_n \right\}, \mu \right) = \sum_{k=2}^{\infty} \frac{1}{k} V_{n_1, \, ..., \, n_k}^{(k)} \, \delta c_{n_1} \, ... \, \delta c_{n_k} \qquad (2.11)$$

where $\delta c_{n_i} = p_{n_i} - c$ and $V_{ni, \, ...n_k}^{(k)}$, defines the kth order effective cluster interaction (ECI) involving sites $n_1, n_2, ... n_k$

$$V_{n_1, \, n_2, \, ..., \, n_k}^{(k)} = -\frac{Im}{\pi N} \int_{-\infty}^{+\infty} d\varepsilon \, \Theta \, (\varepsilon - \mu) \, \mathrm{Tr} \left(\Delta X_{n_1} \, \tau^{c, n_1 n_2} \, \Delta X_{n_2} \, ... \, \Delta X_{n_k} \, \tau^{c, n_k n_1} \right) .$$

$$(2.12)$$

Thus, to the lowest order in the perturbation, as far as the small parameter $|\Delta X \tau^c|$ of the theory is less than unity up to the Fermi energy, the quantity Ω'_e which is commonly called the ordering energy is given by

$$\Omega'_e = \frac{1}{2} \sum_s V_{os}^{(2)} \delta c_o \, \delta c_s \, , \tag{2.13}$$

where V_{os} is the effective pair interaction between sites o and s:

$$V_{os} = V_{os}^{AA} + V_{os}^{BB} - 2 \, V_{os}^{AB} \, . \tag{2.14}$$

Therefore, at the Fermi energy, as far as the 2nd order terms of the expansion are the predominant ones, the positive (negative) sign of V_{os} clearly indicates a tendency toward ordering (phase separation).

The ECM

We consider Eq. (2.7) in connection with a system consisting of a cluster of n sites, C_n, embedded in a disordered material. An exact treatment of this system would require a complete configurational average over all configurations of the material surrounding the cluster C_n. Computational considerations, however, require the use of approximation schemes in computing this average. At the first level of approximation, all sites outside the cluster C_n are taken as being occupied by effective medium or "CPA atoms". Calling this the coherent potential approximation for the configurational energy, we can write

$$\Omega_e\left(\{p_n^i\}, \mu\right) = E_{CPA} + \sum_i V_i^{(1)} \delta c_i + \frac{1}{2} \sum_{i,j}{}'' V_{ij}^{(2)} \delta c_i \delta c_j +$$

$$\frac{1}{3} \sum_{i,j,k}{}'' V_{ijk}^{(3)} \delta c_i \, \delta c_j \, \delta c_k + \ldots \tag{2.15}$$

where E_{CPA} is the contribution of the CPA medium and the $V_{i_1 i_2 \ldots i_n}^{(n)}$ are the renormalized ECI as defined in the ECM,[5]

$$V_{i_1 i_2 \ldots i_n}^{(n)} = \left[V_{i_1 i_2 \ldots i_{n-1}}^{n-1} \right]_n^A - \left[V_{i_1 i_2 \ldots i_{n-1}}^{n-1} \right]_n^B$$

$$= -\frac{2}{\pi} \, \mathrm{Im} \int_{-\infty}^{\mu} d\varepsilon \, \mathrm{Tr} \ln \left[\Pi_j \, \mathbf{Q}^{j(even)} \right] \left[\Pi_j \, \mathbf{Q}^{j(odd)} \right]^{-1} \tag{2.16}$$

Here, j(even) and j(odd) denote cluster configurations with even and odd numbers of minority (B) atoms, respectively, $\mathbf{Q}$ is the matrix $(1 - \mathbf{X} \, \hat{\tau}^c)$ occurring in Eq. (2.7), and the double prime on the summations in Eq. (2.15) denotes sums over distinct sets of sites. In Eq. (2.16), the symbol $[V_{i_1}^{n-1}, i_2 \ldots i_{n-1}]_n^i$ denotes an effective multisite interaction among the n sites of C_n under the restriction that site n is occupied by an atom of type α, ($\alpha = A$ or B).

Expanding the quantities $Tr \ln \mathbf{Q}$ in Eq. (2.16), Eq. (2.15) can be written[5] in the form,

$$\Omega\left(\{p_n^i\},\mu\right) = E_{CPA} + \frac{1}{2}\sum_{i,j}{}' V_{ij}^{(2)}\, \delta c_i\, \delta c_j + \frac{1}{3}\sum_{i,j,k}{}' V_{i,j,k}^{(3)}\, \delta c_i\, \delta c_j\, \delta c_k + ... \qquad (2.17)$$

in which the sums are taken over consecutively distinct sites. Since Eq. (2.17) is precisely the GPM expansion[4] for the configurational energy, it follows that the ECM corresponds to complete summations of sets of terms (diagrams) of the GPM expansions.

NUMERICAL RESULTS

In this section we exhibit the results of numerical calculations of ECI's using the CWM, the GPM and the ECM. In Fig. 1a the pair, triangle and tetrahedron cluster interactions, V_2^T, V_3^T and V_4^T, respectively, calculated[11] within the CWM in a TB description of the Hamiltonian are shown as functions of the number of d-electrons of the B species for fixed $\Delta N = N_A - N_B$. T refers to the tetrahedron truncation of Eq. (2.1) which takes into account the [100] family of ordered structures, namely LI_2, A_3B and AB_3, LI_0 and AB, in addition to the pure metals A and B.

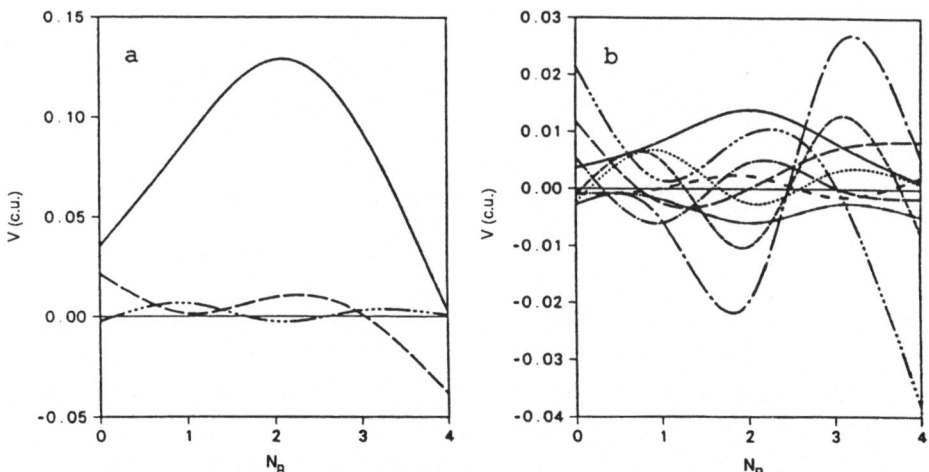

Figure 1. CWM cluster interactions in canonical units (c.u.): a) T-CWM —— V2, — — — V3, –···– V4; b) TO-CWM —— first neighbor pair divided by 10, — — — second neighbor pair, –···–···– equilateral nearest neighbor triangle, —·—· isoceles triangle, ······ regular tetrahedron, ------ irregular tetrahedron, —·—·— pyramid, — — — — octahedron.

As can be seen here, V_3^T and V_4^T are much smaller than V_2^T, except at large values of N_B. Figure 1b shows the V_γ^{TO}, $\gamma = 2, 3, ..., 10$, computed within the tetrahedron-octahedron (TO) truncation of Eq. (2.1) which includes a fit to the input information for the $[1\frac{1}{2}0]$

family of ordered structures, i.e., DO_{22}, A_3B and AB_3, MoPt$_2$, A_2B and AB_2, A_2B_2, (phase 40 in Kanamori's notation), Cu Pt type, AB, in addition to the structures used in the tetrahedron truncations. Although the V_γ^{TO}, $\gamma = 2, 4, 6$ compare correctly with the results of the V_γ^T for $\gamma = 2, 3, 4$, higher order TO cluster interactions can be significant (note those corresponding to the isosceles triangle). Thus, the V_γ appear to depend strongly on the cluster truncation used in their determination (tetrahedron or tetrahedron-octahedron).

Table 1 indicates the V_γ^T for various sets of ordered structures. Note that especially the values for the triangle and tetrahedron can vary significantly. Surprisingly, the [100] family gives the smallest values of V_3^T and V_4^T. Therefore, we can conclude that: i) the rapid convergence of expansion of Eq. (2.1) is not completely fulfilled and ii) that the V_γ, within the same cluster truncation, are not unambiguously defined. By contrast, in the GPM and the ECM the concentration dependent effective cluster interactions do not suffer from such an ambiguity.

Effective cluster interactions calculated within the GPM and the ECM for the alloys $Pd_{0.75}V_{0.25}$, $Pd_{0.5}V_{0.5}$ and $Pd_{0.5}Rh_{0.5}$, using either a TB description of the Hamiltonian or a first-principles, multiple scattering approach are shown in Figs. 2 through 6, and 7 through 9, respectively. Several important characteristic features of the ECI's can be distinguished in these figures. First, as is seen in Fig. 2, the various terms $V_{ij}^{(n)}$, $n > 2$ calculated within the GPM are several factors smaller than the leading term, $V_{ij}^{(2)}$, which in turn is extremely similar to the ECM result (heavy solid curve).

Table 1.   T-CWM cluster interactions (in c.u.) for various sets of ordered structures: a) $\{A, L1_2, L1_0, B\}$, b) $\{A, DO_{22} - A_3B, L1_0,$ $L1_2 - AB_3, B\}$, c) $\{A, DO_{22}, 40, B\}$.

V	a	b	c
0	-.13665	-.13953	-.13437
1	-.33290	-.33866	-.34127
2	.09455	.09455	.08647
3	.00118	.00694	.00956
4	.00299	.00587	.00841

These higher order terms enter[5] the summations for the configurational energies multiplied by numerically small, concentration-dependent factors and their overall significance in the final summation decreases rapidly with increasing order n. Thus, the lowest order term, calculated in the GPM, gives a reasonably accurate representation of the sum, calculated in the ECM, indicating that the GPM is a valid, rapidly convergent perturbation expansion. Second, the effective pair interactions for near-neighbor and more-distant terms (see following figures), exhibit considerable structure as a function of band filling (Fermi energy). Because of this structure, phase stability depends greatly on the value of the Fermi energy, with the most stable structure often being determined by the predominant pair interaction at E_F.

The last point is amply illustrated by the result depicted in Fig. 3, which shows effective pair interactions for second (thick solid curve), third (dashed curve), and fourth

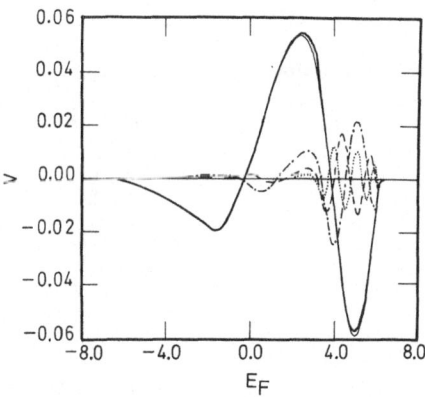

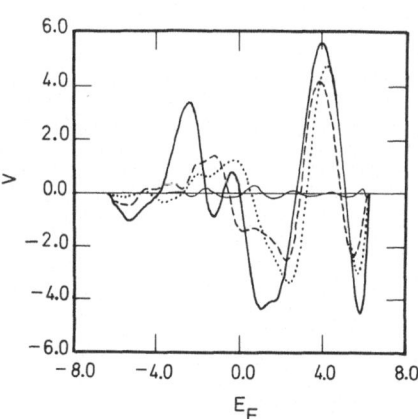

Figure 2. Effective pair interactions, for the $Pd_{0.75} V_{0.25}$ alloy calculated in the GPM and the ECM. The GPM contributions associated with $(\delta t)^n$ for $n = 2, 4, 6,$ and 8 are shown by the thin solid, dashed-dotted, dashed, and dotted curves, respectively. The renormalized ECM effective pair interaction is shown by the thick solid line.

Figure 3. Renormalized effective pair interactions in the $Pd_{0.75} V_{0.25}$ alloy nearest neighbors (thick solid curve), third nearest neighbors (dotted curve), fourth nearest neighbors (dashed curve) calculated in the ECM. These curves are practically indistinguishable from those obtained to second order in the GPM expansion. The vanishingly small fourth-order GPM contribution to second nearest-neighbor interaction is shown by the thin solid curve.

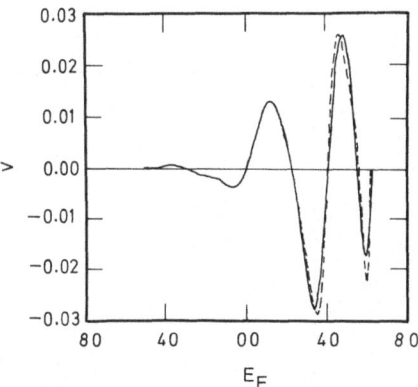

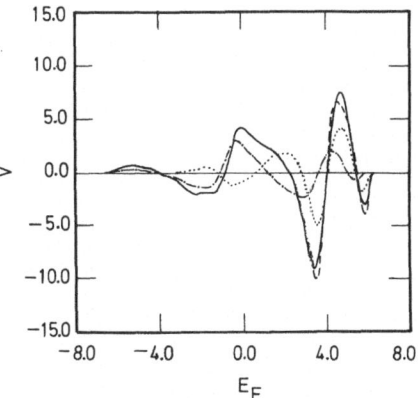

Figure 4. Cluster interactions for three sites aligned along the (110) direction in the fcc $Pd_{0.75} V_{0.25}$ alloy. The dashed curve represents the summation of the GPM expansion to sixth order, while the solid curve shows the renormalized ECM effective interation.

Figure 5. Cluster interactions for three sites forming an equilateral nearest-neighbor triangle in the fcc alloy $Pd_{0.75} V_{0.25}$. The solid curve depicts the renormalized effective interaction calculated in the ECM, and is to be compared to the summation of the lowest order in the GPM expansion (dashed curve). The contributions of two of those order, corresponding to $(\delta t)^3$ and $(\delta t)^4$ are shown by the dashed-dotted and the dotted curves, respectively.

25

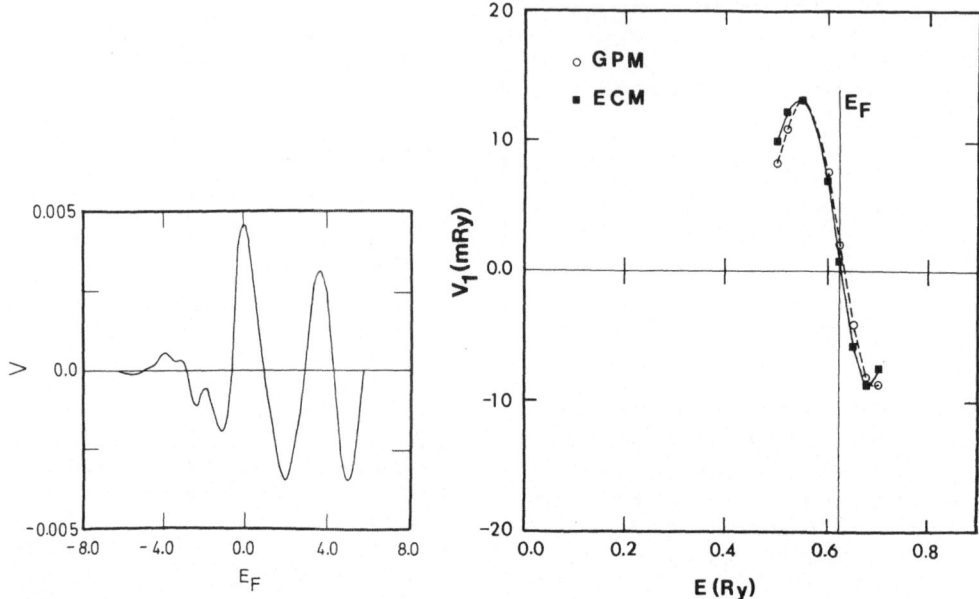

Figure 6. Rebormalized cluster interactions for a tetrahedral cluster of sites in the $Pd_{0.75}V_{0.25}$ alloy calculated in the ECM.

Figure 7. Effective near-neighbor pair interaction in $Pd_{0.75}$ $V_{0.25}$ alloys calculated in the GPM and the ECM.

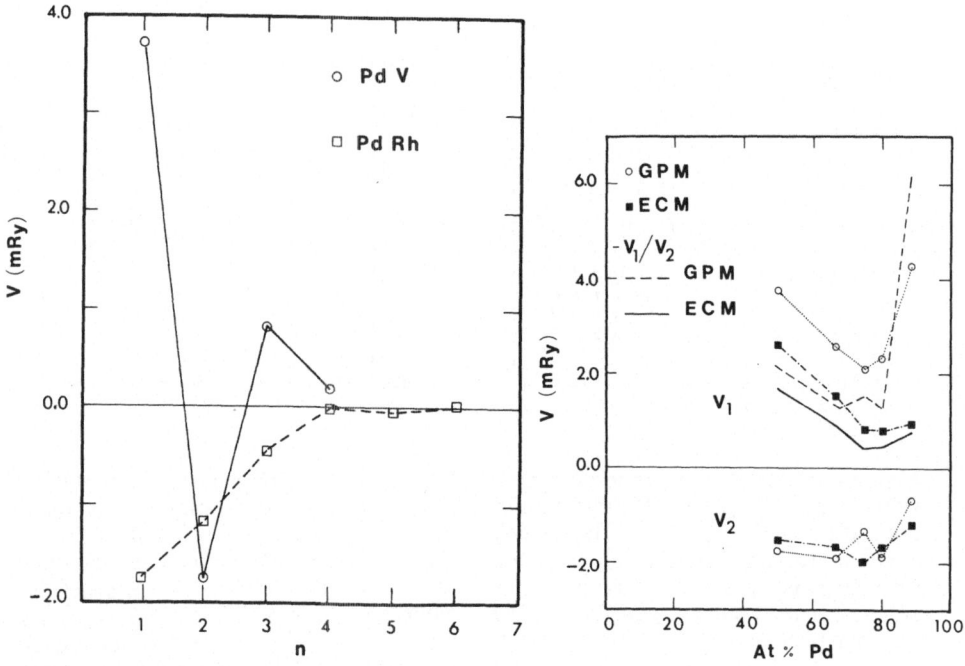

Figure 8. Variation of nth neighbor effective pair interactions in Pd_cV_{1-c} and $Pd_c Rh_{1-c}$, c = 0.5, alloys calculated in the GPM.

Figure 9. The concentration dependence of the first, V_1, and second, V_2, neighbor effective interactions and the ratio - V_1/V_2, in $Pd_c V_{1-c}$ alloys as obtained in the GPM and the ECM.

the search of the ground states of the Ising model,[23] it is possible to predict the most probable stable ordered structures at various concentrations. For example, the negative values of all effective pair interactions in $Pd_{0.50} Rh_{0.50}$ indicate a tendency toward phase separation at $T = 0K$, as is in fact observed experimentally. On the other hand, the variation in the sign of V for the case of Pd V alloys leads to much richer behavior. Taking account of pair interactions up to and including fourth neighbors, we predict a stable ground state in the DO_{22} ordered structure at c = 0.75 as is observed experimentally. Considering only first and second neighbors yields an (incorrect) LI_2 ordered structure. It follows that the convergence of the energy expansions in terms of long-range pair interactions is an important ingredient in the calculation of alloy phase diagrams. The variation of first- and second-neighbor effective interactions in $Pd_c V_{1-c}$ as functions of concentration, c, is shown in Fig. 9.

DISCUSSION

An accurate determination of multisite effective interactions is an indispensible ingredient in the study of alloy phase stability, and in the construction of alloy phase diagrams. The present lecture was concerned with the determination of such interactions through the use of electronic structure calculations based on the TB-CPA and the first-principles KKR-CPA formalism, and two methods, the GPM and the ECM, devised to treat local statistical fluctuations away from the CPA-determined effective medium. These two methods were shown to be essentially equivalent, the ECM corresponding to the complete summation of certain terms in the GPM expansion. In contrast to the fitting procedure used in the method of Connolly and Williams, both the GPM and the ECM yield uniquely defined concentration-dependent multisite interactions.

A number of problems remain in our quest toward a feasible computational method for the calculation of alloy phase diagrams. As alluded to above, lattice entropy, lattice relaxation effects, and the configurational dependence of vibrational properties must be included in the calculation of the total free energy. Further, accurate self-consistent calculations of the total energy (disorder energy) must become available, to supercede the rather simplified use of "band" energies discussed above. This latter problem is currently under studious attack by a number of workers in the field, and alloy total energy calculations have begun to emerge.[20] The statistical-mechanical methods presently available would then allow the construction of both coherent and incoherent alloy phase diagrams from a first-principles approach utilizing the ECI's determined from such a generalized scheme. It is hoped that the future holds pleasant developments in these admittedly difficult studies.

ACKNOWLEDGMENT

This work was supported by the U.S. Department of Energy, under contract No. DE-FG02-84ER45 with Northwestern University, No. W-7405-ENG-48 with the Lawrence Livermore National Laboratory (University of California), and No. DE-AC05-84ORR21400 with Martin Marrietta Energy Systems.

REFERENCES

1. R. Kubo, "Statistical Mechanics", North Holland, Amsterdam (1965).
2. V. Heine and D. Weaire, Sol. St. Physics 24:249 (1970).
3. J.W.D. Connolly and A.R. Williams, Phys. Rev. B, 27:5169 (1983).
4. F. Ducastelle and F. Gautier, J. of Phys. F, 6:2039 (1976).
5. A. Gonis, X.-G. Zhang, A.J. Freeman, P.E.A. Turchi, G.M. Stocks, and D.M. Nicholson, Phys. Rev. B, 36:4630 (1987), and references therein. See also, P.E.A. Turchi, A. Gonis, X.-G. Zhang, and G.M. Stocks, Proceedings of NATO/ASI on "Alloy Phase Stability", A. Gonis and G.M. Stocks, eds., KLUWER Publications, (1989).
6. B.L. Gyorffy and G.M. Stocks, Phys. Rev. Lett. 50:374 (1983).
7. W. Schweika, private communication.
8. R. Kikuchi, Phys. Rev. 81:998 (1951); J.M. Sanchez, F. Ducastelle and D. Gratias, Physica, 128A:334 (1984); D. de Fontaine, Sol. St. Physics, 34:73 (1979).
9. K. Binder, J.L. Lebowitz, M.H. Phani, and M.H. Kalos, Acta Metall., 29:1655 (1981); K. Binder, in: "Monte Carlo Methods in Statistical Physics", Topics in Current Physics, Vol. 7, K. Binder, ed., Springer-Verlag (1986).
10. A.A. Mbaye, L.G. Ferreira, and A. Zunger, Phys. Rev. Lett. 58:49 (1987); K. Terakura, T. Oguchi, T. Mohri, and K. Watanabe, Phys. Rev. B, 35:2169 (1987); T. Mohri, K. Terakura, T. Oguchi, and K. Watanabe, submitted to Acta Metall.; A.E. Carlsson, Phys. Rev. B, 35:4858 (1987).
11. M. Sluiter and P. Turchi, NATO/ASI summer institute on "Alloy Phase Stability", G.M. Stock and A. Gonis, eds., Maleme, Greece, June 14-27, 1987, KLUWER Publications (1989).
12. P. Soven, Phys. Rev. 156:809 (1967).
13. B. Velicky, S. Kirkpatrick, and H. Ehrenreich, Phys. Rev. 175:747 (1969).
14. B.L. Gyorffy and G.M. Stocks, in: "Electrons in Disordered Metals and Metallic Surfaces", P. Phariseau, B.L. Gyorffy, and L. Scheire, eds., Plenum, New York, p. 89 (1978); G.M. Stocks and H. Winter, in: "Electronic Structure of Complex Systems", P. Phariseau and W.M. Temmerman, eds., Plenum, New York, p. 463 (1985); J.S. Faulkner, in: "Progress in Materials Science", J.W. Christian, P. Hassen, and T.B. Massalski, eds., Pergammon, New York, Vols. 1 and 2 (1982).
15. J. Korringa, Physica, 13:392 (1947).
16. W. Kohn and Rostoker, Phys. Rev. 94:1111 (1954).
17. B.L. Gyorffy and M.J. Stott, in: "Band Structure Spectroscopy of Metals and Alloys", D.J. Fabian and L.M. Watson, eds., Academic, New York, p. 385 (1973).
18. J.S. Faulkner and G.M. Stocks, Phys. Rev. B21:3222 (1980).
19. W.H. Butler, Phys. Rev. B31:3260 (1980).
20. D.D. Johnson, D.M. Nicholson, F.J. Pinski, B.L. Gyorffy, and G.M. Stocks, Phys. Rev. Lett. 56:2088 (1986).
21. B.L. Gyorffy and G.M. Stocks, J. de Phys. (Paris), Colloq. 35:C4-75 (1974).
22. P.E.A. Turchi, G.M. Stocks, W.H. Butler, D.M. Nicholson, and A. Gonis, Phys. Rev. B, 37:5982 (1988).
23. J. Kanamori and Y. Kakehashi, J. de Phys. Collog. 38:C7-274 (1977).

SIMULATION OF ISOVALENT IMPURITIES IN MAGNESIUM OXIDE USING HARTREE-FOCK CLUSTERS

Jun Zuo, Ravindra Pandey, and Albert Barry Kunz

Department of Physics
Michigan Technological University
Houghton, MI 49931

INTRODUCTION

An unrestricted Hartree-Fock (UHF) approach has been developed by Kunz and his coworkers[1,2,3] to study the optical excitation properties of atomic, molecular, and solid-state systems and the results have been sufficiently satisfactory[4]. In the present study, three computational models are used to simulate the sulphur-doped magnesium oxide (MgO). The first model is the seven-ion cluster (S-6Mg): one sulphur ion at the center with six nearest magnesium ions. The second one is the same cluster plus some surrounding ions considered as point charges at the sites of the nuclei. The third is the one used in "Ionic Crystal with Electronic Cluster, Automatic Program" (ICECAP)[5]. In this model, the (S-6Mg) cluster is embedded in a shell-model lattice. The ions in this lattice are taken as dipole polarizable point charges coupled harmonically to uniformly charged massless spherical shells. The interactions among the ions are in terms of the Coulomb and short-range potentials. In fact, ICECAP is developed especially for determining the defect properties of crystalline materials arising from excess electrons, holes, or impurities in the crystal. A detailed description of this program package will be given by A.B. Kunz at this conference.

The low-lying excitonic states of sulphur-doped MgO are dominated by the $3p^6 \rightarrow 3p^5 4s$ transition in the sulphur anion and are described by electron-hole pairs. The excitation energy is obtained from taking the difference between the excited state energy and the ground state energy which are calculated separately.

Since there is no experimental data available to us for sulphur-doped MgO, we calculate the corresponding energies of pure MgO, for which we have some experimental results to compare with, in the thought that oxygen and sulphur are isovalent elements so that the calculations would yield results of approximately equal quality. In the calculations of pure MgO, the seven-ion cluster (S-6Mg) in all the three models is replaced by the (O-6Mg) cluster.

RESULTS AND DISCUSSION

The Gaussian basis set for magnesium is the $Mg(7,7/4)$[6] set. The basis sets for oxygen and sulphur are $O(4,3/4)$[7] and $S(4,3,2/4,2)$[7], respectively. To make these basis sets flexible for excited state calculations, additional S-type and P-type Gaussian primitives are added to them.

The calculated excitation energies of pure MgO are listed in Table 1 and the corresponding results of sulphur-doped MgO are listed in Table 2. In the tables, ΔT, ΔS, and ΔST stand for the energy difference between the triplet state and the ground state, the energy difference between the singlet state and the ground state, and the triplet-singlet splitting, respectively.

It can be seen that the simulation of the environment of the seven-ion cluster has greatly improved the result, as we expected. For the ground-triplet excitation of pure MgO, the energy difference is 9.65 eV from the "cluster only" model, 7.37 eV from the "cluster with additional point charges" model, and 7.73 eV from the ICECAP model ("cluster with shell-model lattice"). Comparing with the experimental value 7.69 ± 0.01 eV, the second is in good agreement while the third can be said to be excellent. For ground-singlet excitation energy and triplet-singlet splitting, similar improvements are obtained. For sulphur-doped MgO, the triplet-ground excitation energy and singlet-ground excitation are 13.24 and 13.77 eV from the "cluster only" model, 8.99 and 9.00 eV from the "cluster with additional point ions" model, and 7.09 and 7.12 eV from the ICECAP model, respectively. And, the triplet-singlet splittings turn out to be 0.53 eV, 0.01 eV, and 0.03 eV, respectively from the three models.

Table 1

Results of Pure MgO and Experimental Data (eV)

	Cluster	Cluster + Ions	ICECAP	Experiment[8]
ΔT	9.65	7.37	7.73	7.69 ± 0.01
ΔS	11.05	7.34	7.77	7.76 ± 0.01
ΔST	1.40	0.02	0.04	0.07 ± 0.02

Table 2

Calculated Results of Sulphur-Doped MgO (eV)		
Cluster	Cluster + Ions	ICECAP
ΔT	13.24	8.99	7.09

Calculated Results of Sulphur-Doped MgO (eV)		
Cluster	Cluster + Ions	ICECAP
ΔT	13.24	8.99	7.09
ΔS	13.77	9.00	7.12
ΔST	0.53	0.01	0.03

In the calculations of pure MgO, the distance between the oxygen ion and the nearest magnesium ion is taken to be 2.106 Angstrom. In ICECAP calculations, the magnesium ions are found to have relaxed inward slightly (less than 1%). But, in the case of sulphur-doped MgO, the relaxation is outward about 6%, which has been expected because the sulphur ion is larger than the oxygen ion so that there is a considerably large distortion in the lattice.

CONCLUSION AND FUTURE PLANS

The ICECAP method developed by Kunz and co-workers again gives sufficiently satisfactory predictions of some of the excitation properties of ionic crystal. The inclusion of the response of the surrounding shell-model lattice to the defect cluster appears to be significant in obtaining the reliable energies of pure and sulphur-doped MgO.

We now plan to extend our investigation to the cases of hydrogen-doped MgO, selenium-doped MgO and tellurium-doped MgO. The second-order electronic correlation correction will be incorporated in our calculations. In addition, we propose to examine, using ICECAP, the transport properties of these impurity anions by generating interatomic potentials for sulphur in MgO.

ACKNOWLEDGMENT

This study was supported by the U.S. Department of Energy under grant DE-FG02-85-ER45224.

REFERENCES

1.   A.B. Kunz and D.L. Klein, Phys. Rev. B17:4614 (1978).
2.   A.B. Kunz and P.W. Goalwin, Phys. Rev. B34:2140 (1986).
3.   A.B. Kunz and J.M. Vail, Phys. Rev. B38:1058 (1988).
4.   For example, S.T. Pantelides, D.J. Mickish, and A.B. Kunz, Solid State Comm., 15:203-205 (1974);
       A.B. Kunz and P.W. Goalwin, Phys. Rev., B34:2140 (1986);
       A.B. Kunz, J. Meng, and J.M. Vail, Phys. Rev., B38:1064 (1988);
       A.B. Kunz and R. Pandey, Phys. Rev., in press.
5.   J.H. Harding, A.H. Harker, P.B. Keegstra, R. Pandey, J.M. Vail, and C. Woodward, Physica, 131B:151 (1985).
6.   R. Pandey and J.M. Vail, J. Phys : Condensed Matter, in press (1989).
7.   S. Huzinaga, "Gaussian Basis Sets for Molecular Calculations", Elsevier, New York, 1984.
8.   D.M. Roessler and W.C. Walker, Phys. Rev. Lett., 17:31 (1966); Phys. Rev., 159, 733 (1967).

DEFECT ABUNDANCES AND DIFFUSION MECHANISMS
IN DIAMOND, SiC, Si AND Ge

J. Bernholc,* A. Antonelli,* C. Wang* and R. F. Davis**

Departments of Physics* and Materials Science and Engineering**
North Carolina State University, Raleigh, NC 27695-8202

S. T. Pantelides

IBM T. J. Watson Research Center, Yorktown Heights, NY 10598

I. INTRODUCTION

Many of the important properties of semiconductors are affected by the presence of native defects. These include diffusion of native atoms and impurities, electronic properties, structural rearrangements during growth or processing, etc. We have used local density theory and non-local pseudopotentials to study trends in the relative abundance of native defects and in self-diffusion (diffusion of native tracer atoms) in diamond, SiC, Si and Ge. Both point-defect-mediated diffusion and direct exchange of lattice atoms were considered. All calculations were carried out using the same methodology and similar convergence criteria.

Previous theoretical work[1,2,3,4] on the mechanisms of diffusion has concentrated mainly on Si. The competing mechanisms of diffusion in Si involve vacancies, various interstitial configurations (e.g. tetrahedral, bond-centered, or split, see Fig. 1), and a direct exchange of atoms on a Si lattice. Since diamond has the same lattice structure as Si, the same mechanisms should be considered for diamond.

The interest in diamond has increased substantially with the development of techniques for the growth of thin film diamond,[5] since high quality diamond films could form a basis for high-temperature and/or high speed electronics. This is due to diamond's exceptional thermal stability and heat conductivity as well as high electron and hole mobilities. To date, however, epitaxial growth has only been achieved on diamond substrates, while growth on a non-diamond substrate results in the formation of a continuous polycrystalline film. Since the growth of the film proceeds simultaneously at many nucleation centers, the development of a continuous, single crystal film must necessarily involve atomic motion of both surface and subsurface atoms. The determination of the mechanisms of atomic motion in the bulk is thus also important for the understanding and improvement of the methods of growth of thin film diamond.

Cubic (zinc-blende) SiC is also suitable for high temperature applications and has high electron mobility. Good quality crystals of cubic SiC can readily be grown by CVD techniques, but p-type doping is a problem due to a large concentration of compensating defects. Since SiC is a two-component material, deviations from stoichiometry may affect its electronic properties. Indeed, analyses of lattice parameters and densities of several SiC polytypes indicate that the material is not stoichiometric and Si-rich. Although the deviations from stoichiometry are believed to be quite small, they are large when compared to the typical donor or acceptor concentrations.

Turning to Si, previous calculations[1,2,3,4] have shown that the vacancy, the interstitial and the concerted exchange mechanisms have comparable activation energies for self-diffusion. Our own calculations for Ge have found essentially identical results. For that reason our further work on diffusion mechanisms has concentrated on Si. In particular, we have focused on finding ways to determine the relative contributions of the various diffusion mechanisms in Si. Since the diffusion coefficient depends exponentially on the activation energies, differences of even a few kT in the activation energies will have a substantial effect. However, it is very difficult to determine these contributions directly, either experimentally or theoretically. We will show below, however, that the pressure dependence of the three diffusion mechanisms is strikingly different, so that a combination of calculations and experiments investigating pressure dependence may lead to the determination of their relative contributions. To this end we have calculated the pressure coefficients for the three diffusion mechanisms in Si.

CALCULATIONS

The calculations were carried out using local density theory, Hamann-Schluter-Chiang non-local pseudopotentials and the plane wave supercell method.[6] For diamond and SiC, the quoted results correspond to supercell sizes of 32 atoms for the substitutional defects, while a 16-atom unit cell was used for the higher energy interstitial defects. The 32-atom unit cells were used for Si. Comparisons between 16-atom and 32-atom results have shown that the present results are converged with respect to supercell size.

Supercell plane wave calculations for diamond and SiC required special attention due to the lack of p core orbitals in the C atom. We have used a soft-core Hamann-Schluter-Chiang pseudopotential from Ref 7, generated by enlarging the pseudopotential matching radii. With this pseudopotential, good convergence was obtained with cutoffs of 14 and 28 Ry, for the the plane waves included directly and by Löwdin's perturbation theory, respectively. Even with these relatively small cutoffs the number of plane waves was rather large, especially for SiC, where 1880 plane waves were included directly and 5540 by perturbation theory. For Si, the cutoffs were 6 and 12 Ry, respectively. Symmetry was used to reduce the calculations to a manageable level. The calculations are described in more detail in Refs 8, 9, and 10, for the cases of diamond, SiC, and Si, respectively.

RESULTS AND DISCUSSION

Self Diffusion in Diamond

We have considered the vacancy, the tetrahedral, bond-centered, and <100>-split interstitials (see Fig 1), and the direct exchange as possible mechanisms for self-diffusion. The calculated formation energies and the saddle point energy for the direct exchange are given in Table I. The tabulated results include energy gains due to radial relaxation of nearest neighbor atoms and an estimate of the relaxation effects of further neighbors using the Keating model.

Table I Calculated formation energies of neutral point defects and the saddle point energy for the direct exchange in diamond [eV].

vacancy	tetrahedral interstitial	<100> split-interstitial	bond-centered interstitial	direct exchange
7.2	23.6	16.7	15.8	13.2

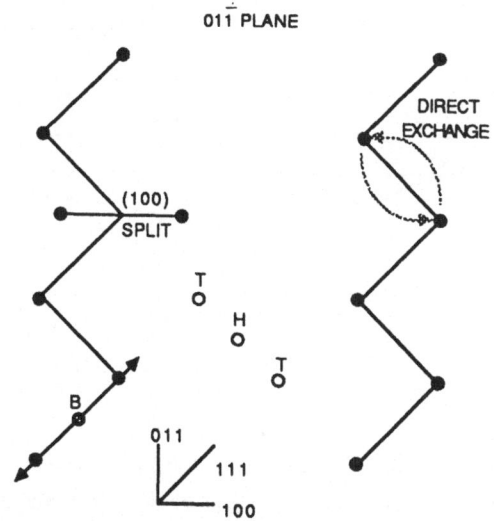

Fig. 1 Schematic view of the native defect and direct exchange configurations which can be involed in in self-diffusion in a diamond lattice. T, H, and B denote the tetrahedral, hexagonal and bond-centered sites.

From Table I it is clear that the vacancy has by far the lowest formation energy and, as such, dominates self-diffusion in diamond. This result may seem surprising given the fact that in Si all three self-diffusion mechanisms have similar activation energies. However, bonding in diamond is quite different than in Si. In particular, the diamond bonds are very stiff, which prevents bond twisting and large reconstructions, and results in a large saddle point energy for direct exchange. For the self-interstitials, the reasons for their high formation energies are: (i) the lack of low lying d-orbitals in carbon atoms which makes overcoordination unfavorable; (ii) the high electron density which leads to a substantial increase in the kinetic energy of the extra electrons; and (iii) the large band gap which increases the energy of the two T_2 electrons near the conduction band edge of the tetrahedral interstitial.

The migration barrier of the vacancy in diamond is substantial (1.7-1.9 eV depending on the charge state), resulting in the total activation energy for self-diffusion via the neutral vacancy mechanism of 9.1 eV. The size of this activation barrier may explain the difficulties in grain boundary annealing during growth of thin film diamond despite the high substrate temperature in most methods of preparation. However, the formation energy of the vacancy and thereby the activation energy for self-diffusion depends strongly on the Fermi level position (see Fig 2a). Although these results suffer from the usual uncertainties associated with the use of local density theory to calculate band gaps and electronic excitations, they still show that the vacancy formation energy is lowered substantially in p- and n-type diamond. They suggest doping of diamond films during growth (e.g. with boron) in order to enhance the annealing of defects and grain boundaries.

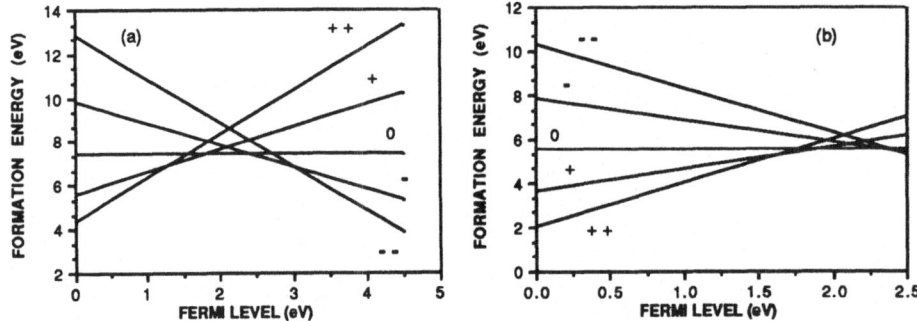

Fig. 2   Formation energy of the carbon vacancy as a function of Fermi level in a) diamond; and b) cubic SiC.

<u>Defect Abundances and Self-Compensation Mechanisms in Cubic SiC</u>

We have calculated the formation energies of all eight elementary point defects with tetrahedral symmetry (see Table II), namely the two vacancies, two antisite defects and the tetrahedral interstitials surrounded by the Si and C atoms (denoted by TSi and TC, respectively). We have also carried out calculations for the bond-centered interstitials and for the V_C-V_{Si} and Si_C-C_{Si} nearest neighbor pairs. All the formation energies are quoted relatively to the binding energy per atom in the perfectly stoichiometric compound. Note that, as in diamond, the formation energies of interstitials are higher than those of substitutional defects. For Si interstitials, this is mainly due to the small size of the interstitial cavity in SiC compared to the size of the Si atom. C interstitials surrounded by C nearest neighbors have a substantially higher binding energy than those surrounded by Si neighbors, due to the inability of C atoms to take advantage of overcoordination.

Since SiC is a two-component system, stoichiometry plays a very important role in determining the most abundant native defects. In a perfectly stoichiometric SiC, the lowest energy defect is the Si_C-C_{Si} nearest neighbor antistructure pair, since it is stoichiometry-preserving. Its formation energy is 5.9 eV. For non-stoichiometric SiC, the most abundant defect is the one which has the lowest formation energy per accommodated atom of the excess component. Furthermore, the nature of the compensating defect may change depending on the Fermi-level, since the formation energies of charged defects are Fermi-level dependent.

In Si-rich n-type and intrinsic material the most abundant defect is the Si_C antisite, since it is twice as effective in accommodating the excess Si atoms than the carbon vacancy. It is electrically inactive, regardless of the position of the Fermi level. In p-type material, however, the reaction $Si_C \rightarrow 2V_C^{++}$ is exothermic (see Fig. 2b and Table II). Since the carbon vacancy is a double donor, it counteracts the p-type doping. The present results are consistent with the observed low doping efficiency of acceptors,[11,12] and they explain the substantial self-compensation effects occurring while doping cubic SiC during growth with Al.[12] They support the diffusion model of Birnie,[13] whose analysis of experimental data suggested that self-diffusion of carbon atoms occurs via carbon vacancies, while Si atoms should diffuse via carbon vacancies and Si_C antisites.

In the C-rich material, the electrically inactive C_{Si} antisite is the dominant stoichiometry-compensating defect regardless of the position of the Fermi level. The slightly C-rich cubic SiC is thus a more suitable material for p-type doping. Its hole mobilities should be substantially higher, due to a lower concentration of acceptors necessary to achieve a given concentration of holes. The feasibility of growing C-rich material has been recently demonstrated via MBE techniques.[14]

Table II   Formation energies for neutral T_d symmetry point defects relative to the binding energy per atom in a stoichiometric cubic SiC.

defect	V_C	V_{Si}	Si_C	C_{Si}	Si_{TSi}	Si_{TC}	C_{TSi}	C_{TC}
formation energy [eV]	5.9	6.8	7.3	1.1	15.0	14.7	8.6	11.0

Pressure Dependence of Self-Diffusion Activation Energies in Si

In contrast to diamond and SiC, the activation energies for the various self-diffusion mechanisms in Si are very similar (within 0.5-1.5 eV), making the direct determination of their relative contributions very difficult. However, the pressure dependence of the diffusion coefficient can be used to distinguish between the contributions of the competing mechanisms.

For a crystal under pressure, the diffusion coefficient can be written as:

$$D = D_0 \exp[-(\Delta E^* + p\Delta V)/kT] \qquad (1)$$

where ΔE^* is the sum of the internal energies of formation and migration of the defect, and ΔV is the volume added to or subtracted from the crystal during the formation of the defect. The pressure dependence of the activation enthalpies for the three mechanisms should be different, since the volume of the crystal decreases upon the formation of an interstitial, increases with the formation of the vacancy and remains the same during the direct exchange. Since the migration energies are normally much smaller than the formation energies, the pressure dependence will be mainly determined by the changes in the formation energies. In fact, the use of pressure measurements for studies of diffusion mechanisms was suggested long ago, first for metals,[15] and then for semiconductors.[16]

In Fig. 3 we plot the energies and enthalpies of formation as a function of pressure for the tetrahedral and bond-centered interstitials, the vacancy, and the concerted exchange in their neutral charge states. The qualitative trends predicted on the basis of eq. (1) are evident in the figure, demonstrating that the $p\Delta V$ term dominates the pressure-induced changes of the activation enthalpy for diffusion. Although only the results for neutral defects are shown in the figure, their charge state dependence is minimal.

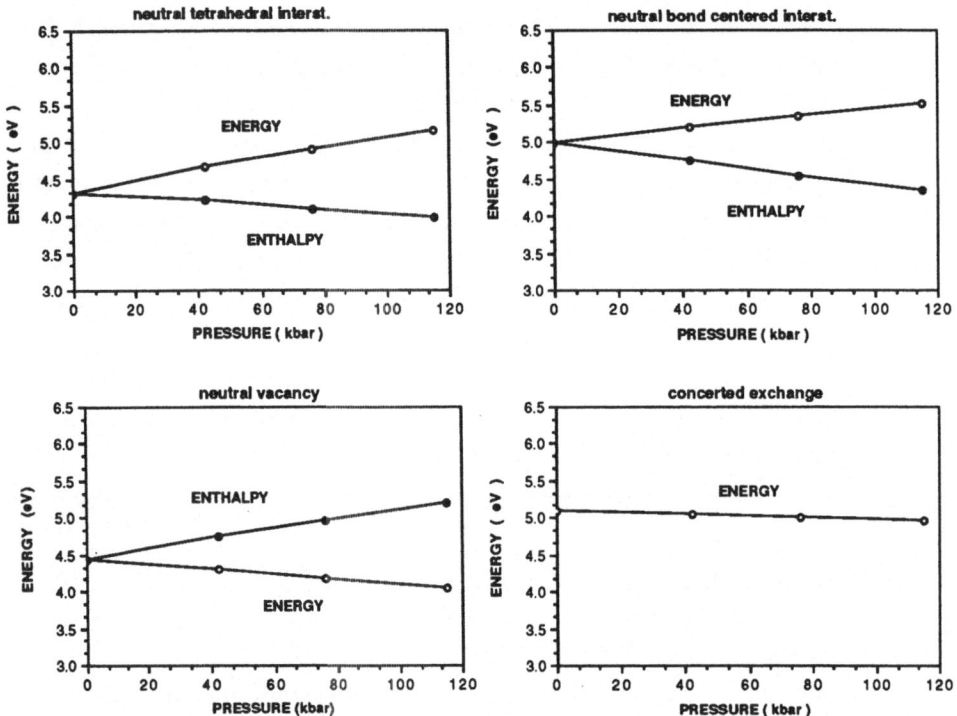

Fig. 3 Formation energies and the saddle point energy for the direct exchange as function of pressure for Si.

In analyzing the results and comparing to experiment one should consider several points: (i) Only the variation in the formation energies was taken into account in the present calculations. The entropy-dependent prefactor will thus have to be extracted from the temperature dependence of the experimental data. If a strongly non-Arrhenius behavior is found, it would indicate that several mechanisms are contributing in this temperature range and possibly a change in the dominant mechanism is occurring with pressure. (ii) Eq. 1 shows that if the activation energies of the competing mechanisms as well as their entropies are very similar (within a few kTs), the pressure induced changes in vacancy and interstitial mechanisms will cancel to a significant extent and the apparent contribution of the direct exchange will be exaggerated. However, for this to happen, both the activation energies *and* the entropies have to be nearly equal (within a few kT). This is an unlikely possibility at the temperatures of interest. Therefore, either a single mechanism will dominate in a given temperature range or a transition to a different mechanism will occur, manifesting itself in a non-Arrenius behavior or a large change in the pre-exponential factor. For example, the pre-exponential factor for As diffusion in Si changed by about three orders of magnitude at the pressure of 30 kbar.[17]

The only experimental data available so far[18] have shown a tenfold increase in the self-diffusion coefficient at a pressure of 35 kbar and temperature of 1000° C. Assuming no changes in the pre-exponential factor, this would correspond to a 0.2 eV decrease in the activation enthalpy. However, since the experiment was done at only one temperature and the pressure was not truly uniform, no firm conclusions can be drawn at this time. In general several measurements at various temperatures and pressures, as well as a careful comparison with the theoretical results, will be necessary before the dominant mechanism is uncovered and/or the relative contributions of the three competing mechanisms are determined.

REFERENCES

1. R. Car, P.J. Kelly, A. Oshiyama, and S.T. Pantelides, Phys. Rev. Lett. 52:1814 (1984).
2. R. Car, P.J. Kelly, A. Oshiyama, and S.T. Pantelides, Phys. Rev. Lett. 54:360 (1985).
3. Y. Bar-Yam and J. Joannopoulos, Proc. 13th Intern. Cont. on Defects in Semi conductors, edited by L. C. Kimerling and J. M. Parsey Jr., in: "The Metallurgical Society of AIME", New York, p. 261 (1985).
4. K.C. Pandey, Phys. Rev. Lett. 57:2287 (1986).
5. See, for example, the *Presentation Summaries* of the Diamond Technology Initiative Symposium, 12-14 July 1988.
6. Y. Bar-Yam and J. Joannopoulos, Phys. Rev. Lett. 52, 1129 (1984); Phys. Rev. B30:1844 (1984).
7. Y. Bar-Yam, J. Joannopoulos, and S. T. Pantelides, to be published.
8. J. Bernholc, A. Antonelli, T. Del Sole, Y. Bar-Yam, and S. Pantelides, to be published.
9. C. Wang, J. Bernholc, and R. F. Davis, to be published.
10. A. Antonelli and J. Bernholc, to be published.
11. H. J. Kim and R. F. Davis, J. Electrochem. Soc.: Solid State Sci. and Techn. 133:2250 (1986).
12. M. Yamanaka, H. Daimon, E. Sakuma, S. Misawa, and S. Yoshida, J. Appl. Phys. 61: 599 (1986).
13. D. P. Birnie, III, J. Am. Ceram. Soc. 60:C-33 (1986).
14. S. Kaneda, Y. Sakamoto, T. Mihara, and T. Tanaka, J. Cryst. Growth 81:536 (1987).
15. D. Lazarus and N. H. Nachtrieb, in: "Solids under Pressure", D. Paul and D. M. Warschauer, ed., McGraw-Hill, New York (1963).
16. S. M. Hu, in: "Atomic Diffusion in Semiconductors", D. Shaw, ed, Plenum, New York (1973).
17. E. Nygren, M. J. Aziz, D. Turnbull, J. M. Poate, D. C. Jacobson, and R. Hull, Appl. Phys. Lett. 47:105 (1985).
18. M. J. Aziz, E. Nygren, W. H. Christie, C. W. White, and D. Turnbull, Mat. Res. Symp. Proc. 36:101 (1985).

A COMPUTATIONAL METALLURGICAL APPROACH TO THE ELECTRONIC PROPERTIES AND STRUCTURAL STABILITY OF INTERMETALLIC COMPOUNDS

A.J. Freeman and T. Hong

Department of Physics and Astronomy
Northwestern University
Evanston, Illinois 60208-3112

J.-h. Xu

Shanghai Institute of Metallurgy
Chinese Academy of Sciences
Beijing, People's Republic of China

INTRODUCTION

There is more and more the growing recognition that technological advances depend strongly on a thorough understanding of the thermodynamic, mechanical, and electronic properties of materials. These, in turn, depend on our predictive ability regarding these properties. Understanding the structure and stability of the phases of compounds and alloys is an area of vital importance to materials science and technology. Such an understanding is starting to emerge from our first steps into carrying out all-electron quantum mechanical investigations on materials of aerospace interest. Briefly put, our emphasis has been on obtaining an understanding of the effects of alloying on bonding, crystal structure and phase stability of structural materials and to use this information to help design new alloy systems - in close collaboration with experimental efforts at a number of laboratories. In this paper, a brief indication is given of the progress made in a number of illustrative areas.

THEORETICAL/COMPUTATIONAL APPROACH: SELF-CONSISTENT LOCAL DENSITY FUNCTIONAL THEORY

The fundamental quantities of density functional theory are two observables, the total electron density (charge density) and the total energy of the system (due to the kinetic energy of electrons and the potential energy due to electrostatic interactions). An important theorem of density functional theory [1] states that given the correct ground state density, $n(r)$, this density rigorously determines all electronic properties of the system, and in particular its total energy. Furthermore, the total energy of a system can be expressed as a functional of the density n'(r) and this functional, E[n'(r)], has a minimum when n'(r) becomes equal to the ground state density.

The total energy as a functional of the density can be written in the form

$$E[n] = T[n] + U[n] + E_{xc}[n] \tag{1}$$

where T[n] is the kinetic energy of a system of non-interacting particles of density n; U[n] is the classical electrostatic energy due to the Coulomb interactions of a negative charge distribution of density n with itself, the interation of the charge density n with the positive nuclei, and the Coulomb interactions between the nuclei, and $E_{xc}[n]$ includes all many-body contributions to the total energy, in particular the exchange and correlation contributions.

To obtain single-particle Schrodinger equations, we decompose the total density into single particle densities which are expressed by single particle wave functions as

$$n(r) = \Sigma_i \, |\psi_i(r)|^2 \tag{2}$$

where the sum goes over all occupied states. Using this decomposition into single particle wave functions, the condition for E[n] to have its minimum leads to the so-called Kohn-Sham equations of the form

$$[-\frac{\nabla^2}{2} + V_{eff}(r)] \, \varphi_i = \varepsilon_i \, \varphi_i \tag{3}$$

where the effective potential is written as a sum of the Coulomb potential and the exchange-correlation potential:

$$V_{eff}(r) = V_C(r) + V_{xc}(r) \tag{4}$$

The Coulomb potential is related to the charge density via Poisson's equation

$$-\Delta V_C(r) = 4\pi \, e^2 \, n(r) \tag{5}$$

and the exchange-correlation potential is given by

$$V_{xc}(r) = \delta E_{xc}(r)/\delta n(r) \tag{6}$$

So far, the expressions given are formally rigorous. Clearly, the last term in equation (1) requires approximation. From a great number of calculations it became obvious that the following approximation, called the local density approximation (LDA), works surprisingly well: Using the known many-body energy, (n), of an electron in a homogeneous, interacting electron gas of density n, this exact result can be used to approximate the inhomogeneous case by setting

$$E_{xc}[n] \cong n(R) \, \epsilon[n(r)]dr \qquad\qquad (7)$$

In a sense, this approximation assumes that the leading terms in the exchange and correlation effects between electrons are of a fairly short range nature. Practical applications of the method as discussed below provide evidence that this may be indeed a realistic assumption. Using the LDA, the effective potential in the Kohn-Sham equations (3) can be evaluated explicitly via equations (4)-(6), provided the charge density is known. In turn, the charge density is defined by the solutions of the Kohn-Sham equations (3) via equation (2). Thus, the Kohn-Sham equations can be solved self-consistently: an initial charge density is constructed, e.g., by using overlapping atomic charge densities. From this ("input") charge density, the effective one-particle potential is calculated by solving Poisson's equation (5) and by evaluating the exchange-correlation potential (6) which is given by an analytic function of the density, n.

In the next step, which is numerically the most demanding, the differential equations (3) are solved. One possible way using a variational expansion of the single particle wave functions, ψ_i, in augmented plane waves is described below. Alternatively, an expansion in atomic orbitals can be employed for localized systems. After the eigenvalues and eigenfunctions are found, a new ("output") charge density can be constructed using Fermi-Dirac statistics in equation (2). This closes one iterative step. The output charge density is fed back into the evaluation of the effective-potential energy operator and the procedure is repeated until the output charge density equals the input charge density to within a given tolerance. To construct the input density for a given iteration, the input and output charge densities from previous iterations are combined to ensure damping and an efficient extrapolation. Depending on the system, between 10 and 50 iterations are necessary (magnetic systems such as Ni are more difficult to converge than systems with a low density of states at the Fermi level).

Once the self-consistent charge density is evaluated for a given choice of nuclear coordinates, the corresponding total energy can be found using equations (1)-(7). Next, the nuclei can be displaced and the procedure can be repeated. In this way, the energy hypersurface of the system can be established point by point. In principle, derivatives of the total energy with respect to displacements of atomic nuclei could be evaluated. However, this has not yet been fully implemented in the approaches discussed here.

ILLUSTRATIVE RESULTS

Structural Phase Stability of Ni3Al and Ni3V

To begin with, the advent of supercomputers has had the added benefit of permiting new methodologies and computational approaches to be implemented. In particular, it has led to a demonstration of the high precision and reliability of our total energy calculations. As a first example of our recent work, we describe briefly extensive studies of the structural phase stability and properties of Ni3Al with and without V additions. These studies may be considered as illustrating our basic approach to understanding the aluminides.

1. <u>Structural Stability and Structural Properties of Ni₃Al</u>. The nickel-rich aluminide, Ni_3Al, has well-known attractive properties for structural applications at elevated temperatures [2]. The flow stress increases with increasing temperature to a maximum value, which occurs in the temperature region between 600-800 C; Ni_3Al is the most important strengthening (γ') phase of the Ni-base superalloys; further, the density of Ni_3Al is significantly lower than that of the Ni-base superalloys. However, it has been widely recognized that unlike its single crystal form, polycrystalline Ni_3Al is extremely brittle; hence, to be useful as an engineering material, one has to overcome its severe embrittlement and to improve its ductility.

Since embrittlement in the ordered intermetallic compounds is considered to be mainly caused by an insufficient number of slip systems or poor grain-boundary cohesion, considerable effort has been made to increase the grain-boundary cohesion or to change the crystal structure from low symmetry to high symmetry (more slip systems, and hopefully, greater ductility). The results obtained showed that the brittleness associated with ordered intermetallic compounds (such as Ni_3Al) can be solved using physical metallurgical methods: for example, ternary additions (such as boron) into Ni_3Al greatly promotes its ductility [3]. However, since the crystal structure may be changed due to ternary additions, the central task is to keep the aluminide in a high symmetry (cubic) structure. In other words, understanding crystal stability may have important significance in order to develop materials with sufficient ductility. Over the last decade, the electronic structure and the magnetic properties of Ni_3Al in its cubic $L1_2$ structure have been extensively studied both experimentally [4] and theoretically [5-7]; still, a rather poor understanding of the structural stability of Ni_3Al based on the electronic structure remains [5,7,8].

In our studies of the structural stability of Ni_3Al, the total energy of the three different crystal structures - cubic $L1_2$, tetragonal DO_{22} and hexagonal DO_{19} - were calculated by means of the all-electron semi-relativistic linear muffin-tin orbital (LMTO) method [9] based on the local density functional approach. We find that of these three different structures, the (weakly ferromagnetic) $L1_2$ structure is the most stable in Ni_3Al (cf. figure 1). It is characterized by a small density of states at the Fermi level (4.42 states per eV - formula unit)), and shows a weak ferromagnetic instability with a magnetic moment of about 0.15 μ_B per atom. Further, the exchange energy gained by inducing the magnetic moment in $L1_2$ is much smaller than the band energy gain due to the structural transition into DO_{22} (or DO_{19}). The higher order coupling between the Ni-d and Al-p states may play an important role in accounting for the structural stability of Ni_3Al. The calculated lattice constant (3.55A), the bulk modulus (2.1 Mbar), and the heat of formation (44.4 kcal/mole) are in good agreement with experiment (3.56 A, 2.1 Mbar and 36.6 and 37.5 kcal/mole, respectively).

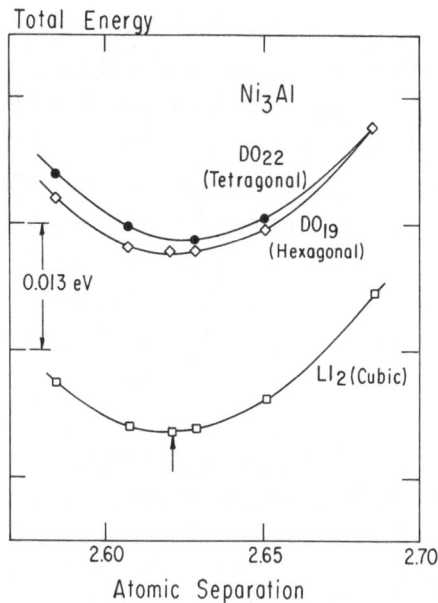

Figure 1.   The total energy as a function of the Wigner-Seitz radius and its structure dependence for Ni3Al: open and closed circles, triangles indicate Ll2, DO22 and DO19 structures, respectively.

In order to make use of our understanding of the phase stability in a practical way, we have studied the density of states for the different structures. We find that there is a correlation between the stability and the density of states at the Fermi energy in that the energetically favorable atomic arrangement of Ni3Al leads to a low density of states at the Fermi energy. This appears to be a universal feature in all the aluminides we have studied and, as we shall see later (cf. Figs. 2 and 3), permits us to make predictions about the possible stabilization of the cubic (L12) phase by means of ternary additions in other aluminide systems.

2. Site Preference and Solid Solution Strengthening: Ni3Al,V). Early on [8], we focused on solid solution strengthening as one of the technologically important features of Ni3Al: Ni3Al can accept into solution substantial alloying additions without losing its long range order. Clearly, knowledge of the preferential occupancy by ternary additions in Ni3Al is fundamental for understanding strengthening affects in Ni3Al and for the development of superalloys for practical use [11]. We therefore set out [8] to understand the effects of ternary additions in Ni3Al focussing, in particular, on the effect of V on the stability of the L12 structure of Ni3Al. (This was combined with separate studies of [10] Ni3Al and [12] Ni3V. For these pure systems [10,12], we found that the lattice constants and bulk modulus were in good agreement with experiment and other calculations). In

these all-electron total-energy local density studies, we compared the total energies of two different structures [$L1_2$-like and DO_{22}-like of $Ni_3(Al_{0.5}V_{0.5})$]. We found that ternary additions of V in the $L1_2$-like structure (i.e., V substitution on Al) are more stable than for V additions in the DO_{22}-like structure (i.e. V substitution on Ni sites). Thus, these model calculations indicate that the $L1_2$-like Ni_3Al can dissolve up to 50% V replacing Al in an ordered way and still retain the $L1_2$-like structure with a lower energy than in the DO_{22}-like structure. In addition, we found a slight hardening effect caused by the V addition in Ni_3Al; this hardness increment agrees with the experimental measurement on a Ni - 12.3% Al - 12.2% V ternary alloy studied by Westbrook and by Dimiduk [11] and earlier by Decker and Mihalisin [13].

3. Role of Density of States in Stabilizing Intermetallics. We have also carried out a careful analysis of the total and the Ni-d and V-d partial density of states for $Ni_3(Al,V)$ in the $L1_2$- and DO_{22}-like structures [12]. We found that the d-d hybridization between V and the host (Ni) appears in both the $L1_2$-like and DO_{22}-like structures; however, the d-d interactions exhibit different features for the $L1_2$-like and DO_{22}-like structures of $Ni_3(Al,V)$. The strong hybridization between Ni-d and V-d in the $Ni_3(Al,V)$ $L1_2$-like structure has a special prominent feature: a well separated bonding and antibonding region. This also turns out to be a noticeable common feature for the $L1_2$ structures of Ni_3Al, Ni_3V and $Ni_3(Al,V)$ [- cf. Figs. 2 and 3]. A deep valley separates the bonding and antibonding parts of the DOS. Obviously, the stability of the compound depends on the position of the Fermi level. It is thus expected, and found from our total energy studies, that the cubic Ni_3V phase, in which the Fermi level is located in the antibonding region, is an unstable compound and Ni_3Al will be a stable compound in the $L1_2$ structure, because the Fermi level lies in the bonding region of the density of states. For $Ni_3(Al,V)$ in the $L1_2$-like structure there are enough valence electrons due to the V addition to bring the Fermi level to a position just below the antibonding d-d hybrid states. Thus, the valence electrons fill all the bonding states and leave all the antibonding states empty; therefore, the strongest bonding effect occurs in $Ni_3(Al,V)$ in the $L1_2$-like structure. For this reason, it is also expected, from the rigid band sense, that Ti, Zr, Hf, Nb, Ta, Si and Ge will stabilize the $L1_2$ structure in Ni_3Al in much the same way as V, and possibly preferentially occupy Al sites.

Stabilization of the $L1_2$ Phase of $ZrAl_3$

As a further effort to understand the stabilization of high temperature structural materials in highly ductile phases, we have investigated the electronic structure and the cohesive properties of $ZrAl_3$ in the metastable $L1_2$ structure using the total energy local density linear muffin-tin orbital approach [10]. ($ZrAl_3$ was suggested to us by M. Fine of Northwestern based on his extensive work on this material). For this metastable structure, the calculated lattice constant (4.073A) is in excellent agreement with experiment (4.0731±0.0008A) and the bulk modulus (1.0 Mbar) is consistent with the observed

Young's modulus (196 GPa). The calculated heat of formation (39.0 kcal/mole) for cubic $ZrAl_3$ is consistent with the experimental value (34.0 kcal/mole) for the similar compound $TiAl_3$.

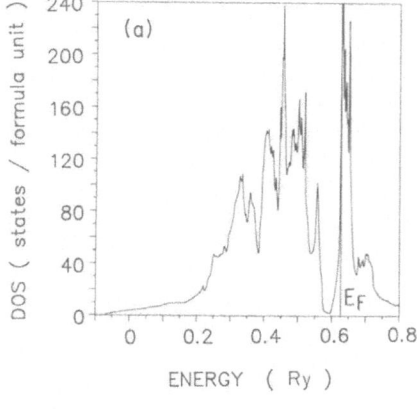

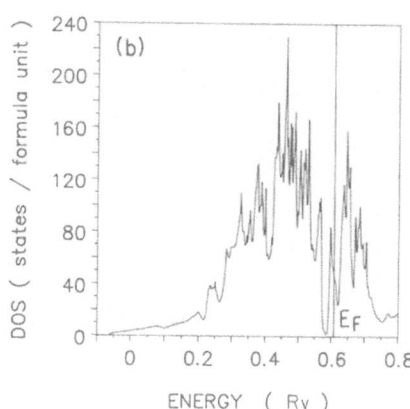

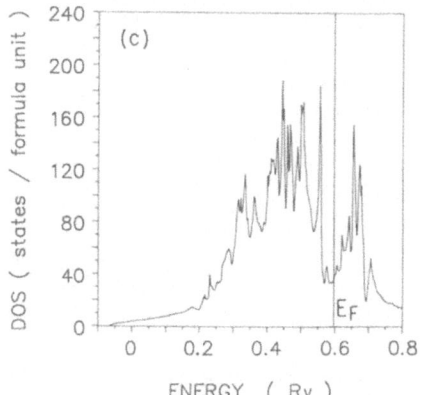

Figure 2. The total density of states for Ni3Al in the three different structures (a) $L1_2$, (b) DO_{22}, (c) DO_{19}.

Encouraged by these results, we set out to determine how to stabilize the cubic ($L1_2$) phase of $ZrAl_3$. Following the above described procedure we first determined the density of states (DOS) at the Fermi energy, $N(E_F)$. We found that $N(E_F)$ was high because E_F occurs in the antibonding peak in the DOS. This suggested that ternary additions, like Li, would lower E_F into the minimum in the DOS and hence stabilize the $L1_2$ phase. This prediction was studied with detailed calculations designed to study site preference for Li substitution. In these calculations, a supercell approach is employed for the $L1_2$-like structure of (i) $ZrLiAl_6$ (i.e., Zr substitution) and (ii) Zr_2LiAl_5 (Al substitution). The results of our total energy studies show clearly a preference for Al substitution and a lower $N(E_F)$ which coincides with the prediction of $L1_2$ stabilization. Thus, these supercell calculations support the simple rigid band concept and thus may be more generally applied to the aluminides.

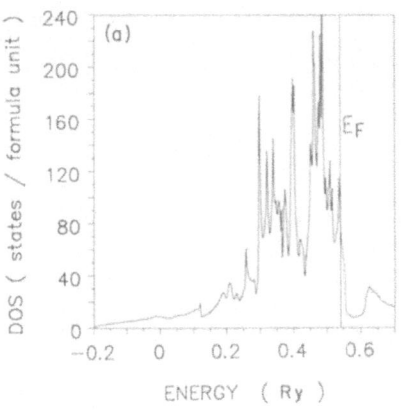

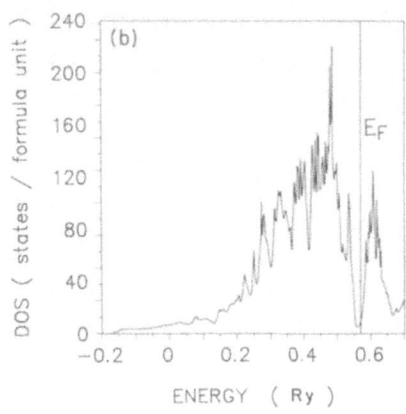

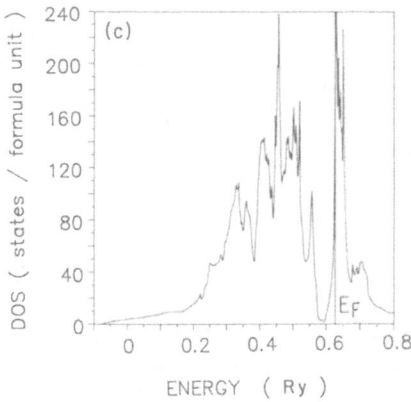

Figure 3. Comparison of the DOS for the L1$_2$ structure of (a) Ni$_3$Al, (b) Ni$_3$ (Al, V), and (c) Ni$_3$V

Structural Phase Stability of Titanium Aluminides

Most recently, we have initiated an in-depth study of the structural phase stability and electronic properties of several TiAl intermetallics (Ti$_3$Al, TiAl, and TiAl$_3$). To begin with, we studied the properties of pure Ti metal and have demonstrated that the total energy of the hcp phase is lower than that of the bcc and fcc structures. An examination of the density of states at the Fermi energy for the three structures shows, once again, that the N(E$_F$) is lowest for the hcp structure.

For the three TiAl compounds studied using our all-electron total energy approach, we find calculated equilibrium lattice constants in quite good agreement with experiment; the heat of formation and the bulk modulus also agree fairly well with the experimental results. In all cases for the stable phase of the compounds, the Fermi energy is on the bonding side of the DOS curve and a lower DOS value appears again at the Fermi energy - in agreement with our earlier studies cited above. This means that we may now have a good method for predicting which ternary additions would stabilize the L1$_2$ phase in TiAl$_3$ and in Ti$_3$Al. Some of these insights and predictions are now being fed to our experimental collaborators.

Combined Statistical Mechanical and Electronic Structure Approach To First Principles Determination of Alloy Phase Diagrams

Stimulated by the complexity of alloy phase diagrams of the intermetallics, notably Ti-Al, we have undertaken a first step towards the calculation of the phase diagrams of binary alloys entirely from first principles [15]. This has been a long-sought and important goal (which started with the work of Kikuchi [16]). In our approach, we start from a local density total energy supercell formulation. After constructing the grand partition function which permits the determination of the entropy from our calculated total electronic energies, we are, in principle, able to obtain all thermodynamic quantities. While this approach faces the same problem as obtaining the entropy in cluster variational methods [17], since one is presently restricted to relatively small unit cells, there are a number of important advantages: (1) solid solution and ordered phases can be calculated with the same numerical method and precision; (2) local environment effects such as charge transfer and chemical bonding are accurately described; (3) any crystal structure for which the total energy has been calculated can be easily included in the grand partition function; (4) the model is self-contained and only needs the total energy; and (5) no fitting and no breaking into pairwise interactions is necessary.

The result is a combined statistical mechanical and electronic structure approach which shows some promise. It should be emphasized that the grand partition function is constructed from volume dependent internal energies obtained from local density total energy supercell calculations. The illustrative results of first calculations for the Al-Li system show: (i) structural properties versus concentration in very good agreement with experiment and (ii) features on the Al rich side of the phase diagram of the fcc solid solution which are important for alloy formation. Since we are limited to small supercells at present, our method should only be applied to regions in experimentally known phase diagrams where complicated structures do not occur and where concentration waves with long wave lengths are not important. Within these restrictions it appears that our combined statistical, mechanical and electronic structure method shows promise for studying alloy phase diagrams entirely from first principles.

Interfacial Properties of Intermetallics: (001) APB Energy of Ni_3Al

We have started some first principles studies of the mechanical properties of the intermetallic compounds [18,19]. Because of their importance in understanding the high temperature properties of Ni_3Al, we have calculated from first principles the energy of formation of an antiphase boundary (APB). Two different approaches were used: (i) a thin film method [18] and (ii) a supercell method [19]. In the supercell approach, this required 16 atoms per supercell and in the thin film approach we set up a comparison of the total energy of the ordered single phase Ni_3Al with that for a 13 layer film representing the APB interface. Surprisingly, both methods gave the same result, namely $E = 140 \pm erg/cm^2$ - in very good agreement with the experimental result of Veyssiere et al.[20]. We are presently also analyzing these results in terms of our ability to compare charge densities so as to better understand the interface. The next step, i.e., undertaking a similar calculation of the (111) APB in Ni₃Al, is now in progress.

Interfacial Properties of Intermetallics: <111> {110} and <111>{112} APB energies of NiAl

Recently, we have been working on mechanical properties of NiAl based alloys [21]. Using the supercell method, the <111> {110} and <111> {112} APB energies of pure B2 NiAl and selected NiAl based alloys with ternary additions have been calculated. The APB energy values for pure B2 NiAl are extremely high, namely about 800 ergs/cm^2 which is several times larger than those for (111) APB in Ni₃Al. However, with some selected ternary additions, the APB energy values can be remarkably lowered to one third of those for pure B2 NiAl. These results indicate that the <111> slip systems may be activated by some alloying process in B2 NiAl.

The intermetallic compound NiAl has been studied intensively because of its potential aerospace applications at high temperatures: it has a high melting point, high ordering energy, good oxidation resistance and is relatively light [22-24]. However, its intrinsic brittleness at room temperature has been an unsolved problem. Deformation experiments have shown that the nature of slip is <001>(110) in [25] NiAl, which implies that only 3 independent slip systems exist. As the independent slip systems are orthogonal, there are no cross slips and therefore, von Mises criterion for ductility is not satisfied [26]. Since the Burgers vectors in the disordered bcc structure are 1/2 <111>, it is quite reasonable to believe that the ductility might be improved if 1/2<111> type slip was shown by an alloying process. Although some mixed results have been obtained [27,28] with the addition of Cr or Mn, the chance of achieving positive results by adding other elements has not been excluded.

As a first step, we have calculated anti-phase boundary (APB) energies for two most probable APB's, namely, 1/2<111> on {110} and 1/2<111> on {112} using the linear muffin-tin orbital (LMTO) method. For simplicity, we considered three smallest APB cells with 4, 6, 8 and 10 layers of (110) and (112) planes, corresponding to unit cells with 8, 12, 16, and 20 atoms, respectively. In these first calculations, no relaxation was allowed at the APB interface. Figure 4 shows such a structure for the 8 layer case. The cells for the 1/2<111>{110} calculations are orthorhombic, while the cells for the 1/2<111>{112} calculations are monoclinic. The quite different nearest neighbors and second nearest neighbors in the 1/2<111>{110}, 1/2<111>{112} APB's and in the original B2 cell make the lattices behave differently.

We started calculations on B2 symmetry lattice by using seven different unit cells: the original one, three orthorhombic and three monoclinic cells. The latter six are chosen to have the same size as those for the APB calculations except that no APB was introduced. The calculated results from the three different symmetry cells are exactly the same within the precision of the method and give us confidence in the accurracy of the APB energy calculations. By comparing the total energy for the cells with and without an APB, we found the APB energy to be around 800 ergs/cm^2 for all the different layer calculations (listed in Table I). As the distance between two consecutive APB layers is increased, the APB energy values decrease gradually, as expected. To test the convergence of our calculations, we carried out an APB calculation using 10 layers for the 1/2<111> on {112} case and the APB energy value is decreased from that of 8 layers value by only a small

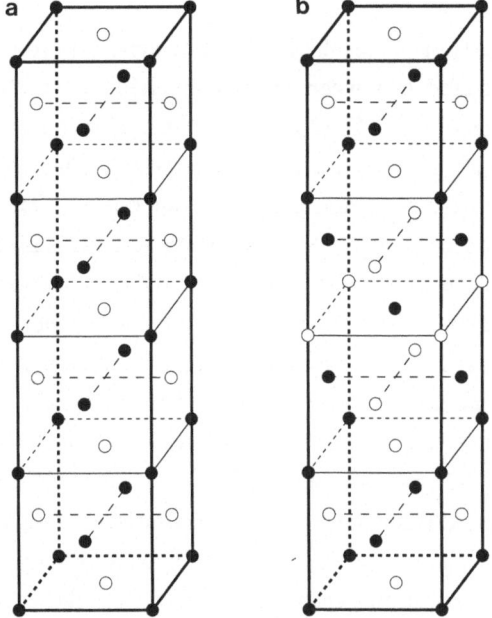

Figure 4. Eight layer cells for (a) pure NiAl and (b) with a 1/2 <111> (110) APB introduced.

amount. This result suggests that the interactions between two APB's only extend to a certain range. After this range, the two consecutive APBs can be considered as totally independent, and the APB energy would become independent of the number of layers. From our tests on 1/2<111> on {112}, it is quite clear that the value obtained for 10 layers would be a very good approximation for the extrapolated APB energy. Similarly we can expect that the APB energy value also becomes layer-independent for 1/2<Ī11> on {110} when the number of layers exceeds 10 layers, as that would affect only high order nearest neighbor configuration.

These calculated APB energies are extremely high, especially when compared to the result for APB in [29] Ni$_3$Al [140 ergs/cm^2 for (100) APB]. They are also considerably larger than the simple estimate made by D.I. Potter [30] for NiAl. The APB energies for 1/2<111>{112} are comparable with those of the 1/2<111>{110}. The energies for breaking a Ni-Al bond {1/2(E_{Ni-Ni}+E_{Al-Al})-E_{Ni-Al}] are also listed in Table I. As in the APB energy case, they also showed a decreasing trend with increase of number of layers. These bond breaking energies for 1/2<111>{112} are a little larger than those for 1/2<111>{110}. [Note that when we calculated the bond breaking energies, we only considered the first nearest neighbors. Remembering that the second and higher nearest neighbors also contribute to the bond breaking, these values are systematically underestimated.]

From an analysis of the results (including calculated layer by layer densities of states, we can understand that when introducing APB's, the strongest Ni d - Al p hybridization DOS peak in B2 NiAl is substantially reduced, while the Ni d - Ni d hybridization DOS peak at higher energy is increased to a much greater height. And the hybridization between the Ni d and the Al p component is much stronger than the corresponding combination of Ni d - Ni d and Al p - Al p. This gives the very high APB energy. Therefore, to decrease the APB energy, we need to either weaken the stronger Ni d - Al p bond or strengthen the Ni d - Ni d and the Al p - Al p bonds. From a consideration of the DOS, we expect to get a weaker transition metal d and Al p hybridization effect if we have some transition metals with fewer d electrons as the d peaks of the transition metal will shift upward in that case. As a consequence, we have calculated the APB energy for the 6 layer 1/2<111>{110} case with substitution of Al or Ni by Cr, Mn, V.

Since the upper APB surface and lower APB surface are on the same plane in the 4 layer APB cases, the interaction between APB's should be very strong. To avoid this situation and still retain reasonable calculational speed, we selected a 6 layer APB cell for the calculations with ternary additions. As we have seen in the pure NiAl APB case, we expect that the APB energy values for the 1/2<111>{110} and 1/2<111>{112} cases should have the same trend. Therefore, we concentrated our calculations only on 1/2<111>{110}. In order to have maximum effect on the APB interface, the ternary additions are put on two APB interface planes to replace either Ni or Al on those two planes.

Table I. APB energies for 1/2<111>{110} and 1/2<111>{112} lattices of pure B2 NiAl.

number of layers in supercell	APB (ergs/cm^2)	
	1/2<111>{110}	1/2<111>{112}
4-layers	1130	1050
6-layers	1000	950
8-layers	880	890
10-layers		885

The calculated APB energy values are listed in Table II. From the results we can see clearly that if the ternary addition substitutes for Ni, the APB energy would be dramatically lower with a ternary addition which has fewer d electrons. This is exactly what we expected from the consideration of the calculated DOS for the pure NiAl case. However, if the ternary addition substitutes for Al, things are different. Although, the APB energy values with ternary additions are substantially decreased from that of pure NiAl, the APB energy with the addition of V is twice as large as for addition of Cr. Our results show that the V atom would be a very favorable choice to lower the APB energy if V were to substitute for Ni. Arguments made by R. Darolia et al. [31] and our site preference

Table II.   APB energies of B2 NiAl with ternary additions (6 layers 1/2<111>{110}) to the interface APB planes.

composition Al_6Ni_6	APB energies (ergs/cm^2) 1000
$Al_6Ni_4Mn_2$	740
$Al_6Ni_4Cr_2$	510
$Al_6Ni_4V_2$	250
$Al_4Cr_2Ni_6$	250
$Al_4V_2Ni_6$	550

calculations show that V substitutes for Al in NiAl. Unfortunately, our calculated APB energy for this case is more than twice as large as that of substituting V for Ni, even if it is only about half of the value for pure NiAl. Recently, Darolia et al.[31] performed some experiments for testing the role V atoms play in NiAl. Somewhat disappointing results were obtained as the V containing specimens exhibit higher flow stresses and decreased ductility compared to pure stoichriometric NiAl. Their results indicate that the effect of the addition of V to decrease the APB energy may have been offset by some other factors (i.e., strengthening effects). Hence, the simple consideration of only the APB energy may not be enough to the ductility problem.

Finally, we note that the addition of Cr to the APB interface plane lowers the APB energy substantially for Cr substitution for Al (its site preference). From the mixed results [27,28] obtained to date, it is not clear, however, if these predictions on an unrealistic model system may be indicative of the actual behavior of the real material.

ACKNOWLEDGEMENTS

We are grateful to Wei Lin for carrying out the site preference calculations and to D. Dimiduk, R. Darolia and R. Fields for helpful discussions and collaboration. This work was supported by the Air Force Office of Scientific Research under contract #F 49620-88-C-0052, and grant #88-0346, and a computing grant at the Wright-Patterson AFB Supercomputing Center.

REFERENCES

1.   W. Kohn and L.J. Sham, Phys. Rev., 140: A1133 (1965).
2.   J.H. Westbrook, ed. Intermetallic Compounds (Wiley, New York, 1967)
3.   K. Aoki and O. Izumi, Nippon Kinzok Takkaishi, 43:1190 (1979);
          C.T. Liu and C.C. Koch, in Proceedings of a Public Workshop
          on Trends in Critical Materials Requirements for Steels of the Future:
          Conservation and Substitution Technology for Chromium (NBSIR-
          83-2679-2, Washington, DC, June 1983)
4.   N.R. Bernhoeft, I. Cole, G.G. Lonzarich, and G.L. Squires, J. Appl.
          Phys. 53:8207 (1982); T.I. Sigfusson, N.R. Bernhoeft, and G.G.
          Lonzarich, J. Appl. Phys. 53:8207 (1982); T.I. Sigfusson,
          N.R. Bernhoeft, and G.G. Lonzarich, J. Phys., F14:2141 (1984)
          and references therein.
5.   D. Hackenbracht, and J. Kubler, J. Phys., F10: 427 (1980).
6.   J.J.M. Buiting, J. Kubler, and F.M. Mueller, J. Phys. F13:L179 (1983).
7.   B.I. Min, T. Oguchi, H.J.F. Jansen, and A.J. Freeman, J. Mag. Magn. Matls.,
          54-57: 1091 (1986)..
8.   J.-H. Xu, T. Oguchi, and A.J. Freeman, Phys. Rev. 36:4186 (1987).
9.   O.K. Andersen, Phys. Rev., B12:3060 (1975).
10.   J.-H. Xu, A.J. Freeman and T. Oguchi, (to be published)
11.   J.H. Westbrook, Ordered Alloys, Physical Metallurgy and Structural
          Applications, Claitors, Baton Rouge (1970), p. 1; D.M. Dimiduk,
          Solid Solution Strengthening of Ordered Ni_3Al, (unpublished).
12.   J.-H. Xu, T. Oguchi, and A.J. Freeman, Phys. Rev., B35: 6940 (1987).
13.   R.F. Decker and J.R. Mihalisin, Trans. ASM, 62: 481 (1969).
14.   T. Hong, T.J. Watson-Yang, A.J. Freeman and T. Oguchi, Bull. Am. Phys.
          Soc., 32:No. 3, p. 413 (1987).
15.   R. Podloucky, R., H.J.F. Jansen, X.Q. Guo and A.J. Freeman, Phys. Rev. B.,
          (to appear)
16.   R. Kikuchi, Phys. Rev., 81:988 (1951).
17.   See, eg., D. de Fontaine in "Alloy Phase Diagrams", eds. L.H. Bennett,
          T.B. Massalski, and B.C. Giessen (North-Holland, 1983), p. 149

18.   A.J. Freeman , C.L. Fu, and J.I. Lee, Bull. Am. Phys. Soc., 32: No.
          3, (1987) p. 772.
19.   J.-H. Xu, A.J. Freeman, and T. Oguchi, (to be published)
20.   P Veyssiere, Phil. Mag. 50:189 (1984).
21.   T. Hong and A.J. Freeman, (to be published).
22.   D.M. Shah et al. (unpublished).
23.   H.L. Lipsitt (unpublished).
24.   D.E. Polk (unpublished).
25.   A. Ball, et al., Acta Metallurgica 14: 1349 (1966); 14: 1527 (1966).
26.   R. von Mises, Z. Angew. Math. Mech. 8:161 (1928).
27.   C.C. Law et al. (interim report).
28.   D.B. Miracle et al., Materials Research Society Symp. Proc. on High
          Temperature Ordered Intermetallic Alloys (Boston, 1988).
29.   J. Xu and A.J. Freeman (to be published).
30.   D.I. Potter, Mater. Sci. Eng. 5, 201 (1969/1970).
31.   R. Darolia et al., Materials Research Society Symp. Proc. on High Temperature
          Ordered Intermetallic Alloys (Boston, 1988).

THEORY OF DEFECTS IN SOLIDS AND THEIR INTERACTIONS

A. Barry Kunz

Department of Physics
Michigan Technological University
Houghton, MI 49931

INTRODUCTION

Materials lie at the core of our technologically oriented society. Increasingly, it becomes necessary as the search for more efficient products or lower cost fabrication intensifies, to be able to design specific materials to achieve the needed design goals. It would be most desirable to be able to use the tools of theoretical materials science combined with advanced computational methodology to accomplish this task.

The rapid advances in computational resources during the last two decades have made it possible to use computationally intense theoretical methods to predict the detailed properties of materials. Furthermore, it may be possible in a limited number of cases to use these predictive capabilities to design materials intended for specific purposes. Such design may transcend the normal range of even such a broad discipline as traditional materials science and may also involve such varied areas as biology, catalysis, or electronic materials.

Computer simulation is adding a new dimension to scientific investigation in the ever-growing disciplines of materials science and technology. In particular, computer experimentation is establishing its greatest value in those situations where the gap between laboratory measurement and theoretical explanation is large. The experimentalist is concerned with obtaining factual information concerning states and processes. Challenged by the need to explain measured physical phenomena, the theorist invents model behavior which sometimes owes its existence to a chosen symmetry or to an assumed process. As is common in scientific investigation, the investigator may find that the theoretical analysis is very complex and demands 'arbitrary' simplification. As a result, the validity of a comparison between a particular theoretical prediction and experimental interpretation is frequently questioned due to the complexity of the experimental interpretation and/or the simplicity of the theoretical model.

In many ways, computer simulations, or "computer experiments", are beginning to alleviate these bottlenecks to progress. The computer experimentalist invents mathematical models for the physical world in an attempt to 'simulate' the behavior of a particular complex observation. Nonlinearity, lack of symmetry, and a large number of degrees of freedom are much less severe obstacles to investigation by simulations in contrast to typical theoretical approaches. In many instances, information not readily accessible to laboratory experiment is obtainable.

The aim of point-defect simulations is to provide detailed knowledge of the local electronic and atomic structure near defects present in real solids. From this knowledge the aim is to predict those macroscopic properties which are defect controlled. In addition, such methods when applied to perfect as well as defective crystals permit one to obtain interatomic potential from total energy calculations. The quantitative modelling of actual systems, rather than computer experiments on idealized models, is the goal at this level as well as to generate potentials for atomistic modelling.

The major objectives include technological links of three main types:
(a) The prediction of defect behavior beyond existing data bases.
(b) The limiting of searches for specific characteristics to short lists.
(c) The prediction of behavior for processes with time scales which are inconveniently short.

These links normally involve defect properties (such as thermodynamic quantities, or atomic geometries associated with specific electronic structure) which depend on total energies, rather than on one-electron eigenvalues. This emphasis, in turn, includes a significant role for interatomic potentials.

Therefore, the essential content of such a fundamental theory must be such that it permits the computation of quantities of essentially chemical interest such as bond strengths, bond lengths, and other associated chemical-type data. Furthermore, the basic theory chosen must not depend upon the existence of periodic symmetry (e.g. energy band theory), and in full generality should not depend upon high point symmetry either. Basically, one must be able to perform total energy type calculations. The necessary computations are computationally intensive. Fortunately the basic theory elements needed for this project have been developed and largely implemented in a computer code called ICECAP. This is a result of the interactions between the MTU physics group, the solid state theory group at the University of Manitoba, and the solid state group at Harwell AERE, UK. The state of ICECAP implementation is seen in the next section.

ICECAP: CONCEPT AND IMPLEMENTATION

Lattice defects in crystalline materials, often in combinations that are difficult to resolve experimentally determine many technologically important properties. Reliable computer simulation of such defects is therefore of potential value. For point defects, an attractive approach is to use quantum mechanics to describe the response of the region containing the defect to that defect, and to embed this region in a potential derived from that of a perfect lattice, allowed to respond to the defect. The response of the embedding crystal may then be described in terms of a much simpler model, applicable to a weakly-perturbed perfect lattice. Such models either exist, or may be expected to emerge, for ionic, molecular, covalent, and simple-metallic crystals. In the present case, we refer specifically to non-metals. In many of these materials the shell model[1], based on classical point charges and masses, interacting by simply-parameterized potentials, has been successful in correlating perfect-lattice equilibrium configurations, and static and dynamic elastic and dielectric properties within the harmonic approximation[2]. We shall therefore think of the embedding lattice in terms of the shell model. The region about the defect called the cluster will be referred to in terms of the unrestricted Hartree-Fock self-consistent field approximation[3], augmented by explicit correlation corrections[4].

For a cluster embedded in a classical lattice, special care must be taken to ensure that mathematical and physical consistency are maintained in dividing the system into segments. This topic has been discussed by Kunz and Klein[5], who also introduced a formal procedure based on localizing potentials to deal with the short-range quantum-mechanical cluster-lattice interaction. The present method at the self-consistent field level is an extension of that of Kunz and Klein.

Simulation of an infinite lattice containing a point defect represented by a cluster may be carried out by minimizing the total energy with respect to parameters that define the lattice configuration, including the defect. For shell-model ions, the configuration parameters are core and shell positions. For a cluster in the Hartree-Fock approximation they are variational parameters of the electronic wave function, and the nuclear positions. The one-electron Hartree-Fock functions are frequently based[6] on a linear combination of atomic-type orbitals centered on ionic and other suitable sites, consisting of spherical harmonics with gaussian radial dependences of given ranges. States other than ground states may be analyzed by applying symmetry constraints to the calculation. The total-energy minimization procedure described above is a practical proposition, and is, in fact, implemented in a general, user-friendly program, named ICECAP[7]. Recent applications[8] of ICECAP, incorporating Kunz-Klein Localizing Potentials (KKLP) to color centers, are demonstrating that calculations in which lattice configuration and defect structure are simultaneously self-consistent are possible[9,10].

ICECAP: One-Electron Problem

We begin by relating the problem of a small cluster in the Hartree-Fock approximation, containing N_A electrons, to the problem of the large crystal, containing N electrons. The cluster-Fock equation depends in part on occupied states of the embedding lattice. This dependence, in our formalism, is due to a subsidiary condition termed the Kunz-Klein Localizing Potential (KKLP). KKLP is considered a localizing potential because the cluster electrons see lattice ions as weak perturbing potentials.

The short-range potential is evaluated once and for all for a given crystal lattice by applying the same localization condition to each species in a perfect-lattice calculation. The use of a perfect-lattice short-range potential with a defect cluster is consistent with the idea that the cluster should contain all significant deviations from perfect-lattice electronic structure. The N-electron Hamiltonian is:

$$H = \sum_{i=1}^{N} (-h^2/2m)\, \nabla_i^2 - \sum_{i=1}^{N} \sum_{I} Z_I\, e^2\, |\vec{r}_i - \vec{R}_I|^{-1}$$

$$+ \sum_{i>j} e^2\, |\vec{r}_i - \vec{r}_j|^{-1} + \sum_{I>J} Z_I Z_J\, e^2\, |\vec{R}_I - \vec{R}_J|^{-1}.$$

We obtain the Bohr-Rydberg atomic units used throughout this work by setting $e^2 = 2$, and $(h^2/2m) = 1$. We use a single Slater determinant to approximate solutions to the Hamiltonian; and, the one-electron functions of the Slater determinant satisfy the Fock equation:

$$F(\underline{r})\, \varphi_k\, (\underline{r}) = \varepsilon_k\, \varphi_k\, (\underline{r}), \quad k = 1,2,...N, \tag{1}$$

where the Fock operator F is:

$$F(\underline{r}) = -\nabla^2 - 2 \sum_{j} Z_j\, |\vec{r} - \vec{R}_j|^{-1}$$

$$+ 2 \int d\underline{r}'\, |\vec{r} - \vec{r}|^{-1} \int d\underline{y}\, \delta\, (\underline{y} - \underline{r}')[1 - P(\underline{r},\underline{y})\, \rho(\underline{r}', \underline{y}) \tag{2}$$

$$\rho(\underline{r}', \underline{y}) = \sum_{k'=1}^{N} \varphi_{k'}\, (\underline{r}')\, \varphi_{k'}^*\, (\underline{y}) , \tag{3}$$

where $\underline{r} \equiv (\vec{r},s)$, $\vec{r}$ is position and s is spin, $\vec{R}_j$, and Z_j are nuclear positions and charges, and $P(\underline{r},\underline{y})$ is the electron pairwise interchange operator. In the unrestricted Hartree-Fock approximation (UHF), φ_k is a spin eigenstate. ρ is the Fock-Dirac one-particle density. Now consider a cluster A within the crystal, having N_A electrons. These N_A electrons are assumed to occupy a manifold of states, denoted k(A), localized in the cluster's vicinity. The N_A occupied states k(A) satisfy a Fock equation that is obtainable by unitary transformation from the orthonormal set φ_k of Eq. (1). The remaining $N_B = (N - N_A)$ new occupied basis functions, denoted k(B), along with k(A) are complete over the original manifold. One sees:

$$(F_A + U_A)\, \varphi_k = \varepsilon_k \varphi_k, \quad k = 1,2,\ldots N_A, \, . \tag{4}$$

where F_A is that part of F, Eqs. (1) and (2), that involves nuclei and electrons in the cluster, and U_A^S is due to those N_B electrons and the nuclei in the environment. Thus:

$$F = F_A + U_A \tag{5}$$

In Eq. (4), U_A clearly depends on k(B). Now the purpose of using the cluster is so that we will not have to solve for the states k(B) of the embedding lattice simultaneously with the states k(A) of the cluster.

In most non-metals, the electronic density of k(B) is particularly well-localized about lattice nuclei or possibly bond centers. For ionic systems, in zeroth order, the lattice potential U_A has been thought of as Madelung potential, arising from point-charge ions:

$$V_A^M = -2 \sum_{j(B)} I_j \, | \vec{r} - \vec{R}_j |^{-1} \tag{6}$$

where $I_j = (Z_j - N_j)$ is the ionic charge of ion j, and N_j is the number of electrons associated with ion j. The remainder of U_A is the short-range part V_A^S, arising from the quantum-mechanical nature of the electrons. Thus:

$$U_A = (V_A^M + V_A^S) \tag{7}$$

$$V_A^S = -2 \sum_{j(B)} N_j \, | \vec{r} - \vec{R}_j |^{-1}$$

$$+ \int d\underline{r}' \, | \vec{r} - \vec{r}' |^{-1} \; d\underline{y} \, \delta \, (\underline{y} - \underline{r}') \, [1 - P(\underline{r}, \underline{y})] \, \rho_B \, (\underline{r}', \underline{y}) \, . \tag{8}$$

A subsidiary condition can now be applied to the cluster-Fock Equation (4) to modify the localization of V_A^S relative to the cluster.

The procedure for applying a subsidiary condition to the Fock equation has been given generally by Gilbert.[11] It is based on the fact that the Fock-Dirac one-particle density operator ρ_A is a projection operation onto the manifold of occupied states k(A):

$$\rho_A = \sum_{k'=1}^{N_A} |k'><k'|, \tag{9}$$

where |k'> is the state vector corresponding to orbital $\varphi_{k'}(\underline{r})$. Using the theory due to Gilbert' one may add a term to Eq. (4) and obtain a modified Fock equation which leaves the total energy unchanged.

$$(F_A + U_A + \rho_A W \rho_A)|k\rangle = \Pi_k |k\rangle, \quad k = 1,2,...N_A. \tag{10}$$

Unoccupied (virtual) orbitals |k'⟩ are unaffected by the procedure.

It is not necessary to use the same W for all orbitals. However, if one changes W, a non-orthogonal solution-set emerges. Since W is arbitrary, we pick it to be $-V_A^S$. This results in the equation which we solve:

$$[F_A + V_A^M + V_A^S - \rho_A V_A^S \rho_A] \,|\, j\rangle = \pi_k |k\rangle \tag{11}$$

The present formulation assumes that there is some distance away from a defect such that the first order density matrix for atoms/ion/molecules (building blocks) located farther from the defect than this distance are, in the Hartree-Fock limit, the same as for the perfect system. Having obtained this solution the long and short range parts of the potential surrounding the cluster are known. Let us examine how the total energy of an infinite crystal containing a molecular cluster is related to the modified cluster-Fock equation (11). Now between Eq. (4) for ε_k and Eq. (11) for π_k there has been a unitary transformation in the cluster manifold k(A), so φ_k in Eq. (4) does not correspond to |k⟩ in Eq. (11). Nevertheless, the trace is invariant, so it follows that:

$$\sum_{k(A)} \varepsilon_\kappa = \sum_{k(A)} \langle k|F_A + U_A|k\rangle = \sum_{k(A)} (\pi_k + \langle k|V_A^S| k\rangle), \tag{12}$$

where we have used the projection property of ρ_A on k(A). This plus the fact that the total energy of the cluster is

$$E = \frac{1}{2} \sum_{k=1}^{N_A} (\varepsilon_k + f_f) + V_c. \tag{13}$$

where V_c is the classical cluster-lattice interaction term, yields a convenient expression for the total system energy.

Electron Correlation

The simple way to include correlation is now presented. The use of Hartree-Fock for obtaining energy band results has been seen highly inaccurate for both alkali halides and also rare gasses since at least 1970[12]. One needs to account for correlation effects if quantitative accuracy is to be achieved. It is known that these consist of three principal parts[13]: electron-electron dynamic correlation, orbital relaxation, and electron-hole attraction effects. The state by state self-consistent Hartree-Fock method employed by us and described previously incorporate directly the latter two effects. Therefore one need only incorporate the electron-electron dynamic correlation effect. This effect is well known to be significant even for tightly -bound situations.[12]

Correlation methods acceptable for solid state calculations of total energy are constrained to be size-consistent.[14] The statement that a correlation method need be size consistent simply implies that if one were to consider a homogeneous system containing a large number N, similar building block units, that the total energy of this system would be directly and linearly proportional to N. This property is also termed extensive. In a formal many body sense an extensive (size-consistent) approach is one which obeys the link-cluster theorem.[14]

One may choose a method based on many-body perturbation theory (MBPT).[14,15] In the normal single-reference application that is appropriate here, one divides the many-electron Hamiltonian H into two parts, namely a zero-order Hamiltonian H_o whose eigenvalues and eigenfunctions are known, and a perturbation V:

$$H = H_o + V. \tag{14}$$

Usually H_0 would be chosen to be the sm of one-electron Fock operators,[14] but that is not convenient here. Rather, we pick H_0 to be the sum of the one-electron localized-orbital (modified Fock) operators from Eq. (11).

$$H_o = \sum_{i=1}^{N_A} [F_A(\underline{r}_i) + V_A^M + V_A^S - (\rho_A V_A^S \rho_A) \underline{r}_i] \tag{15}$$

where ρ_A is defined in Eq. (9).

Now from having solved the local orbital equation, we know the zero-order eigenvalue and eigenfunction of H_0, Eq. (15). We now estimate the corresponding eigenvalue and eigenfunction of the full Hamiltonian H:

$$H\Psi_I = E_I \Psi_I. \tag{16}$$

Formally:

$$\Psi_I = [1 - (H_o - E_{Io})^{-1} (1 - p_I)(E_I - V - E_{Io})]^{-1} \Phi_I, \tag{17}$$

where p_I is the projection operator onto the state Φ_I. If size-consistency is maintained, this leads to the perturbation series:

$$E_I = E_{Io} + \langle \Phi_I |V| \Phi_I \rangle$$
$$+ \langle \Phi_I |V(H_o - E_{Io})^{-1} (1 - p_I) (-V)| \Phi_I \rangle + ... \tag{18}$$

It follows that, to second order:

$$E_I = (E_{Io} + V_{II}) + \sum_{J=I} V_{IJ}V_{JI} (E_{Io} - E_{Jo})^{-1},$$ (19)

where

$$V_{IJ} = <\Phi_I |V| \Phi_J >.$$ (20)

Example of A Charged Defect: V_k Center

The V_k center, or self-trapped hole, in alkali halides has long been considered as a negatively charged diatomic halide molecular-ion (X_2^-; X is the halogen) oriented in one of twelve equivalent [110] directions. Consider LiF, for which V_k center is a well characterized defect system. Its geometry, transition energies for hole excitation and hyperfine interaction with neighboring nuclei have been well-established [16,18]. On the theoretical front, Jette et al.[19] have tried to interpret the experimental results in terms of the so-called 'molecule in crystal' approximation. They have taken self-consistent-field molecular-orbital results for the Free F_2^- molecular-ion wavefunction (obtained by Gilbert and Wahl[20] using the restricted Hartree-Fock method) and have combined the results with a classical description of the surrounding lattice to include lattice-distortion and polarization. Their calculated results show an order-of-magnitude agreement with the experimental results for the hole excitation and the hyperfine constant emphasizing the need for more sophisticated calculations. With similar approximations, Cade et al.[21] have then used the HADES methodology to simulate the V_k center in alkali halides. HADES[22] uses pair potentials to mimic lattice interactions and a shell model to account for polarization in the lattice.[23] Therefore, HADES is inadequate in dealing with electronic states. Cade et al. have used the potential energy curves of the Free F_2^- molecular-ion (obtained by using a valance bond pseudopotential calculation) to calculate the transition energies for the hole excitation. The calculated results show only qualitative agreement with the observed transition energies.

With the availability of the UHF[3] and ICECAP[8] program packages, we have embarked upon such a detailed analysis of the V_k center in alkali halides involving improvements in the treatment of lattice distortion and polarization and a more accurate description of the electronic structure of the molecular-ion. In this paper, we report the results of our investigation of the ground state of the Vk center in LiF and assess the role of accurate solid state modelling for a successful description of charged defects in ionic crystals. We anticipate significant level distortion will be caused by the X_2^- ion defect.

We use the unrestricted Hartree-Fock self-consistent-field (UHF-SCF) approximation to describe the electronic structure of the molecular cluster consisting of the Vk center. We note here that the UHF-SCF approximation has the advantage of allowing for a proper description of spin-density at a nucleus by including the spin-polarization effects due to the self-trapped hole. Correlation corrections are incorporated using Rayleigh-Schrodinger perturbation theory in an implementation described by Goalwin and Kunz.[9]

In the present work, the V_k center in LiF is simulated in several well-defined models. The molecular cluster is then embedded either in a limited point-ion array simulating only the Madelung field or in a shell-model lattice using ICECAP which treats the defect electronic structure and the lattice response to the defect in a self-consistent way. Thus the simulation models considered are (i) [F_2^-] molecular-ion, (ii) [F_2^-] in point-ion array, (iii) [F_2^-] in shell model lattice, (iv) [$F_2^- + 10\ Li^+$] cluster in point-ion array and (v) the same cluster in a shell-model lattice.

Basis sets associated with fluorine in the F_2^- molecular-ion and near-neighbor Li^+ ions used in our calculations are of gaussian type. They are originally taken from Huzinaga's compilation[14] and are split to a contraction of (4,2,1/3,1) from (4,3/4) for F^- and to (3,1) from (4) for Li^+ ions. Shell-model parameters for LiF used in ICECAP calculations are taken from Caltow et al.[24] which were referred to as set I in the reference.

Referring to Table I, the internuclear separation and stabilization energy for the free F_2^- molecular-ion simulating the V_k center comes out to be 3.65 bohr and 1.04 eV respectively. In comparison, Gilbert and Wahl[20] and Woon[25] have calculated the separation of about 3.60 bohr and 3.65 bohr respectively using a different basis set than the present one for the F_2^- molecular-ion. Now, the introduction of the crystalline environment by using a point-ion array to the F_2^- molecular-ion increases the separation to 3.90 bohr. However, when we allow the surrounding lattice to relax (ICECAP calculation) the separation decreases to 3.73 bohr. A similar trend has been observed when we have taken account of the ion-size effect of the nearest-neighbor Li^+ ions by including them in the Hartree-Fock cluster. Thus, the internuclear separation in the F_2^- molecular-ion does depend on the accuracy of our treatment of crystalline environment. The calculated value with our best model comes out to be 3.74 bohr as compared to 3.65 bohr for the free F_2^- molecular-ion. The lattice therefore seems to have an effect, although small, on the internuclear separation. On the other hand, the stabilization energy is found to be 2.76 eV, a significant increase of about 1.7 eV from free molecular-ion state. This is due to lattice polarization.

The addition of correlation correction to UHF-SCF results does not seem to change the internuclear separation. The internuclear separation is also found to be insensitive to the ion-size effect of near-neighboring Li^+ ions in ICECAP calculations. This may be due to the fact that these Li^+ ions move outward by as much as 19% in the relaxed configuration and thereby provide negligible ion-size effect to the V_k center in our model.

Commencing to Table 1, it is interesting to note that our calculated value of the internuclear separation, 3.74 bohr comes out to be in between the values, 4.0 bohr and 3.67 bohr obtained by Jette et al.[19] and Cade et al.[21] respectively. Both the calculations have suggested that these values of the internuclear separation are either too large (Jette et al.[19]) or too small (Cade et al.[21]) to predict the correct transition energies for the hole excitation.

Theoretical prediction of the hyperfine constants has long been recognized as the most critical test of a defect model. The isotropic hyperfine constant is proportional to the total spin-density at the nucleus and thereby furnishes an exact determination of the wavefunction amplitude in the ground state. We have therefore calculated the isotropic hyperfine constant to test our defect model for the V_k center. For the F_2^- molecular-ion, the calculated value comes out to be 41.77 MHz as compared to the measured ESR value of 46.74 MHz[16], showing reasonably good agreement. However the agreement becomes poor for the neighboring Li^+ ions. Here, the calculated values are about 20%, 33% and 75% of the measured ENDOR values[17] for Class A, C and E ions, respectively. This poor agreement may be due to the relatively large displacements of Li^+ ions in the lattice calculated by our model as mentioned earlier or the limited basis sets used on the Li ions.

In summary, the aim of this study was to assess the calculations in a series of models as the assumptions were being improved for the ground state of the V_k center in LiF. What we have found is that this system shows strong relaxation of the ions immediately surrounding the V_k center itself. These distortions appear to extend well into the surrounding lattice. The stabilization energy of the resultant V_k center is almost entirely controlled by the presence of the distorted lattice. The internuclear separation appears to be affected by the crystalline environment. However, a more elaborate simulation model is needed involving the near-neighbor F^- ions in the cluster and associating Kunz-Klein localizing potentials with the ions at the cluster boundary before anything can be said conclusively.

TABLE 1

Internuclear separation and stabilization energy for the F_2^- molecular-ion in LiF.

Source		Internuclear Separation (bohr)	Stabilization Energy (eV)
(a)	**This work**:		
(i)	[F2] - UHF	3.65	1.04
	UHF + Correlation	3.65	0.75
(ii)	[F2] in point-ion array		
	UHF	3.90	-
	UHF + Correlation	3.90	-
(iii)	[F2] in shell-model lattice		
	ICECAP	3.73	-
(iv)	[F2 + 10 Li$^+$] in point-ion array		
	UHF	3.96	-
	UHF + Correlation	4.10	-
(v)	[F2 + 10 Li$^+$] in shell-model lattice		
	ICECAP	3.74	2.76
(b)	**Jette et al.,**[19] [F2] in lattice	4.0	-
(c)	**Daly and Mieher,**[17] fitted to hyperfine constants	3.91/3.80	-
(d)	**Cade et al.,**[21] HADES calculation	3.67	-

REFERENCES

1. B.G. Dick and A.W. Overhauser, <u>Phys. Rev</u>. 112:90 (1958).
2. H. Bilz and W. Kress, "Phonon Dispersion Relations in Insulators" (Springer-Verlog, Berlin, 1979).
3. A.B. Kunz, "Theory of Chemisorption", J.R. Smith Ed. (Springer-Verlag, Berlin, 1980).
4. P.W. Goalwin and A.B. Kunz, <u>Phys. Rev</u>. B34:2140 (1986).
5. A.B. Kunz and D.L. Klein, <u>Phys. Rev</u>. B17:4614 (1978).
6. S. Huzinaga, "Gaussian Basic Sets of Molecular Calculations", (Elsevier, New York ,1984).
7. J.H. Harding, A.H. Harker, P.B. Keegstra, R. Pandey, J.M. Vail and C. Woodward, <u>Physica</u>, 131B:151 (1985).
8. A.B. Kunz, J. Meng, and J.M. Vail, <u>Phys. Rev. B38</u>:1064 (1988).
9. R. Pandey and A.B. Kunz, <u>Phys. Rev</u>.,B38:2460 (1988).
10. J. Meng, A.B. Kunz and C. Woodward, <u>Phys. Rev</u>., B38:2480 (1988).
11. T.L. Gilbert, "Molecular Orbitals in Chemistry Physics and Physics and Biology", P.O. Lowdin and B. Pullman ed. (Academic Press, New York, 1964).
12. A.B. Kunz, <u>Phys. Rev</u>. B6:606 (1972).
13. A.B. Kunz, D.J. Michish and T.C. Collins, <u>Phys. Rev. Lett</u>., 31:756 (1973).
14. D.J. Thouless, "The Quantum Mechanics of Many Body Systems", (Academic Press, New York 1961); E.R. Davidson and D.W. Silers, <u>Phys. Lett</u>. 403 (1977).
15. R.J. Bartlett, I. Shavitt and G.D. Purvis III, <u>J. Chem. Phys</u>., 71:281 (1979).
16. D. Shoemaker, <u>Phys. Rev</u>. B7:786 (1973).
17. F. Daly and R.L. Mieher, <u>Phys. Rev</u>. 183:368 (1969).
18. For a review, see A.M. Stoneham, "Theory of Defects in Solids" (Oxford University Press, London, 1975), Chap. 18, pp. 653-669.
19. A.N. Jette, T.L. Gilbert and T.P. Das, <u>Phys. Rev</u>. 186:919 (1969).
20. T.L. Gilbert and A.C. Wahl, <u>J. Chem. Phys</u>. 55:5247 (1971).
21. P.E. Cade, A.M. Stoneham, and P.W. Tasker, <u>Phys. Rev</u>. B30:4621 (1984).
22. J. Meng, Ph.D. Thesis, Michigan Technological University, unpublished (1987).
23. Ravindra Pandey and John M. Vail, <u>J. Phys. : Condensed Matter</u>, in press (1989).
24. C.R.A. Catlow, K.M. Diller, M.J. Norgett, J. Corish, B.M.C. Parker and P.W.M. Jacobs, <u>Phys. Rev</u>. B18:2739 (1978).
25. D.E. Woon, M.Sc. Thesis, Michigan Technological University, unpublished (1987).

THE ATOMISTIC STRUCTURE OF SILICON CLUSTERS AND CRYSTALS:

FROM THE FINITE TO THE INFINITE

James R. Chelikowsky

Department of Chemical Engineering and Materials Science
Minnesota Supercomputer Institute
University of Minnesota
Minneapolis, MN 55455

INTRODUCTION

One of the most fundamental problems of materials science is to describe the microscopic structure of condensed matter. For crystalline matter, this problem can be viewed as "solved in principle", as one can now accurately determine the structure of simple crystals. For example, the structure of virtually all elemental solids can now be determined from first principles[1]. However, for large clusters of atoms without periodic symmetry, few good methods, either experimental or theoretical, exist for describing the structure of such systems.

Several factors have motivated a vigorous effort to develop techniques to determine theoretically the properties of clusters. For example, one would like to understand in a fundamental way the progression from the *atomic -* to *molecular-* to the *solid-* state. In addition, numerous technological issues are centered on cluster structures such as the catalytic activity and electronic properties of clusters. Here we will review recent progress made for describing the theoretical structure of large clusters. We will concentrate on large clusters of silicon atoms, although many of the basic ideas and concepts presented here are more generally applicable.

BUILDING UP A SOLID FROM SCRATCH

One could imaging building up a crystalline solid by adding one atom at a time, at least conceptually. For example, one could start by building small clusters and adding atoms to such a cluster. If one were to ask what the ground state structure would be at various stages in the process, one might find that the small clusters had a very different type of bonding than large clusters and most certainly different from the crystalline material. In the case of silicon, based on highly accurate quantum calculations[2-4], one would expect that the small clusters of silicon would be metallic like, both in their structural and electronic properties. On the other hand, we know in the limiting case that this can not be correct as crystalline silicon possesses a covalent semiconducting structure: the diamond structure. Thus, as we pass from a finite cluster to the infinite crystal, at some point the structural and electronic properties must undergo a transition[5].

One of the fundamental problems facing condensed matter theorists is to predict at what critical cluster size this transition will occur. State-of-the-art "*ab initio*" methods for describing such problems are woefully inadequate. At present, if we consider a system such as Si_n, $n \approx 10$ is probably the largest system for which one has hope for determining the precise structure. Therefore, new methods are required to describe systems where we lack periodicity, but $n \gg 10$. It seems clear that new methods will involve approximations to the actual quantum mechanical solution.

Unfortunately, our knowledge of bonding properties is quite limited in that it is not obvious how to translate the effects of quantum mechanical concepts such as hybridization, covalency and Jahn-Teller distortions into effective valence force fields, i.e. tractable algebraic expression which will replicate true quantum mechanical interatomic forces[6-11]. This "primitive state" of our physical intuition is especially so for covalent systems. Consider that in graphite and diamond covalent bonding takes place through sp^2 and sp^3 hybrids, but as we descend the Group IV column from C to Si to Ge to Sn and finally to Pb this hybridized orbital description of tetravalent structures becomes progressively inadequate. Although some theoretical descriptions of these systems can be made without explicit quantum mechanical calculations, e.g., the dielectric theory, these methods by and large are limited to equilibrium configurations at low temperatures and pressures.

SEMI-EMPIRICAL TOTAL ENERGY EXPRESSIONS

One fairly obvious approach is to use a very approximate quantum mechanical description of clusters and improve this approximation by the addition of valence force descriptions[3,4,12]. For example, one approach might be to write an expression of the form:

$$E(\{\mathbf{R}_i\}) = \sum_i E_i + E_{2B} + E_{3B} + U \sum_i (q_o - q_i)^2 \tag{1}$$

where

$$E_{2B} = \sum_{i<j} e_{2B}(R_{ij})$$

$$e_{2B}(R_{ij}) = A_o exp(-2\beta (R_{ij} - R_o))[1 - 2exp(-\beta (R_o - R_{ij}))] - \sum_i E_i^d$$

and

$$E_{3B} = B_o \sum_{i<j<k} cos(\theta_{ijk} - \theta_o) \qquad\qquad 0° \le \theta_{ijk} \le 180°$$

The first term in Eq. (1) is a sum over the eigenvalues obtained from semi-empirical pseudopotentials and a Gaussian basis as outlined elsewhere[1,5]. The second term, E_{2B}, is a two body interaction. This term contains the parameters: A_0, β and R_0 which are fit to the silicon dimer using Eq. (1), i.e. $e_{2B}(R) + \Sigma E_i^d$ where ΣE_i^d is the sum of eigenvalues for the dimer and where R is the interatomic distance. By construction, the experimental values of the binding energy, bond length and vibrational frequency for the silicon dimer will be exactly reproduced. The three body term, E_{3B}, is similar to one used elsewhere[7]. θ_{ijk} is the angle formed by vectors from atom i to the nearest neighbor atoms j and k (we cut this term off for distances greater than 10% of the crystalline bond length). (B_0 and θ_0) are determined from the trimer and larger clusters; effectively these parameters are fit to reproduce more rigorous calculations[2-4] for small n. This term is not allowed to grow without bound as the cluster size increases. We saturate E_{3B} so that each term is less than some maximum value, i.e. $B_0 \cos(\theta_{ijk} - \theta_0) < E_{bond}$ where the bond energy is taken to be $E_{bond} \cong -2.5$ eV. Finally, with respect to Eq. (1), we have a term which accounts for the lack of a self-consistent solution. U is adjusted so that the silicon trimer dissociates properly in terms of the total energies known for the monomer and the dimer. q_0 is taken as 4 to suppress ionic configurations.

We can compare this procedure to several classical models[6-10] and more sophisticated quantum mechanical results[2-4]. Our form contains quantum mechanical effects via the first term in Eq. (1), yet it is far easier to evaluate than a full self-consistent field solution of the electronic structure problem. The number of basis functions which we use are minimal in number. Typically, 8 Gaussians which have the form $\{1,x,y,z\}$ $\exp(-\beta r^2)$ with two sets of β's, i.e. two s- and six p- symmetry Gaussians with $\beta = 0.25$ and 1.0 are used so that for the largest cluster ($n \approx 50$) considered here, the matrix size was on the order of 400 x 400. Computational details have been presented elsewhere[1,4,5].

The chief deficiency of this method compared to a fully self-consistent solution is its failure to account for charge transfer properly and the need to parameterize significant interactions. Compared to other tight binding models or linear combination of atomic-like orbitals (LCAO) methods[3,12] we have the advantage of not employing bond counting terms. Also, we do not have to scale matrix elements as we calculate explicitly the required matrix elements. With respect to valence force field descriptions[6-9], we have the advantage of including some quantum effects, e.g. hybridization and coordination effects, which are difficult to include with classical valence force fields. Most notably those classical approaches which fail to account for incomplete coordination have been unreliable for the structural energies of clusters and interstitial defects[7-9].

In determining the structural properties of Si_n, we adjusted the parameters Eq. (1) to fit a pseudopotential-local density approximation (LDA) calculation[3]. A comparison for small silicon clusters is given in Figure 1. Overall the agreement is quite good; however, notable disagreements occur at n=2 and to a lesser extent at n=8. Our dimer is fit to experiment; the pseudopotential calculation is not. Another point to emphasize is the excellent agreement between our semi-empirical calculation of the energy of the n = 6 and

10 structures both for the lowest energy structures (which are metallic like) and the diamond structure fragments corresponding to a six-fold ring and an open cage structure. If one intends to compare the diamond fragments with metallic structures, it is important to verify that one obtains structural energies in regimes which are known to be correct.

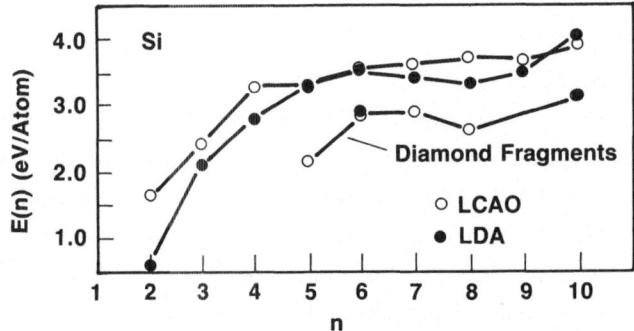

Figure 1. Binding energy in eV for silicon clusters. The curve labelled diamond fragments represents clusters which correspond to fragments of the diamond crystal. The other curves are for the lowest energy structure for a given cluster size. Open circles are from the present work (LCAO) and the filled circles are from Ref. 3 (LDA).

In determining which cluster structure is the ground state, one can use simulation techniques to optimize the lowest energy structure. One of the simplest methods uses Monte Carlo simulated annealing approaches[13]. A random assemblage of atoms is considered at a high temperature. Each atom is then moved a small step distance. If the energy change for the step is such that the cluster's energy is lowered, then this step is accepted and the structure is so altered. If the energy change raises the energy of the cluster, then the step is accepted with a probability given by a Boltzmann factor. At high temperatures, the probability for any step to be accepted is quite high. However, as the temperature is lowered only steps which result in small increases in the energy, or lower the energy, of the cluster become feasible. This process avoids metastable solutions provided one anneals slowly, i.e. by taking enough steps and by lowering the energy in a slow fashion. Clearly, this is realizable for large clusters only in the limit of an infinite number of steps and an infinitely slow anneal. In practice, we consider a number of topologically different clusters and anneal each cluster. If a random starting cluster also results in one of the "known" structures, we have some reassurance that this cluster is the lowest energy one. We can also alter the annealing schedule and verify that we get the same structure. Typically 10^4 steps per atom are used with an initial temperature of $T_i \approx$

3000 K and a final temperature of $T_f \approx 300$ K. This annealing schedule is similar to that used for amorphous silicon[14].

DETERMINING WHEN COVALENCY BEGINS

We can illustrate how one might consider going from a small cluster to the limiting case of an infinite cluster by considering two types of structures: a prototypical metallic structure (fcc) and a prototypical covalent structure (diamond). One could then consider an expansion of the cluster energy as a function of size:

$$E(n)/n = E_b \left(1 - a / n^{1/3} - b / n^{2/3}\right) \qquad (2)$$

where $E_b = E(n) / n$ in the limit $n \to \infty$, i.e. the binding energy for the solid state. E_b (diam) and E_b (fcc) are fixed by experiment and theory, respectively[15]. We may interpret (a,b) as decreasing the binding energy of the solid by the creation of surface and edge atoms. Initially for small n, we expect the bulk binding energy to be dramatically reduced and by different amounts for the two types of structures. Metallic structures with their greater coordination are able to support more readily the presence of surfaces, whereas open structures such as the diamond structure are not.

We present the structural energy as a function of n for diamond and fcc fragments in Figure 2. The structures we used for the diamond fragments come from the work of Saito *et al.*[16] They assumed perfect sp^3 bonding and determined the optimal structure for diamond fragments via a valence force field calculation. While we do not feel this procedure is accurate in terms of comparing different types of structural configurations, we feel for the restricted set of diamond-like structures it is reliable. Saito *et al.*[16] do not give structures for $n > 20$ so here we have simply taken diamond fragments formed by taking nearest neighboring shells, e.g., if we include 5th nearest neighbors to Si atom in the diamond lattice, this would generate a cluster of 29 atoms. For the closed packed structures, we have taken an fcc fragment of ten atoms and considered symmetric structures built up from this core.

From Figure 2, the cross-over point at which diamond and fcc silicon fragments are comparable in energy is about 40-50 atoms. Clearly, this is only an estimate. We have not fully optimized the structure in the fcc or diamond fragments. However, one can emphasize two points here. First, the diamond fragments are expected to have greater relaxation, or reconstruction energies, than the fcc clusters. This effect would tend to push the cross-over point to lower values of n. Second, even if one has not achieved a highly accurate energy for the fcc and diamond fragments, the estimate should not be off by more than a few tenths of an eV which would bracket the cross-over point between 20-100 atoms. Thus we estimate a much lower transition point in terms of a critical size than previous studies[3,7].

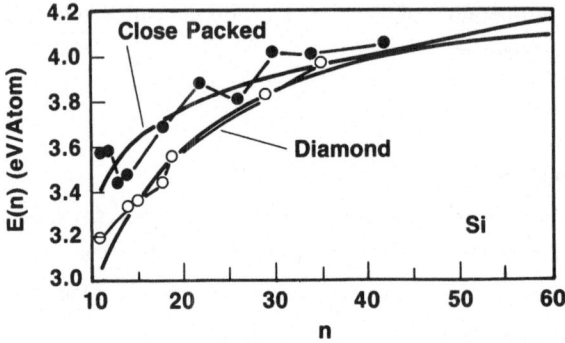

Figure 2. Binding energy of large silicon clusters as a function of cluster size. The open circles correspond to fragments of a diamond crystal; the filled circles correspond to fragments for an fcc crystal. The solid curves are from Eq. (2).

One can also compare to the recent experimental work of Smalley *et al.*[17] which lends support to our estimate. Smalley *et al.* find for clusters of Si_n^+ undergoing reaction with NH_3 that dramatic oscillations in reactivity occur with n. They observe for clusters above n = 47 the oscillations subsided and a slow monotonic increase in activity occurs. Phillips[18] has suggested that the periodicity can be explained by a cylindrical morphology for $n \leq 47$. We have not examined charged clusters, but for neutral clusters we find that cylindrical structures can be competitive with metallic structures, at least for $n \leq 24$, supporting Phillips' model. Moreover, the lack of oscillations for $n \geq 47$ is certainly suggestive of a morphology change occurring near $n \cong 50$ which is consistent with this work.

THERMODYNAMIC INTERATOMIC FORCE FIELDS FOR CLUSTERS AND BULK PHASES

In this subsection we consider the issue of using only algebraic expressions to describe the total energy of a cluster or a bulk phase of silicon. Several attempts have been made to model the results of *ab initio* pseudopotentials calculations for silicon polytypes[7-10]. For example, one might use classical two and three body forces to fit the theoretical structural data on silicon in the diamond, bcc, fcc, β-Sn, hcp, sc, etc. structures and then use such forces to describe cluster structures. Most of these attempts have met with little or no success. The classical force fields (CFF) depend on the interatomic vectors R_{ij} through R_{ij}^2 and $R_{ij} \cdot R_{jk}$ and thus utilize the angular factor cos (θ_{ijk}) as in Eq. (1). A key difference between this approach and our efforts here is that we focus from the outset on an angular function which is designed to describe the free energy change at a first order covalent-metallic transition which is discontinuous. An energy function which is macroscopically discontinuous becomes "S"-shaped on an atomic scale. The smallest interatomic angle varies from π/3 in close-packed structures to 2π/3 in graphite and so an appropriate angular factor is actually given by cos $(3\theta_{ijk})$. If we sum with equal weighting over all nearest neighbor three-body forces, the multi-valued nature of cos $(3\theta_{ijk})$ will produce undesirable cancellations. However, we know from a CFF point of view that it is necessary to introduce radiál weighting factors to describe rapidly decreasing interatomic wave function overlaps. For consistency we do the same here with an angular weighting factors which suppress such cancellations.

To be more specific, we chose the following form for our thermodynamic interatomic force field (TIFF) for the bulk phases of a covalent material like silicon:

$$E(\{R_{ij}\}) = \sum_{ij} (A\exp(-\alpha R_{ij}^2) + g_{ij} \exp(-\beta R_{ij}^2) / R_{ij}) \tag{3}$$

g_{ij} is defined to be

$$g_{ij} = g_0 + g_1 S_{ij} S_{ji} \tag{4}$$

where

$$S_{ij} = 1 + < \cos(3\theta_{ijk}) >$$

$$< f(\theta_{ijk}) > = [f] / [1]$$

$$[f(\theta_{ijk})] = \sum_{k \neq i,j} f(\theta_{ijk}) \exp(-\gamma_1 \theta_{ijk}^4) \exp(-\gamma_2 R^4)$$

We define $R = (R_{ij} + R_{jk}) / 2$. The results depend essentially on the sharp cut-off functions with $(\gamma^1)^{-1} = (\pi/2)^4$ and $(\gamma_2)^{-1} = (R_0)^4$ where R_0 is the nearest neighbor distance in the diamond structure. Our strength function, g_{ij}, describes increasing covalent bonding strength as the bond angle θ_{ijk} increases from $\pi/3$ to $2\pi/3$.

There are seven parameters which describe our TIFF expression. To determine these parameters we use a data base provided by *ab initio* pseudopotential calculations. This data base includes cohesive energies, lattice parameters and the bulk moduli for several silicon polytypes: diamond, simple cubic, face-centered cubic and body centered cubic structures. We attempted to fit these theoretical values using Eqs. (3-4) as indicated in Figure 3. The improvement over previous efforts which used cos (θ_{ijk}) instead of cos $(3\theta_{ijk})$ are quite apparent.

When the TIFF expression in Eqs. (3-4) is used to calculate equilibrium energies and structures of small silicon clusters (n = 5-10), comparison with molecular orbital theory (MO) shows that the binding energies are too small and the bond lengths too large[2]. The cluster topologies agree well with those previously obtained using a CFF based on cos (θ_{ijk}), but are much more open than those obtained quantum mechanically[19,20]. In particular, the transfer of bond strength to back bonds can produce more tightly bound structures. This transfer depends on the angle between the dangling bond and the back bond. Such an effect is completely missing from most CFF discussions.

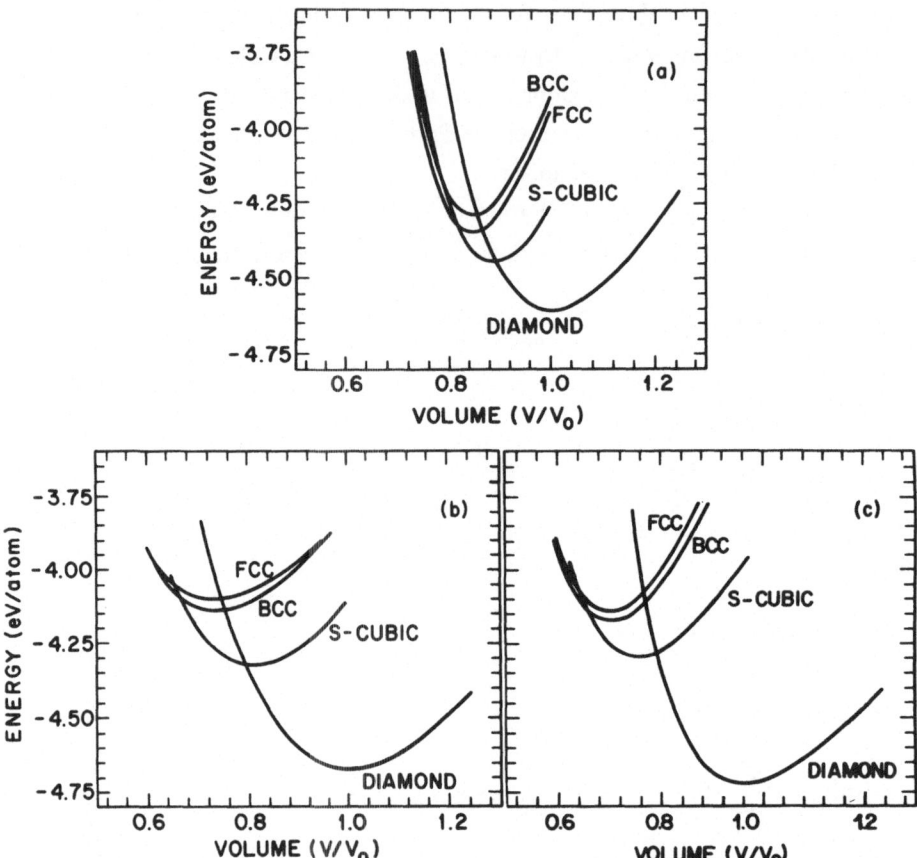

Figure 3. Energy versus volume for several bulk silicon phases. In (b) are shown *ab initio* pseudopotential results (Ref. 15), in (a) a previous fit without an angular cut-off (Ref. 7) and in (c) present work.

To mimic this quantum mechanical effect, we define a "dangling bond" vector

$$\mathbf{D}_i = \sum_j \mathbf{R}_{ij} \exp(-\gamma_2 R_{ij}^4) / \sum_j \exp(-\gamma_2 R_{ij}^4) \tag{5}$$

We expect that the covalent effect of back-bond strengthening is small when the bonding is metallic ($\theta_{ijk} = \pi/3$). Thus, we describe back-bond strengthening by the factor:

$$T_{ij} = 1 + z |\mathbf{D}_i| \sin(\alpha (\theta_{ijk} - \pi/3)) \tag{6}$$

In crystals, $\mathbf{D}_i = 0$ and it will be small in clusters except for surface atoms, i.e. miscoordinated atoms.

Changes in back bonding strengths affect the relative contributions of the attractive metallic and covalent interactions in opposite fashions, i.e. it should increase the role of "two" body interactions and the decrease the role of hybridization. To minimize the number of parameters, we assume that for each bond ij:

$$\Delta \, g_o/g_o = -\Delta \, g_1/g_1 = u \, (T_{ij}T_{ji} - 1) \tag{7}$$

To evaluate our results for the structures of small silicon clusters, we compare them to those obtained from MO calculations[2] and those obtained from an earlier force field without dangling bond corrections[20]. The MO structures can include Jahn-Teller (J-T) or surface reconstruction effects. Especially for small Si_n clusters with n < 10 these effects can be significant (i.e. energy changes of order 1 eV per cluster) and depend on valence electron filling of frontier orbitals. With increasing n steric hindrance reduces J-T distortion energies and surface reconstruction transition temperatures toward bulk values (of order 10^3 K or less), so that we expect that our formalism to best describe clusters where n > 10.

In Figure 4 we illustrate the structures obtained for n $\leq$ 10. It is particularly gratifying that we obtain a bicapped pentagon for n = 7 which is the same structure obtained by MO calculations. In general, our structures are similar to the ground states calculated quantum mechanically[2,3], but our TIFF description omits π bonding energies which favor uniaxial structures so that the results are not identical. The disagreements are generally in the range of ~0.2 eV per atom which we take to represent the limiting accuracy achievable without quantum methods for small clusters. As opposed to CFF calculations, we have much more metallic-like structures, i.e. higher-coordination. For example, the average coordination per atom of our n = 7 structure is about 4.5. The CFF value is about 2.5.

CONCLUSIONS

In summary we have illustrated two approaches to determining the structural properties of large clusters. One approach is to attempt to combine elements of a quantum mechanical description of clusters with a valence force field description. This approach has several advantages. First, larger clusters can be examined than can be handled by fully quantum mechanical methods. Here we have illustrated the method for Si_n clusters up to n $\approx$ 50. Also, simulation techniques such as simulated annealing can be used to determine the cluster sizes as one can rapidly evaluate the structural energy. Second, the method does include explicitly quantum mechanical effects such as hybridization.

Unfortunately, while this semi-empirical method is clearly superior to classical valence force fields, it is still too slow for realistic descriptions of large clusters, e.g. n ~ 100 atoms or larger. For larger clusters it is clear than an approach such as our TIFF method will be required. This approach has the advantage of describing both small clusters and crystals of silicon correctly by its formulation. We believe our TFF model contains two vital improvements over previous efforts: (1) an angular cut-off in three body interactions and (2) a back-bonding term which represents an n+1 body force for an n-fold coordinate atom. Applications to many problems such as the structure of amorphous silicon appear promising.

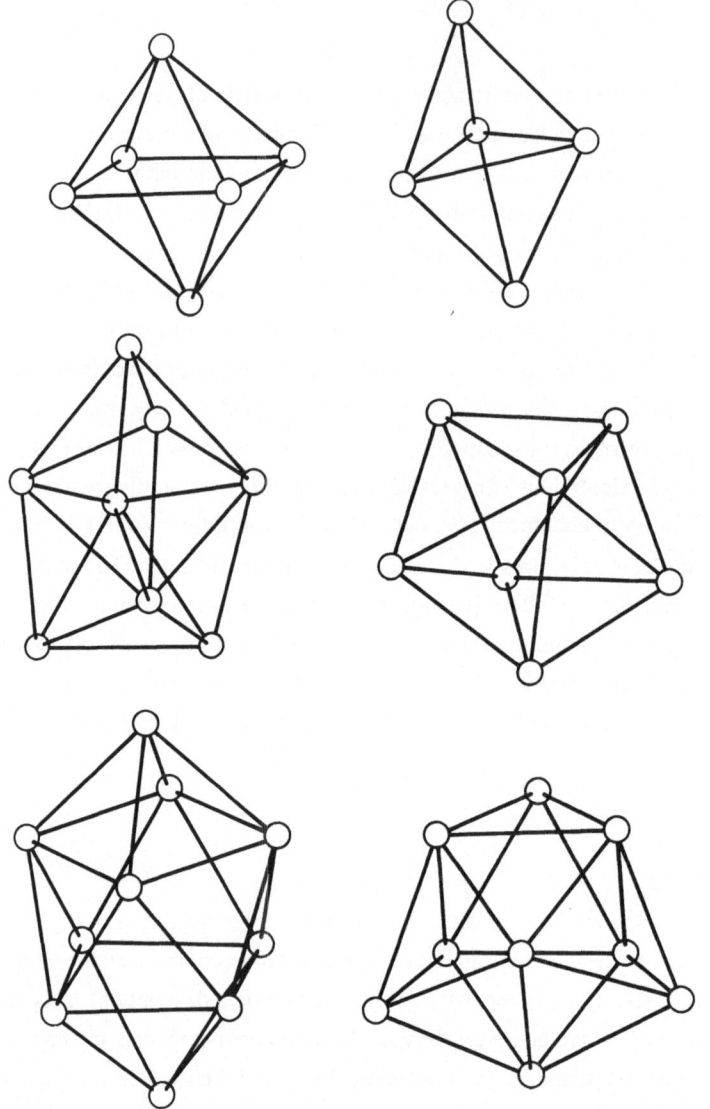

Figure 4. Lowest energy structures for Si_n where $5 \leq n \leq 10$ from the TIFF formalism. For n = 3 we obtain an equilateral triangle and for n = 4 we obtain an ideal tetrahedron.

ACKNOWLEDGMENTS

Computational support from the Minnesota Supercomputer Institute is greatly appreciated. Helpful discussions with J.C. Phillips are also acknowledged.

REFERENCES

1.   J.R. Chelikowsky and S.G. Louie, Phys. Rev. B29:3490 (1984), C.T. Chan, D. Vanderbilt and S.G. Louie, Phys. Rev. B33:2455 (1986). J.R. Chelikowsky and M.Y. Chou, Phys. Chem Minerals, 14:308 (1987) and references therein.

2. K. Raghavachari, J. Chem. Phys. 83:3520 (1985), *ibid*, 84:5672 (1986) and to be published.

3. D. Tomanek and M. Schlüter, Phys. Rev. Lett. 56:1055 (1986), Phys. Rev. B, 36:1208 (1987).

4. J.R. Chelikowsky and R. Redwing, Solid State Comm. 64:843 (1987).

5. J.R. Chelikowsky, Phys. Rev. Lett. 60:2669 (1988).

6. F. Stillinger and T. Weber, Phys. Rev. B31:5262 (1985).

7. R. Biswas and D.R. Hamann, Phys. Rev. Lett. 55:2001 (1985); Phys. Rev. B36:6434 (1987).

8. J. Tersoff, Phys. Rev. Lett. 56:632 (1986).

9. B.W. Dodson, Phys. Rev. B35:2795 (1987).

10. M.I. Baskes, Phys. Rev. Lett. 59:2666 (1987).

11. J.R. Chelikowsky, J.C. Phillips, M. Kamal and M. Strauss, Phys. Rev. Lett. 62:292 (1989).

12. See D.J. Chadi, Phys. Rev. B29:785 (1984) and references therein for a general description of this technique applied to semiconductor surfaces.

13. For details see Refs. 7, 9 and 14.

14. P.C. Kelores and J. Tersoff, Phys. Rev. Lett. 61:562 (1988).

15. M.T. Yin and M.L. Cohen, Phys. Rev. B26:5668 (1982).

16. S. Saito, S. Oshishi and S. Sugaro, Phys. Rev. B33:7036 (1986).

17. J.L. Elkind, J.M. Alford, F.D. Weiss, R.T. Laaksonen, and R.E. Smalley, J. Chem. Phys. 87:2397 (1987).

18. J.C. Phillips, J. Chem. Phys. 88:2090 (1988).

19. E. Blaisten-Barojas and D. Levesque, Phys. Rev. B34:3910 (1986).

20. B. Feuston, R.K. Kalia and P. Vashista, Phys. Rev. B37:6297 (1988).

APPLICATIONS OF SIMULATED ANNEALING IN ELECTRONIC STRUCTURE STUDIES OF METALLIC CLUSTERS

Mark R. Pederson, Michael J. Mehl, and Barry M. Klein

Condensed Matter Physics Branch
Naval Research Laboratory
Washington, DC 20375-5000

Joseph G. Harrison

Department of Physics
University of Alabama at Birmingham
Birmingham, AL 35294

INTRODUCTION

With the advent of the density functional formalism[1], improved numerical schemes, and the steady increase in computational power; researchers are now confidently studying a wide variety of technologically important materials properties which in some cases are not amenable to laboratory observation. In this paper, we survey an all-electron, full potential computational algorithm which employs a compact basis set of Gaussian type function for such studies. In Sec. II, the computationally intensive steps of this problem and recent work-toward reducing the computational complexity, are briefly reviewed[2,3]. By incorporating a simulated annealing algorithm we are able to simultaneously vary both the linear and nonlinear "electronic coordinates" and, when necessary, bypass the direct diagonalization step. With this formulation, the computational cost increases linearly with the number of atoms in the regime of tens to hundreds of non-identical atoms[3,4]. This method enables the accurate evaluation of Hellmann-Feynman (HF) forces. In Sec. III through V we present static and dynamic simulations on a variety of lithium clusters ranging in size from two to twenty seven atoms. By way of these examples, we explicitly show how to predict vacancy formation energies, defect induced lattice relaxation, cohesive energies and vibrational phenomena in many-atom systems. In addition, by carrying out calculations on successively larger crystal fragments, we are able to simulate crystal growth and observe the transition from atomistic to bulk phenomena. The cohesive energies, bulk moduli and lattice constants are presented as a function of cluster size and are found to agree favorably with other theoretical and experimental perfect crystal results.

COMPUTATIONAL CONSIDERATIONS

The Hohenberg-Kohn theorem[1] relates the ground state electronic total energy to the ground state density through a unique but unknown energy functional. Although the energy functional must be approximated, local density approximations[5] (LDA) have been introduced which provide reasonably accurate first principles electronic structure information. By coupling the LDA with the Born-Oppenheimer approximation, we approximate the total energy for a paramagnetic system of 2M electrons and N nuclei by

$$E_\gamma = 2 \sum_i \left\langle \varphi_i \left| -\frac{1}{2} \nabla^2 + V_{nuc} \right| \varphi_i \right\rangle + \frac{1}{2} \int drdr' \; \frac{\rho(r)\rho(r')}{|r - r'|} + \int dr\rho(r) \, \epsilon_{xc} [\rho]$$
$$+ \frac{1}{2} \sum_{\nu\mu}{}' \frac{Z\nu Z\mu}{|R_\nu - R_\mu|} \tag{1}$$

$$V_{nuc} = - \sum_\nu \frac{Z_\nu}{|r - R_\nu|} \; , \tag{2}$$

$$\rho(r) = 2 \sum_{i=1}^{M} |\varphi_i (r)|^2 . \tag{3}$$

In the above equations, $\rho(r)$, the electronic charge density, is constructed from the occupied orbitals ($\varphi_i(r)$) and Z_ν is the charge on the νth nucleus which is centered at R_ν. The total energy [Eq. (1)] consists of a sum of the electronic kinetic energy, the nuclear-electronic, electronic-electronic and nuclear-nuclear Coulomb energies and the exchange correlation energy[5]. In this work the occupied orbitals are parametrized in terms of a linear combination of Gaussian type functions according to:

$$\varphi_i(r) = \Sigma_j \, a_{ij} \, e^{-\alpha_j (r - A_j)^2} \bullet P_{lj} (r - A_j) . \tag{4}$$

In the above equation, $P_l(r)$ represents a polynomial such as 1, x, xy etc. Static properties, such as ground state geometries and cohesive energies, are obtained by variationally adjusting the occupied orbitals and the nuclear positions so as to minimize the total energy. In addition to ground state total energies and equilibrium geometries, other information such as bulk moduli and vibrational phenomena, may be determined by evaluating the dynamical matrix (of second derivatives with respect to the nuclear positions) at the ground state geometry.

The occupied orbitals which minimize the total energy for a particular set of nuclear positions are adjusted so as to satisfy

$$\left\langle \delta\varphi_n | H_o | \varphi_n \right\rangle - \Sigma_m^{occ} \, \lambda_{mn} \left\langle \delta\varphi_n | \varphi_m \right\rangle = 0 . \tag{5}$$

with Ho, the LDA Hamiltonian, determined by the first variation of Eq. (1) with respect to the electron density. Providing that Eq. (4) is satisfied and that the parameterization of the wavefunctions contains no explicit dependence on the nuclei, the Hellmann-Feynman (HF) theorem[6] states that the force on a given nucleus may be determined by evaluating the classical electric field due to the electronic charge density and the remaining N-1 nuclei. Hence, equilibrium geometries, including metastable minima and saddle points, may be found by adjusting the nuclear positions until all of the HF forces vanish. Further, by incorporating force options into electronic structure codes, the path to equilibrium may be traversed substantially faster by simultaneously calculating the total energy and the Hellmann-Feynman forces as one steps through parameter space. For example, assuming that one is near the equilibrium geometry, the number of calculations required to find the exact minimum scales as the square of the number of non-identical atoms if one calculates only total energies. In contrast, with a force option, the number of calculations scales linearly. Aside from a faster approach to equilibrium and a more efficient determination of the dynamical matrix, there are other advantages to electronic structure codes with force options which are now discussed.

In the past, the solution of Eq. (5) has generally been carried out by first expanding the wavefunctions in terms of a linear combination of auxiliary functions, and transforming Eq. (5) to an equivalent diagonal form which leads to an algebraic eigenvalue problem. The computer time required for the solution of an algebraic eigenvalue problem increases with the cube of the dimension of the matrix which scales with the number of atoms. Aside from the pessimistic scaling law, another shortcoming of this method is that, to date, it has been limited to linear parameterizations of the wavefunctions. Recently, Car and Parrinello[7] (CP) have derived an indirect approach to the solution of Eq. (4) by introducing a fictitious Lagrangian. In addition to a more favorable scaling, methods akin to that of CP are more general since they only rely on the capability of calculating first derivative information. As such, they may be used to <u>simultaneously</u> vary linear and non-linear electronic parameters and, providing HF forces can be calculated, the nuclear positions. For example, in Ref. (2), a slight variation of the CP method was used to simultaneously vary both the linear expansion coefficients (a_{ij}) as well as the Gaussian decay constants (α_j) and centers of gravity (A_j) in static applications to the Li_2 molecule. More recently[3], the conjugate gradient method[8], has been implemented for the minimization. This method requires the same information per parameter move as the fictitious Lagrangian approach but appears to be more suitable for localized basis calculations. For example, the results presented in the next section were performed 10.2 times faster with the conjugate gradient method than with the fictitious Lagrangian method discussed in Ref. (2).

A DYNAMICAL SIMULATION OF THE Li_2 MOLECULE

In Ref. (2), we presented both formal and numerical proof that Hellmann-Feynman forces could be determined by systematically varying all parameters (a_{ij}, α_j, A_j) in Eq. (5). This opens up the possibility of displacing a set of atoms from their equilibrium positions and then allowing the nuclei to evolve in time under the action of their Hellmann-Feynman forces. We illustrate this method on diatomic lithium. We start the simulation off with the atoms at rest but displaced 6.05 a.u. from one another. This corresponds to a temperature of 460 Kelvin. Prior to each nuclear move, a complete SCF quench is carried out with the total energy converged to 10^{-6} a.u. The resulting motion for a half a period is shown in Fig. 1. By Fourier transforming this motion, the Harmonic decomposition [R(t) = R_0 + $\Sigma_n R_n cos(nwt)$] of the nuclear motion has been found. The classical period of oscillation is found to be 0.105 picoseconds which corresponds to a frequency shift of 10 cm^{-1} from the zero temperature result. In addition, the time averaged separation of the two nuclei increases by approximately 0.1 a.u. due to the increased temperature. For further details on this calculation see Ref. (3).

Before continuing, we note that the calculation discussed in this section could also have been, and was, carried out by first performing several calculations near the equilibrium separation, using the force and energy data to obtain, by least square fitting, a representation of the potential curve in terms of a power series, and thirdly calculating analytic forces from the parametrized version of the potential. This approach was substantially faster and as accurate as the former approach. It is likely that for low-temperature dynamical studies of all systems which are near a convex quadratic minimum, it will be more efficient to generate the dynamical matrix rather than performing a real time simulation. However, for studies of higher temperature processes in many atom systems, where anharmonic effects become important, dynamical simulations similar to that of the preceding paragraph may be the only alternative.

SIMULATION OF CRYSTAL GROWTH

In this section, we discuss our results on calculations of successively larger crystal fragments of lithium atoms. In addition to extracting information about isolated clusters, we wish to determine to what extent it is possible to predict bulk characteristics as well. For results of this section and, the remainder of the paper, we have not employed floating Gaussians and defer questions concerning basis set completeness to a later paper. We start with a discussion of a one dimensional simulation of crystal growth.

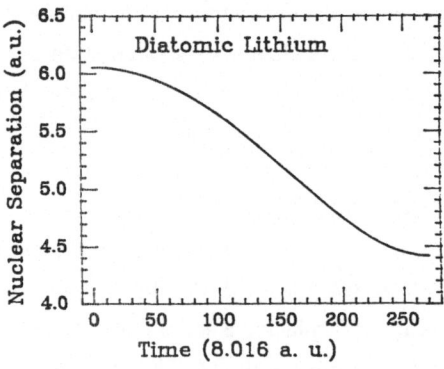

Fig. 1.   Nuclear separation as a function of time for Li$_2$.

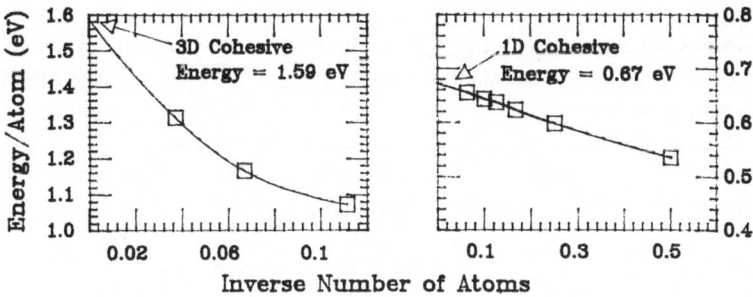

Fig. 2.   Cohesive energy of lithium clusters for a variety of Li clusters.

We have carried out SCF calculations on one dimensional chains of lithium atoms of size N = 2, 4, 6, 8, 10 and 16. In Fig. 2, we plot the cohesive energy per atom as a function of 1/N. The resulting curve is found to be quite smooth. By fitting the cohesive energy to a polynomial (degree 3 or 4) we find the asymptotic cohesive energy to be 0.673 eV. However, it is interesting to note that using only the first three data points (N = 2, 4, 6), we may fit the cohesive energy to a polynomial of degree 2 and obtain an asymptotic cohesive energy of 0.679 eV which is in excellent agreement with the previous results. Hence, certain bulk properties such as cohesive energies may be obtainable by studying successively larger crystal fragments.

In addition, we have carried out the three dimensional analog of this calculation by performing calculations on BCC crystal fragments containing 9, 15 and 27 atoms respectively. In Fig. 3, the cohesive energy per atom is plotted as a function of (1/N) for the BCC case. Once again, the data appears to be rather smooth. By fitting the data to a parabola in (1/N), the large N cohesive energy extrapolates to 1.5 eV which agrees favorably with the experimental measurement[9] of 1.63 eV and a variety of bulk LDA results which range from 1.65 to 1.77 eV[10-12]. While the agreement is good, we feel that in order to draw definitive conclusions, an additional 51-atom calculation should be performed. Work on this is in progress. In Table 1 we compare the lattice constants, cohesive energies and bulk moduli obtained from our cluster calculations to the experimental bulk values and the theoretical values of Moruzzi et al. While the lattice constants and bulk moduli show some oscillations for cluster sizes that we have studied, the close agreement to the bulk results is very encouraging.

VACANCY FORMATION ENERGIES AND VACANCY INDUCED LATTICE RELAXATION

In this section, we study vacancies in large lithium clusters and calculate the vacancy formation energy and vacancy induced lattice relaxation as a function of cluster size. Here, we define the vacancy formation energy as the energy difference between an

Table 1. Lattice constants (ao), bulk modulus (B), cohesive energies (U), Vacancy formation energies (V), and vacancy induced inward relaxation (ΔR) of nearest neighbor shell obtained from calculations on large lithium clusters as compared to theoretical and experimental bulk counterparts.

No. of Atoms	a_0(au)	B (Mbar)	U(eV)	V(eV)	ΔR (%)
8/9[a]	5.62	0.15	1.07	1.5	5.6
14/15[a]	6.36	0.11	1.17	2.1	6.4
26/27[a]	6.16	0.13	1.31	1.7	7.9
Infinity[b]			1.59		
Bulk-Theory[c]	6.42	0.15	1.65		
Bulk-Theory[d]	6.52	0.14	1.65		
Bulk-Theory[e]	6.36	0.15	1.77		
Bulk-Expt.[f]	6.58	0.12	1.63		

[a]Results from this work for N+1 atom BCC fragments and N atom BCC vacancy fragments. See Secs. (3) and (4).
[b]Extrapolation to infinite limit as described in Sec. (3).
[c]Muffin Tin bulk values of Moruzzi, Janak and Williams [Ref. (10)].
[d]Linear Combination of Gaussian Orbitals of Callaway et al. [Ref. (11)].
[e]Linear Augmented Plane Wave of Mehl and Krakauer [Ref. (12)].
[f]Experimental bulk values [Ref. (9)].

isolated (N+1) atom BCC fragment and the sum of the energies of an atom and isolated N-atom cluster. In reality, one expects that the removed atom would migrate to the surface rather than infinity. We reserve this question for later studies. While it is clear that the vacancy formation energy would be substantially reduced by the latter process, we expect the effect on the lattice relaxation to be negligible. In order to calculate these quantities, we surround a complex of either 8 or 9 lithium atoms by either 0, 6, or 18 nearest neighbor atoms. The surrounding neighbors have been constrained to reside at the LDA ground state positions obtained in Sec.. IV. We then allow the 8 nearest neighbor atoms to relax to minimize the total energy. The vacancy induced lattice relaxation may then be determined by comparing the positions of the 8 atom complex to those of the 9 atom complex. In all cases we find that the presence of a vacancy causes the nearest neighbor sh ell of atoms to relax inward by 5-8%, partially filling the void. In Table 1, the vacancy formation energy and nearest neighbor shell relaxation is presented as a function of cluster size. In contrast to the cohesive energies obtained in Sec. IV, there is no obvious way of extrapolating to an infinite result. However, the inward trend is readily apparent.

In order to assess the importance of the placement of the outer shells, we have repeated the 26 and 27 atom calculation with the outer shells frozen at the experimental lattice points rather than our theoretical lattice points. This lower density structure leads to an inner shell lattice relaxation of 3% inward. In principle, one should allow each shell to breath independently. Work on this problem is in progress. All indications are that these calculations will lead to inward relaxation as well. However, the appearance of multiple shallow minima, particularly in the intermediate sized clusters, may compound the problem.

ACKNOWLEDGMENTS

Thanks to Dr. J.Q. Broughton and D.D. Johnson for discussions on various aspects of this work. This work was supported in part by the Office of Naval Research. The calculations were performed on the MULTIFLOW at NRL, and CRAY computers at the University of Alabama and the University of Pittsburgh.

REFERENCES

1. P. Hohenberg and W. Kohn, Phys. Rev. B 136:864 (1964); W. Kohn and L.J. Sham, Phys. Rev. A 140:1133 (1965).
2. M.R. Pederson, B.M. Klein, and J.Q. Broughton, Phys. Rev. B 38:3825 (1988).
3. M.R. Pederson, "Proceedings of the Third International Conference on Supercomputing", Vol. I, 179 (1988).
4. W.E. Pickett, "Proceedings of the Third International Conference on Supercomputing", Vol. I, 172 (1988).
5. We use the parametrization of the Ceperley-Alder exchange correlation potential of J.P. Perdew and A. Zunger, Phys. Rev. B 23:5048 (1981).
6. H. Hellmann, Einfuhrung in die Quantenchemie, (Deutick, Leipzig, 1937); R.P. Feynman, Phys. Rev. 56:340 (1939).
7. R. Car and M. Parrinello, Phys. Rev. Lett. 55:2471 (1985).
8. W.H. Press, B.P. Flannery, S.A. Teukolsky, and W.T. Vetterling, "Numerical Recipes: The Art of Scientific Computing", (University Press, Cambridge, 1986).
9. C. Kittel, "Introduction of Solid State Physics", 5th Ed., (John Wiley and Sons, Inc. 1976).
10. V.L. Moruzzi, J.F. Janak, and A.R. Williams, "Calculated Electronic Properties of Metals", (Pergamon Press, 1974).
11. J. Callaway, X. Zou and D. Bagayoko, Phys. Rev. B 27:634 (1983).
12. M. Mehl and H. Krakauer (unpublished data). For a discussion of the method see S.H. Wei and H. Krakauer, Phys. Rev. Lett. 55:1200 (1985).

AB-INITIO MOLECULAR DYNAMICS SIMULATION OF

ALKALI-METAL MICROCLUSTERS

W. Andreoni and P. Ballone

R. Car and M. Parrinello

I.B.M. Research Division
Zurich Research Laboratory
8803 Rüschlikon
Switzerland

International School for Advanced
Studies
Strada Costiera 11, 34100 Trieste
Italy

In the rapidly growing field of cluster physics alkali metals have played a key role. Due to their relative simplicity of preparation, alkali-metal clusters have in fact been studied experimentally for a longer time and more extensively than aggregates of other materials[1,2,3]. In particular, abundance spectra have revealed the existence of a regular sequence of "magic numbers" which can be understood with a simple jellium shell model[2]. The interrelation between stability and structure and electronic properties has been investigated with model calculations which include the effect of realistic electron-ion interactions[4] and also with calculations following the standard methods of quantum chemistry[5] and solid state theory[6].

In spite of the availability of a number of theoretical studies, it is fair to say that the description of alkali-metal clusters so far achieved is far from being complete. In fact, on one hand, due to their numerical complexity, *ab initio* approaches have been restricted to very small aggregates and to few (usually highly symmetric) atomic configurations. On the other hand, the field of dynamical and thermal properties is still largely unexplored. This is in contrast to the case of homogeneous alkali metals where the latter have been successfully studied in terms of effective pair potentials[7]. No simple and reliable prescriptions have yet been proposed to extend these methods to finite and strongly inhomogeneous systems such as small clusters.

In this work we have adopted the Ab-Initio Molecular Dynamics scheme introduced by two of us[8]. In this approach the forces acting on the ions are derived from the electronic ground state, described within the Local Density Approximation (LDA) of Density Functional Theory. The method has been described elsewhere[8,9]. Here we recall only the points relevant to our discussion.

For a given set of ionic positions $\{\vec{R}_I\}$ the energy of the system in its electronic ground state is given by the minimum of a functional of the electronic density $\rho(\vec{r})$[10]:

87

$$E = min\ \varepsilon\ (\rho\ ,\ \{\vec{R}_I\}) \tag{1}$$

The introduction of one-electron orbitals $\{\varphi_i\ (\vec{r}\ )\}$ allows one to isolate the independent particle contribution to the kinetic energy[10]:

$$E = -\frac{1}{2} \sum_i f_i \left\langle \varphi_i / \vec{\nabla}^2 / \varphi_i \right\rangle + V\ (\ \rho\ ,\ \{\vec{R}_I\}\ ) \tag{2}$$

where

$$\rho(\vec{r}) = \sum_i f_i\ /\ \varphi_i\ (\vec{r})\ /^2 \tag{3}$$

(f_i are occupation numbers).

We adopt the pseudopotential formulation, so that the $\{\varphi_i\}$ refer only to the valence electrons. The functional $V\ (\ \rho,\ \{\vec{R}_I\}\ )$ contains four contributions: the electron-electron Hartree energy, the electron-ion interaction energy, the electronic exchange and correlation energy and the Coulomb energy of the ions. We assume a supercell geometry and expand the orbitals in a plane wave basis set:

$$\varphi_i\ (\vec{r}\ )\ =\ \sum_{/\vec{G}/\ <\ /\vec{G}_{max}/} C_i\ (\vec{G}\ )\ \exp\ (i\vec{G}\ .\ \vec{r}\ ) \tag{4}$$

where the sum is truncated as usual by an energy cutoff $E_{cut} = 1/2|G_{max}|^2$. The functional ε can be optimized with the constraint of orthonormality for the $\{\varphi_i\}$ by treating the partial derivatives:

$$\vec{F}[C_i(\vec{G})] = -\vec{\nabla}_{C_i^*(\vec{G})}\ (\ E - \sum_k \Lambda_{ik} \left\langle \varphi_i\ /\ \varphi_k \right\rangle) \tag{5}$$

as classical forces acting on the Fourier coefficients $C_i(\vec{G})$, and by solving at each $\{\vec{R}_I\}$ the equations of motion:

$$\mu \ddot{C}_i(\vec{G}) = \vec{F}\ [C_i(\vec{G})] \tag{6}$$

The Λ_{ik} are Lagrange multipliers introduced to guarantee the orthonormality constraints of the wavefunctions:

The extremum condition $\vec{F}[C_i(\vec{G})] = 0$ defines a point on the Born-Oppenheimer (BO) surface for which the forces on the ions are computed as:

$$M_I \ddot{\vec{R}}_I = -\vec{\nabla}_{\vec{R}_I}\ E \tag{7}$$

The two coupled equations (6) and (7) can be used to generate the time evolution of the system: Eq. (7) describes the dynamics of the ions while the "pseudoforces" (5) are used to keep the electronic structure close to its instantaneous ground state[8,9].

Following this scheme we have performed a Molecular Dynamics simulation of Sodium and mixed Sodium-Potassium clusters with a number of ions up to 20. The electron-ion interaction was described by *ab-initio* norm-conserving non-local pseudopotentials[11]. The exchange-correlation LDA-functional was computed using the Perdew and Zunger parametrization of electron gas data[12]. The equations of motion for the Fourier coefficients $C_i(\vec{G})$ and the ionic positions $\{\vec{R}_I\}$ have been integrated using the Verlet algorithm with a time step of $1.44 * 10^{-16}$ sec. The other parameters of the calculation are: $\mu = 400$ *a.u.*, $E_{cut} = 5$ Ry., $\Omega \approx 40000$ *a.u.*^{3}. Tests of convergence are reported in Ref. 13, where also other details of the results are described.

Molecular Dynamics was used with two different purposes: (i) as a tool to perform a Dynamical Simulated Annealing (DSA) search for the ground state, i.e. to perform a global optimization of functional ε with respect to both the electronic and the ionic degrees of freedom; (ii) to study the ionic dynamics of the clusters at finite temperature. We describe separately the results of (i) and (ii).

We have considered neutral clusters with a number of electrons $N = 8, 18, 20$ which correspond to the filling of energy levels in the spherical jellium model. As we verified at the end of the computation, even in the presence of a realistic electron-ion interaction, these numbers identify "closed shell system", i.e. systems with filled multiplets of quasi-degenerate levels.

The DSA search for the ground state has been performed for Na_8, Na_{18}, Na_{20} and the mixed cluster $Na_{10}K_{10}$. The DSA cycles was composed by two parts. First, we thermalized the system at a temperature of 350-400K during a time interval Δt of about 0.3 ps. At this temperature, $\Delta \tau$ was sufficiently long to observe the ions diffuse over a significant fraction of the diameter of the cluster. Secondly, the system was gradually cooled down to very low temperatures by rescaling the ionic velocities. At each time step we decreased the kinetic energy by one part over 10^4. Therefore each annealing was completed after a number of steps of the order of 10^4, corresponding to an average cooling rate of about 10^{14} K/sec. Although this is very large on the scale of experimental cooling rates, our DSA schedule has proven to be very efficient in finding atomic configurations of very low energy. These structures are very different from quenched liquid structures and possess a high degree of symmetry.

Na_8 has been used mainly as a test of the method and of the ability of DSA to reach the global minimum of the functional (1). For this system we find a structure of D_{2d} symmetry in analogy with a previous LDA study[6]. Of the four electronic levels the lowest in energy has s angular character, the others have p character and are split into a singlet and a doublet separated by 0.4 eV.

The results for Na_{18} and Na_{20} were much less predictable. The ionic configuration of lowest energy found by DSA for Na_{20} is illustrated in Fig. 1a). Despite some minor distortions in the bond lengths and bond angles it is easy to recognize in it some remarkable properties. It has a well defined layered structure, shown in detail by Fig. 1b); the ionic

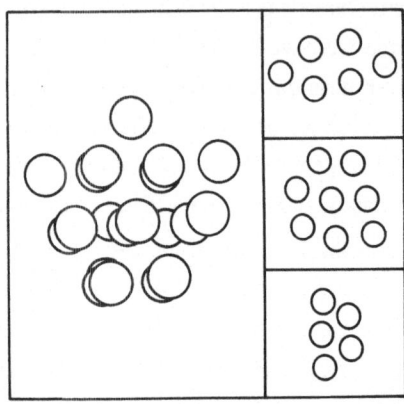

Figure 1.   Configuration of lowest energy produced by DSA for Na_{20} (a) and its decomposition in planes (b).

distribution is fairly spherical with the 3 principal momenta of inertia differing from their average by 6% at most; it is closely packed with the majority of bond angles very close to 60°; each atom has at least 4 neighbors; close to the center of mass there is a twelve-fold coordinated atom. The nearly spherical ionic configuration and the relative weakness of the pseudopotential preserve the shell model feature in the electronic structure of this cluster. The electronic density is remarkably spherical, with only negligible dipole and quadrupole components. A decomposition of the $\{\varphi_i\}$ in spherical harmonics around the center of mass of the cluster shows that each orbital has a well-defined angular character with only a limited hybridization between the s and d states of highest energy. Also the energy ordering (1s,1p,1d,2s) and the (quasi) degeneracy of the levels are as expected on the basis of the shell model. Both structural and electronic properties of Na_{18} are very similar to those of Na_{20}.

By repeated quenches from intermediate configurations of the annealing run and by relaxing regular structures suggested by intuition we have explored the shape of the energy surface of Na_{20} close to the minimum. It turns out that even in an interval of few tenths of an eV above the ground state there are many distinct local minima. Around them the slope of the energy surface is very gentle and the barriers separating different valleys are probably quite small. The implications of this shape of the BO surface on the dynamical properties of the system are discussed below.

The DSA search for the ground state of $Na_{10}K_{10}$ presented an intrinsically more demanding task than in the case of pure sodium clusters, since the interchange of position between ions of different kind is very slow in the time scale of our simulation. In order to speed up the search for the global minimum, we have added to the basic DSA strategy described above a sampling of particle interchanges. Near the end of annealing we selected couples of ions of different species and interchanged their position. Each attempt was followed by relaxation of the ionic and electronic structure. The move was accepted only if

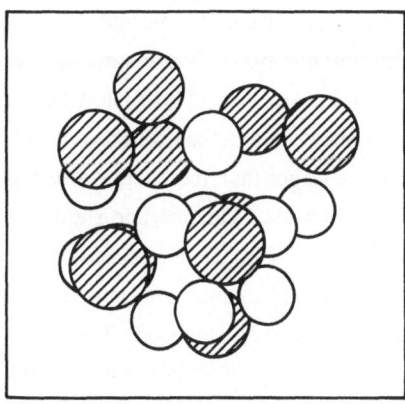

Figure 2.   Structure of lowest energy for $Na_{10}K_{10}$. Empty and filled circles represent Na and K atoms, respectively.

it lowered the total energy. The final structure produced by this procedure is shown in Fig. 2.

If we neglect the difference between Na and K ions it is possible to recognize some similarity with the structure of Na_{20}. Here again there is layered structure with some large distortion than in Na_{20}, due to the mismatch in the ionic dimension of Na and K. The structure is close packed and nearly spherical. In spite of the bigger perturbation due to the K pseudopotential it is still possible to assign to each orbital a dominant angular character and the resulting electron density is nearly spherical. However, the structure is not simply given by a random substitution of potassium ions in the Na_{20} configuration. Instead, we observe a clear tendency to segregation, with the Na ions grouping together close to the center of the cluster in a region of high electron density. This tendency can be quantified: the average distance of K ions from the center of mass of the cluster is 2.5 a.u. larger than for Na. As described in Ref. 13 we explicitly checked that this result is not an artifact due to the initial configuration and to the short time of the annealing. Segregation is not completely unexpected on the basis of the low solubility of Na and K in the solid homogeneous phase. The tendency to segregate is likely to be strongly enhanced by the large inhomogeneity characteristic of small clusters, where the larger K ions prefer to sit in a region of lower density. It is remarkable to see this effect taking place in the short time of our observation.

Starting from an intermediate configuration of an annealing cycle we have performed a long simulation of Na_{20} at finite temperature. We have thermalized the system for 0.4 ps, keeping the temperature in the range 150-200K by rescaling the ionic velocities. At the end we removed the temperature control and followed the microcanonical evolution of the system for 1.44 ps, corresponding to 10000 time steps of our MD. The average temperature of the run turned out to be slightly lower than 200K. Although this temperature is well below the melting temperature of the homogeneous system, the motion of the cluster is far from resembling that of a rigid body, nor can it be described as given by small oscillations around the equilibrium positions. In all cases the ionic displacements turned out to be a significant fraction of the interatomic distances. We also observed few ions crossing from an equilibrium position to a new one. The time of the simulation was too short to allow for an accurate analysis of diffusion and for an estimate of energy barriers. However, the low temperature at which we observe these jumps already suggests that barriers are not significantly larger than the energy differences among local minima, i.e. of the order of few tenths of an eV. The average static structure appears to be very different from that of the ground state. Peaks in the one-particle distribution function reveal that the central region is now occupied by two ions, surrounded by a shell of 16 with the two remaining ions confined in an outermost region. The electronic density maintains its spherical symmetry, but the radial shape changes substantially and it now extends over slightly larger distances from the center of the cluster. The angular character predicted by the spherical shell model is still recognizable in most of the orbitals with, however, two significant changes from the low T situation. The 1d orbitals acquire a sizeable 2s component whose energy is lowered by the removal of the central ion. The highest occupied level, formerly a 2s, has many angular components (also with $l > 2$), none of which is clearly dominant.

Trajectories generated during the simulation have been used to compute an average Velocity Velocity Autocorrelation Function (VVAF). Its Fourier transform, proportional to the vibrational density of states, is displayed in Fig. 3 where it is compared to the same quantity measured in the liquid[14]. The two curves extend over the same interval of frequencies and below the vibrational frequency of the molecule. As expected the VVAF of the cluster is more structured than that of the liquid. However, the conspicuous continuous background is already reminiscent of the disordered, extended system.

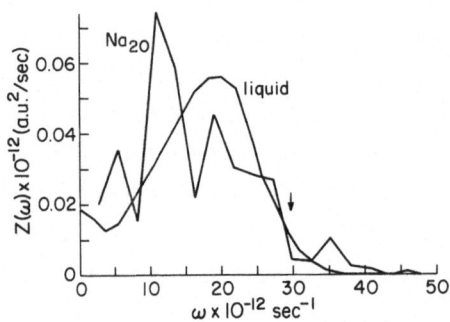

Figure 3.   Fourier transform of the VVAF for Na$_{20}$ compared to that of the liquid[14]. The arrow indicates the vibrational frequency of the molecule.

The study presented here is part of an extensive investigation of the properties of atomic clusters using the Ab-Initio Molecular Dynamics scheme. As already pointed out in previous papers[15,16] this method represents a major improvement over traditional computational techniques: it allows the treatment of larger systems and extends the investigation to dynamical and thermal properties. As far as the subject of the present calculation is concerned, i.e. alkali metal clusters, our conclusions are: (i) In the search of the ground state DSA has produced structures with remarkable properties for both the pure and mixed clusters; (ii) The general picture of the electronic structure, at least for the clusters corresponding to the magic numbers, is in good agreement with the spherical jellium shell model. Sizeable but small variations are introduced by the effect of disorder in the ionic structure due to either thermal motion or alloying; (iii) Our simulations at finite temperatures suggest that rigid body dynamics is restricted to very low temperatures.

REFERENCES

1. W.A. de Heer, W.D. Knight, M.Y. Chou, and M.L. Cohen, in: "Solid State Physics", edited by H. Ehrenreich, F. Seitz and D. Turnbull (Academic, New York, 1988), Vol. 40, p. 93.
2. M.L. Cohen, in: "Microclusters", edited by S. Sugano, Y. Nishina, and S. Ohnishi, Springer Series in Material Sciences (Springer-Verlag, Berlin, 1987), Vol. 4, p.2.
3. W.A. de Heer, K. Selby, V. Kresin, J. Masu, M. Vollmer, A. Châtelain and W.D. Knight, Phys. Rev. Lett. 59:1805 (1987).
4. M. Manninen, Solid State Commun. 59:281 (1986); Phys. Rev. B34:6886 (1986).
5. I. Boustani, W. Pewestorf, P. Fantucci, V.. Bonacic-Koutecky and J. Koutecky, Phys. Rev. B35:9437 (1987); V. Bonacic-Koutecky, P. Fantucci, and J. Koutecky, Phys. Rev. B37:4369 (1988).
6. J.L. Martins, J. Buttet and R. Car, Phys. Rev. B31:1804 (1985).
7. J.E. Inglesfield in: "Computer Simulation of Solids", ed by C.R.A. Catlow and W.C. Mackrodt (Springer-Verlag, Berlin, 1982), p. 115 and references therein.
8. R. Car and M. Parrinello, Phys. Rev. Lett. 55:2471 (1985).
9. R. Car and M. Parrinello in: "Proceedings of the NATO ARW: Simple Molecular Systems at Very High Density", Les Houches (France), March 29 - April 7, 1988 (NATO ASI Series - Plenum Publishing Co.).
10. W. Kohn and L.J. Sham, Phys. Rev. 140:A1133 (1965); P. Hohenberg and W. Kohn, Phys. Rev. 136:B864 (1964).
11. D.R. Hamann, M. Schlüter, and C. Chiang, Phys. Rev. Lett. 43:1494 (1979).
12. J.P. Perdew and A. Zunger, Phys. Rev. B23:5048 (1981).
13. P. Ballone, W. Andreoni, R. Car, and M. Parrinello, to be published.
14. P.A. Egelstaff, "An Introduction to the Liquid State" (Academic Press, New York, 1967) p. 142.
15. R. Car, M. Parrinello, and W. Andreoni in: "Microclusters", edited by S. Sugano, Y. Nishina, and S. Ohnishi, Springer Series in Material Sciences (Springer-Verlay, Berlin, 1987), Vol. 4, p. 134; P. Ballone, W. Andreoni, R. Car, and M. Parrinello, Phys. Rev. Lett. 60:271 (1988).
16. D. Hohl, R.O. Jones, R. Car, and M. Parrinello, Chem. Phys. Lett. 139:540 (1987); and to be published.

A SIMPLIFIED FIRST-PRINCIPLES TIGHT-BINDING METHOD FOR MOLECULAR DYNAMICS SIMULATIONS AND OTHER APPLICATIONS

Otto F. Sankey and David J. Niklewski

Department of Physics
Arizona State University
Tempe, AZ 85287

INTRODUCTION

The growth of a crystal or interface, the interaction between an adatom and surface, defect reactions in crystals, migration of atoms in solids, and a wide range of other phenomena can be simulated by the technique of molecular dynamics. Here the many-body classical equations of motion are solved as a function of time, and the physical process is studied in real time. The equations of motion are prescribed once the instantaneous forces are given. In covalent solids, bonds are formed between atoms which share electrons. The strength of the bond depends on the local environment, making the forces between atoms more complicated than a sum of two-body forces. Potentials have been devised [1] which mimic these many atom effects. However, the many-body effects are clearly rooted in the electronic structure of the material and a superior method is to obtain these forces directly from the electronic structure.

Two electronic structure methods relevant here are the empirical tight-binding model [2,3] and the more rigorous plane wave expansion. [4]. The empirical tight-binding model is relatively simple to use and includes many-body effects, but since it is empirical and primarily fit to bulk properties, using it far from equilibrium is uncertain. The plane wave method does not have this problem, but is exceedingly difficult to implement, making many systems impossible to treat.

The purpose of this paper is to develop an approximate first-principles electronic structure method which approximates very closely a more rigorous calculation, yet is simple enough to be used for a wide variety of purposes, including molecular dynamics simulations of covalent materials. The method employs a first-principles tight-binding approximation, making it versatile and easy to use, and is executed entirely in real space, so periodicity is not necessary and there are no plane wave expansions. It can be used for supercells, slabs, clusters, or within a Green's function technique. The method has been thoroughly tested in Si, and only a brief outline will be given in this manuscript. A complete description is forthcoming [5].

Our approach is to use a number of physically motivated approximations within the framework of a well established first-principles theory, to retain accuracy and transferability, yet still achieve the goal of simplifying the computation of the total energy and atomic forces. The theoretical foundation that we use is density functional theory within the local density approximation [6], with the full atomic potential replaced by a pseudopotential operating only on the valence electrons. The pseudopotentials we use are of the Hamann-Schluter-Chiang type [7].

THE PSEUDO-ATOMIC-ORBITALS

Our first major approximation is to use pseudo-atomic-orbitals (PAO's). Here the atomic orbitals are wavefunctions of the atom in the non-local pseudopotential approximation. These wavefunctions are nodeless, and the pseudo-atom itself contains only the valence electrons.

These pseudo-atomic-orbitals have been tested in ten different covalent semiconducting materials [8]. They were found to yield a quite accurate picture of the bonding and energetics. Comparing with experiment, it was found that lattice constants are within about 2%, bulk moduli are within about 10%, and optic phonon frequencies are within about 1%. The band structures of the bulk material, as well as substitutional and interstitial defect levels and total energies, are also quite satisfactory [9].

The total energy and tight-binding-like Hamiltonian matrix elements were computed exactly in the work of Ref. 8. It was found that the one-electron Hamiltonian matrix elements were quite long range, sometimes going out as far as fifth or sixth nearest neighbors. Neglecting these distant overlaps was found to lead to substantial error.

In the present work, we seek an approximation to the tight-binding-like Hamiltonian and demand that it have a shorter range. To shorten this range, we restate the atomic problem, and impose the boundary condition that the atomic orbital vanish outside and at a predetermined radius r_c,

$$\phi^{PAO}(\vec{r})_{r=r_c} = 0 .$$
(1)

The limit of $r_c \to +\infty$ gives the true pseudo-atomic-orbital. The boundary condition of Eq. (1) has the physical effect of mixing in slight amounts of excited orbitals of the atom inside r_c; this can be seen by adding a small amount of the 2S wavefunction to the hydrogen atom 1S ground state.

The precise value of r_c chosen is not critical, as long as it is well past the peak of the wavefunction. In Figure 1, we show the slightly excited s-orbital of Si for various values of r_c. The bonding region (defined as the half bond length of bulk Si) is ~$2.2a_B$. For values of r_c greater than $5a_B$, there is very little change of the atomic function in the bond region, but the long wavefunction tail is eliminated. Thus, for example, when $r_c = 5a_B$, the tight-binding-like Hamiltonian matrix elements in bulk Si rigorously form a third neighbor model.

The LDA total energy reproduces quite accurately the equilibrium lattice constant and bulk modulus of Si using these excited orbitals. We show the total energy of bulk Si in Fig. 2, with $r_c = 5.5a_B$. The agreement with experiment is excellent. The expected increase in total energy due to confinement of the electrons is not significant here since the electron can hop from atom to atom. Lowering r_c to $5a_B$ gives similar results, but for $r_c <$ $4.5a_B$ serious errors begin to emerge. Thus the introduction of excited PAO's is a controllable approximation.

THE TOTAL ENERGY

The total density approximation to the total energy is given by

$$E_{tot} = E_{BS} + [U_{ii} - U_{ee}(n,n)] + \delta U_{xc}(n) \tag{2}$$

Here E_{BS} is the "band-structure" energy, $E_{BS} = \Sigma_{i\ occ}\ \varepsilon_i$ where ε_i are the one-electron energy eigenvalues, U_{ii} is the Coulomb ion-ion interaction, and the electron-electron repulsion $U_{ee}(n,n)$ is

$$U_{ee}(n_1, n_2) = (e^2/2) \int d^3r n_1(\vec{r}) \int d^3r' n_2(\vec{r}')/|\vec{r} - \vec{r}'| . \tag{3}$$

The electron-electron repulsion is subtracted due to the double counting in the band structure. The term $\delta U_{xc}(n)$ is a correction to the band structure energy from the exchange/correlation interaction, $\delta U_{xc}(n) = \int (\varepsilon_{xc}(n) - \mu_{xc}(n))n(r)d^3r$, where $\varepsilon(\mu)$ is the LDA approximation of the exchange/correlation energy (potential).

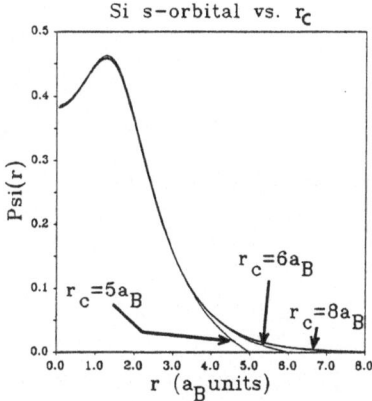

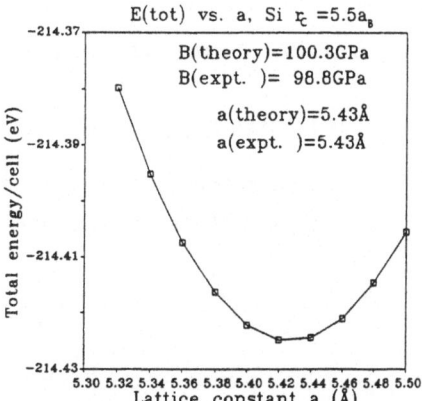

Figure 1. The pseudo-atomic-orbital ϕ_s^{PAO} of Silicon at different values of r_c.

Figure 2. A plot of the total energy versus lattice constant for bulk Si using pseudo-atomic-orbitals with $r_c = 5.5a_B$. No other approximations to the total energy were made.

The single particle energy eigenvalues are solutions of the tight-binding-like matrix equations:

$$\sum_\beta h_{\alpha\beta} C_\beta^i = \varepsilon_i \sum_\beta S_{\alpha\beta} C_\beta^i . \tag{4}$$

Here the energy eigenstates are written as a linear combination of pseudo-atomic-orbitals $\varphi_i = \Sigma_\alpha C_\alpha^i \phi_\alpha^{PAO}$. The Hamiltonian matrix elements $h_{\alpha\beta} = < \phi_\alpha^{PAO} |h| \phi_\beta^{PAO} >$ and overlap $S_{\alpha\beta}$ are zero when the atoms corresponding to α and β are further than $2r_c$ of each other.

The problem is now in the standard form for a self-consistent iteration loop. This path is viable, but at this point we wish to keep the theory as simple as possible without substantially sacrificing accuracy or reliability. We thus use an approximation first suggested by Harris [10], where only changes in the electron number density to first order are kept. We write

$$n(\vec{r}) = n_0 (\vec{r}) + \delta n(\vec{r}) \tag{5}$$

where $n_0 (\vec{r}) = \Sigma_l n^{atomic} (\vec{r}_l)$, i.e. the sum of the spherically symmetric neutral pseudoatomic charge densities. Keeping terms to first order in δn, it is easy to show that

$$E_{tot} = E_{BS}^0 + U_{SR}^0 + \delta U_{xc}^0 (n_0) . \tag{6}$$

where the "bandstructure" energy E_{BS}^0 is obtained by summing the eigenvalues from the one-electron Hamiltonian constructed from a sum of neutral spherically symmetric atomic charges. The term U_{SR} is a "short-ranged" repulsive potential given by $U_{SR} = U_{ii} - U_{ee}(n_0, n_0)$, and $\delta U_{xc}^0 (n_0)$ is the exchange/correlation correction from a sum of neutral atoms. We will outline briefly how we handle only the "bandstructure" term as this is by far the most difficult term. A full account can be found in Ref. [5].

THE SINGLE-PARTICLE HAMILTONIAN AND THE HELLMANN-FEYNMAN

The bandstructure energy is given by summing the eigenvalues of the tight-binding-like matrix (Eq. 4) with the neutral atom Hamiltonian given by,

$$h^0 = T + \int n_0 (\vec{r}\,') / |\vec{r} - \vec{r}\,'| \, d^3r + \sum_l (V_{ion} (\vec{r} - \vec{r}_l) + V_{NL} (\vec{r} - \vec{r}_l)) + \mu_{xc} (n_0) \tag{7}$$

The eigenvalue equation (4) can be either in a slab, cluster, or supercell geometry with N atoms/cell, or using a perturbed Green's function subspace of N atoms. For a system with sp^3 orbitals, this leads to a secular equation of 4N x 4N.

Given the eigenfunctions and energies ε_i^0, the band structure energy can be written

$$E_{BS}^0 = \sum_{l \, occ} \mathcal{E}_l^0 = \sum_{l \, occ} < \varphi_l /h/ \, \varphi_l > = \sum_{l \, occ} \rho_{\mu\nu}^{ij} \, h_{\mu\nu}^{ij} \, . \tag{8}$$

The density matrix $\rho_{\mu\nu}^{ij}$ gives the contribution to the total charge of the bondcharge $\phi_\mu^{PAO} \, (\vec{r} - \vec{r}_i) \, \phi_\nu^{PAO} \, (\vec{r} - \vec{r}_j)$ (μ, ν refer to orbital type s, p_x, p_y, or p_z) in the sense that

$$\rho(\vec{r}) = - e \sum_{\mu\nu ij} \rho_{\mu\nu}^{ij} \, \phi_\mu^{PAO} \, (\vec{r} - \vec{r}_i) \, \phi_\nu^{PAO} \, (\vec{r} - \vec{r}_j) \, . \tag{9}$$

The density matrix $\rho_{\mu\nu}^{ij}$ can be obtained from the Green's function $G(E) = (E - h^0)^{-1}$ as

$$\rho_{\mu\nu}^{ij} = -1/\pi \, Im \int_{-\infty}^{E_F} < \phi_\mu^{PAO} \, (\vec{r} - \vec{r}_i) \, // \, G(E) \, // \, \phi_\nu^{PAO} \, (\vec{r} - \vec{r}_j) > dE \, , \tag{10}$$

where E_F is the Fermi energy. We also define an energy density matrix

$$E_{\mu\nu}^{ij} = -1/\pi \, Im \int_{-\infty}^{E_F} < \phi_\mu^{PAO} \, (\vec{r} - \vec{r}_i) \, // \, EG(E) \, // \, \phi_\nu^{PAO} \, (\vec{r} - \vec{r}_j) > dE. \tag{11}$$

The double bar is a reminder that the matrix elements are taken as if the orbitals on different atoms are orthogonal. This last form is particularly useful for Green's function approaches [3].

A generalization of the Hellmann-Feynman Theorem can now be used to evaluate the derivative with respect to an atomic coordinate $\vec{r}_\gamma$ and hence the band-structure contribution to the force on atom γ:

$$-f_\gamma^{BS} = \partial \, E_{BS}^0 \, / \, \partial \vec{r}_\gamma = \Sigma \, [\rho_{\mu\nu}^{ij} \, (\partial_{\mu\nu}^{ij} \, / \, \partial \vec{r}_\gamma) - E_{\mu\nu}^{ij} \, (\partial S_{\mu\nu}^{ij} \, / \, \partial \vec{r}_\gamma)] \tag{12}$$

The second term is the "repulsive overlap" coming from the non-orthogonality of the PAO's and is quite trivially calculated. The important feature is of course that the derivatives do not operate on the wavefunctions (which are contained in ρ and E) but only on matrix elements of the single-particle Hamiltonian and overlap.

We thus need a fast and efficient method for calculating the Hamiltonian matrix elements and overlap. The Hamiltonian contains kinetic, the local core pseudopotential, the Hartree electron repulsion, the non-local pseudopotential, and the exchange potential. We exactly compute the matrix elements of the kinetic and non-local pseudopotential [5]. We approximate matrix elements of the core potential V_{ion} ($\sim 1/r$ for large r), hartree potential V_e, and exchange/correlation potential.

The local core pseudopotential V_{ion} and the neutral atom hartree repulsion V_e are added together to give a strong short ranged potential of a neutral atom $V_{NA}(\vec{r})$. The matrix elements of $\phi_\mu(\vec{r} - \vec{r}_1)$ and $\phi_v(\vec{r} - \vec{r}_2)$ of the neutral atom potential are obtained by considering the bond charge $\phi_\mu(\vec{r} - \vec{r}_1)\phi_v(\vec{r} - \vec{r}_2)$ as a classical charge distribution centered at $(\vec{r}_1 + \vec{r}_2)/2$. We expand this charge distribution in multipole moments including $l = 0, 1, 2, 3,$ and 4. This distribution is integrated against the neutral atom potential which has only spherically symmetric monopole-like terms, whose monopole moment (charge) exactly vanishes outside r_c. As an example; the matrix element formed by the bondcharge of two Si s-orbitals separated by a distance d along the z axis is [5].

$$< \phi_s^{PAO}(\vec{r} + d/2\hat{z}) \, | \, V_{NA}(\vec{r} - \vec{r}_{NA}) \, | \, \phi_s^{PAO}(\vec{r} - d/2\hat{z}) > =$$

$$q_{ss}(d, r_{NA}) + P_2(\cos\theta) Q_{ss}(d, r_{NA}) + P_4(\cos\theta) N_{ss}(d, r_{NA}) + \ldots \ldots \tag{13}$$

This particular matrix element contains no dipole or octupole moments by symmetry. The matrix elements depend on d, r_{NA}, and cos θ where d is the distance between the two orbitals, r_{NA} is the distance of the neutral atom from the center of the bond charge and θ is the angle between $\vec{r}_{NA}$ and the "molecular" z axis of the bond charge. The important point is that the functions q, Q, and N can be tabulated for a two-dimensional grid of distances up to $2r_c$ once and for all before the simulation begins. The matrix element for any geometry can be simply reconstructed from Eq. (13). The derivative is also easily constructed from the appropriate derivatives of q, Q, and N and of the Legendre polynomials P_l.

The other matrix elements which are computed approximately are those due to the exchange/correlation energy $\varepsilon_{xc}(n)$ is a non-linear function of n, so that the total exchange/correlation energy is not given by a simple sum over neighboring atoms.

We evaluate the matrix elements approximately by expanding the exchange function $\varepsilon(n)$ about an effective density δ for each matrix element,

$$< \phi_\mu^{PAO}(\vec{r} - \vec{r}_i) \, | \, \varepsilon_{xc}(n) | \phi_v^{PAO}(\vec{r} - \vec{r}_j) > = < \phi_\mu^{PAO}(\vec{r} - \vec{r}_i) \, | \, \varepsilon_{xc}(n - \delta + \delta) \setminus \phi_v^{PAO}(\vec{r} - \vec{r}_j) >$$

$$= \varepsilon_{xc}(\delta) < \phi_\mu^{PAO}(\vec{r} - \vec{r}_i) \, | \, \phi_v^{PAO}(\vec{r} - \vec{r}_j) >$$

$$+ (\partial\varepsilon_{xc}/\partial n)_{n=\delta} < \phi_\mu^{PAO}(\vec{r} - \vec{r}_i) \, | \, (n - \delta) \, | \, \phi_v^{PAO}(\vec{r} - \vec{r}_j) >$$

$$+ (1/2!) (\partial^2 \varepsilon_{xc}/\partial n^2)_{n=\delta} < \phi_\mu^{PAO}(\vec{r} - \vec{r}_i) \, | \, (n - \delta)^2 | \phi_v^{PAO}(\vec{r} - \vec{r}_j) > + \ldots \ldots \tag{14}$$

Our choice of δ is such that the second term vanishes identically, and the contribution from the third term is minimized. When the overlap is non-zero, the value of δ which accomplishes this is:

$$\delta = \, <\phi_\mu^{PAO}(\vec{r}-\vec{r}_i) \,|n|\, \phi_\nu^{PAO}(\vec{r}-\vec{r}_j)> \,/\, <\phi_\mu^{PAO}(\vec{r}-\vec{r}_i) \,|\, \phi_\nu^{PAO}(\vec{r}-\vec{r}_j)> \,. \qquad (15)$$

The calculation of δ is linear in the neighboring atoms. However, the first term of the series for the matrix element, $\varepsilon_{xc}(\delta)\, S_{\mu\nu}^{ij}$, uses δ in a non-linear way, since ε is non-linear in δ. The remainder of the series is approximately included by writing the matrix element as $S_{\mu\nu}^{ij}\, \varepsilon_{xc}(\delta)(1 - C_\infty)$, where C_∞ is fit to the exact result in bulk matrial. C_∞ is small, being ~0.03 in Si.

We have tested the method in bulk Si. No adjustable parameters were used other than the fitting of C_∞ using a first-principles planewave calculation. A value of $r_c = 5a_B$ was used. We find the minimum energy configuration at lattice constant 5.45 Å (Expt. 5.43 Å), a bulk modulus of 101 GPa (expt. 99 GPa). and a $\vec{k} = 0$ optic phonon frequency $\omega = 0.93\ 10^{14}\ s^{-1}$ (expt. $\omega = 0.99\ 10^{14}\ s^{-1}$) which overall is in very good agreement with experiment.

CONCLUSION

The approximate electronic structure method we have developed is physically motivated, fast and easy to use, and is entirely first principles. It requires no experimental input, and avoids complexities of more rigorous methods. It is executed in real-space and requires no periodicity, and has been developed with the application of molecular dynamics in mind. The method has been successfully tested on the physical properties of bulk materials, such as lattice constants, bulk moduli, and optical phonon frequencies. Molecular dynamics simulations using the method are currently in progress.

ACKNOWLEDGMENTS

We wish to thank the Office of Naval Research for their support under contract No. *ONR - N*00014 - 85 - *K* -0442.

REFERENCES

1. F.H. Stillinger and T.A. Weber, Phys. Rev. B31:5262 (1985); M.I. Baskes Phys. Rev. Lett. 59:2666 (1987); M.I. Biswas and D.R. Hamann Phys. Rev. Lett. 55:2001 (1985); J. Tersoff, Phys. Rev. Lett. 56:632 (1986).
2. O.F. Sankey and R.E. Allen, Phys. Rev. B33:7164 (1986).
3. R.E. Allen and M. Menon, Phys. Rev. B33:5611 (1986); M. Menon and R.E. Allen, Phys. Rev. B33:7099 (1986); Superlattices and Micro-structures 3:295 (1987); Solid State Commun. 64:53 (1987).
4. R. Car and M. Parrinello, Phys. Rev. Lett. 55:2471 (1985); M.C. Payne, J.D. Joannopolous, D.C. Allen, M.P. Teter, and D.H. Vanderbilt,

Phys. Rev. Lett. 56:2656 (1986); M. Needles, M.C. Payne, and
J.D. Joannopolous, Phys. Rev. Lett. 58:1765 (1987); D.C. Allan
and M.P. Teter, Phys. Rev. Lett. 59:1136 (198).

5.   D.J. Niklewski and O.F. Sankey, to be published.
6.   P. Hohenberg, W. Kohn, Phys. Rev. 136:B864 (1964); W. Kohn and
L.J. Sham, Phys. Rev. 140:A1133 (1965).
7.   D. Hamann, M. Schluter, and C. Chiang, Phys. Rev. Lett. 43:1494 (1979);
G.B. Bachelet, D.R. Hamann, and M. Schluter, Phys. Rev. B26:4199
(1982).
8.   R.W. Jansen and O.F. Sankey, Phys. Rev. B36:6250 (1987).
9.   R.W. Jansen and O.F. Sankey, Solid State Commun. 64:197 (1987); O.F. Sankey
and R.W. Jansen, J. Vac. Sci. Technol. B6:1240 (1988); R.W. Jansen,
D.S. Wolde-Kidane, and O.F. Sankey, J. Appl. Phys. 64:2415 (1988).
10.   J. Harris, Phys. Rev. B31:1770 (1985); H.M. Polatoglou and M. Methfessel,
Phys. Rev. B37:10403 (1988).

ANGULAR FORCES IN TRANSITION METALS AND DIAMOND STRUCTURE SEMICONDUCTORS

A. E. Carlsson

Department of Physics
Washington University
St. Louis, Missouri 63130

INTRODUCTION

This paper describes the motivation for developing angular forces in transition metals and semiconductors, and discusses the calculation of such forces and their relation to observed structures. The successes and failures of radial force schemes are first briefly described. Subsequently, existing methods for generating angular forces are discussed, with emphasis on methods based on tight-binding analysis. Plots of the angular potentials are given for model transition metals and semiconductors. The features in the potentials are used to interpret observed bond angles in these systems.

MOTIVATION

The need to use angular forces in simulations of properties of diamond structure semiconductors is fairly obvious from their open crystal structure. If radial forces dominated, it would in all likelihood be preferable to fill in the holes in the structure, thereby increasing the coordination and lowering the total energy. Therefore existing simulations of their properties have included such forces. In contrast, the more closely-packed structures of transition metals can be described by a radial forces, without having to assume a very bizarre form for the interactions. A large number of simulations of transition metal properties have, in fact, been performed with radial forces. Recently, schemes have been employed which, although based on radial interactions, are more general than descriptions based on radial pair potentials.[1,2] We will denote the generalized schemes pair *functionals,* as opposed to pair potentials. These include the "embedded atom" method[3] and "N-body" potentials.[4] Pair functionals assume the following form for the configurational energy:

$$E_c = \frac{1}{2} \sum_{i,j} V_2\left(R_{ij}\right) + \sum_i U\left[\sum_j g_2\left(R_{ij}\right)\right] \tag{1}$$

where V_2 is a radial pair potential, g_2 is a radial function, and $R_{ij} = |\vec{R}_i - \vec{R}_j|$. The sum in the argument of U describes some property of the local environment of atom i; this property is assumed to depend additively, via g_2, on the positions of the neighbors. The function U describes the dependence of the part of the configurational energy associated with atom i, on the environmental property. It could, for example, describe the dependence of the local configurational energy on the background electron density due to the other atoms.[3] A linear dependence of U on its argument would lead to a pair potential description. In real metals, U is typically non-linear, with a positive curvature; a form proportional to $-(\sum_j g_2 (R_{ij}))^{1/2}$ is commonly used. Some of the more important advantages of pair functionals over pair potentials are the following:

1) A better fit to the dependence of E_c on the configurational energy is obtained.[5] For example, using pair functionals matched to the cohesive energy one obtains[4,6] reasonable estimates of vacancy formation energies and surface energies.[7-9] Pair potentials matched to the cohesive energy give vacancy[10] and surface energies at least a factor of two too high.

2) The estimates of relaxations[7-9] occurring around broken-bond defects, such as vacancies and surfaces, are much more plausible than those obtained by pair potentials. Usually, negative vacancy relaxation volumes, contractions in the first interlayer spacing at surfaces, and positive surface stresses are obtained; these results are consistent with typical experimental values. The basic physics underlying these results is that the strength of a bond connecting two atoms increases if the coordination number of these atoms is reduced. Nearest-neighbor equilibrium pair potentials, in contrast, obtain vanishing results for these quantities; other types of pair potentials often give the wrong sign.[11]

3) It is possible[12] to obtain non-zero values for the Cauchy pressure C_{12}-C_{44} (here C_{12} and C_{44} are elastic constants). Equilibrium pair potentials always produce a vanishing Cauchy pressure in cubic structures.

However, pair functionals fail in describing a variety of structural energies. For example, the anisotropic parts of the relaxations around vacancies in bcc metals are obtained poorly.[13] In addition, surface reconstructions (the *symmetry-breaking* part of the surface relaxations) are obtained unreliably. The chemical trends in reconstruction energies for noble and near-noble metal (110) surfaces are obtained by some schemes[14] but not by others;[8] the (2x1) reconstruction of the W(100) surface is not obtained.[9] A standard test for accuracy in calculating structural energy differences is the bcc-fcc energy difference in transition metals. Pair functionals produce much too small a value[15] for this energy difference, and do not even necessarily obtain the correct sign.

QUANTUM-MECHANICAL ORIGINS OF ANGULAR FORCES

While a number of recently proposed fitted potentials[16-20] and other simplified energy functionals[21-25] for semiconductors have included angular forces, none have derived the form of the *angular* part of the forces from quantum-mechanical analysis (although some[21-25] have included the bond-strengthening effects mentioned above in connection with defect relaxations). Simplified quantum-mechanical analyses of the bonding in semiconductors have been made using "bond-orbital"[26,27] and "bond-charge" models,[28-30] but these have not yet been developed into interatomic potential schemes suitable for atomistic simulations.

In transition metals, angular forces have been developed using both nearly-free-electron treatments and tight-binding models. In the nearly-free-electron treatment[31,32] one uses as a starting point a uniform electron gas with d-impurities that are not coupled to each other; the s-p electrons are assumed to "see" a weak pseudopotential. By expanding in powers of the pseudopotential and various couplings involving the d-orbitals, one obtains a series of pair, triplet, and higher-order potentials. We will call these "constant-volume" potentials. The pair potential corresponds, roughly to an interaction between screened ions with an additional d-bonding contribution. All the triplet level, angular forces are included, and, where results are available, the bcc-fcc structural energy difference and the phonon spectrum are obtained well.[31] However, because "constant-volume" potentials are based on a uniform electron gas starting point, it is difficult to apply them to inhomogeneous defect environments; one is faced, for example, with the problem of suitable defining a local electron density at a surface. Attempts have been made[33,34] to define such a local density, but at present there is no universally accepted definition. Somewhat similar in spirit to "constant-volume" potentials are interactions based on "bond-charge" model,[35] which replace part of the screened ion contribution with charges at the bond centers. These have, however, not been developed into interatomic potential schemes for treating defect geometries.

A method that holds promising results for both transition metals and semiconductors is based on a moment analysis of the density of states. In this type of analysis, one obtains an r-space description of the electronic band energy associated with a tight-binding Hamiltonian of the form

$$H = \sum_{i,\mu} \varepsilon_i^\mu \mid i,\mu >< i,\mu \mid + \sum_{i \neq j} \sum_{\mu,\nu} h_{ij}^{\mu\nu} \mid i,\mu >< j,\nu \mid \tag{2}$$

Here the $\mid i,\mu>$ are localized basis orbitals (assumed orthogonal) located on the atom i. The index μ denotes the angular momentum of these orbitals. In transition metals one often keeps only the d orbitals, and one has five values of μ ($-2 \leq m \leq 2$); in Si one typically has four values of μ corresponding to s, p_x, p_y, and p_z basis orbitals. The on-site matrix elements ε_i^μ in Eq. (2) correspond roughly to atomic energy levels, but are often fitted to quantitative band structure results. The interatomic couplings $h_{ij}^{\mu\nu}$ are usually fitted as well.

Given H, one seeks an r-space description of the electronic band energy

$$U_{TB} = 2 \int_{-\infty}^{\varepsilon_F} E\rho\,(E)\,dE = \sum_i U_{TB}^i \ ,$$

where

$$U_{TB}^i = 2 \int_{-\infty}^{\varepsilon_F} E\rho_i(E)\, dE.$$ (4)

Here $\rho(E)$ is the electronic density of states, and $\rho_i(E)$ is the projection of this density of states on the orbitals on site i; the factor of two accounts for spin degeneracy. The crucial observation that allows an r-space description is that the moments

$$\mu_n^i = \int_{-\infty}^{\infty} E^n \, \rho_i(E)\, dE$$ (5)

of $\rho_i(E)$ can be given[36] rigorously in r-space:

$$\mu_2^i = \sum_j \sum_{\mu,\nu} \left(h_{ij}^{\mu\nu} \right)^2 ,$$ (6a)

$$\mu_3^i = \sum_{j,k} \sum_{\mu,\nu,\gamma} h_{ij}^{\mu\nu} h_{jk}^{\nu\gamma} h_{kl}^{\gamma\nu} ,$$ (6b)

$$\mu_4^i = \sum_{j,k,l} \sum_{\mu,\nu,\gamma,\eta} h_{ij}^{\mu\nu} h_{jk}^{\nu\gamma} h_{kl}^{\gamma\eta} h_{li}^{\eta\mu} ,$$ (6c)

and the higher-order moments are given by parallel expressions. Thus to obtain μ_n one needs to perform n-atom sums (although for the higher-order moments techniques[37] exist for calculating the μ_n which are faster than direct summation).

Knowledge of all the μ_n^i determines $\rho_i(E)$ uniquely. Our focus, however, will be on low-order descriptions, typically with n≤4. In this case, one guesses[36] a parametrized form for the density of states, and fits the parameters in this form to the known μ_n. Typical forms are continued fractions[38] and the "maximum-entropy" form,[39-42] which is an exponential of a polynomial in E. Once $\rho_i(E)$ is known, one obtains U_{TB}^i via Eq. (4). The total configurational energy is then given by

$$E_c = \frac{1}{2} \sum_{ij} V_2\left(R_{ij} \right) + \sum_i U_{TB}\left(\mu_2^i, \mu_3^i, \ldots \right) ,$$ (7)

If one keeps only μ_2^i, then by Eq. (6a) a pair functional description (cf. Eq. (1)) is obtained, with $g_2\left(R_{ij} \right) = \sum_{\mu\nu} (h_{ij}^{\mu\nu})^2$. When higher-order moments are included, one obtains a more sophisticated description which we will denote a cluster functional.

Cluster functionals explicitly include information about the bond angles in the system. For example, in a d-band model, in only σ-type couplings (those having $l=0$ character about the bond axis) are included, the contribution from a four-body path (*cf.* Fig. 1) to μ_4 is given[43] by $P_2(\cos\theta_i)\,P_2(\cos\theta_j)\,P_2(\cos\theta_k)\,P_2(\cos\theta_l)$, where P_2 is the second-order Legendre polynomial and θ_i–θ_l are the bond angles at the sites i–l; analogous results also hold for different angular momenta, and paths of different lengths. It is convenient to visualize this angular dependence via effective potentials,[44,45] given by

$$V_n^{\text{eff}}\left(\vec{R}_i,\vec{R}_j,\ \ldots\right) = \sum_{m=n}^{M} \mu_m^{ij\ldots}\left(\frac{\partial U_{\text{TB}}}{\partial\mu_m}\right). \qquad (8)$$

Here $\mu_m^{ij\ldots}$ is the contribution to μ_m from paths including all of the atoms $i,j\ldots$ and no others; M is the order of the highest moment included in the description. The $\partial U_{\text{TB}}/\partial\mu_m$ term is typically obtained by numerical differentiation of U_{TB}, at values of the μ_m in a "reference environment" which is not too dissimilar from the atomic configurations of interest.

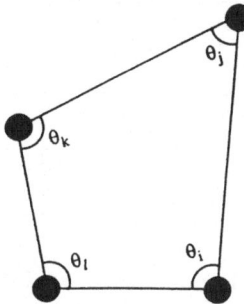

Figure 1.   Sketch of geometry of four-atom cluster contributing to m_4.

From Eq. (8), we see that the angular dependence of V_n^{eff} simply reflects that of the $\mu_m^{ij\ldots}$. The coefficients $\partial U_{\text{TB}}/\partial\mu_m$, which determine the sign and magnitude of the V_n^{eff}, are strongly dependent on the filling of the band of interest and are thus important in determining the chemical trends in the angular forces. We will denote the fractional band filling by N_v, so that $N_v = 0$ corresponds to an empty band and $N_v = 1$ to a filled band. We will be particularly interested in $\partial U_{\text{TB}}/\partial\mu_4$. While μ_2 determines the effective width of the band, μ_4 determines some aspects of its shape. As shown in Figs. 2a and 2b, a small value of μ_4 (μ_4 is the smallest possible value) corresponds to a density of states dominated by two peaks, which would typically correspond to bonding and antibonding states; usually, if μ_4 is larger, the absolute bandwidth increases, and the density of states at the center of the band grows. (The densities of states in Fig. 2 both have $\mu_0 = \mu_2 = 1$, so that only the effects of changing μ_4 are seen.) One can thus begin to see the effects of a gap through μ_4. If the band is nearly empty ($N_v \ll 1$), then the electronic band energy per electron for the rectangular band (Fig. 2b) is approximately $-(3\mu_2)^{1/2}$. That for the δ-function density of states (Fig. 2a), which has a smaller value of μ_4, is only $-(\mu_2)^{1/2}$. Thus $(\partial U_{\text{TB}}/\partial\mu_4) < 0$. On the other hand, if $N_v = 1/2$, then U_{TB} (rectangular band) = $-1/2(3\mu_2)^{1/2}$ while U_{TB} (δ-functions) = $-(\mu_2)^{1/2}$. Therefore $(\partial U_{\text{TB}}/\partial\mu_4) > 0$, and a small value of μ_4 is preferred. The behavior of a nearly filled band is identical to that of a nearly empty band, resulting in the behavior indicated in Fig. 2c. The two zero crossings seen in the Figure are seen in more sophisticated calculations of U_{TB} as well; in fact, their presence is required by an exact theorem[46-48] for tight-binding models.

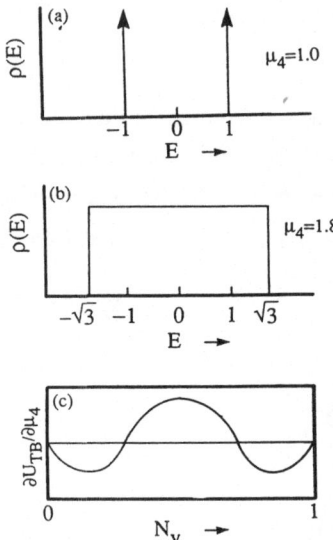

Figure 2.   a) and b) Model electronic densities of states, with $\mu_0 = \mu_2 = 1$. Arrows in a) denote δ-functions. c) Schematic illustration of dependence of $(\partial U_{TB}/\partial \mu_4)$ (*cf.* Eq. (8)) on fractional band filling N_v.

ANGULAR DEPENDENCE OF EFFECTIVE TRIPLET AND QUADRUPLET POTENTIALS

As mentioned above, the angular dependence of the effective potentials (*cf.* Eq. (8)) is determined by that of the μ_m. We will be concerned with three- and four-body potentials in a model of the electronic density of states based on $\mu_0 - \mu_4$, so that M=4. Then

$$V_3^{eff}\left(\vec{R}_i, \vec{R}_j, \vec{R}_k\right) = \mu_4^{ijk}\left(\partial U_{TB}/\partial \mu_4\right) + \mu_3^{ijk}\left(\partial U_{TB}/\partial \mu_3\right) \tag{9}$$

and

$$V_4^{eff}\left(\vec{R}_i, \vec{R}_j, \vec{R}_k, \vec{R}_l\right) = \mu_4^{ijkl}\left(\partial U_{TB}/\partial \mu_4\right). \tag{10}$$

In diamond structure semiconductors, μ_3^{ijk} and μ_4^{ijkl} both involve second-neighbor hops and are therefore much smaller than μ_4^{ijk}, which has contributions from paths containing only nearest neighbors. Therefore, we analyze only μ_4^{ijk}. As shown in Fig. 3, μ_4^{ijk} involves a self-retracing path. We have evaluated the contribution from this path in an sp³ tight-binding model. Neglecting angle-independent terms, one obtains the functional form given in the Figure, with

$$g_1\left(R_{ij}\right) = \sqrt{2}\ h_{ij}^{sp}\left(h_{ij}^{ss} - h_{ij}^{pp\sigma}\right) \tag{11}$$

and

$$g_2\left(R_{ij}\right) = \left(h_{ij}^{sp}\right)^2 + \left(h_{ij}^{pp\sigma}\right)^2 - \left(h_{ij}^{pp\pi}\right)^2 ; \tag{12}$$

here all the couplings have $l=0$ symmetry about the bond axis except $h_{ij}^{pp\pi}$, which has $l = 1$ symmetry. The angular dependence of the resulting effective potentials is shown in Fig. 4, for two models. The sign of V_3^{eff} corresponds to a half-filled band. In the first model (solid line), we use sp³ tight-binding parameters[49] appropriate for Si; in the second (dashed line) only p orbitals are included. The latter case would be more appropriate for group V elements in which the s-orbitals are more core-like, and the p-band can be regarded as roughly half-filled. For comparison, we include a plot (dotted line) of the "Stillinger-Weber" (SW) empirical three-body potential[17] for Si.

PATH	MOMENT	FUNCTIONAL FORM

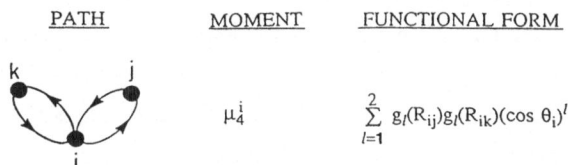

| μ_4^i | $\displaystyle\sum_{l=1}^{2} g_l(R_{ij})g_l(R_{ik})(\cos\theta_i)^l$ |

Figure 3.   Sketch of three-body contribution to μ_4 in sp³ model semiconductor. Functions g_l are radial functions determined by interatomic couplings. θ_i denotes angle between i-j bond and i-k bond.

EQUATION

Comparison of the sp³ curve with the SW curve shows close agreement in shape; the sp³ curve minimizes at $\theta = 115°$, while the SW potential is constrained to minimize at $\theta = 109°$, the nearest-neighbor bond angle in the diamond structure. (The latter constraint is not necessary for the stability of this structure.) For $\theta < 90°$, both of these curves rise rapidly, making a repulsive contribution to the energies of more closely packed hypothetical structures, such as fcc ($\theta = 60°$) and bcc ($\theta = 71°, 55°$). It should be noted, however, that the closely packed structures also have nearest-neighbor four-body paths, which contribute to the energy differences between these structures and the diamond structure. Thus V_3^{eff} alone cannot account for these energy differences.

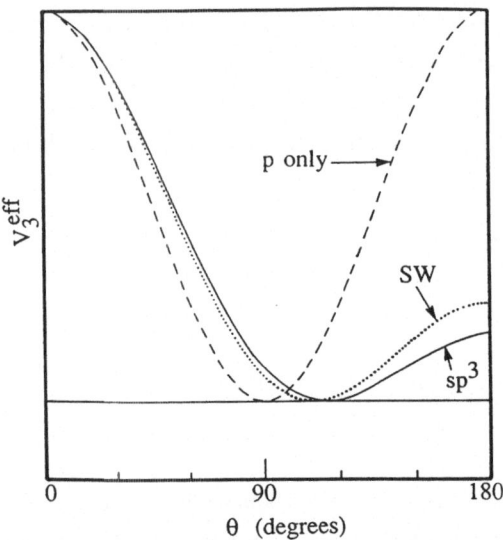

Figure 4.   Angular parts of effective three-body potentials for Si (arbitrary units). Shown for clusters of three atoms i,j,k, with $R_{ij} = R_{ik} = 2.35$Å (=bulk nearest-neighbor bond length). θ denotes angle between i-j and i-k bonds. Curve labeled "sp^3" obtained from s-p tight-binding model with parameters described in text. Curve labeled "p-only" obtained from tight-binding model containing only p-orbitals. Curve labeled "SW" is empirical potential due to Stillinger and Weber (Ref. 17). A constant shift is added to "sp^3" and "p-only" curves, so that their minimum value is zero.

The position of the minimum in V_3^{eff} is strongly influenced by the s-p hybridization terms. In the p-only model, V_3^{eff} has no contributions linear in cos θ and thus minimizes at 90° (*cf*. Fig. 4). The shift relative to the sp^3 case is probably in part due to the fact that p orbitals are orthogonal to each other only if their directions are 90° apart, whereas sp^3 orbitals are orthogonal if their directions are 109° apart.

As mentioned above, the p-only model is more appropriate for group V elements. Therefore, one would expect the shift in the minimum of the potential to result in smaller bond angles for the group V elements than for the group IV elements. This trend is, in fact, present in the observed crystal structures.[50] Starting with the group IV elements, C has bond angles of either 120° (graphite) or 109° (diamond). Si and Ge, of course, have the diamond structure angles of 109°. Turning to the group V elements, black P has an average bond angle of 100°, while α-As, α-Sb, and α-Bi all have bond angles less than 97°. Thus the trends in the observed bond angles are consistent with the calculated potentials. A similar, but less pronounced trend is also seen in the group VI elements. Their p-bands are probably somewhat more than half-filled, but V_3^{eff} is still expected to have the same sign as in the half-filled case.[52] Here the bond angles are less than 109°, with typical values being 102°-103°. In addition, an empirical angular potential, similar to the "SW" potential for Si discussed above, has recently been developed[53] for S. This potential minimizes at θ=95°, 14° smaller than the angle for Si.

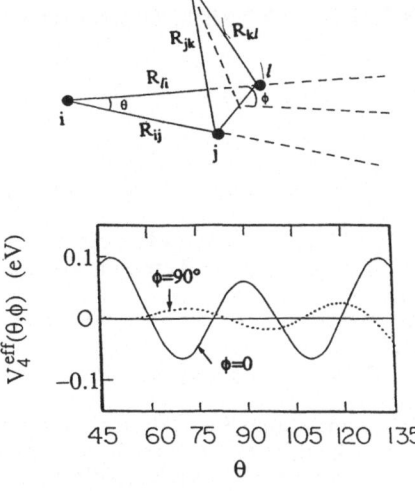

Figure 5.   Four-body effective potential for a model transition metal with a nearly half-filled band (for which the bcc structure is favored). Geometry, which is illustrated, has $R_{ij} = R_{jk} = R_{kl} = R_{li}$.

In more closely-packed transition metal structures, the four-body terms[15,43,54] cannot be ignored. Figure 5 shows the calculated[55] V_4^{eff} for a four-atom cluster of transition metal atoms, with a roughly half-filled d-band; for this band filling the bcc structure is preferred. For simplicity, only configurations with four equal bond lengths are considered. The angles θ and ϕ are as illustrated; the two curves correspond to different values of ϕ. The angular dependence of V_3^{eff} is very rapid, with a strong maximum at $\theta = 90°$ and minima at $\theta \approx 70°$ and $110°$. For $\phi=90°$ V_4^{eff} is much weaker than for $\phi = 0°$. Thus a particularly large penalty is associated with square nearest-neighbor paths, which have $\theta=90°$ and $\phi=0°$. There are many such paths in the fcc structure, but none in the bcc structure, which is favored here. In addition, the nearest-neighbor bond angles in the bcc structure are 71°, 109°, and 180°, all of which occur in planar clusters; the first two fall very close to the minima in V_4^{eff}. Thus we have the beginnings of a picture of the preferability of the bcc structure based on effective interatomic potentials.

CONCLUSION

A qualitative picture of the shape of angular forces in transition metals and semiconductors is now beginning to emerge from tight-binding theory. Effective potentials based on low-order moments of the electronic density of states display angular features which correlate well with the observed structures. Since the form of these potentials is derived from quantum-mechanical analysis, much of the ambiguity that is present when one uses an assumed angular dependence to fit an empirical potential can be eliminated. It is hoped that this analysis will lead to useful *ab-initio* interatomic potential schemes, and will improve the accuracy of empirical schemes.

ACKNOWLEDGEMENTS

I appreciate support for this work from the Department of Energy under Grant No. DE-FG02-84ER45130.

REFERENCES

1. M. W. Finnis, A. T. Paxton, D. G. Pettifor, A. P. Sutton, and Y. Ohta, Phil. Mag. A58:143 (1988).
2. A. E. Carlsson in: "Solid State Physics: Advances in Research and Applications", edited by H. Ehrenreich and D. Turnbull, Volume 43, (Academic, New York) (in press).
3. M. S. Daw and M. I. Baskes, Phys. Rev. Lett. 50:1285 (1983); Phys. Rev. B 29: 6443 (1984).
4. M. W. Finnis and J. M. Sinclair, Phil. Mag. A50:45 (1984); Phil. Mag. A53:161 (1986).
5. V. Heine and D.Weaire, in: "Solid State Physics: Advances in Research and Applications", edited by H. Ehrenreich, R. Seitz and D. Turnbull, Volume 35, (Academic, New York, 1970), p. 1.
6. J. M. Harder and D. J. Bacon, Phil. Mag. A54:641 (1986).
7. S. M. Foiles, M. I. Baskes, and M. S. Daw, Phys. Rev. B 33:7983 (1986); Phys. Rev. B 37:10378 (1988).
8. G. J. Ackland, G. Tichy, V. Vitek, and M. W. Finnis, Phil. Mag. A56:735 (1987).
9. G. J. Ackland and M. W. Finnis, Phil. Mag. A54:301 (1986).
10. C. P. Flynn, "Point Defects and Diffusion", (Clarendon, Oxford, 1972), p. 6.
11. R. Benedek, J. Phys. F8:1119 (1978).
12. F. Ducastelle, J. Phys. (Paris) 31:1055 (1970).
13. Y. Ohta, M. W. Finnis, D. G. Pettifor, and A. P. Sutton, J. Phys. F17:L273 (1987).
14. S. M. Foiles, Surf. Sci. 191:L779 (1987).
15. P. Turchi and F. Ducastelle, in: "The Recursion Method and Its Applications", edited by D. G. Pettifor and D. L. Weaire, (Springer, New York, 1985), p. 104.
16. E. Pearson, T. Takai, T. Halicioglu, and W. A. Tiller, J. Cryst. Growth 70:33 (1984).
17. F. H. Stillinger and T. A. Weber, Phys. Rev. B 31:5262 (1985).
18. R. Biswas and D. R. Hamann, Phys. Rev. Lett. 55:2001 (1985).
19. R. Biswas and D. R. Hamann, Phys. Rev. B 36:6434 (1987).
20. D. W. Brenner and B. J. Garrison, Phys. Rev. B 34:1304 (1986).
21. J. Tersoff, Phys. Rev. Lett. 56:632 (1986).
22. J. Tersoff, Phys. Rev. B 37:6991 (1988).
23. M. I. Baskes, Phys. Rev. Lett. 59:2666 (1987).
24. B. W. Dodson, Phys. Rev. B 35:2795 (1987).
25. K. E. Khow and S. Das Sarma, Phys. Rev. B 38:3318 (1988).
26. W. A. Harrison, "Electronic Structure and the Properties of Solids", (W. H. Freeman, San Francisco, 1980), Chapter 7.
27. M. van Schilfgaarde and W. A. Harrison, Phys. Rev. B 33:2653 (1986).
28. J. C. Phillips, Phys. Rev. 166:832 (1968).
29. R. M. Martin, Phys. Rev. 186:871 (1969).
30. Reference 26, Chapter 9.
31. J. A. Moriarty, Phys. Rev. Lett. 55:1502 (1985).
32. J. A. Moriarty, Phys. Rev. B 38:3199 (1988).
33. R. J. Harrison, Surf. Sci. 144:215 (1984).
34. K. W. Jacobsen, J. K. Nørskov, and M. J. Puska, Phys. Rev. B 35:7423 (1987).
35. C. C. Matthai, P. J. Grout, and N. H. March, J. Phys. Chem. Solids 42:317 (1981).
36. F. Cyrot-Lackmann, J. Phys. Chem. Solids 29:1235.
37. See articles by R. Haydock, (p. 216) and M. J. Kelley (p. 296), in Ref. 5.
38. E. T. Jaynes, Phys. Rev. 106:620 (1957); Phys. Rev. 108:171 (1957).

40. R. Collins and A. Wragg, J. Phys. A10:1441 (1977).
41. L. R. Mead and N. Papanicolaou, J. Math. Phys. 25:2404 (1984).
42. R. H. Brown and A. E. Carlsson, Phys. Rev. B 32:6125 (1985).
43. K. Hirai and J. Kanamori, J. Phys. Soc. Jpn. 50:2265 (1981).
44. A. E. Carlsson and N. W. Ashcroft, Phys. Rev. B 27:2101 (1983.
45. A. E. Carlsson, Phys. Rev. B 32:4866 (1985).
46. F. Ducastelle and F. Cyrot-Lackmann, J. Phys. Chem Solids 32: 285 (1971).
47. V. Heine and J. H. Samson, J. Phys. F10:2609 (1980).
48. V. Heine and J. H. Samson, J. Phys. F13:2155 (1983).
49. W. A. Harrison, "Solid State Table of the Elements," Ref. 26.
50. J. Donohue, "The Structures of the Elements", (Wiley, New York, 1974),
    Chapter 8.
51. Reference 50, Chapter 9.
52. In typical calculations (see Ref. 42, for example), the $\partial U_{TB}/\partial \mu_4$ contribution to the V_n^{eff} changes sign at $N_v \approx 0.25$ and 0.75. In group VI elements $N_v < 0.67(=4/6)$, since not all the electrons reside in the p-band.
53. F. H. Stillinger, T. A. Weber, and R. A. LaViolette, J. Chem. Phys. 85:6460 (1986).
54. F. Ducastelle and F. Cyrot-Lackmann, J. Phys. Chem. Solids, 31:1295 (1970).
55. R. B. Phillips and A. E. Carlsson (unpublished).

PSEUDOPOTENTIAL STUDIES OF STRUCTURAL PROPERTIES FOR TRANSITION METALS

E. J. Mele, M. H. Kang and I. A. Morrison

Department of Physics, Department of Materials Science
and Laboratory for Research on the Structure of Matter
University of Pennsylvania
Philadelphia, PA 19104-6396

INTRODUCTION

Modern applications of the density functional theory within the "local density approximation" have demonstrated that the theory provides a powerful approach for studying the structural properties of solids. This method is generally regarded as a nonempirical theory, since once the density functional method is adopted, the only remaining input required for the calculation of ground state structural properties is the atomic number of the elements in the structures to be studied. In most applications to date, the formalism has proven exceedingly useful for ground state structural properties (i.e. predictions of equilibrium crystal structures, lattice constants and compressibilities). The prediction of absolute cohesive energies remains somewhat problematic, limited by the accuracy of the local density functional for calculating the absolute internal energy of the interacting many electron system. In this paper I will review some recent progress in extending the "first principles pseudopotential" approach to complex structures involving elements from the 3d transition series. For several reasons mentioned below, these systems have been resistant to well controlled studies on complex structures using available computational techniques. Here, I will briefly present a computational scheme which allows for accurate work on these systems, and will illustrate it with a brief review of results we have obtained on structural studies in Cu and NiAl.

THEORETICAL METHODS

As is well known, the minimization of the energy of the interacting many electron system, within the local density functional theory, requires the self consistent solution of a set of coupled integro-differential equations for the effective one electron "states" in the system. A number of computational techniques for studying this effective one electron problem have been developed; in our work, we have been concerned with the pseudopotential approach to this problem. In this approach, the bare ionic "all electron" potential, in which both core and valence electronic states may be calculated, is replaced by an effective ionic potential, with a short ranged repulsive core. Due to the repulsive core potential, the pseudopotential binds no core states, and the lowest valence solutions are then nodeless "images" of the all electron valence wavefunctions. These pseudowavefunctions are images of the all electron valence wavefunctions in the sense that they have the same binding energy as the all electron state in a self consistent solution for the neutral pseudoatom, and retain the shape and normalization of the all electron function outside a critical core radius which encloses the repulsive potential. These resulting effective ionic pseudopotentials are nonlocal by virtue of their dependence on the orbital angular momentum of the valence state. It has been shown formally, and demonstrated in calculations on physical systems, that the choice of the repulsive potential satisfying the criterion noted above leads to an effective potential which preserves the scattering properties of the true all electron potential over a very wide energy range (typically of order 20 eV)[1]. The resulting potentials are often referred to as "first principles" pseudopotentials.

There are a number of practical advantages of the pseudopotential approach to the calculation of ground state structural energies. The first is that the effective valence pseudopotential is considerable weaker (i.e. smoother) than the all electron potential, especially for systems which bind many core levels. Thus the valence electrons may be regarded as weakly scattered by the effective ionic cores in the problem. In fact for a wide variety of systems, the electronic structure can be studied by means of a rapidly convergent plane wave expansion for the one electron levels. Additionally, since the effective potentials bind no core states, the absolute total energy of the interacting system is reduced by a factor typically of order 10^2 relative to the all electron energy. Thus the accurate calculation of small energy differences between competing condensed phases is considerably simplified within the pseudopotential approach. This has allowed a number of nontrivial studies of equilibrium structural properties for relatively complex systems[3].

There are a class of systems which have been resistant to the computations within the pseudopotential method. These are "light" elements, i.e. s-p systems in the first row of the periodic chart, or transition metal systems in the 3d series. In these systems the effective pseudopotential is "unscreened" in at least one angular momentum channel, so that the effective pseudopotential is not weak relative to the bare ionic potential. As an example, the Bachelet Hamann Schluter[1] d-wave ionic pseudopotential for Cu is nearly 200 Rydbergs deep in the core (r<0.2 a.u.). It should be noted that traditional approaches to the calculation of the electronic structure of condensed phases involving the 3d transition elements have made use of cellular methods, i.e. KKR scattering method, or the APW expansion[4]. In these methods the deep core potential is in fact exploited so that the rapidly varying part of the wavefunction is evaluated by a real space integration on a radial mesh. Alternatively expansion of the valence wavefunction in gaussian orbitals has also frequently been employed in solid state calculations[5]. A "mixed" basis has also been developed for this problem[6]. This approach attempts to preserve the advantages of the plane wave expansion for the nearly free component of the wavefunction for the transition elements, by augmenting the expansion set with a set of site centered gaussian expansion functions to describe the "localized" part. This has proven reasonably successful for the heavier (i.e. 4d and 5d) transition elements, but accurate work in the 3d series has proven impractical using this scheme.

We have developed a modification of this mixed basis approach which exploits a useful technique from the APW approach to this problem[7]. We seek an expansion in terms of a small number of plane waves, augmented by a small number of site centered "localized functions" to describe the rapidly varying component of the valence wavefunctions. We choose these on site functions so that they are nonzero only within a critical radius around the atomic core. Thus they are rigorously nonoverlapping in the solid state. This tremendously simplifies the calculation of various matrix elements, screening densities, etc. in solid state calculations involving these orbitals, since most numerical manipulations involving them can be studied by very fast and accurate on dimensional radial quadrature just as in the cellular schemes. In our method the plane waves in the theory are *not* constrained to describe the wavefunction only in the interstitial region, but are allowed to penetrate the core regions. Thus the basis set consists of a small number of plane waves which propagate through the entire structure, and a small number of localized functions which are nonzero only within a small cell around each site.

We have found that to keep the plane wave expansion well conditioned (i.e. rapidly convergent) care must be exercised in choosing these localized "core functions." For example an unphysical truncation of the core function on the surface of the core sphere produces a singularity in a Fourier expansion of the function which ultimately must be healed by a plane wave "correction" with very high spatial Fourier components. This obviates the usefulness of the mixed basis approach. Consequently we have derived a set of "optimized" localized functions for this purpose, which in essence are atomic functions with a set of long wavelength Fourier components projected out of the functions. The strategy for this projection is to choose the long wavelength components to eliminate any significant amplitude for the projected function in the "tail" region, i.e. outside the core radius (typically of order 0.2 a.u. in Cu). Since the tail is relatively smooth, this projection can be very accurately accomplished with very few spatial Fourier components, provided that the plane waves are permitted to penetrate into the core region (i.e. we do not truncate the projection on the surface of the core spheres). The projected localized functions are tabulated numerically on a radial mesh. They offer the computational advantage of describing *only* the short wavelength behavior of the localized functions; variational freedom for the state applications are then carried entirely by the plane wave basis set. Explicit techniques for numerically choosing these optimized real space function, and numerical manipulations involving them are presented in reference 7. We should note that the real space functions derived here bear some resemblance to the quasiatomic orbitals derived by Bendt and Zunger for a similar purpose[8].

In the applications involving the transition metals, the resulting optimized local basis functions are used to model the d-wave component of the valence wavefunctions. Interestingly, the resulting dispersion in the d bands is entirely due to resonant mixing of these spatially nonoverlapping d functions through the propagating plane waves. The model thus ultimately resembles a kind of refined "Anderson" model, and appears to be a natural vehicle for studying systems which exhibit a well localized component in the valence one electron eigenfunctions.

RESULTS

In this section I will briefly review a series of structural studies on systems containing elements from the 3d transition series using the scheme described above.

(a) Bulk Elemental Cu

As a test of the method we consider first the ground state structural properties of elemental Cu[7]. This is a natural starting point, since the structural properties of bulk elemental Cu have been studied by more traditional local density methods[4,9], providing a benchmark for the accuracy of our approach. Our calculations employ a single radial d-functions multiplied by the five independent azimuthal components of the $l = 2$ harmonics, and typically 60-80 plane waved in the basis set (corresponding to an energy cutoff of 15 Rydbergs). The charge density is sampled on 29 wavevectors in the irreducible Brillouin zone and the screening potential is accumulated on a set of 3K plane waves.

The self consistent band structure obtained by this method is presented in reference 7. Of interest here is that the dispersion of the d-bands agrees quite well with that obtained from KKR and APW methods. Thus the dispersion in the d bands provided by resonant mixing of the truncated d functions with the plane waves provides a good account of the valence one electron eigenfunctions. It should be noted that there remains a discrepancy between the computed binding energies of these states and the energies as deduced from angle resolved photoemission[10] (typically of order 0.5 eV in the d bands), a difficulty which almost certainly requires a proper treatment of the quasihole self energy in excited state to remedy.

We have studied the structural properties of Cu in the observed FCC structure, and the results are summarized in table I where they are compared with several other reported results for the ground state structure of Cu. We note that the calculations account well for both the observed lattice constant and the bulk modulus. Our cohesive energy is underestimated in the theory. However, this result must be regarded as less rigorous than the ground state structural properties since it requires a comparison of the solid state energy with a pseudoatomic total energy where the calculation of the electronic structure does not involve the same basis set constraints.

Table I. Ground state structural properties for Cu in the BCC structure:

	KKR[9]	ASW[4]	Present	exp
lattice constant a.u.	6.76	6.81	6.85	6.81
cohesive energy (Ryd/atom)	0.301	0.298	0.246	0.257
bulk modulus (Mbar)	1.55	1.29	1.50	1.42

We have also studied Cu in a hypothetical BCC crystal structure. Cu does not exist in the BCC crystal structure in bulk form, and indeed we find that over a wide range of densities the energy of the BCC structure exceeds the FCC structure, but by a very small amount (on order 5 mRyd per atom at the equilibrium FCC density). [Our original calculations on this system also suggested an anomalous bimodal dependence of the BCC structural energy on cell volume[7]. While we ultimately traced this to a nonspherical distortion of the d-like charge density around the Cu sites under compression of the BCC structure, the precise physical origin of this effect remains unexplained.]

More recently experiments have suggested that it is possible to produce a metastable BCC phase of Cu by epitaxial growth on Fe (001)[11]. BCC Cu would appear to be an ideal epitaxy partner to BCC Fe, since the estimated BCC equilibrium lattice constant of Cu is 2.87 Å (as compared to 2.82 Å for Fe). To study this possibility more closely we studied the total energy of Cu in a range of structures within the body centered tetragonal crystal system which spans both the BCC and FCC cubic phases as limiting high symmetry structures[12]. Interestingly, we find that the free energy with a constant lateral latice constant in the (100) plane (as constrained by epitaxy on Fe with a=2.82 Å) reveals that the most stable epitaxial phase for this situation corresponds to an intermediate metastable BCT configuration in between the cubic BCC and FCC reference phases. In fact the original experimental reports of BCC Cu on Fe suggest the appearance of a "somewhat distorted" BCC phase which we suspect is this metastable intermediate BCT structure.

(b) Nonpolar NiAl Surfaces

We have applied the scheme to study the bulk and surface structural properties of NiAl. NiAl is a bimetallic alloy in the CsCl crystal structure. Simple chemical arguments predict a charge exchange from the Al sites to the Ni sites, saturating the Ni d-band and yielding a composite system which in a zeroth order approximation resembles Cu more closely than pure Ni or Al. Unlike Cu however, NiAl is tremendously stable with a melting point exceeding 1600 K. The pseudopotential description of the bulk NiAl system is particular interesting since the Al pseudopotential is known to be exceptionally weak. The compound is then well described as a simple cubic array of strong d-wave scatterers (Ni) embedded in a nearly uniform electron gas (Al). Our bulk structural studies yield structural parameters with agreement with experiment comparable to the previous Cu studies. We find a cubic lattice constant a=5.401 Å and a bulk modulus B=1.71 Mbar in good agreement with experiment (a=5.457 Å) and a LMTO calculation (B=1.69 Mbar)[13]. A very important structural point in these studies is that the charged transfer from the Al to the Ni leaves the free volume around the Al sites nearly 20% smaller than in elemental Al. The Al sites are thus under compressive stress, and several of the surface properties may be interpreted as a microscopic mechanism to release this stress.

We have studied the (110) surface of NiAl. This is a nonpolar termination of the crystal, with alternating Ni and Al sites on a two dimensional centered rectangular lattice. Theoretical interest in this surface originated in the early observation that the clean surface is not atomically flat, but exhibits a rippled structure, with the Al forced away from the crystal and the Ni sites contracting towards the bulk. The relative displacements are uncharacteristically large for a metallic surface, involving relative displacements of order 7% of the (110) bulk interplanar spacing. To study the surface we construct a model thin film consisting of five NiAl layers in a supercell geometry, and fixed the atomic geometry in the three layers at the center of the structure. The atoms in the two bounding layers were allowed to relax freely along the (110) direction. A potential energy surface for the relaxation is given in Figure 1. The theory predicts an equilibrium structure with the Ni contracting towards the bulk by 6.9% of an interlayer spacing and the Al withdrawing from the surface by 6.6% of an interlayer spacing. The result is in reasonably good agreement with the LEED and MEIS scattering geometries indicated by the points. The agreement is especially good for the relative Ni position, which appears to dominate the relaxation energy. The potential for the Al relaxation is softer, and the error compared to experiment is slightly larger.

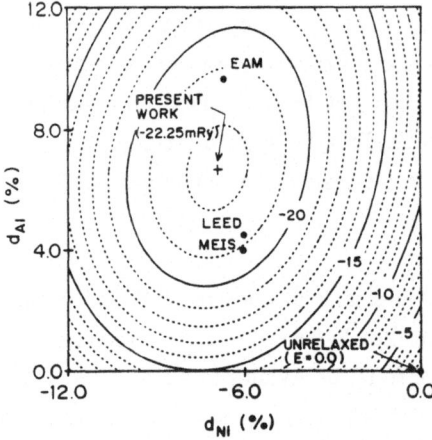

Figure 1. Ground state potential surface for surface relaxation of NiAl (110)

The mechanism underlying the rippled relaxation is relief of bulk compressive stress on the surface Al sites. More precisely, we find that the charge transfer from the Al to the Ni is slightly quenched at the surface. As the free electron charge flows back to the Al sites, the Al are forced out of the compressed bulk positions away from the crystal. This also has the effect of smoothing the free electron density profile parallel to the surface, and lowers the electronic kinetic energy. Electrostatic effects then draw the Ni back towards the bulk. There is a rich spectrum of surface states associated with the surface Ni within 1-2 eV of the Fermi energy, these have been characterized theoretically, and observed by angle resolved photoemission (a complete report is given in reference 13). Unfortunately, we find that the binding energies of these surface states depend relatively weakly on the surface relaxation, so that the experimental measurements do not precisely pin down the equilibrium surface structure.

The potential energy surface of Figure 1 was also used to calculate restoring forces for surface atomic motions polarized along the surface normal, these in turn were used to estimate intrinsic surface phonon frequencies for the relaxed surface[15]. Interestingly, we find an intrinsic surface mode at 27.2 meV, polarized parallel to the surface normal, and strongly coupled to the charge fluctuation implicated above in the surface relaxation. This mode occurs within a projected gap of the bulk phonon spectrum, and describes a strongly localized surface phonon. Crudely, this mode may be described as a harmonic oscillation of the collective coordinate which leads from the ideal termination to the relaxed surface structure. Experimentally a strong dipole active feature, with perpendicular polarization is observed at 27 meV, undoubtedly associated with this structural degree of freedom.

(c) Polar (111) NiAl Surfaces

Recent experimental measurements on the polar (111) NiAl surface have led to some puzzling conclusions about the structure surface. Unlike the (110) surface the (111) surface is a polar surface, with two inequivalent terminations on a Ni layer or an Al layer. Interestingly, LEED measurements indicate that the surface is unreconstructed, and exhibits a sharp 1x1 LEED pattern. In analysis of the intensities of the diffracted beams however, it has been concluded that an optimized structural configuration is not obtained with a pure Ni or Al termination, but is better described as an incoherent superposition of the two[16]. This has led to the speculation, that the surface, locally 1x1, exhibits a domain structure separating regions terminated on a Ni or an Al layer. On the contrary, measurements on vicinal surfaces near the (111) orientation suggest a mean step size of two interatomic layers, which quite strongly suggests that one termination is preferred[17]. This conclusion is consistent with one's physical intuition that the free energies of the two surfaces are not well matched (and thus that one is preferred energetically).

We investigated this problem by studying the equilibrium structural configurations, and by comparison of valence one electron binding energies with the bands measured in angle resolved photoemission for this surface. The calculations were performed on a slab of 9 atomic layers for NiAl (111) with a Ni termination. The Al termination was studied by adding an additional Al film on either side, yielding a film with 11 atomic layers. In our surface relaxation studies, relaxations of the atoms in the outer two layers were allowed to relax, with the remaining sites fixed at their bulk positions. We considered both the equilibrium structures within our theory, and the structures deduced from the LEED measurements[18-19].

A striking result emerges from this study. The equibilibrium structure we deduced for the Ni termination describes an inward relaxation of both the Ni sites in the outermost layer and the Al atoms in the second layer. The relaxation provides a substantial lowering of the total energy (26 mRyd relative to the ideal surface), and furthermore agrees very well with the LEED optimized geometry for this termination. For the Al termination, we find a similar relaxation pattern with both the Ni and Al (now in the outermost layer) contracting towards the bulk. This yields a modest reduction of the surface energy (5.5 mRyd). Moreover the relaxation pattern disagrees qualitatively with the LEED geometry for this surface, which suggests an inward relaxation of the Al, but a substantial (5%) outward relaxation of the second layer Ni. Upon closer study of the surface structure, the experimentally suggested geometry for the Al terminated structure is difficult to understand; the surface layer is relatively loosely packed, so that the Ni relaxation pattern should be comparable on both surfaces. In fact we find a slightly smaller inward relaxation of the Ni on the Al terminated surface. As a check we find that the LEED structure for the Al terminated surface increases the surface energy by a modest amount (3.5 mRyd).

Analysis of the angle resolved photoemission data is, unfortunately, not so unambiguous. We find that the Ni terminated structure accounts well for an observed dispersionless surface band, bound by 1 eV relative to the Fermi energy, which is strongest in a projected bulk gap at the M point. This feature is associated with a Ni d-like surface state with xy symmetry, i.e. lobes directed in the surface plane. A second feature bound by 3 eV (relative to E_f) in an even symmetry gap at the Γ point does not appear on the Ni terminated surface, and surprisingly does appear on the Al terminated surface. It is characterized as an Al p-like surface state, hybridized with second layer d-like z^2 functions directed along the (111) direction. The appearance of this feature is the only evidence in this study favoring the "two termination" model.

If the equilibrium surface does exhibit coexisting domains with both terminations, this raises an interesting question about the underlying dynamics favoring this configuration. It is unlikely that the "bare surface" energies of these two structures are equivalent. An intriguing possibility is that either structure separately is under lateral stress, so that the macroscopic elastic energy is minimized by a long range ordering of the structure with inequivalent terminations. Such a possibility has been recently explored for clean surfaces of semiconductors such as Si^{20}. The applicability of this mechanism on the NiAl surface remains to be explored.

SUMMARY

We have developed a modification of the mixed basis approach to electronic structure calculations on complex systems containing elements from the transition series. The method has allowed for well controlled studies on complex structures involving both Cu and Ni which were previously inaccessible by the pseudopotential method. The method appears to provide a natural approach to such problems, and should be readily generalizable to other systems.

ACKNOWLEDGMENTS

We thank S. C. Lui and E. W. Plummer for their contributions to this work. The research was supported jointly by NSF through the MRL program under grant DMR85-19059, and by DOE through grant DE FG02 84ER45118.

REFERENCES

1.   G. B. Bachelet, D. R. Hamann, and M. Schluter, Physical Review B 26:4199 (1982).
2.   G. Kerker, J. Phys. C 13:L 189 (1980).
3.   see e.g. M. L. Cohen, Physics Reports 110:293 (1984).
4.   see e.g. V. L. Moruzzi, J. F. Janak and A. R. Williams, in: "Calculated Electronic Properties of Metals" (Pergamon Press, 1978).
5.   J. R. Chelikowsky and S. G. Louie, Physical Review B 29:3470 (1984).
6.   S. G. Louie, K. M. Ho and M. L. Cohen; Physical Review B 28:5480 (1983).
7.   M. H. Kang, R. C. Tatar, E. J. Mele and P. Soven, Physical Review B 35:5457 (1987).
8.   P. Bendt and A. Zunger, Physical Review B 26:3114 (1982).
9.   M. R. Norman, Physical Review B 29:2956 (1984).
10.   J. A. Knapp, F. J. Himpsel and D. E. Eastman, Physical Review B 19:4952 (1978).
11.   Z. Q. Wang, S. H. Lu, Y. S. Li, F. Jona and P. M. Marcus, Physical Review B 35:9322 (1987).
12.   I. A. Morrison, M. H. Kang and E. J. Mele, Physical Review B 1988).
13.   M. H. Kang and E. J. Mele, Physical Review B 36:7371 (1987).
14.   S. C. Lui, M. H. Kang, E. J. Mele and E. W. Plummer, Physical Review B (submitted, 1988).
15.   M. H. Kang and E. J. Mele in: "The Structure of Surfaces" (F. van der Veen, ed.; Springer Verlag, 1988).
16.   J. R. Noonan and H. L. Davis, Physical Review Letters 59:1714 (1987).
17.   J. Wendleken (unpublished).
18.   M. H. Kang (Ph. D. Thesis, University of Pennsylvania, unpublished).
19.   M. H. Kang, S. C. Lui, E. J. Mele and E. W. Plummer (in preparation).
20.   D. Vanderbilt, in reference 15.

THEORY OF GROUND- AND EXCITED-STATE PROPERTIES OF SOLIDS, SURFACES, AND INTERFACES: BEYOND DENSITY FUNCTIONAL FORMALISM

Steven G. Louie

Department of Physics, University of California
Materials & Chemical Sciences Division
Lawrence Berkeley Laboratory
Berkeley, CA 94720

INTRODUCTION

A number of methods, going beyond pair-potentials, now exist for calculation of the structural and electronic properties of materials. These range from the use of classical many-body atomic potentials to the semiempirical quantum approaches to parameter-free *ab initio* methods.[1] Among the first-principles methods, basically one of the following two approaches is employed in treating many-electron correlations: (1) the Hartree-Fock plus correlation corrections approach or (2) the local density functional formalism (LDA). The LDA is by far the more commonly employed approach. The Hartree-Fock-plus-corrections calculations have been mostly restricted to the lighter elements and to binding energies.

The LDA[2] has been applied to a wide range of materials with many successes.[3,4] Among them are some striking results on the structural properties, vibrational properties, electron-phonon and phonon-phonon interaction parameters, and structural phase transitions. There are, however, problems with the LDA. The binding energies are significantly too large for virtually all molecular and solid-state systems. The ground-state properties of many magnetic and other highly correlated electron systems are not well described. Finally, since the density functional formalism is a ground-state theory, the eigenvalues from the LDA do not give accurate band gaps or in general electron excitation energies.[5]

In this paper, we give a short review of two recent theoretical developments in first-principles calculation of the ground- and excited-state properties of real materials. The calculated results are shown to be in significantly better agreement with experiment than those from state-of-the-art LDA or Hartree-Fock methods. The paper is organized as follows. First, a quasiparticle approach[6] for excitation energies in solids is discussed. Results for bulk solids as well as for semiconductor surfaces and heterojunctions are presented to illustrate the general applicability of the method. Next, an approach[7] which combines the use of variational quantum Monte Carlo techniques with nonlocal pseudopotentials for ground-state properties is presented. Results from calculations on diamond and graphite are discussed. Finally, we present a summary and some conclusions.

QUASIPARTICLES AND EXCITED-STATE PROPERTIES

It is well known now that, since the density functional formalism[2] is a ground-state theory, LDA calculations do not provide in principle direct information on the excitation spectra of an electronic system. Indeed, in most cases, the practice of comparing LDA eigenvalues to spectroscopic data has led to rather severe discrepancies.[4] For example, the band gaps of semiconductors and insulators are drastically underestimated in the LDA by as much as 50% or more (see Table I). The optical and photoemission spectra of both bulk and surface states are similarly not well-described. The same problem exists even for the band gaps and widths of the simple metals such as sodium.[8] The reason for these discrepancies is that the interpretation of excitation spectra such as those measured in optical, photoemission or transport experiments requires the concept of quasiparticles, the particle-like excitations in an interacting many-body system.[5,9]

Recently, several many-body approaches have been proposed for calculating (either semi-empirically or from first principles) the quasiparticle properties of solids going beyond the one-particle picture.[6,10-14] For the special case of the minimum gap in insulators, an explicit correction to the Kohn-Sham minimum gap has also been derived and calculated.[14] For a complete description of the excitation spectrum, direct quasiparticle calculations are however necessary. First-principles methods are of special importance for predicting the properties of new materials and those of defects or surfaces and interfaces.

In this section, we present a brief review of a first-principles quasiparticle approach[6] which has shown to give very accurate excited-state properties for real materials with basically the atomic number of the constituent elements as the only input. The method is based on the so-called GW approximation[15] in which the electron self-energy operator is calculated to first order in the dressed Green's function G and the screened Coulomb interaction W.

TABLE 1. Comparison of calculated band gap E_g (in eV) with experiment. The results for Ge include relativistic effects.

	LDA	Present Theory	Experiment
Diamond	3.9	5.6	5.48[a]
Silicon	0.52	1.29	1.17[a]
Germanium	< 0	0.75	0.744[a]
LiCl	6.0	9.1	9.4[b]

[a]Ref. 21   [b]Ref. 22

The Self-Energy Approach

In the present approach, the quasiparticle energies are given by [6,9]

$$(T + V_{ext} + V_H) \, \varphi_{n\vec{k}} (\vec{r}) + \int d\vec{r}' \, \Sigma \, (\vec{r}, \vec{r}'; E_{nk}) \, \varphi_{n\vec{k}} (\vec{r}') = E_{n\vec{k}} \varphi_{n\vec{k}} (\vec{r}) \tag{1}$$

where T is the kinetic energy operator, V_{ext} is the external potential due to the ions, V_H is the average (Hartree) Coulomb potential, and the exchange and correlation contributions are included in the self-energy operator, Σ. In general, Σ is nonlocal, energy-dependent, and nonHermitian with the imaginary part giving the lifetime of the quasiparticles.

The self-energy operator Σ can be systematically expanded in a series in terms of the screened Coulomb interaction W and the fully dressed Green's function G.[15] In the GW approximation, Σ is taken to be the first-order term with

$$\Sigma \, (\vec{r}, \vec{r}'; E) = i \int (d\omega/2\pi) \, e^{-i\delta\omega} \, G(\vec{r}, \vec{r}'; E - \omega) \, W(\vec{r}, \vec{r}', \omega), \tag{2}$$

where δ is a positive infinitesimal. The major components of the theory are then the fully interacting Green's function for which we use a quasiparticle approximation

$$G(\vec{r}, \vec{r}'; E) = \sum_{n\vec{k}} \frac{\varphi_{n\vec{k}} (\vec{r}) \, \varphi^*_{n\vec{k}} (\vec{r}')}{E - E_{n\vec{k}} - i\delta_{n\vec{k}}} \, , \tag{3}$$

and the dynamically screened Coulomb interaction

$$W(\vec{r}, \vec{r}'; \omega) = \Omega^{-1} \int d\vec{r}'' \, \varepsilon^{-1} (\vec{r}, \vec{r}''; \omega) \, V_c (\vec{r}'' - \vec{r}') \, , \tag{4}$$

where ε is the time-ordered dielectric matrix whose off-diagonal elements in Fourier space describe the local fields (variation in the screening in the unit cell due to charge inhomogeniety) and V_c is the bare Coulomb interaction. Our approach[6] is to take Eq. (2) as the basic approximation and proceed to calculate Σ from first principles with minimum further approximations.

The quasiparticle energies together with Σ and G must be obtained in a self-consistent fashion. In the calculations, the electron Green's function is constructed initially using the LDA Kohn-Sham eigenfunctions and eigenvalues and is subsequently updated with the quasiparticle spectrum from Eq. (1). The static dielectric matrix $\varepsilon(\vec{r}, \vec{r}', \omega = 0)$ is obtained as a ground-state quantity from the LDA calculation and extended to finite frequencies using a generalized plasmon pole model with exact sum rules.[6,16]

This approach has been applied successfully to a variety of solids including semiconductors,[6] ionic insulators,[17] and metals[18] as well as surfaces[19] and interfaces.[20] For a full physical description of the quasiparticles, it is found that the use of the crystalline Green's function and inclusion of both local fields (the full dielectric matrix) and dynamical screening effects are important factors. For semiconductors and insulators, local field effects are of crucial importance. For the alkali metals, the quasiparticle bandwidths are shown to be also sensitive to the treatment of exchange-correlation effects in the dielectric screening.[18]

Band Gaps and Excitation Spectra of Crystals

Very accurate band gaps, optical transition energies, and photoemission spectra have been obtained using the quasiparticle approach.[6,17,18] For the case of the minimum gap, this is illustrated in Table I where the calculated results for several selected crystals are compared to the experimental values. As seen from the table, the gaps open up dramatically as compared to the LDA eigenvalues. Similar results have been obtained for the ionic semiconductors, e.g. GaAs and AlAs.[20,23] This level of accuracy is achieved only when local fields and dynamical screening effects are included in the evaluation of the electron self-energy operator.

For the optical properties of semiconductors, the agreement between theory and experiment is generally at the same level as for the minimum gaps. A comparison between experimental direct optical transitions[21,24-27] and calculated results are given in Table II for diamond, Si and Ge. The theoretical results are within 0.1-0.2 eV of the experimental values for all transitions except for the very high energy ones in diamond where the experimental uncertainties are large. This excellent agreement is a dramatic improvement over previous theories.[10-13] In fact, the present quasiparticle results are comparable to results from the Empirical Pseudopotential Method[28] in which the band structure is obtained by fitting to optical data using several parameters.

In addition to the minimum band gaps and optical transition energies, the theory yields excellent band dispersions. Figure 1 depicts the calculated Ge quasiparticle valence band structure together with data from angle-resolved photoemission measurements.[29] The agreement between theory and experiment is generally within experimental errors. Similar levels of agreement with experiment have been obtained for other semiconductors and for empty band states as measured in inverse photoemission experiments.[6] A comparison of the calculated band structure with photoemission data for Na[30] is given in Fig. 2. The surprisingly large observed bandwidth reduction is explained by the self-energy effects although the origin of the dispersionless feature near the Fermi energy remains a subject of debate.

Another important result from the calculations is that the quasiparticle wavefunctions are found to be remarkably well represented by the LDA wavefunction. For the materials considered, the quasiparticle wavefunction, in general, has better than 99.95% overlap with the corresponding LDA wavefunction although the energy spectrum is altered significantly as seen in Table I. The approximation of using only the diagonal element $<nk \mid \Sigma \mid nk>$ in evaluating the quasiparticle energies leads to errors ranging from less than 0.01 eV for Si to less than 0.05 eV for LiCl. This approximation thus may be used with only negligible loss of accuracy in the final quasiparticle energies.

TABLE II. Comparison between theory and experiment for optical transitions in Ge, Si, and diamond.

	LDA	Present Work	Experiment
Ge			
$\Gamma_{7v} \rightarrow \Gamma_{8v}$	0.30	0.30	0.297[a]
$\Gamma_{8v} \rightarrow \Gamma_{7c}$	-0.07	0.71	0.887[a]
$\Gamma_{8v} \rightarrow \Gamma_{6c}$	2.34	3.04	3.006[a]
$\Gamma_{8v} \rightarrow \Gamma_{8c}$	2.56	3.26	3.206[a]
$X_{5v} \rightarrow X_{5c}$	3.76	4.45	4.501[a]
Si			
$\Gamma_{25'v} \rightarrow \Gamma_{15c}$	2.57	3.35	3.4[b]
$\Gamma_{25'v} \rightarrow \Gamma_{2'c}$	3.26	4.08	4.2[c]
$L_{3'v} \rightarrow L_{1c}$	2.72	3.54	3.45[b]
$L_{3'v} \rightarrow L_{3c}$	4.58	5.51	5.50[b]
Diamond			
$\Gamma_{25'v} \rightarrow \Gamma_{15c}$	5.5	7.5	7.3[d]
$\Gamma_{25'v} \rightarrow \Gamma_{2'c}$	13.1	14.8	$15.3 \pm .5$[e]
$X_{4v} \rightarrow X_{1c}$	10.8	12.9	12.5[c]

[a]Ref. 24   [b]Ref. 25   [c]Ref. 21   [d]Ref. 26   [e]Ref. 27

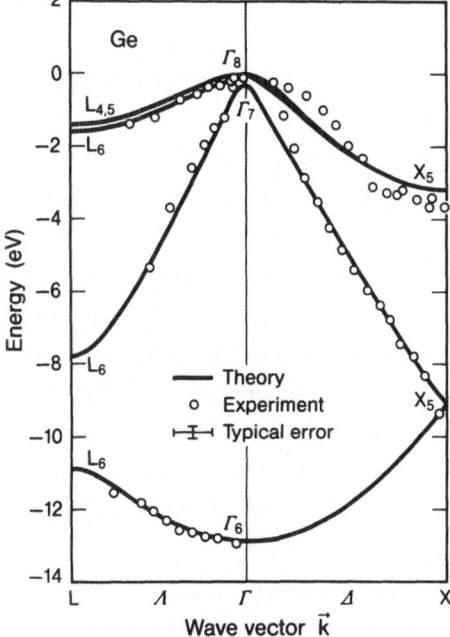

Fig. 1   Photoemission data from Ref. 29 vs. theoretical quasiparticles energies.

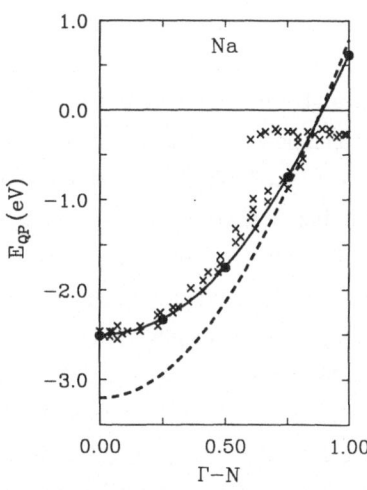

Fig. 2   Quasiparticle energies for Na: Experimental data from Ref. 30 (crosses), LDA eigenvalues (dashed line), and calculated quasiparticle energies (filled circles).

Surface and Interfaces

The quasiparticle theory has been extended to surfaces and interfaces. We discuss here results for two prototypical systems: the As-capped Si (111) and Ge (111) surfaces[19] and the GaAs-AlAs (001) heterojunction.[20] The major objective is to provide a predictive theory for surface-state energies and band offsets at interfaces.

At saturation coverage, the chemisorbed As atoms are found to substitute for the outermost-layer atoms on the Si (111) and Ge (111) surface.[31-33] The surface becomes chemically inactive and is stable against reconstruction showing a 1x1 periodicity. These are good prototype systems for the many-body calculations because of their geometric simplicity and the availability of detailed experimental data.[31-33] They are also of intrinsic importance as an initial stage of growth of GaAs on Si and Ge. The calculations[19] were carried out using a repeated slab geometry with a 12-layer thick slab. The surface geometry was determined by a LDA total energy minimization. After the structure has been determined, the quasiparticle energies for both the bulk and surface states are then calculated using the same formalism as in the case of bulk crystals.

Figure 3 depicts the calculated quasiparticle energies for the As-capped Si (111) system. The quasiparticle surface-state bands (solid lines) together with the LDA surface-state bands (dashed lines) are plotted against the projected quasiparticle band structure of Si. The zero of the energy scale is set at the valence band maximum. As expected, the fully occupied surface band corresponds to the lone-pair states of the As adatoms. These surface states have been studied in detail using angle-resolved photoemission techniques.[31] The theory also predicted an empty surface state band in the gap. These empty states correspond to backbond surface states. Very similar results are obtained for the Ge (111)-As surface.

For both the Si (111)-As and Ge (111)-As surface, the occupied quasiparticle surface-state band is lower in energy and has a broader dispersion as compared to the LDA results. These changes bring the theoretical energies into close agreement with experiment. Figures 4 and 5 compare the quasiparticle surface-state band with angle-resolved photoemission data. The agreement is excellent in both the placement and the width of the band and is well within the estimated errors of ± 0.1 eV with experiment and theory. As seen in Fig. 3, the effect of many-body correction to the LDA values on the empty surface states is even more dramatic. These states are substantially shifted upwards in energy, opening up the gap between empty and occupied surface states by nearly an extra 1 eV at some k-points. A scanning tunneling microscopy study[34] has recently been done for the Si (111)-As surface to search for these states. Not only is the 1x1 surface structure seen directly, the observed surface-state gap of 1.9 - 2.3 eV in the normalized differential conductivity agrees very well with the predicted value of 2.2 eV.

The surface quasiparticle results, therefore, address the difficulties associated with the LDA surface-state energies in ways very similar to the bulk excitation energies. In particular, the gap between the empty and filled surface states is substantially opened in comparison with the LDA surface-state gap. Analysis of the calculated results,[19] however, shows that the size of the correction depends on the detailed character of the surface states which can have a substantially different character than the bulk states. This means that the shortcomings of the LDA surface-state spectra cannot easily be corrected by a simple rigid shift derived from bulk data.

Using the same theoretical techniques, we have computed the band offsets of the GaAs-AlAs (001) interface.[20] The band discontinuities at a semiconductor interface are simply the differences in the quasiparticle energy E_{qp} across the junction for the band edge states. Since, as mentioned above, the quasiparticle wavefunctions are virtually identical to the LDA wavefunctions, we may write E_{qp} near an interface as

$$E^{qp}_{n\vec{k}} = \varepsilon^{LDA}_{n\vec{k}} + \Sigma_{n\vec{k}} - V^{n\vec{k}}_{xc} \tag{5}$$

where $\Sigma_{n\vec{k}}$ and $V^{n\vec{k}}_{xc}$ are, respectively, the expectation value of the self-energy operator and the LDA exchange-correlation potential for a given state. Then, the valence band offset ΔE_v becomes

$$\Delta E_v = \Delta E^{LDA}_v + \delta_{vbm} \tag{6}$$

where ΔE^{LDA}_v is the LDA calculated valence band offset and δ_{vbm} is a many-body correction given by

$$\delta_{vbm} = (\Sigma - V_{xc})^{vbm}_{GaAs} - (\Sigma - V_{xc})^{vbm}_{AlAs} . \tag{7}$$

Since the quasiparticle energies in determining the band offsets should be evaluated at a distance away from the interface and both Σ and V_{xc} are short range interactions, $(\Sigma - V_{xc})^{vbm}$ can be replaced by their bulk values. To calculate the LDA band offsets, we perform a 12-layer superlattice calculation using an approach similar to that of Van de Walle and Martin.[36]

We find that for the GaAs-AlAs (001) interface the many-body correction to the valence band offset is quite significant; that is, $\delta_{vbm} = 0.12$ eV. It is about 30% of the LDA result of $\Delta E^{LDA}_v = 0.41$ eV. Equation (6) thus gives a calculated value of $\Delta E_v = 0.53$ eV which is in good agreement with recent experimental values of 0.53 - 0.56 eV.[37] The sign of the many-body correction can also be understood in terms of a more localized valence band wavefunction for AlAs which leads to a more negative self-energy for the valence band states of AlAs as compared to GaAs, and, hence, a positive value for δ_{vbm} as defined by Eq. (7). We expect that similar corrections will be even more important for junctions that are made of materials with lesser chemical similarities than GaAs and AlAs.

In this section we describe a method[7] of performing variational quantum Monte Carlo calculations for solids and atoms using nonlocal pseudopotentials. The goal is to provide a first principles method which would be capable of treating accurately the properties of some of the strongly correlated electronic systems.

<u>Variational Quantum Monte Carlo Method with Nonlocal Pseudopotentials</u>

The variational quantum Monte Carlo (QMC) method was pioneered by McMillan[38] to study liquid He^4 and first applied to fermion liquid problems by Ceperley, Chester, and Kalos.[39] However, a straightforward application of the method to the electronic properties of real materials has been severely hampered by a number of conceptual and technical problems. Among these include the treatment of the single-particle orbitals in the presence of electron correlations and the very rapid growth of computation effort with increasing atomic number caused by fluctuations in the energies of electrons in the core region.[40] This has prompted our development of a quantum Monte Carlo pseudopotential approach, which incorporates the effects of the core electrons in an ionic potential. The ionic pseudopotentials used are the norm-conserving pseudopotentials.[41] The integral operator which arises in the nonlocal pseudopotential makes the present problem different from previously considered QMC problems. However, this operator can be evaluated statistically within the variational QMC method with a computational effort comparable to that for the kinetic energy.[7,42]

In the present approach, a correlated trial wavefunction of the Jastrow-Slater form,

$$\psi(\vec{r}_1,...,\vec{r}_N) = \exp\left\{\sum_{i=1}^N \chi(\vec{r}_i) - \sum_{1<i<j<N} u(\vec{r}_{ij})\right\} D(\vec{r}_1, ..., \vec{r}_N),$$ (8)

is employed where D is a Slater determinant of LDA single-particle wavefunctions. For this wavefunction we evaluate the expectation value of the exact many-electron Hamiltonian

$$H = \sum_{i=1}^N -\left\{-\frac{h^2}{2m}\nabla_i^2 + V_{ext}(\vec{r}_i) + \frac{1}{2}\sum_{j\neq1}\frac{e^2}{r_{ij}}\right\},$$ (9)

which consists of the usual three terms: the kinetic energy of the valence electrons, an external potential due to the ions, and the Coulomb interactions between the electrons. The many-body integrals are evaluated using the Metropolis Monte Carlo algorithm[43] for importance sampling with the importance function $|\psi|^2$. In the calculation, the kinetic and electron-electron energies are evaluated as in Ref. 39. The external potential is the sum of the ionic pseudopotentials which have a local and a (short-range) nonlocal part. The value of the local potential at each configuration of the random walk is evaluated using Ewald summation techniques. The value of the nonlocal potential is evaluated using a statistical method with a special point scheme.[7,42]

The two-particle correlation term, $u(\vec{r}_{ij})$, in the Jastrow factor lowers the energy by reducing the probability of two electrons coming close together. In the solid, this term is taken to be of the standard form,[39] $u(r) = A[1 - \exp(-r/F)]/r$, with A and F as spin-dependent variational parameters. For diamond and graphite, the optimum values for A and F are found to be very close to the values given by the physical considerations of the so-called "tail" and "cusp" conditions.[39] In the atom, we have used both the form of $u(r)$ for the solid and $u(r) = -ar/(1 + br)$ and obtained essentially identical total energies.

The one-particle term $\chi(\vec{r})$ allows a variational adjustment of the single-particle orbitals in the presence of the two-particle term which tends to make the charge density overly diffuse. Although not relevant in liquids, the one-particle term is quite important for atoms and solids. There are several possible implementations of the one-particle term including the use of an Euler-Lagrange equation for χ to minimize the energy.[42] In the calculation discussed here, we simply set $\chi(\vec{r}) = \alpha\log[\rho_{\chi,u=0}(\vec{r})/\rho_{\chi=0}(\vec{r})]/2$, where $\rho(\vec{r})$ is the charge density and α is a variational parameter. The optimum value of α is close to 1, as expected, since the LDA charge density is generally quite good.

TABLE III. The theoretical and experimental values of electron affinity and first ionization potential (in eV) for carbon. Statistical error in the last digits is in parentheses.

	E.A.	1st I.P.
LDA	C⁻ unbound	11.76
QMC	1.05 (10)	11.43 (5)
Experiment	1.27	11.26

Application to Atomic Carbon, Diamond, and Graphite

The ionization energy and electron affinity of atomic carbon are obtained by performing calculations on $C(^3P)$, $C^+(^1P)$, and $C^-(^4P)$. In each case, we fixed the parameter a in $u(r)$ using the cusp condition and searched the b, α parameter space to determine the optimal energy. Since the atoms are spin-poplarized, different χ-functions for different spin types were used. The QMC results, together with LDA results and experimental values, are presented in Table III. The approach given carbon ionization energy and electron affinity in agreement with experiments within $\pm$ 0.2 eV (C⁻ is unbounded in LDA). Since the number of three-body interactions is very different for the three cases, this result suggests that the three-body terms in the Jastrow factor can only lower the energy by no more than approximately 0.2 eV/atom.

TABLE IV. Terms in the total energy of diamond (64 electron simulation at a lattice constant a = 3.63 Å) for a single Slater determinant of LDA wavefunctions and for a Jastrow-Slater function with a two-body term only in the Jastrow factor, as discussed in the text, and with LDA wavefunctions in the determinant. Energies in eV/atom.

	Slater Det.	Jastrow-Slater (u only)
Local Pot.	-87.1	-73.6
Elec-Elec	-29.2	-39.0
Kinetic	121.3	116.8
Non-local Pot.	15.8	12.8
Ewald Sum	-171.0	-171.0
Total	-150.2	-154.0

The method has been applied to study the binding energy and structural properties of diamond. In the calculation, simulation cells containing up to 54 atoms (or 216 electrons) have been used. We find that the size dependence for larger simulations is almost entirely determined by the convergence of the single-particle terms in the total energy, as given within band theory by the $\vec{k}$-point sampling of the Brillouin zone. The various contributions to the total energy for a specific simulation are given in Table IV illustrating the effects of electron correlation. As is shown in Table IV, the introduction of a Jastrow factor with only the two-particle term $u(r_{ij})$ lowers the total energy of the solid by approximately 3.8 eV/atom. With the introduction of the Jastrow factor, the electron-electron energy is substantially reduced. However, unlike uniform systems, the kinetic energy also decreases. It is the electron-ion (local plus nonlocal potential terms) which is greatly increased. The general trends in the atom are similar. Without the one-particle term χ in the Jastrow factor, the presence of a nonzero $u(r_{ij})$ alters the charge density from that of the Slater determinant alone. Because $u(r)$ is positive and a decreasing function of r, its effect is to remove charge from the high density regions leading to both an increase in the electron-ion energy and the decrease in the kinetic energy.

TABLE V. Total energies (in eV/atom) of the carbon atom and of diamond (with finite-size correction) for (a) LDA calculation, and for Monte Carlo calculations with (b) single Slater determinant of LDA wavefunctions and (c) Jastrow-Slater function with one-and two-body terms in the Jastrow factor. The expected statistical error in the last digits is in parentheses.

	Carbon Atom	Diamond	Cohesive Energy
(a) LDA	-146.79	-155.42	8.63
(b) Slater Det.	-145.55 (7)	-151.3 (2)	5.85 (25)
(c) Jastrow-Slater	-147.93 (3)	-155.38 (6)	7.45 (7)
(d) Experiment[a]	------	------	7.37

[a]see Ref. 46.

The total energies obtained with and without the full Jastrow factor (i.e., including both the u and c terms) in the wavefunction are presented in Table V for diamond at the minimum-energy lattice constant together with the atomic results. The correlation energies for the valence electrons in the atom and the solid are thus found to be 2.4 ± 0.1 eV and 4.1 ± 0.2 eV/atom, respectively. These values are in agreement with results for the carbon valence electrons in a recent calculation using a similar ansatz for the many-body wavefunction but evaluating the energy by diagrammatic techniques.[44] Our value for the Hartree-Fock cohesive energy obtained using LDA wavefunctions in a single Slater determinant is 5.85 ± 0.25 eV/atom.

The final results for the binding energy of diamond in the present approach are also given in Table V and compared with the LDA result obtained using the Ceperley-Adler form for the exchange-correlation energy.[45] We have included the zero-point energy of the phonons in the energy for the solid. The quantum Monte Carlo calculation gives a cohesive energy of 7.45 ± 0.07 eV, as compared to the experimental value[46] of 7.37 eV/atom. This result is in significantly better agreement with experiment than the value of 8.63 eV/atom computed using the LDA formalism. To obtain the structural properties of diamond, the QMC total energies as a function of lattice constant are fitted with a Murnaghan equation of state, as shown in Fig. 6. We obtain a fitted equilibrium lattice constant of 3.54 ± 0.03 Å and bulk modulus of 420 ± 50 GPa, compared with experimental values of 3.567 Å and 443 GPa, respectively.[46]

To assess further the accuracy of the method, we performed a calculation for graphite. We obtained a cohesive energy for graphite which is identical to the diamond cohesive energy within the statistical noise of 0.07 eV/atom. This result is in excellent agreement with experiment since experimentally, the binding energy of graphite is only 0.025 eV larger than that of diamond.[47]

SUMMARY AND CONCLUSIONS

We have discussed, in this brief overview, two new approaches for calculating the properties of materials going beyond density functional formalism. A self-energy approach for calculating the quasiparticle energies from first principles is presented. The electron self-energy operator is evaluated to first order in the dressed Green's function and the screened Coulomb interaction including local field effects. This method allows an *ab initio* determination of electronic excitation energies in crystals and at surfaces and interfaces which can be directly compared with spectroscopic measurements. Excellent results have been obtained for a wide-range of materials systems. A method of calculating total energies of solids using nonlocal pseudopotentials in conjunction with the variational quantum Monte Carlo approach is also presented. Electron-electron correlation effects are treated using the exact interaction and a correlated Jastrow-Slater many-electron wavefunction. We demonstrated the computational feasibility of the method for solids and obtained excellent results for the cohesive properties of diamond and graphite. This development opens an exciting new theoretical avenue for studying the properties of solids, especially for the highly correlated electron systems.

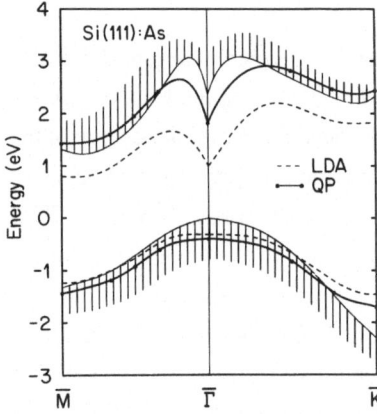

Fig. 3   Quasiparticle surface-state energies compared to LDA surface-state energies for the Si(111):As surface.   Also shown is the quasiparticle bulk projected band structure.

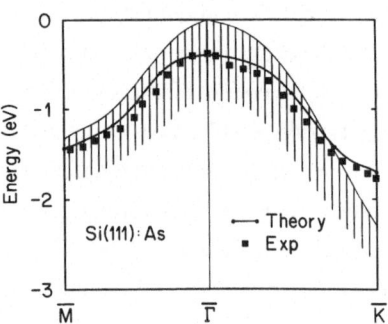

Fig. 4   Calculated filled quasiparticle surface-state energies for Si(111):As compared to data from photoemission (Ref. 31).

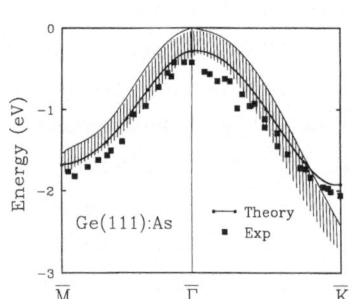

Fig. 5   Same as Fig. 4 except for Ge (111):As. Experiment data are from Ref. 32

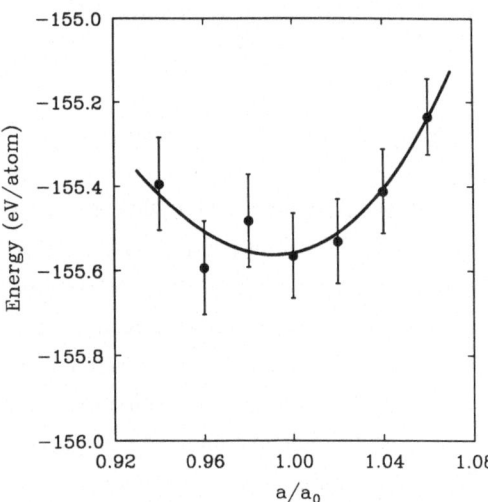

Fig. 6   Calculated total energy of diamond as a function of the ratio of the lattice constant to the measured lattice constant. The error bars indicate the standard deviation of the mean in each QMC calculation.

ACKNOWLEDGEMENT

This work was supported by National Science Foundation Grant No. DMR-8818404 and by the Director, Office of Energy Research, Office of Basic Energy Sciences, Materials Sciences Division of the U.S. Department of Energy under Contract No. DE-AC03-76SF00098.

REFERENCES

1. See the various articles in this volume.
2. P. Hohenberg and W. Kohn, Phys. Rev., 136:B864 (1964); W. Kohn and L.J. Sham, Phys. Rev., 140:A1133 (1965).
3. "Theory of the Inhomogeneous Electron Gas", S. Lundqvist and N.H. March, eds., Plenum, New York (1983), and references therein.
4. S.G. Louie, in "Electronic Structure, Dynamics and Quantum Structural Properties of Condensed Matter", J. Devreese and P. van Camp, eds., Plenum, New York (1985), p. 335.
5. M.S. Hybertsen and S.G. Louie, Comments Cond. Mat. Phys., 13:223 (1987).
6. M.S. Hybertsen and S.G. Louie, Phys. Rev. Lett., 55:1418 (1985); Phys. Rev. B, 34:5390 (1986).
7. S. Fahy, X.W. Wang, and S.G. Louie, Phys. Rev. Lett., 61:1631 (1988).
8. E.W. Plummer, Surf. Sci., 152/153:162 (1985).
9. See, for example, the excellent review by L. Hedin and S. Lundqvist, Solid State Phys., 23:1 (1969).
10. For references on earlier work, see Ref. 6.
11. C.S. Wang and W.E. Pickett, Phys. Rev. Lett., 51:597 (1983).
12. C. Strinati, H.J. Mattausch, and W. Hanke, Solid State Commun., 51:23 (1984), and references therein.
13. S. Horsch, P. Horsch, and P. Fulde, Phys. Rev. B, 29:1870 (1984).
14. L.J. Sham and M. Schluter, Phys. Rev. Lett., 51:1888 (1983); J.P. Perdew and M. Levy, Phys. Rev. Lett., 51:1884 (1983); R.W. Godby, M. Schluter, and L.J. Sham, Phys. Rev. Lett., 56:2415 (1986).
15. L. Hedin, Phys. Rev., 139:A796 (1965).
16. It is shown recently that an accurate ε^{-1} ($\vec{r}$, $\vec{r}'$, ω) for the quasiparticle problem may also be obtained in a more simplified model with one parameter for the semiconductors. See M.S. Hybertsen and S.L. Louie, Phys. Rev. B, 37:2733 (1988).
17. M.S. Hybertsen and S.G. Louie, Phys. Rev. B, 32:7005 (1985).
18. N.E. Northrup, M.S. Hybertsen, and S.G. Louie, Phys. Rev. Lett., 59:819 (1987).
19. M.S. Hybertsen and S.G. Louie, Phys. Rev. Lett., 58:1551 (1987); Phys. Rev. B, 38:4033 (1988).
20. S.B. Zhang, D. Tomanek, S.G. Louie, M.L. Cohen, and M.S. Hybertsen, Solid State Commun., 66:585 (1988).
21. "Landolt-Borstein: Zahlenwerte und Funktionen aus Naturwissenscharten und Technik", Springer, New York (1982), Vol. III, pt. 17a.
22. G. Baldini and B. Bosacchi, Phys. Sat. Sol., 38:325 (1970).
23. R.W. Godby, M. Schluter, and L.J. Sham, Phys. Rev. B, 35:4170 (1987).
24. D.E. Aspnes, Phys. Rev. B, 12:2797 (1975).
25. R.R.L. Zucca and Y.R. Shen, Phys. Rev. B, 1:2668 (1970).
26. R.A. Roberts and W.C. Walker, Phys. Rev., 161:730 (1967).
27. F.J. Himpsel, J.F. van der Veen, and D.E. Eastman, Phys. Rev. B, 22: 1967 (1980).
28. M.L. Cohen and J.R. Chelikowsky, "Electronic Structure and Optical Properties of Semiconductors", Springer-Verlag, Berlin (1988).
29. A.L. Weeks, T. Miller, T.C. Hsieh, A.P. Shapiro, and T.-C. Chiang, Phys. Rev. B, 32:2326 (1985).
30. E. Jensen and E.W. Plummer, Phys. Rev. Lett., 55:1912 (1985).

31. For the Si (111) surface, see M.A. Olmstead, R.D. Bringans, R.I.G. Uhrberg, and R.Z. Bachrach, Phys. Rev. B, 34:6041 (1986); R.I. Uhrberg, R.D. Bringans, M.A. Olmstead, R.Z. Bachrach, and J.E. Northrup, ibid, 35:3945 (1987).

32. For the Ge (111) surface, see R.D. Bringans, R.I.G. Uhrberg, R.Z. Bachrach, and J.E. Northrup, Phys. Rev. Lett., 55:533 (1985); R.D. Bringans, R.I.G. Uhrberg, R.Z. Bachrach, and J.E. Northrup, J. Vac.Sci. Tech.A, 4:1380 (1986); R.D. Bringans, R.I.G. Uhrberg, and R.Z. Bachrach, Phys. Rev. B , 34:2373 (1986).

33. J.R. Patel, J.A. Golovchenko, P.E. Freiland, and H.J. Grossmann, Phys. Rev. B, 36:7715 (1987).

34. R.S. Becker, B.S. Swartzenruber, J.S. Vichers, M.S. Hybertsen, and S.G. Louie, Phys. Rev. Lett., 60:116 (1988).

35. This expression assumes that the electrostatic potential at the interface is well-represented in the LDA. This should be a very good approximation since LDA is known to give excellent charge densities.

36. C.G. Van der Walle and R.M. Martin, J. Vac. Sci. Technol. B, 3:1256 (1985).

37. P. Dawson, K.J. Moore, and C.T. Foxon, in "Quantum Well and Superlattice Physics", Proceedings of SPIE, 792, G.H. Dohler and J.N. Schulman, eds., SPIE, Washington (1987), p. 208; D.J. Wolford, private communications.

38. W.L. McMillan, Phys. Rev. 138:A442 (1965).

39. D. Ceperley, G.V. Chester, and M.H. Kalos, Phys. Rev. B, 16:3081 (1977).

40. The amount of computation time increases approximately as $z5.5$; see D.M. Ceperley, J. Stat. Phys., 43:815 (1986).

41. D.R. Hamann, M. Schlüter, and C. Chiang, Phys. Rev. Lett., 43:1494 (1979).

42. S. Fahy, X.W. Wang, and S.G. Louie, to be published

43. N. Metropolis, A.W. Rosenbluth, M.N. Rosenbluth, A.H. Teller, and E. Teller, J. Chem. Phys., 21:1087 (1953).

44. G. Stollhoff and K.P. Bohnen, Phys. Rev. B, 37:4678 (1988).

45. J.R. Perdew and A. Zunger, Phys. Rev. B, 23:5048 (1981).

46. See S. Fahy and S.G. Louie, Phys. Rev. B, 36:3373 (1987), and references therein.

47. L. Brewer, Lawrence Berkeley Laboratory Report No. LBL-3720 (unpublished).

MOLECULAR DYNAMICS SIMULATION OF THE PHYSICS OF THIN FILM GROWTH ON SILICON: EFFECTS OF THE PROPERTIES OF INTERATOMIC POTENTIAL MODELS

W. Lowell Morgan+

Joint Institute for Laboratory Astrophysics
National Bureau of Standards
University of Colorado
Boulder, CO 80309-0440

+1987-1988 JILA Visiting Fellow
Permanent Address: Chemistry and Materials Science Department
Lawrence Livermore National Laboratory
Livermore, CA 94550

INTRODUCTION

During the past several years numerous papers have been published on simulations of properties of silicon. There have been a number of papers published that develop multibody interatomic potentials for silicon[1-4] and many more that investigate a variety of surface phenomena via molecular dynamics simulation. Examples relevant to this discussion include modeling of epitaxial growth;[5-8] and surface reconstruction.[9-11] Reference [4] provides a good, but not exhaustive, summary of the wide variety of atomistic simulations that have been performed on silicon as well as on other materials.

It is well known that silicon surfaces are rarely found in unreconstructed form. The usual structure of the Si (100) face is a 2x1 surface net and 2x1, 5x5, and 7x7 for the Si (111) face. Hence, if one is going to simulate surface phenomena on silicon, these would be the surfaces of greatest interest. Several authors[9-11] have shown that some of the available silicon model potential functions allow the Si (100) - 1x1 surface to spontaneously reconstruct into the 2x1 and 2x2 structures thereby reducing the number of dangling bonds on the surface and lowering the surface energy. In this paper I present some results obtained for silicon using the potential energy function developed by Stillinger and Weber[1] of AT&T Bell Laboratories. This potential has become a workhorse for such simulations. I have used it to compute the surface energies of the Si (111) 2x1, 5x5, and 7x7 reconstructions. I also discuss features of this potential function that affect simulations of surface diffusion and epitaxial growth. Finally, I present some preliminary results using variations on this potential to simulate growth of a Ga thin film on Si (100) -2x1.

SIMULATION ON Si(111) AND ITS RECONSTRUCTIONS

The silicon interatomic potential function developed by Stillinger and Weber[1] consists of the sum of two and three-body potentials. The pair potential has the form:

$$V_2(r) = A\varepsilon[B(\sigma/r)^P - 1] \exp[\sigma/(r - r_0)]$$

and the three body potential energy is the sum over all triplets of:

$$V_3(r_{ij}, r_{ik}, \theta_{jik}) = \lambda\varepsilon\exp\{\gamma\sigma[r_{ij} - r_0)^{-1} + (r_{jk} - r_0)^{-1}]\} \times (\cos\theta_{jik} + 1/3)^2$$

The minimum of the pair potential is $-\varepsilon$ at $r = 2^{1/6}\sigma$. The three body potential is zero at the tetrahedral angle $\theta \simeq 109°$ and is positive for other angles; this destabilizes the non-diamond cubic crystal structures. The exponential factor yields a smooth cutoff of the potential and all derivatives at $r = r_0 = 3.77$ A for Si. This potential mimics the vibrational properties of silicon well, as can be seen from the phonon spectrum in Fig. 1. The main TO peak is only about 10% too high in frequency.

Surface Energies of Reconstructions

To evaluate the utility of using the potential energy function of Stillinger and Weber for simulations of Si surface processes I applied it to several reconstructed Si (111) surface structures. The surface energy is computed from:[13]

$$E_s = (V_T - NE_0)/N_L$$

where V_T is the total potential energy of the slab, N is the number of atoms in the slab, $E_0 = 4.335$ eV is the cohesive energy per bulk atom, and N_L is the number of atoms per layer. The calculated surface energies are:

	1 x 1	2 x 1	5 x 5	7 x 7
Energy (ev)	2.17	2.54	2.94	2.71
ΔE/atom	0.0	0.37	0.77	0.54
N	160	160	200	396

We see that the 1 x 1 surface has the minimum energy and that the energies of the reconstructions are larger. These reconstructions are metastable. However, as they do not spontaneously rearrange themselves to a different structure. This can be seen in the melting curve, Fig. 2, for the 7 x 7 surface. Fig. 3 shows a side view of the 7 x 7 slab. We see that although the surface retains the 7 x 7 structure, there is substantial buckling or strain of the subsurface layers. This is why the computed energy is larger than that of the 1 x 1 despite a reduction in the number of dangling bonds. These results are in contrast to those

140

obtained from simulations on the Si (100) surface. Results[9-11] using the Stillinger and Weber potential demonstrate that the energies of the 2 x 1 and 2 x 2 reconstructions are lower (2 x 1 being the most stable) and that the 1 x 1 spontaneously reconstructs to 2 x 1 and 2 x 2.

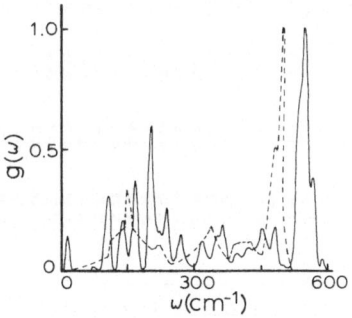

Fig. 1   Relative density of states function for silicon computed by molecular dynamics using the S&W potential (solid line) compared to that derived from experiment (dashed line) in Ref. [12].

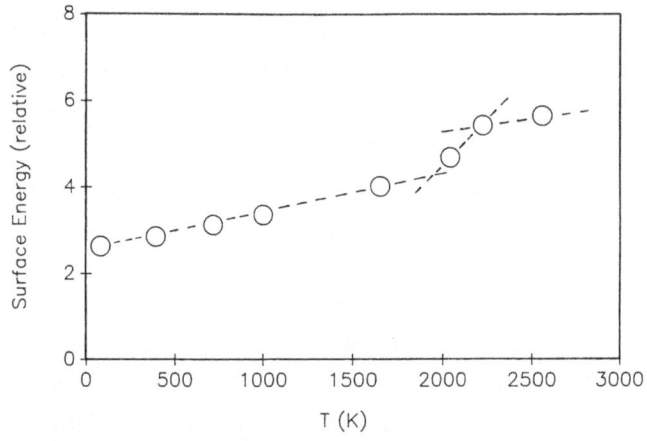

Fig. 2   Surface energy of Si (111) - 7 x 7 as temperature is increased past the melting point.

Surface Diffusion and Epitaxial Growth

Molecular dynamics simulation of Si slab and sampling of the mean square displacement of the (111) surface atoms (Debye-Waller factor) yields $\sqrt{<x - x_0)^2>} = 0.22A$ for 1000 K, which is nearly twice the published value.[14] This is attributable to the atoms of the (111) surface being second neighbors while the potential, having a cutoff of only 3.77 A, includes only first neighbors. The surface atoms are effectively tied to each other by only the angular dependence of the three body potential.

Noorbacha et al.[15] have carried out surface diffusion simulations for Si atoms on Si

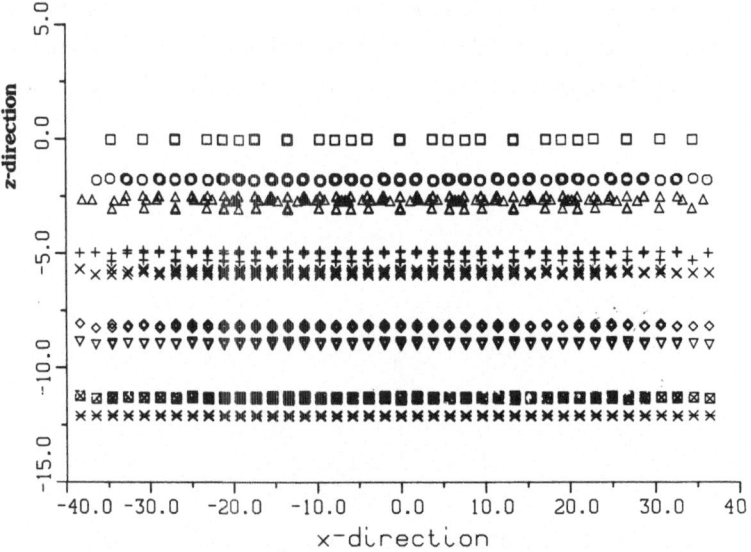

Fig. 3   Side view of Si (111) - 7 x 7 slab at steady state.

(111) - 1 x 1 using Keating's potential. They obtained D > 8.53 x 10^{-4} cm^2/s. In MD simulations using the Stillinger and Weber potential, Das Sarma, et al.[7] observed very slow silicon atom diffusion on the (111) surface even at temperatures of 1411 K! They estimate the D $\simeq$ 3x10^{-5} cm^2/s. An investigation of the barrier height between sites on the (111) surface using the S&W potential shows that it is approximately 1 eV, which would yield ~1.5 x 10^{-5} cm^2/s for the diffusion coefficient at 1411K. The Noorbacha, et al. results give a barrier height of only 0.21 eV, which clearly is responsible for their much larger diffusion coefficients. My own results suggest that Si adatoms on the (111) surface lie too deeply in the sites because the surface atoms relax too much, as we have seen in looking at the Debye-Waller results. This can be overcome by adjusting the two and three body parameters, which affects other properties of the potential, however. Another possibility is to develop a second neighbor potential. I have looked into modifying the S&W pair potential to increase its range but retaining the three body potential largely unmodified. I find with this approach that the wurtzite structure tends to be more stable than the diamond structure. Recently Biswas and Hamann[2] have developed a new potential that extends to several neighbors. This may eliminate some of these difficulties and is worthy of further exploration.

The effects that the issues discussed above might have on the simulation of epitaxial growth on silicon were foreseen by Dodson[16] who stated that they make "...modeling of the epitaxial growth of (111) silicon impossible" using the S&W potential. On the other hand, Schneider, et al.[5] were able to simulate such growth with the S&W potential on (111) and Gawlinski and Gunton[6] performed a similar simulation on (100). The Schneider et al. MD calculation used a time step of 1.5 fs and deposited a Si atom on the surface every 90 time steps or 0.135 ps, which is a deposition rate many orders of magnitude faster than experimental. Temperature in the simulation was controlled by rescaling the velocities of the substrate atoms. I have found in performing similar calculations that, although the thermal diffusion rate of the Si adatoms is negligible even at the 1058 K temperature used by Schneider et al., the large deposition rate leads to substantial diffusion due to collisions between incoming atoms and adatoms. This, in conjunction with the high vibrational temperature of the lattice, tends to keep the adatoms on the move with a diffusion rate much larger than thermal until they settle into a low energy site and, hence, promotes epitaxial growth.

SIMULATION OF Ga DEPOSITION AND GROWTH ON Si (100) - 2 x 1

In a series of recent papers Leone and his coworkers[17-19] have presented various results from their experimental research on the growth of Ga thin films on the Si (100) - 2 x 1 surface. Among these results are LEED determinations of the Ga structures formed on the silicon surface over a range of partial coverages (as determined by Auger spectroscopy). They observe, upon deposition, 3x2, 5x2, 2x2, and 8x1 patterns for coverages $\theta < 0.9$. For $\theta > 1$ they observe the 2x1 LEED pattern.

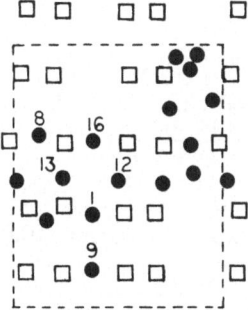

Fig. 4   MD simulation of Ga atoms • deposited on Si(100) - 2x1 surface. The symbols are the Si atoms on the surface.

I have attempted to model Ga growth on silicon using molecular dynamics simulation. I developed Ga-Si two and three body potentials using the S&W form and published theoretical[20] and experimental[21,22] information on Ga on Si (111). I also developed Ga-Ga two and three body potentials based on theoretical potential curves[23] and liquid Ga pseudopotential and structure factor information[24,25] (note that Ga metal is liquid at T > 303 K). An example of the kind of structure that I observe in the MD simulations of Ga deposition on Si (100) - 2x1 is shown in Fig. 4. The calculations represent the deposition of 16 Ga atoms having 0.05 eV kinetic energy onto a 4 x 4 silicon slab at 300 K. Although the temperature is far too low for there to be any thermal surface diffusion on the 30 ps time scale of the simulation, there is substantial diffusion due to collisions, as I have discussed above. Adatoms 1 and 9 form a 2x1 unit cell having the structure proposed by Leone and coworkers. We see, however, that adatoms[1, 13, 16] and 12 form form another 2x1 structure. Note that the bonding is unrealistic indicating that a constraint on the potential that saturates the number of allowed bonds is required. We also see that adatom 8 has broken the Si-Si dimer; we expect this to happen more readily at higher temperatures and with larger Ga atom kinetic energies.[8] Finally, similar simulations for Al atoms on silicon indicate a different structure having a pair of Al atoms bridging the gaps between the rows of Si-Si dimers. Further work in this area is under way.

ACKNOWLEDGMENTS

This research began while I was a visitor at the AT&T Bell Laboratories. I am greatly indebted to Dr. John Tully of AT&T for his help in getting me started in the simulation of surface phenomena. This work was performed in part under the auspices of the U.S. Department of Energy at the Lawrence Livermore National Laboratory under Contract No. W-7405-Eng-48. Finally, many of the calculations were performed at the San Diego Supercomputer Center.

REFERENCES

1. F.H. Stillinger and T.A. Weber, Computer simulation of local order in condensed phases of silicon, Phys. Rev. B31:5262 (1985).
2. R. Biswas and D.R. Hamann, New classical models for silicon structural energies, Phys. Rev. B36:6434 (1987).
3. J. Tersoff, New empirical approach for the structure and energy of covalent systems, Phys. Rev. B37:6991 (1988).
4. M. Baskes, M. Daw, B. Dodson, and S. Foiles, Atomic scale simulation in materials science, MRS Bulletin 13(2):28 (1988).
5. M. Schneider, I.K. Schuller, and A. Rahman, Epitaxial growth of silicon: a molecular dynamics simulation, Phys. Rev. B36:1340 (1987).
6. E.T. Gawlinski and J.D. Gunton, Molecular dynamics simulation of molecular beam epitaxial growth of the silicon (100) surface, Phys. Rev. B36:4774 (1987).
7. S. Das Sarma, S.M. Paik, K.E. Khor, and A. Kobayashi, Atomistic numerical simulation of spitaxial crystal growth, J. Vac. Sci. Technol. B5:1179 (1987).
8. B.J. Garrison, M.T. Miller, and D.W. Brenner, Kinetic energy enhanced molecular beam epitaxial growth of Si(100), Chem. Phys. Lett. 146:553 (1988).
9. T.A. Weber, private communication.
10. F.F. Abraham and I.P. Batra, A model potential study of the Si(001) 2x1 surface, Surf. Sci. 163:L752 (1985).
11. K.E. Khor and S. Das Sarma, Model potential based simulation of Si(100) surface reconstruction, Phys. Rev. B36:7733 (1987).
12. R. Tubino, L. Piseri, and G. Zerbi, Lattice dynamics and spectroscopic properties by a valence force potential of diamond-like crystals: C, Si, Ge, Sn, J. Chem. Phys. 56:1022 (1972).
13. G.A. Somorjai, "Principles of Surface Chemistry", Prentice-Hall, Englewood Cliffs, NJ (1972).
14. T. Soma and H. Matsuo, Debye-Waller factor of Si and Ge, Phys. Stat. Sol. B111:K93 (1982).
15. I. NoorBatcha, L.M. Raff, and D.L. Thompson, A phenomenological approach to the calculation of the diffusion coefficient for Si on Si (111) using classical trajectories, J. Chem. Phys. 82:1543 (1985).
16. B.W. Dodson, Evaluation of the Stillinger-Weber classical interaction potential for tetragonal semiconductors in non-ideal atomic configurations, Phys. Rev. B33:7361 (1986).
17. K.L. Carleton and S.R. Leone, Laser probing of gallium interactions with silicon (100) surfaces, J. Vac. Sci. Technol. B5:1141 (1987).

18. B. Bourguignon, K.L. Carleton, and S.R. Leone, Surface structures and growth mechanism of Ga on Si(100) determined by LEED and Auger electron spectroscopy, Surf. Sci., 199:455 (1988).

19. B. Bourguignon, R.V. Smilgys, and S.R. Leone, AES and LEED studies correlating desorption energies with surface structures and coverages for Ga on Si(100), Surf. Sci., 199:473 (1988).

20. T. Thundat, et al., Experimental and theoretical investigation of chemisorbed Ga on Si(111), J. Vac. Sci. Technol. A6:681 (1988).

21. J.M. Nicholls, B. Reihl, and J.E. Northrup, Unoccupied surface states revealing the Si(111) $\sqrt{3}$ x $\sqrt{3}$-Al, -Ga, and -In adatom geometries, Phys. Rev. B35:4137 (1987).

22. A. Kawazu et al., Structural studies of Ga-adsorbed Si(111) $\sqrt{3}$ x $\sqrt{3}$ surfaces by low energy electron diffraction, Phys. Rev. B36:9809 (1987).

23. K. Balasubramanian, Electronic states of Ga, J. Phys. Chem. 90:6786 (1986).

24. W. Schommers, Pair potential in disordered many-particle systems: a study for liquid gallium, Phys. Rev. A28:3599 (1983).

25. J.L. Bretonnet and C. Regnaut, Determination of the structure factor of simple liquid metals from pseudopotential theory and optimized random phase approximation: application to Al and Ga, Phys. Rev. B31:5071 (1985).

SELF-CONSISTENT CLUSTER-LATTICE SIMULATION

OF IMPURITIES IN IONIC CRYSTALS

Jie Meng

Physics Department
Virginia Commonwealth University
P.O. Box 2000
Richmond, VA 23284

A. Barry Kunz

Department of Physics
Michigan Technological University
Houghton, MI 49931

INTRODUCTION

When simulating defects in ionic crystals by a finite cluster model, one always has a problem of taking account of the influence of the lattice beyond the cluster. Some cluster models consider only the defect and the nearest-neighbor ions. Some calculations approximate the lattice potential by a certain number of point charges and effective core potentials. Several approximations to the lattice potential in the region of the cluster were compared to the exact Madelung potential in the recent work of Winter et al.[1] That study emphasized the influence of the surrounding lattice ions on the energy level splitting and geometry of the nearest-neighbor cluster. It was found that the error in the calculated nearest-neighbor distance for the pure host is proportional to the error in the lattice potential. Unlike the finite cluster model, ICECAP (Ionic Crystal with Electronic Cluster; Automatic Program)[2,3] incorporates the polarization and the distortion of the surrounding infinite lattice with the electronic structure of the cluster self-consistently. ICECAP has been used to study the impurity Cu+ and Ag+ ions in alkali halides. The ground state energy, excited state energies and crystal field splitting were calculated. The interatomic potentials for impurities Cu+ and Ag+ in alkali fluorides and alkali chlorides were determined and used to study the transport properties.[4,5]

CALCULATIONS AND RESULTS

The impurity of Cu+ and Ag+ ions are located at the center of an octahedral host alkali site in the ground state in the systems $NaF:Cu^+$, $LiCl:Cu^+$ and $NaCl:Ag^+$. There are two singlets and two triplets associated with the excited state d^9s with the spin-orbit coupling omitted. The defect clusters consist of the impurity ion and the nearest-neighbor halides with 88, 136 and 154 electrons respectively for the system $NaF:Cu^+$, $LiCl:Cu^+$, and $NaCl:Ag^+$. The surrounding infinite lattice is described by the shell model. The shell-model short-range potentials were taken from Catlow[6], and had been fitted to the elastic and dielectric constants.

The defect cluster embedded in the shell-model lattice is treated in Unrestricted Hartree-Fock (UHF) approximation. The correlation corrections were calculated by use of Many Body Perturbation Theory (MBPT) for $NaF:Cu^+$.[7] All the electrons in the defect cluster were explicitly included in the quantum mechanical calculations without any replacing pseudopotential. The contracted Gaussian basis sets (533,53,5) for Cu^+ (d^{10}) and Cu^{2+} (d^9), (5333,53,5) for Cu^+ (d^9s) singlet and triplet were optimized based on the basis sets for the Cu atom in Huzinaga.[8] The optimized contraction coefficients and the total energy of the ion are given in Table I. The Gaussian basis sets (5333,533,53) for Ag^+ taken from Huzinaga[8], with the 5s orbital being omitted, were used for the ground state of $NaCl:Ag^+$. Additional Guassian primitives were optimized by minimizing the total energy for the MBPT calculation.[7]

TABLE I. The contraction coefficients of the Gaussian Basis Sets for Cu^+ and Cu^{2+} ions, with their total energy.

	d^{10}	d^9	d^9s singlet	d^9s triplet
Total Energy (Hy)	-1638.24198	-1637.42623	-1638.085432	-1638.19259
S	-0.005131	0.005122	0.005132	-0.005132
	-0.038944	0.037830	0.038495	-0.038944
	-0.176121	0.175821	0.176152	-0.176121
	-0.468240	0.456329	0.468321	-0.468240
	-0.450701	0.464112	0.450779	-0.450701
S	-0.108983	-0.062013	0.108962	0.108962
	0.638191	0.642130	-0.638187	-0.638187
	0.436235	0.419188	-0.436232	-0.436232
S	0.226176	-0.152061	-0.221699	-0.221699
	-0.733189	0.714811	0.732195	0.732195
	-0.401310	0.390872	0.400765	0.400765
S			-0.006668	0.008424
			0.973426	-0.974200
			0.035648	-0.035676
X	0.009514	0.009486	0.009514	0.009514
	0.070469	0.070314	0.070469	0.070469
	0.266356	0.265505	0.266356	0.266356
	0.510530	0.509915	0.510530	0.510530
	0.323997	0.325331	0.323996	0.323996
X	0.341064	0.330885	0.347471	0.347471
	0.549134	0.561415	0.551447	0.551447
	0.233149	0.225038	0.223027	0.223027
XX	0.034829	0.031120	0.036670	0.036670
	0.175709	0.163897	0.188737	0.188737
	0.389764	0.383815	0.412406	0.412406
	0.458082	0.475384	0.470367	0.470367
	0.314193	0.288259	0.250059	0.250059

The ground state energy is minimized by varying the nearest-neighbor distance, with the whole lattice relaxing self-consistently. Fig. 1 shows the energy curve of LiCl:Cu+.

The energy curve was fitted parabolically. The energy minimum was found with the nearest neighbors at a distance of 5.11 a.u. in LiCl:Cu$^+$, giving a distortion of 5.1% relative to the host lattice, consistent with the larger ion size of the Cu$^+$ ion (0.96 A) compared to the Li$^+$ ion (0.78 A).

Table II. Comparison of the calculated equilibrium nearest-neighbor distance r_0 (a.u.) and the distortions relative to the nearest-neighbor distance, a, in host lattice with the results of the finite cluster model.

		ICECAP		Finite Cluster	a (a.u.)
	r_0 (a.u.)	distortion	r_0 (a.u.)	distortion (%)	
NaF:Cu$^+$	4.37	0.2	4.42[1]	0.9	4.38
			4.62[9]	5.4	
LiCl:Cu$^+$	5.11	5.1	5.63[12]	15.8	4.86
NaCl:Ag$^+$	5.38	1.3	5.95[11]	12.0	5.31

The nearest-neighbor distance for the host lattice were taken as 4.34 a.u. for NaF:Cu$^+$, which is smaller than the experimental value of 4.38 a.u. The energy minimum was found with the nearest neighbors at a distance of 4.37 a.u., with a distortion of 0.2% compared to the experimental Na-F distance for NaF. Consistent with the similar ion size of Na$^+$ and Cu$^+$ (0.95 and 0.96 A, respectively), the Cu$^+$ impurity does not appreciably distort the lattice. In a finite cluster Hartree-Fock (HF) calculation[9] for a 33 ion cluster with effective core potentials on the lattice Na$^+$ ions, the equilibrium separation of Cu$^+$ and F$^-$ was found to be 4.62 a.u., with a distortion of 5.4%. The recent work of Winter et al[1], adjusting the charge of the Na$^+$ ion to include the effects of the remainder of the lattice, got an equilibrium distance of 4.42 au., with a distortion of 0.9%.

In the system NaCl:Ag$^+$, the nearest-neighbor distance for the energy minimum was found at 5.38 a.u., with a distortion of 1.3%. The relative ionic radii of the Ag$^+$ (1.26 A) and Na$^+$ being considered, it seems that the distortion may have been underestimated in this system because the basis sets for Ag$^+$ were not optimized. Table II shows the comparison of the calculated equilibrium distance and the distortions with the results of the finite cluster model.

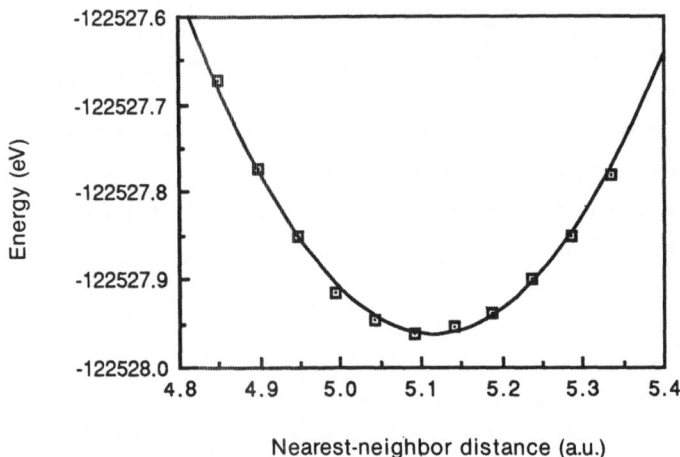

Figure 1.   The Ground State Energy curve of LiCl:Cu$^+$

The excited stated energies of the singlet and triplet T_{2g} and E_g were calculated at the equilibrium distance of the ground state. The d^{10} to d^9s singlet transition energy was calculated by projecting the singlets to pure spin eigenstates and taking the average of the two singlet states. At the UHF level, without correlation correction, the d^{10} to d^9s singlet transition energy was found to be 3.6 eV for NaF:Cu$^+$ and 4.3 eV for LiCl:Cu$^+$. The crystal splitting, 10Dq, of the d orbital was calculated by taking the difference between d10 to d9 (T2g) and to d9 (Eg). At the UHF level, 10Dq was 0.24 eV. With the correlation correction included, the d^{10} to d^9s transition energy was found to be 4.0 eV, versus the experimental result[10] of 4.2 eV, and the 10Dq was 0.31 eV versus the experimental value of 0.35 eV. It is noticed that the experimental 10Dq was the splitting between the twin singlet states. The comparison of the transition energy and the crystal field splitting, 10Dq, with experiment and with the finite cluster HF calculation are given in Table III. The spin-orbit interaction was not included in this work. Winter et al.[1] included the spin-orbit coupling using a semi-empirical approach.

Table III.   Comparison of the transition energy and the crystal fieldsplitting, 10Dq, with experiment and with the finite cluster calculation.

	ICECAP		Experiment[10]	Winter[1]
	(UHF level)	(with MBPT)		
NaF:Cu+				
d to s (eV)	3.6	4.0	4.2	3.5
10Dq (eV)	0.24	0.31	0.35	
LiCl:Cu$^+$				
d to s (eV)	4.3		5.0	

CONCLUSIONS

The calculations presented here demonstrated an advantage of ICECAP over the finite cluster models, with the ability of representing more accurately and easily both the short range repulsive interactions between neighboring ions and the long range potential due to the Coloumb interaction of the cluster with the infinite lattice. They also demonstrated the importance of including the correlation corrections to get more accurate results in transition energies and the crystal field splittings. The use of optimized basis sets was found to be essential to the accuracy of the Hartree-Fock self-consistent calculations. The spin-orbit interactions should be included in future calculations to get fine structure.

ACKNOWLEDGMENTS

This work was supported by the U.S. Department of Energy under Grant No. De-FG02-85ER45224.

REFERENCES

1. N.W. Winter, R.M. Pitzer, and D.K. Temple, J. Chem. Phys. 86 (6):3549 (1987).
2. J.H. Harding, A.H. Harker, P.B. Keegstra, R. Pandy, J.M. Vail, and C. Woodward, Physica B + C 131B:151 (1985).
3. A.B. Kunz, J. Meng, and J.M. Vail, Phys. Rev. B38:1064 (1988).
4. J. Meng, R. Pandey, J.M. Vail, and A.B. Kunz, Phys. Rev. B, 38:10083 (1988).
5. J. Meng, R. Pandey, J.M. Vail, and A.B. Kunz, (submitted).
6. C.R.A. Calow, K.M. Diller, and M.J. Norgett, J. Phys. C. 10:1395 (1977).
7. J. Meng, A.B. Kunz, and C. Woodward, Phys. Rev. B , 38;10870 (1988).
8. "Gaussian Basis Sets for Molecular Calculations", edited by S. Huzinaga (Elsevier, New York, 1984).
9. N.W. Winter and R.M. Pitzer, in: "Tunable Solid State Lassers", edited by P. Hammerling, Springer Series of Optical Science, Vol. 47 (Springer, Berlin, 1985), p. 164.
10. S.A. Payne, A.B. Goldberg, and D.S. McClure, J. Chem. Phys. 81:1529 (1984).
11. C. Pedrini, H. Chermette, A.B. Goldberg, D.S. McClure, and B. Moine, Phys. Stat. Sol. 120:753 (1983).
12. B. Moine, H. Chermette, and C. Pedrini, J. Chem. Phys. 85:2784 (1986).

THE EFFECTIVE MEDIUM APPROACH TO THE ENERGETICS

OF METALLIC COMPOUNDS

Andrew C. Redfield

Physics Department
Williams College
Williamstown, MA 01267

Andrew Zangwill

School of Physics
Georgia Institute of Technology
Atlanta, GA 30332-0245

In this paper we discuss the use of the effective medium approach, an embedded atom method (EAM), to study the cohesion in, and structures of, intermetallic compounds. We show that the method includes important contributions to the cohesion of many such compounds, resulting from the proximity of different types of atoms, that are neglected in simple pair potentials. We also show that important contributions exist of the form calculated from pseudopotential pair potentials. We argue that these contributions are neglected in currently existing EAM schemes, but could be included in an improved effective medium model.

A rich array of metallic compound structures exists in nature. More than a thousand have been identified and determined [1]. Many of these structures are complex, containing on the order of a hundred atoms, or in extreme cases, even a thousand atoms, per unit cell. A complete, detailed understanding of the energetics of such systems is clearly not feasible. On the other hand, various intriguing geometrical patterns and structural motifs recur frequently, and an understanding of them would be both interesting and a major step toward understanding these complex structures. Examples of common patterns include icosahedral arrangements of atoms, various stackings of simple units known as Friauf polyhedra, and layered nets (such as the so-called 3-5-4 nets) [2]. For specificity, in this paper we will limit our attention to aluminum - transition metal (TM) alloys, although we anticipate the discussion to be of much broader relevance. We will concentrate on Al-rich alloys and compounds [3]. In Figure 1, the structures of the most Al-rich binary compound containing each of the transition metals is shown. For TM atoms near the left side of the figure (e.g., Ti) the TM atoms substitutionally replace Al atoms on the sites of an FCC lattice. As one moves somewhat to the right (e.g., Mn), the TM atoms are surrounded by an icosahedral shell of Al atoms, and these icosahedra are packed in a BCC arrangement. Still farther to the right, the number of Al atoms surrounding each TM atom is reduced from the close-packing value of 12 (dropping to 9 for Co and 8 for Ni).

Ti Al_3Ti	V $Al_{10}V$	Cr $Al_{45}V_7$	Mn $Al_{12}W$	Fe $Al_{13}Fe_4$	Co Al_9Co_2	Ni Al_3Ni
Zr Al_3Zr	Nb Al_3Ti	Mo $Al_{12}W$	Tc $Al_{12}W$	Ru $Al_{13}Fe_4$	Rh Al_9Co_2	Pd $Al_{21}Pt_5$
Hf Al_3Zr	Ta Al_3Ti	W $Al_{12}W$	Re $Al_{12}W$	Os $Al_{13}Os_4$	Ir Al_9Co_2	Pt $Al_{21}Pt_5$

Figure 1. Periodic table of the transition metals, showing the crystal structure of the most Al-rich intermetallic compounds. (NOTE: The name of a structure is the chemical formula of one compound that forms the structure. Thus, for example, Al_3Ti is the name of a structure that exists in both Al-Ti and Al-Nb alloys). The Al_3Ti and Al_3Zr structures substitute TM atoms on some of the sites of the Al FCC structure. The $Al_{12}W$ structure has an icosahedral shell of Al around each TM atom.

To gain insight into these trends and other structural properties requires an understanding of the cohesive energies of metallic alloys. Numerous approximation methods exist for computing cohesive energies of solids. The most accurate method for such problems is generally the local density approximation (LDA). Unfortunately, this method is computationally unfeasible for very large systems. Even for somewhat smaller systems, investigation of large numbers of alternative structures or relaxation of structures is not possible. Also, it is frequently difficult to extract simple physical principles from the numerical results. Therefore, we have looked for a simpler, less precise method that would retain the essential physics of the problem and be appropriate for the primarily simple metal systems under consideration. We believe the EAM best meets these requirements, as will be discussed below.

First, however, let us explain the basic idea of the EAM [4]. View the solid as a collection of atoms, each immersed in an electron gas coming from the wavefunction tails of the neighboring atoms. Approximate this gas a uniform background electron gas. (This should be a good approximation for a simple metal system [5]). The energy for an atom in such an electron gas can be calculated, essentially exactly, using the LDA. (This is where the particular method we use, developed by Jacobsen, Norskov, and Puska [6] and known as the effective medium approach, differs from empirical EAMs, which fit to various experimental values). The resultant energy, known as the embedding energy, is shown as a function of background electron density for a variety of atoms in Figure 2. The key point to notice is that each curve has a minimum at some finite value of the density. Thus the atom "prefers" to be surrounded by this density from the tails of the neighboring atoms. Finally, the total cohesive energy of the solid is assumed to be a sum of these embedding energies $E_i(n)$ for each atom i, plus a residual pair interaction term:

$$E_{TOT} = \sum_i E_i(n_i) + \sum_{i,j} V(r_{ij}).$$ (1)

Here, n_i is the background density on site i from the neighboring atoms. Since the LDA calculation for the atom in the background electron gas includes a computation of the density tail, n_i is simply the sum of these tails from each of the neighbors of atom i. Actually, the tails exhibit Friedel oscillations, as would be expected, which in the Jacobsen et al. prescription are smoothed out. The pair potential V(r) will be discussed below.

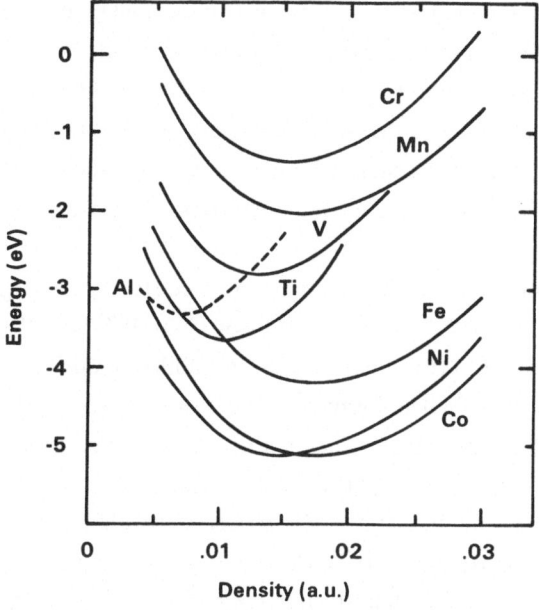

Figure 2. Embedding energies E_i as a function of the background electron density n for Al and the 3d transition metals (from ref. 6).

It is instructive to compare this expression with the corresponding expression derived from pseudopotentials:

$$E_{Tot} = E(n) + \sum_{i,j} V(r_{ij}) + \text{higher order terms.}$$ (2)

In this expression, the zeroth order approximation, E(n), is the energy of a uniform electron gas of density n. The difference in the form of these equations is that the EAM allows a *different* density on each site in the zeroth order approximation. For some alloys, different atoms want and have very different densities (as can be seen in Fig. 2). Requiring a uniform density electron gas in zeroth order will be a poor approximation and therefore large third and higher order terms will be necessary. On the other hand, in elemental solids, each site generally has the same density, so the form of Eq. (1) reduces to that of

Eq. (2). Also, for compounds where the different constituents want similar densities, which are generally atoms in the same or nearby columns of the periodic table, the pseudopotential method should be quite good. Indeed, Hafner has applied such methods [7] and gets good results when both atoms are alkali metals. He gets increasingly more error as the atoms come from more widely separated columns of the periodic table, thus supporting our claim.

We have shown in previous papers [8,9] that the first term of the EAM is the key determining factor of a variety of metallic alloy structures. For example, it explains the transition from the $TiAl_3$ to the WAl_{12} structure as one moves from left to right in Fig. 1, as discussed above. The reason is the following. Looking at Figure 2 one sees that Ti and Al want to be embedded in similar background densities. Ti can therefore be substituted for Al on an FCC lattice with both atoms near their desired embedding densities. On the other hand, Mn wants a much larger density than Al so it would be energetically very costly for a Mn atom to sit in the FCC environment. The neighboring Al atoms therefore rearrange to try to be closer to the Mn (thus supplying more tail density on the Mn site) while maintaining maximum distance from each other (so their densities do not get too high). The optimal way to do this is for the Al atoms to form an icosahedral shell around the Mn.

The same ideas explain various other observed structures, including 54-atom units known as Mackay icosahedra found in the α phase of AlMnSi (and also believed important in the icosahedral phases [10]), the role of Si in stabilizing this structure, different bond lengths and coordination shell uniformities in these structures, etc. [8].

However, there are many structures which cannot be explained simply by this density argument. One example is the transition to the more complicated structures as one moves to the extreme right in Figure 1. On reflection, this failure is not surprising; in fact the argument does not explain at all the structures of pure simple metals (for example, the FCC-BCC transition). This *is* understood in the pair potential picture: the screening of the electron gas produces an oscillatory pair potential and the resultant structure is one that puts the neighboring atoms near the minima of this pair potential (while preserving the overall density n). But this effect has been omitted in the EAM as presently implemented. There are a few ways to see this. First, as mentioned earlier, the Jacobsen et al. method explicitly averages over the Friedel oscillations in the density tails. But this eliminates the effect being sought. Even if this were changed there is a more fundamental problem. At the boundary between cells, the density is matched, but no attempt is made to ensure that the density is smooth there. Normally, the slope at the boundary will be discontinuous. But it is precisely the necessity of smoothing this out, (and thus avoiding its infinite energy) that leads to the oscillatory screening solutions. Thus an improved effective medium approach would need to consider more correctly the density changes across the boundary. On the other hand, in the empirical EAM, the pair potential term is not determined from the density profile, so in principle one could simply supply an oscillatory potential. Unfortunately, however, since the parameters are fit to experimental values, several additional input values would be required. What is worse, much of the curve is not determinable from ground state properties of the system, so the readily available experimental data would not suffice. Present EAMs have generally *assumed* some simple, non-oscillatory form.

What is the effect of such a pair-potential on the Al-TM systems? Qualitatively, the following would be expected. As one moves to the right across the transition metal rows in the periodic table, the d-shell contracts, making the core radius smaller. The radius of the first minimum of the pair potential therefore moves inward. [9]. Toward the left end of the rows, the TM atoms thus want to keep the neighboring atoms fairly far away, while toward the right end they move in closer. As they move in, they can achieve the desired density with less atoms, thus explaining why Co and Ni have fewer Al atoms around them. For TMs farther to the left, the effect of the pair potential term is simply to keep the atoms far apart. Then many atoms need to be around each TM atom to achieve the desired density. But the most that can fit geometrically is the close-packing number, 12. (To fit more requires a much larger distance between the central atom and the surrounding shell, drastically reducing the density on the central atom). That is why, in the discussion of AlMn alloys above, it was tacitly assumed that the number of Al atoms around a Mn remains at 12 [11].

In summary, then, we have shown that the structures of many intermetallic compounds can be understood using the EAM. The preference of different atomic types to have different background electron densities is an important structural determinant, which cannot be obtained from pure pair potential methods. On the other hand, the usual oscillatory pair potential terms are also important, but are not properly accounted for in present embedded atom methods. Inclusion of a better pair potential term in the EAM, including such oscillations, could be expected to provide increased accuracy and the ability to understand many of the structures found in metallic alloy systems.

REFERENCES

1. P. Villars and L.D. Calvert, eds., *Pearson's Handbook of Crystallographic Data for Intermetallic Phases* (American Society of Metals, Metals Park), 1985.
2. W.B. Pearson, *The Crystal Chemistry and Physics of Metals and Alloys* (Wiley, New York), 1972.
3. As we shall see, the computational method we use will only be reliable in this case.
4. The EAM is also discussed in several other papers in this symposium.
5. It would not be a good approximation for a system with many TM-TM nearest neighbors, since directional d-bonding will occur. However, in the Al-rich alloys under consideration, each TM atom is surrounded only by Al atoms, so the background electron gas will be fairly uniform.
6. K.W. Jacobsen, J.K. Norskov, and M.J. Puska, Phys. Rev. B35:7423 (1987).
7. J. Hafner, Phys. Rev. B15:617 (1977) and MRS Symposium Proc., Vol. 19:1 (1983).
8. A.C. Redfield and A. Zangwill, Phys. Rev. Lett., 58:2322 (1987) and Phil. Mag. Lett. 57:255 (1988).
9. A. Zangwill and A.C. Redfield, J. Phys. F18:1 (1988).
10. V. Elser and C.L. Henley, Phys. Rev. Lett. 55:2883 (1985).
11. Of course, the numerical calculations using Eq. 1 properly take these competing effects into account. In particular, the V(r) used (the same as in ref. 6), while not oscillatory, does tend to push atoms to larger distances.

AB-INITIO STUDY OF AMORPHOUS AND LIQUID CARBON

Giulia Galli and Richard M. Martin

University of Illinois, Dept. of Physics
Urbana, IL 61801

Roberto Car and Michele Parrinello
International School for Advanced Studies
Strada Costiera 11
34014 Trieste, Italy

INTRODUCTION

The bonded forms of carbon in solid and liquid states have considerable variations with many characteristics which make them interesting for a wide range of researchers[1-4]. Nevertheless, outstanding questions concerning the properties of both non-crystalline and liquid carbon are yet unanswered; in particular, knowledge of the structural and electronic properties of the low density disordered phases, the nature of the liquid state and the melting mechanism - despite systematic investigations[4] of the carbon phase diagram carried out in different fields, such as condensed matter physics[1], astrophysics[2] and geology[3].

From a technological point of view, the two common forms of disordered carbon at room temperature (T) - glassy (g-C) and amorphous (a-C)[5] - are particularly attractive, since they combine semiconducting properties in some respect similar to graphite with a much higher hardness[6]. A question of primary interest regards the understanding of the atomic structure of a-C films; the determination of the fraction of sp^2: sp^3 sites is, in particular, a problem greatly debated in the literature[1a,6-8] and yet unsolved. Estimates of sp^2:sp^3 sites concentration ratio, in these films, ranging from 5 to 50% have been obtained from radial distribution function (RDF) analyses[7]. This is an example of the need to understand atomic coordination which is key for the phase diagram of carbon.

The knowledge of the liquid state (1-C) properties is even more incomplete than that of the amorphous materials. The carbon phase diagram originally proposed by Bundy[9], and investigations carried out in the sixties and seventies[4a], indicate the occurrence of a triple point in the high-T, low pressure (P) regime. This would mean that 1-C does not exist at atmospheric pressure. Recent results about graphite surface melting by high-energy laser pulses [2b,4b,10,11] cast these early predictions into doubt, by showing that there is at least a small range of T for which 1-C exists at low P. A greatly debated question regards the electronic structure of the liquid, namely whether it is an insulating or a metallic phase.

A very recent experiment[4b] indicates that near T = 4450 K and close to atmospheric pressure carbon undergoes a solid-to-liquid transition, leading to a metallic liquid with a nearly temperature independent electrical resistivity. Previous experiments of the same kind[11] (time resolved refelectivity measurements of graphite samples irradiated by intense laser pulses) had instead brought to the conclusion that the low-P liquid is not metallic, but has an electronic structure intermediate between that of graphite and diamond. Evidences have been cited[9,12] to show that the density of liquid carbon is less than that of the solid at the triple point. An extrapolation from thermodynamical data of the density of the solid at the triple point, and the somewhat arbitrary assumption of a volume expansion of 20% for graphite melting, gives an estimate of $\rho = 1.6$ gr.cm^{-3} for 1-C[13]. On the other hand, the hypothesis that carbon behavior upon melting parallels that of the other group IV elements, would indicate a density larger than that of graphite in the liquid phase. A value of $\rho = 2.7$ gr.cm^{-3}, for example, has been proposed in the literature[1b]. There are no measurements which allow to infer the structure of 1-C, but only evidences of structural transformations as graphite melts[12]. Speculations[13] based on thermodynamic properties of the vapor and liquid phases depict the liquid in the low-P regime as a mixture of C_n chains, similar to those found in carbon vapors, with the simpler species C_1 and C_2 tending to gain in relative importance with raising T. According to the results of several spectroscopic investigations of graphite at high-T[14], Whittaker[15], and other authors[16] proposed that graphite is not stable above 2600K, at any pressure, but transform to the so-called "carbyne" solid, composed of chain-like structures containing triple-bonds. If so, this would support the picture that liquid carbon is a low coordinate compound, with sp-bonded carbon atoms. However, in the last decade, many controversies about the existence of the carbyne region have been reported.[14,16]

In this work we present an analysis of both structural and electronic properties of a room-temperature a-C structure[17] and discuss preliminary results about the liquid state, which we have generated with a computer simulation based on a first-principle Molecular Dynamics (MD) method.[18] Unlike conventional Molecular Dynamics and Monte Carlo methods, the interatomic potential is constructed directly from the electronic ground state and the latter is treated with accurate density functional techniques. This is achieved by constructing a Lagrangian whose dynamical variables are the ionic positions and the continuous degrees of freedom in the Kohn-Sham orbitals for the electrons. In our investigation, we have used the dynamical trajectories obtained from the solution of the equations of motion to generate both a disordered carbon structure at room temperature and a liquid state at $T = 5000\ K$.

We have carried out the MD calculations for 54 carbon atoms with periodic boundary conditions, corresponding to fcc supercell with macroscopic density ρ fixed to be 2 gr.cm^{-3}. The chosen density allows to represent an amorphous phase of carbon, better than a glassy one, at low T[5], and, according to the (ρ,T) diagram proposed in Ref. 13, it is possibly a reliable density for the liquid as well, at high T and low P. The a-C network has been obtained with constant-volume (CV) MD runs. The 1-C state has instead been generated with a constant-volume-constant-temperature (CVT) MD, originally proposed by Nosé[19] for classical systems. To this end, a Lagrangian appropriate for a coupled system of ions and electrons, in thermal equilibrium with an external heat reservoir of fixed temperature, has been defined and equations of motions for ions consistently derived. Unlike CV simulations, which lead to phase-space sampling of the microcanonical ensemble, CVT simulations produce a sampling of the canonical ensemble[19]. For both CV and CVT approaches, the electronic Kohn Sham orbitals at the $k = 0$ point of the Brillouin zone have been expanded in plane waves. A kinetic energy cut-off of 32 Ry and 20 Ry has been used in different runs, the larger cut-off implying the use of 12,000 plane waves. The interaction between valence and core electrons has been described by a non-local pseudopotential of the form suggested in Ref. (20), including s-only non locality. In the MD runs a time step (Δt) equal to 10^{-16} sec. has been used and the fictitious mass parameter introduced in Ref. 18 set at 200 a.u. The optimal mass-like parameter associated with the scaling variable entering CVT equations of motions[19a] have been determined to be $1.2 \cdot 10^5$ a.u. The initial atomic positions for the a-C simulation have been chosen by randomly displacing the atoms from the positions in a perfect diamond lattice and the evolution in time followed for about 3000 Δt, using a plane wave cut-off of 20 Ry. The configuration thus obtained has been heated up to 5000 K, evolved for another 3000 Δt, and finally cooled down at a rate of 10^{16} deg/sec. After an equilibration of about 3000 Δt at room temperature, temporal averages have been taken in order to measure static properties for the amorphous structure. In order to check the convergence of the calculation with respect to the plane wave cut-off (Ecut), starting from a well equilibrated 20 Ry configuration at 300 K, we have increased Ecut to 32 Ry and after a new annealing we have generated a different a-C structure. The results are very similar in the two cases. We have studied the liquid state with Ecut = 32Ry, heating this final annealed a-C structure up to 5000 K.

The computer generated a-C, described in more detail in Ref. 17, consists of 85% sp^2 sites and 15% sp^3 sites[21], at 300 K. The former are graphitic in nature, while the latter are distorted diamond structures. The average bond angle of sp^2 sites is 117 deg., slightly smaller than that of graphite; the angular distribution of sp^3 sites has instead a peak at 105 deg, with a non-negligible contribution from angles near 90 deg. Most measurements[6-8] indicate a sp^3 sites concentration of 10-20%, in agreement with our findings.

As the system is heated up, a change in the atomic coordination is observed. At T around 2500-3000 K, a small proportion of two-fold coordinate atoms starts being present in the network. sp-sites[21] concentration increases as T is again raised, at the expenses of both 3- and 4-fold coordinated atoms. Above T = 4000 K the system begins showing a diffusive behavior, possibly to indicate that a melting transition is taking place. Above 4500 K, the behavior of atomic mean square displacement as a function of time suggests that a liquid state has been generated. Our results are consistent with several experimental findings[6,7] of an increasing graphitic character of a-C samples as the temperature is raised.

They also support the idea that liquid carbon is composed of chain-like structures with sp-bonding[13].

The calculated particle-particle correlation function of a-C, $g(r)$, and the partial correlation functions $g_{ij}(r)$ - $g(r) = \sum_{ij=3,4} g_{ij}(r)$ show interesting features. The first maximum of $g_{33}(r)$ is found to be at a distance equal to the calculated nearest neighbor distance in graphite, while that of g_{44} is 6% larger than the computed first neighbor distance in diamond[22]. It has been argued[71] that the presence of sp^2 and sp^3 sites in a-C films should be revealed by the existence of a double first peak in $g(r)$, which diffraction experiments performed with large wave vectors (q) should be able to resolve. This is not found to be the case in our simulation: the first maximum of $g(r)$ is indeed unique but the presence of sp^3 sites is revealed only by a very weak shoulder. Our calculation leads to the conclusion that this shoulder could be detected only by experiments using extremely high wavelength cut-off ($q \geq 70\text{Å}^{-1}$). As the temperature is raised and the sp sites fraction increases, the position of the first maximum of $g(r)$ moves towards shorter distances and the integral of the $g(r)$ first peak indicates a shell coordination number (n_1) less than 3.0.

In Table 1 we report the calculated positions of the first and second peak of the a-C RDF $J(r) = 4\pi\rho r^2 g(r)$ and the coordination number n_1, together with the corresponding quantities obtained from different experiments. In addition to the good agreement with experiment shown in Table 1, the overall features of $J(r)$ are nicely reproduced in our computer simulation. This can be seen from Fig. (1), where we compare the computed $J(r)$ with the one obtained from electron diffraction by Kakinoki et al.[7b], chosen because it is perhaps the most accurate, i.e. derived from measurements performed with the largest cut-off wave vector (27 Å^{-1}).

TABLE 1. Position of the first (r_1) and second (r_2) peak of the a-C pair-correlation function $J(r)$ and the first (n_1) shell coordination number, as obtained from different experiments (Ref. 7a [1]; Ref. 7b [2]; Rev. 7c [3]; Ref. 7d [4]) and from theory (present results [5], with Ecut = 32R). ρ is the macroscopic density of the system. The theoretical results correspond to the unconvoluted $J(r)$.

Non-Crystalline C	$r_1(\text{Å})$	n_1	$r_2(\text{Å})$	$\rho(\text{gr-cm}^{-3})$
Glassy Carbon[1]	1.42	2.99	2.45	1.49
Evaporated a-C[2]	1.50	3.45	2.53	2.40
Evaporated a-C[3]	1.43	3.30	2.53	2.10
a-C from PTFE[4]	1.46		2.54	
a-C (theory)[5]	1.44	3.20	2.56	2.00

Although the computer generated a-C is a truly three dimensional structure, the atoms are arranged into several "thick planes": two, which are labelled (*a*) and (*b*) in Fig. (2a) are roughly parallel to each other, with a stacking sequence reminiscent of that of graphite (*a,b,a*). They are connected by orthogonal planes (indicated with (*c*) and (*d*) in Fig. (2a)). Most of the bonds formed by sp^2 atoms tend to lie nearly on the same plane, as in graphite, but substantial buckling can occur locally. A ring statistic analysis shows that a-C is mainly composed of 5-, 6-, and 7-fold rings, with several coupled (5 + 7) rings similar to those found in carbon azulene molecules (see Fig. (2b)).

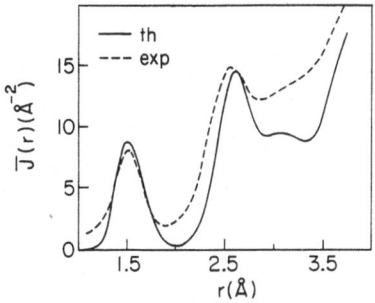

Figure 1 a-C theoretical and experimental (Ref. 7b) correlation function *J(r)*. The theoretical *J(r)* has been convoluted with the experimental resolution function, but not scaled for different theoretical and experimental densities, due to large uncertainties on the latter (See, e.g., Ref. 7a).

The computed Electronic Density of States (EDOS) of a-C, which is the result of the same self-consistent procedure from which the atomic positions were derived, is in good agreement with experiment. In the valence band, the split *s* peak around - 16 eV and the broad σ peak centered at about 7 eV are correctly reproduced, as well as the π state shoulder near the Fermi level[6a]. The conduction band spectrum shows a sharp π* peak and a broader σ* peak, in agreement with experimental findings[6c]. In addition, we find a deep minimum in the EDOS, near the Fermi level. Within the finite resolution of our calculation, this is consistent with the experimental observation[6] of a small gap of about 0.4-0.7 eV. A detailed investigation, still underway, is necessary to understand the liquid state controversial electronic structure. Preliminary results indicate that the high-T carbon structure which we have generated is a metal.

In summary, we have presented an *ab initio* study of both disordered and liquid carbon. Our results are in good agreement with the limited experimental information available. Extension of the first-principles MD method to the study of other properties of the liquid state as well as of other complex phases of carbon are in progress.

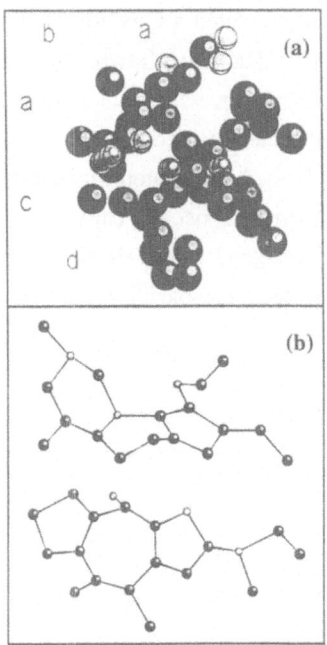

Figure 2   Microscopic structure of the computer-generated a-C network: the entire set of atoms belonging to one MD cell (a) and several 5-, 6-, and 7-fold ring structures the system is composed of (b) are displayed (see text). Black and grey spheres indicate 3- and 4- coordinate atoms, respectively.

ACKNOWLEDGEMENTS

This work was supported by the NSF grant DMR86-12860 and the SISSA-CINECA collaborative project, under the sponsorship of the Italian Ministry for Public Education.

REFERENCES

1.   See, e.g., (a) J. Robertson, Adv. Phys. 35:317 (1986) and references therein;
        (b) J. Steinbeck, G. Braunstein, M.S. Dresselhaus, T. Venkatesan
        and D.C. Jacobson, J. Appl. Phys. 58:4374 (1985).
2.   M. Ross, Nature, 292:435 (1981).
3.   (a) J.S. Dickey, W.A. Bassett, J.M. Bird and M.S. Weathers, Geology, 11:219
        (1983); (b) D.A. Young and R. Groover, Nature (1988).
4.   (a) F.B. Bundy, J. of Geophys. Res. 85:6930 (1980); (b) J. Heremas, C.H. Olk,
        G.L. Eeseley, J. Steinbeck and G. Dresselhaus, Phys. Rev. Lett. 60:453
        (1988).
5.   g-C has low density (ρ = 101.5 gr.cm^{-3}), while experiments (see, e.g., Ref. 1a)
        suggest that a-C has ρ = 1.8-2.7 gr.cm^{-3}.
6.   (a) J. Fink, T. Muller-Heinzerling, J. Pfluger, A. Bubenzer, P. Koidl and
        G. Crecelius, Solid St. Comm. 47:687 (1983); (b) J. Fink, T. Muller-
        Heinzerling, J. Pfluger, B. Scheerer, B. Dischler, P. Koidl, A. Bubenzer
        and R.E. Sah, Phys. Rev. B 30:4713 (1984); (c) D. Wesner, S. Krum-

macher, R. Carr, T.K. Sham, M. Strongin, W. Eberhardt, S.L. Weng, G. Williams, M. Howells, F. Kampas, S. Heald and F.W. Smith, Phys. Rev. B 28:2152 (1983).

7.   (a) D.F.R. Mildner and J.M. Carpenter, J. Non-Cryst. Solids 47:391 (1982); (b) J. Kakinoki, K. Katada, T. Hanawa and T. Ino, Acta Cryst. 13:171 and 448 (1960); and ibid 18:578 (1965); (c) B.T. Boiko, L.S. Palatnik and A.S. Derevyanchenko, Sov. Physics Doklady, 13:237 (1968); (d) L. Cervinka, F.P. Dousek and J. Jansta, Phil. Mag. B 51:604 (1985).

8.   (a) A.L. Ritter, J.R. Dennison and R. Jones, Phys. Rev. Lett. 53:2054 (1984); (b) Y.Y. Wnag, A.L. Ritter, T.J. Fabish and J.A. Nemetz, Bull. Amer. Phys. Soc. 33:354 (1988); (c) C. Gao, A.L. Ritter, T.J. Basish and J.A. Nemetz, Bull. Amer. Phys. Soc. 33:354 (1988).

9.   F.P. Bundy, J. Chem. Phys. 38:631 (1963).

10. T. Venkatesan, D.C. Jacobsin, J.M. Gibson, B.S. Elman, G. Braunstein, M.S. Dresselhaus and G. Dresselhaus, Phys. Rev. Lett. 53:360 (1984).

11. (a) A.M. Malvezzi, N. Bloenbergen and C.Y. Huang, Phys. Rev. Lett. 57:146 (1986); (b) E.A. Chauchard, C.E. Lee and C.Y. Huang, Appl. Phys.Lett. 50:812 (1987).

12. (a) G.J. Schoessow, Phys. Rev. Lett. 21:738 (1968); (b) N.S. Fateeva and L.F. Vereshchagin, JEPT Lett. 13:119 (1971).

13. H.R. Leider, O.H. Krikorian and D.A. Young, Carbon 11:555 (1973).

14. See, e.g., R.B. Heimann, J. Kleinman and N.M. Salansky, Nature, 306:164 (1983).

15. A.G. Whittaker, Nature 276:695 (1978); Science, 200:763 (1978) and ibid, 229:485 (1985).

16. P.P.K. Smith and P.R. Buseck, Science 216:985 (1982) and ibid, 229:487 (1985).

17. G. Galli, R.M. Martin, R. Car and M. Parrinello, Phys. Rev. Lett., 62:555 (1989).

18. R. Car and M. Parrinello, Phys. Rev. Lett. 55:2471 (1985).

19. (a) S. Nosé, Mol Phys. 52:255 (1984) and J. Chem. Phys. 81:511 (1984); (b) W.G. Hoover, Phys. Rev.A 31:1695 (1985).

20. L. Kleinman and D.M. Bylander, Phys. Rev. Lett. 48:1425 (1982).

21. sp, sp^2 and sp^3 sites designates 2, 3 and 4 coordinated atoms, respectively. We have defined the coordination by considering neighbor atoms to lie at a distance less than the first minimum of the pair correlation function $g(r)$.

22. The bond lengths in the crystal structures have been obtained with the same kinetic energy cut-off as that adopted for a-C, which leads to an overestimate of the corresponding experimental values of about 2%, if Ecut = 32 Ry is used.

MODELLING OF INORGANIC CRYSTALS AND GLASSES

USING MANY BODY POTENTIALS

C.R.A. Catlow, R.A. Jackson and B. Vessal

Department of Chemistry
University of Keele
Keele, Staffs. ST5 5BG
United Kingdom

INTRODUCTION

The application of modelling methods to the study of the structures and properties of inorganic materials is a rapidly expanding field. Growth in computer power is allowing systems of increasing complexity to be studied, and refinements in interatomic potentials are leading to greater precision in the calculations. Several reviews of the techniques and achievements of this field have appeared in recent years (see e.g., Catlow, 1986; Mackrodt, 1984; Catlow and Cormack, 1987; Catlow et al., 1988). Our account will therefore be brief and will concentrate on those areas where the use of many body terms has had a significant impact. These concern principally the study of crystalline and glassy silicates, the modelling of complex oxides including the recently discovered high T_c materials, and the development of potentials for the silver halides. Discussion of these systems will follow our general account of potential models for inorganic materials and their use in simulation studies.

MODELLING OF INORGANIC MATERIALS: SCOPE AND POTENTIALS

Modelling methods have been applied to a wide range of inorganic materials; table (1) presents a list of some of the compounds that have been studied. The modelling techniques used include energy minimization, molecular dynamics and Monte-Carlo and they have been applied to the prediction of perfect lattice structure and properties and to the study of impurities, defects and atomic transport properties.

Interatomic potentials for these systems have generally been based on the Born model, using ionic central-force pair potentials. Such 'two-body' potentials have the following common features:

(i) The use of the ionic model generally with formal (but occasionally with partial) ionic charges. The question of the degree of ionicity in oxides and halides has been extensively debated and was reviewed recently by Catlow and Stoneham (1983). In practice it is found that models based on formal charges work well in simulating properties of the perfect and defective crystal even for those compounds such as the silicates, where it is known that there is an appreciable covalent contribution to the bonding.

(ii) The description of the short range interactions in terms of central-force pair potentials, commonly of a simple and analytical form, such as the widely used Buckingham potential:

$$V(r) = A \exp(-r/\rho) - Cr^{-6}. \tag{1}$$

Other analytical functions such as the Lennard-Jones and Morse potentials may be used. In addition, Mackrodt and co-workers (Mackrodt, Colbourn and Kendrick 1981) have made extensive use of numerical potentials.

(iii) Polarizability is described using the shell model originally developed by Dick and Overhauser (1958). This describes polarization in terms of the displacement of a massless shell (of charge Y, simulating the valence-shell electrons) from a core (in which the mass is concentrated), the core and the shell being connected by a harmonic spring constant K. The free ion polarizability α may then be written as

$$\alpha = Y^2/K \tag{2}$$

Table 1. Examples of Inorganic Materials Studied by Modelling Methods

HALIDES

(i)   Simple: NaCl + alkali halides:
CaF$_2$ + alkaline earth fluorides,
MnF$_2$, LaF$_3$, CaF$_2$

(ii)   Complex: KCaF$_3$

OXIDES

(i)   Simple: MgO, CaO, NiO, TiO$_2$, WO$_3$, Nb$_2$O$_5$,
CeO$_2$, UO$_2$, SiO$_2$, Al$_2$O$_3$, ZrO$_2$

(ii)   Complex: BaTiO$_3$, LiNbO$_3$, Silicates

NITRIDES
Li$_3$N

SULPHIDES
CaS

The advantage of the model is that by specifying repulsive forces to act between shells, it is possible to include coupling between polarizability and short-range repulsion, an important physical effect whose omission from earlier point-ion models led to serious short-comings in calculations of both dielectric, lattice-dynamical and defect properties of ionic solids.

Parameterization is achieved by both empirical fitting and theoretical methods. In the latter are included both electron gas techniques based on procedures developed by Wedepohl (1967) and Gordon and Kim (1972), and *ab initio* methods which are being used increasingly and where calculations based on large clusters are now possible owing to expansion in computer power.

At present only empirical methods may be used to derive shell-model parameters (although recent work of Fowler and Pyper (1985) suggests that polarizabilities may be accurately calculated). Measured static and high-frequency dielectric constants must be available if these are to be fitted with any reliability; otherwise extrapolation methods must be used. Recent work of Cormack (1989) has demonstrated the sensitivity of calculated defect energies in complex oxides to shell-model parameterization. The derivation of reliable and general procedures for deriving these prameters is therefore one of the most urgent requirements.

A typical illustration of a current model for an oxide material is provided by the recent work of Jackson, Murray, Harding and Catlow (1986) on UO_2. Their work adapted the older empirical shell-model potential of Catlow (1977) so that it correctly modelled the thermal expansion of the material, which provides an important and sensitive test of a potential model. The resulting parameters are reported in Table 2, while Table 3 gives the calculated and experimental phonon dispersion curves. The agreement is good for the acoustic modes, but less satisfactory for the optical branches. Table 4 gives calculated and experimental enthalpies for the most important defect processes.

Table 2. Parameters of the modified UO_2 potential. Note that the potential minimum is held at $r = 2.1$ and the potential and its first two derivatives are constrained to be continuous across the spline points. The shell model assumes full ionic charges and uses for uranium $Y = 6.54e$ ($K = 94.24$ eVÅ^{-2}) and for oxygen $Y = -4.4e$ ($K = 296.2$eV Å^{-2}).

Short Range Potentials (energy in eV, distances in Å)

$\Phi_{UO}(r)$	=	$1518.92 \exp(-r/0.38208) - (65.41/r^6)$	
Φ_{OO}	=	$11262.6 \exp(-r/0.1363)$	$r < 1.2$
Φ_{OO}	=	fifth-order polynomial	$1.2 < r < 2.1$
Φ_{OO}	=	seventh-order polynomial	$2.1 < r < 2.6$
Φ_{OO}	=	$= 134.0/r^6$	$r < 2.6$

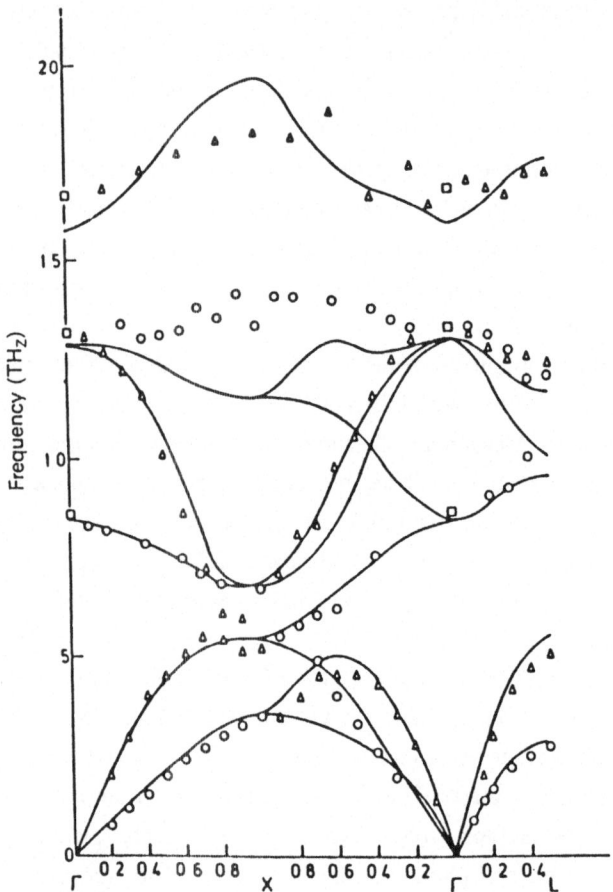

Figure 1.   Phonon dispersion curves for UO_2 at 298K. The solid line shows the calculation. Experimental points are indicated by squares, triangles and circles.

Table 3.   Calculated and experimental lattice properties at 298 K (see Jackson et al. (1986) for details). The calculated values were found using the parameters given in Table 1.

Property	Experimental	Calculated
C_{11} (10^{11} dyne cm^{-2})	38.93	40.06
C_{12} (10^{11} dyne cm^{-2})	11.87	12.94
C_{44} (10^{11} dyne cm^{-2})	5.97	7.22
ε_0	24	21
ε_∞	5.3	5.78

In general, the agreement between theory and experiment is reasonable for both perfect- and defect-lattice properties, although there are definite inadequacies in the calculations regarding the optical phonons and the cation migration energies. Such behaviour is typical for binary oxides, where pair potential models are used.

Table 4. Basic defect parameters in UO_2. The calculated values are from Jackson et al. (1986), the experimental enthalpies are discussed by Catlow (1987).

Defect	Calculated Enthalpy (eV)	Experimental Enthalpy (eV)
Anion Frenkel pair	4.7	3.5 - 4.5
Schottky trio	11.3	~8
Anion vacancy migration	0.6	~0.5
Cation vacancy migration	6.0	~2.5

The inadequacies of pair potential models become increasingly marked as the modelling methods are extended to more covalent materials. These deficiencies are manifested in the following ways:

(i) Inadequate crystal structures may be calculated, as has been found in several studies of silicates using pair potentials.

(ii) Poor elastic constants may be obtained. This is most obviously seen in cubic rock-salt structured materials such as MgO and AgCl where the Cauchy condition requires that $C_{44} = C_{12}$ for a structure at equilibrium in which only central force pair potentials are operative. In practice it is observed that $C_{44} > C_{12}$ for MgO, and the Cauchy condition is violated in the reverse sense for AgCl.

(iii) Inaccurate phonon dispersion occurs especially in the optic branches, as encountered in studies using two body models of SiO_2 (Sanders, 1984) and of AgCl (Baezold et al., 1989).

These problems have stimulated the development of the many-body potential models to be discussed in the next section.

MANY BODY MODELS

Several studies of halide materials attempted to incorporate many body terms using charge transfer models; a good review is available from Singh (1982). However, there has to date been little use of such models in computer modelling studies and we shall therefore concentrate on the following approaches:

(i) Use of bond harmonic terms, employing simple angle dependent energy terms of the type

$$E(\theta) = 1/2 \, k \, (\theta - \theta_0)^2 ,$$

which are applied to specified types of bond angles, e.g., O-Si-O angles in SiO_2 (where θ_0 is the tetrahedral angle). Such methods are obviously most appropriate when modelling systems with an appreciable degree of covalence.

(ii) Use of 'tripole-dipole" terms of the type orginally developed by Axilrod and Teller (1943) by applying 3rd order perturbation theory to the dispersive interaction. Such interactions have the following functional form:

$$E_{TD} = \frac{V_{ijk}(1 + 3\cos\theta_1 \cdot \cos\theta_2 \cdot \cos\theta_3)}{r_{ij}^3 \, r_{jk}^3 \, r_{ik}^3}$$

where i, j, and k are the triplets of atoms or ions and θ_1, θ_2, and θ_3 are the internal angles of a triangle with sides r_{ij}, r_{jk}, and r_{ik}. Such potentials are probably most appropriate in modelling materials comprising ions with high polarizabilities, e.g., the silver halides.

We note that polarizability may be considered to include many body contributions, which are, however, automatically included in treatments such as the shell model.

APPLICATIONS

We will consider three types of materials where many body potentials have been used. The first concerns crystalline and glassy SiO_2 and a range of silicate systems; the second are the newly discovered high T_c oxides; and the third the silver halides which continue to pose challenging problems in the field of interatomic potential development.

Crystalline Silicates and Aluminosilicates

(a) α-quartz

The incorporation of a 3 body bond-bending term into the potential model (see the previous section) enabled a potential to be derived which successfully reproduced the lattice properties of α-quartz (Sanders, Leslie & Catlow, 1984). In addition, this potential gave good agreement with the measured phonon spectrum, and reproduced the variation of the Si-O-Si angle with pressure. Table (5) gives the potential, and the observed and calculated crystal properties.

(b) Zeolites

Zeolites are complex framework structured aluminosilicates with important industrial applications, including catalysis. Recent calculations have shown that the potential derived from α-quartz can be successfully applied to zeolites. A detailed study of Na^+ zeolite A (Jackson and Catlow, 1988) compared experimental and calculated bond lengths and bond angles, and in general, good agreement was obtained. In view of this success, calculations are now being carried out to make predictions about relative stabilities of different zeolite structures. As well as considering the effect on the stability of different structural forms, the effect of variation in the Si/Al ratio can be considered. This is an important factor in zeolite synthesis, in that it is often desired to synthesize a zeolite with a particular Si/Al ratio. It has been observed experimentally that zeolites show a maximum Si/Al ratio, and stability calculations agree well with this maximum (Ooms, van Santen, den Ouden, Jackson and Catlow, 1988). Figure (2) compares measured and calculated maximum Si/Al ratios for a range of zeolites.

Table 5. Potentials and calculated properties of α-quartz. The shell displacements are relative to the core.

Potential parameters and oxygen core/shell displacements

Parameters	Si-O Potential	O-O Potential
A (eV)	1283.9	22764.3
ρ (Å)	0.3205	0.149
C (eVÅ^6)	10.6616	27.88
Oxygen core charge/ lel	+0.8482	
Oxygen shell charge/lel	-2.8482	
Core/shell spring constant (eVÅ^{-2})	74.9204	
x Shell displacement (Å)	+0.0997	
y Shell displacement (Å)	-0.0302	
z Shell displacement (Å)	-0.0666	
Bond-bending constant (eV rad^{-2})	2.097	

Observed and calculated crystal properties of α-quartz

Experimental		Bond-bending Potential	Two-body Potential
Elastic constants (10^{11} dyn cm^{-2})			
c_{11}	8.683	8.815	6.204
c_{33}	10.498	10.605	7.466
c_{44}	5.826	5.296	3.301
c_{66}	3.987	4.269	2.737
c_{14}	-1.8064	-1.666	-1.012
c_{13}	1.193	1.151	1.629
Static dielectric constants			
ε_{11}	4.520	4.452	5.513
ε_{33}	4.640	4.812	6.086
High-frequency dielectric constants			
	2.4	2.04	2.069

Vitreous Silica

Numerical simulation methods give unique information on the structure of glasses and can be used to test approximate theories. Furthermore they can be used to determine certain properties of the glasses which cannot be deduced from experiment e.g., triplet distributions.

In the past few years several glasses including Lennard Jones systems, BeF_2, B_2O_3 and SiO_2 have been simulated, and reasonable success has been enjoyed in reproducing the radial distribution functions (RDF) of the respective glasses studied. However, the bond angle distributions (BAD) have been far broader than those predicted experimentally. This has been attributed to the lack of terms in the potential model representing covalent bonding in glasses under consideration.

173

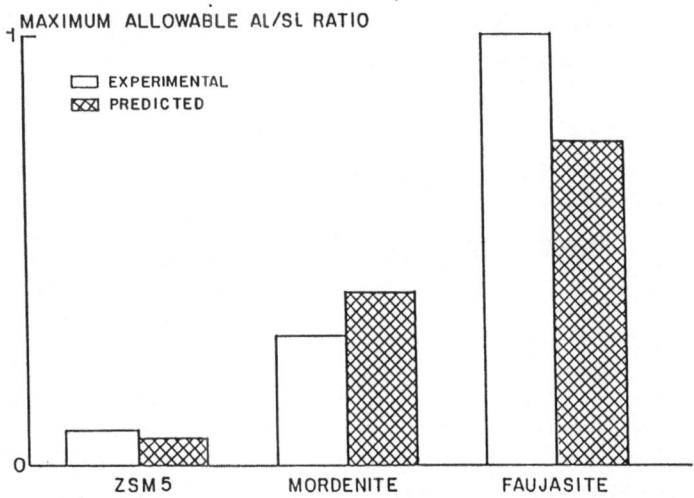

Figure 2. Experimental and predicted maximum allowed Al/Si ratios.

Vessal, Leslie and Catlow (1989) have, using M.D. methods, studied the effect of three body forces on the structure of vitreous silica. In this work we have developed a potential model that gives good structure and experimental properties for α-quartz, and the correct structure for vitreous silica (see Table 6).

The form of the three body potential used is slightly different from that employed previously and takes the form:

$$E_{ijk} = 1/4 \, A_{ijk} \, B_{ijk}^2 \exp(-r_{ij}/\rho_1) \exp(-r_{ik}/\rho_2)$$

where

$$A_{ijk} = k_{ijk}/2 \, (\theta_0 - \pi)^2$$

$$B_{ijk} = (\theta_0 - \pi)^2 - (\theta - \pi)^2$$

k_{ijk} is the three body spring constant and θ_0 is the equilibrium bond angle, i.e., 109°28' for O-Si-O interactions; θ is the calculated bond angle; r_{ij} is the distance between i and the nearest image of j, and ρ_1 and ρ_2 are constants.

Recently, Vessal, Amini, Fincham and Catlow (1989) have successfully undertaken constant pressure simulations of the melting of β-cristobalite and glass formation from the SiO_2 melt using the same potential model. They have improved the structure of the glass obtained by restricting the number of three body interactions to four nearest neighbors for each silicon and simulating at a constant pressure of -40 kbar. It has been shown in their study that three body interactions are indeed essential in modelling vitreous silica especially for getting an accurate BAD.

Table 6.   Comparison of the experimental and simulated vitreous silica structure.

	First Peak $g_{Si-O}(\text{Å})$	Second Peak
Simulated	1.62	4.12
Observed	1.62	4.15
	$g_{O-O}(\text{Å})$	
Simulated	2.65	5.12
Observed	2.65	5.1
	$g_{Si-Si}(\text{Å})$	
Simulated	3.2	5.1
Observed	3.12	5.1

La_2CuO_4

The high T_c superconductors La_2CuO_4 and $Ba_2YCu_3O_7$ have received a great deal of attention recently. However, the defect chemistry and the nature of the charge carriers are still not fully understood. Atomistic simulation procedures yield valuable information about both these problems, which is often unavailable from other theoretical and experimental studies. Little dielectric and elastic data are available for La_2CuO_4. It was therefore necessary to take potential parameters from separate studies of the structure and properties of the binary oxides La_2O_3 and CuO.

The structure of La_2CuO_4 is shown in Figure 3. It may be described as containing alternate layers of perovskite structured $LaCuO_3$ units and rock-salt structured LaO units along the c-axis. The CuO_6 octahedron of the perovskite structured component shows a strong tetragonal elongation. The equatorial Cu-O bond length is 1.907 Å, whereas the axial Cu-O bond length is 2.459 Å. Such a distortion is also shown in CuO, which is itself based on CuO_6 octahedra. In order to reproduce this distortion we had to extend the potential model to include three-body bond-bending interactions (Leslie, 1985). These were applied around the O-Cu-O and O-O-Cu bond angles as shown in Figure 4. Details of the potential model, and the parameters, are reported elsewhere (Islam et al., 1988). Upon equilibration of the La_2CuO_4 structure by energy minimization, the maximum deviation of calculated bond length from that observed is 0.1Å, for the lanthanum to axial-oxygen bond. This error could be eliminated by refinement of the La-O potential. The discrepancy is, however, small and our potential model is clearly an accurate representation of the observed structure. Future refinement will be carried out when more accurate crystal properties are available for this system.

One of the most rigorous tests of a potential model is to calculate phonon dispersion curves for the material. We have done so for a variety of wavevectors (see Figure 5) and found all modes to be positive, which confirms that our potential model yields a stable structure. Moreover, for the $(0,\varepsilon,0)$ direction of the orthorhombic C-centered cell, we find a softening of one of the low-lying optical modes, which we show in Figure 3. This agrees well with the behavior measured by Birgeneau et al. (1987) and is also of interest because phonon mode softening may mediate the favourable lattice-conduction species interaction, which results in superconductivity.

The potential model has been used in studies of defects and hole states in this material (see Islam et al., 1988; Catlow et al., 1988).

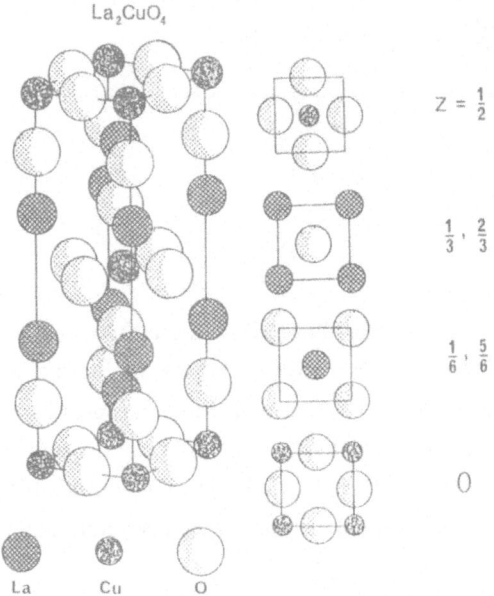

Figure 3.   The structure of La_2CuO_4.

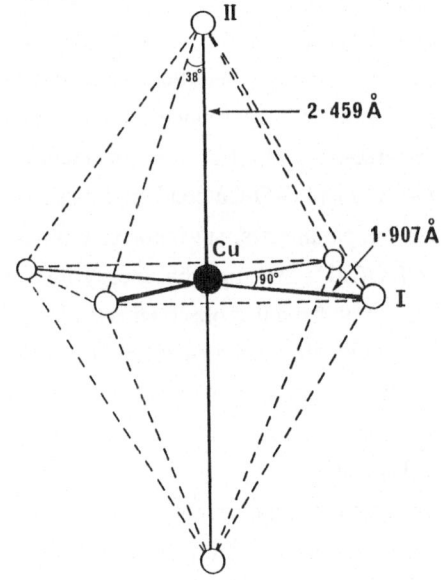

Figure 4.   Details of the structure of La_2CuO_4 showing the Cu position.

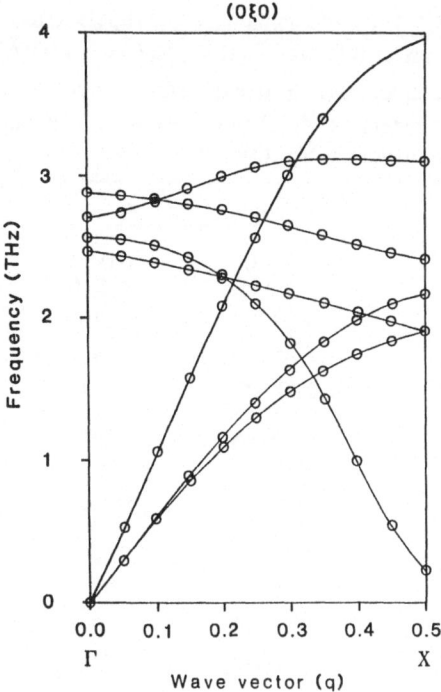

Figure 5. Calculated phonon dispersion curve along the (0 ξ 0) direction for La$_2$CuO$_4$.

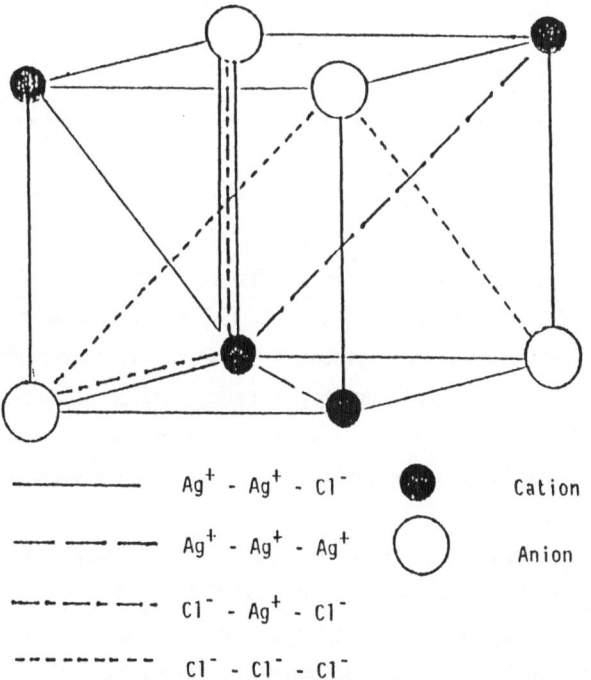

Figure 6. Three body interactions for the triple-dipole model.

Table 7. Potentials derived including triple-dipole terms. The units of each parameter are as follows: A_{ij} (eV), ρ_{ij} (Å), V_{ij} (eVÅ^6) in the Buckingham potential, $\Phi(r) = A_{ij} \exp(-r/\rho_{ij}) - V_{ij}/r^6$; the units of the shell parameters are Y_i (e) and k_i (eVÅ^{-2}). The units of the triple-dipole parameter are V_{ijk} (eVÅ^9).

| | Silver Chloride | | Silver Bromide |
	(I)	(II)	(I)
A_{11}	16528.0	16528.0	16528.0
ρ_{11}	0.2370	0.2370	0.2370
V_{11}	336.0	336.0	268.0
A_{12}	2080.27	2416.4796	4943.00
ρ_{12}	0.33745	0.33017	0.30120
V_{12}	275.0	275.0	228.3897
A_{22}	1227.2	1227.2	2948.20
ρ_{22}	0.3214	0.3214	0.332
V_{22}	75.0	75.0	248.0
V_{212}	15.10933	15.10933	0.0
V_{211}	379.3331	379.3331	7.23312
V_{111}	41.61966	41.61966	100.0303
V_{222}	2996.3194	2996.3194	6764.535
k_1	3000.0	4500.0	4109.339
k_2	29.38	29.38	24.66
Y_1	-13.396	-19.396	-25.967
Y_2	-2.485	-2.485	-2.705

NOTE: $1 = Ag^+$; $2 = Cl^-$ or Br^-)

Silver Halides

Several studies of interatomic potentials for AgCl and AgBr have been reported (see e.g., Catlow et al., 1979; Devlin and Corish, 1987). Until recently these had used exclusively two-body models. These potentials had some success in modelling the unusual defect properties of the materials; but they necessarily fail to describe the large Cauchy violation ($C_{12} > C_{44}$) and they generate poor optic phonon branches. Inadequacies are also found in the calculated defect properties especially regarding the activation energies for Ag migration.

Recently, Baezold et al. (1989) have attempted to obtain improved potentials for these systems by including both bond-bending and triple-dipole terms. The latter which are almost certainly more appropriate for these materials have enjoyed great success. Triple dipole terms were included for the following ion triplets $Ag^+...Ag^+...Ag^+$; $X^-...X^-...X^-$ ($X = Cl$ or Br); $X^-...Ag^+...X^-$; $X^-...Ag^+...Ag^+$, as illustrated diagramatically in Figure 6.

Parameterization was effected by empirical fitting procedures and the resulting parameters are reported in Table (7). We note that two potentials were derived for AgCl.

The calculated elastic constants reported in Table (8) show a clear improvement over the two-body models. The large Cauchy violation is modelled although the calcualted value of C_{44} is still rather high. Significant improvements were also found in the calculated defect energies, although it is still not possible to model accurately cation migration processes. The improvements in the phonon dispersion curves are, however, only marginal.

The remaining inadequacies are almost certainly due to the omission from our models of any representation of the quadrupolar deformativity of the Ag^+ ion - an effect discussed previously by Fischer et al. (1972) and by Kleppmann and Bilz (1976), and whose importance in modelling cation migration in AgCl was shown by Jacobs et al. (1980). Work in progress is including these effects in a systematic way.

Table 8. Comparison of calculated and experimental values of elastic and dielectric constants and polarizabilities for potentials AgClI, AgClII and AgBrI.

	Silver Chloride			Silver Bromide	
	Expt.	(I)	(II)	Expt.	(I)
Elastic Constants					
C_{11}	7.59	7.145	7.546	6.60	7.297
C_{12}	3.908	3.664	3.650	3.49	3.198
C_{44}	0.6894	1.159	1.162	1.00	1.224
Relative Permittivity					
ε_s	9.50	9.187	9.404	10.60	10.922
ε_∞	3.97	2.738	2.999	4.68	4.692
Bulk Lattice Strain	0.0	-0.00248	-0.0013	0.0	0.00449
Polarizabilities					
α_1	2.37	0.861	1.204	2.44	2.363
α_2	2.97	3.023	3.027	4.13	4.273

SUMMARY AND CONCLUSION

The inclusion of many body terms in potentials is greatly extending the range of materials that can be studied using modelling techniques. Bond-bending and triple-dipole functions can be readily incorpated into modelling codes and have proved to be successful in several studies. Future work will concentrate on refining our treatment of these terms and on including more subtle effects such as quadrupolar deformation.

ACKNOWLEDGMENTS

We would like to thank the following with whom we have collaborated in the work summarized in this paper: S.M. Tomlinson, M.S. Islam, D. Fincham, M. Amini, M. Leslie, R. Baezold, Y. Tan, J. Corish, P.W.M. Jacobs, F. Healy, C.J.J. den Ouden.

REFERENCES

Axilrod, B.M. and Teller, E., 1943, J. Chem. Phys., 11:299.

Baezold, R. et al., to be published, J. Phys. Chem. Solids.

Birgenau, R.J., Chen, C.Y. Gabbe, D.R., Jensen, H.P., Kastner, M.A., Peters, C.J., Picone, P.J., Thio, T., Thurston, T.R., Tuller, H .L., Axe, J.D., Boni, P. and Shirance, G., 1987, Phys. Rev. Lett. 59:1329.

Catlow, C.R.A., 1986, Ann. Rev. Mat. Sci., 16:517.

Catlow, C.R.A. and Cormack, A.N., 1987, Int. Rev. Phys. Chem., 6:227.

Catlow, C.R.A., Freeman, C.M., Islam, M.S., Jackson, R.A., Leslie, M. and Tomlinson, S.M., 1988, Phil. Mag. A, 58:123.

Catlow, C.R.A. and Stoneham, A.M., 1983, J. Phys. C., 16:4321.

Catlow, C.R.A., 1977, Proc. R. Soc. Lond. A., 353:533.

Catlow, C.R.A., (ed.)., 1987, J. Chem. Soc. Far. Trans. II, 83.

Catlow, C.R.A., Cox, P.A., Jackson, R.A., Parker, S.C., Price, G.D., Tomlinson, S.M. and Vetrivel, R, 1989, Molecular Simulation, in press.

Catlow, C.R.A., Corish, J., Diller, K.M., Jacobs, P.W.M. and Norgett, M.J., 1979, J. Phys. C., 12:451.

Cormack, A.N., 1987, to be published.

Devlin, B.A. and Corish, J., 1987, J. Phys. C., 20:705.

Dick, B.G. and Overhauser, A.W., 1958, Phys. Rev. B, 112:90.

Fischer, K., Bilz, H., Haberkorn, R., and Weber, W., 1972, Phys. Stat. Solidi b54:285.

Fowler, P.W. and Pyper, N.C., 1985, Proc. R. Soc. Lond. A., 398:377.

Gordon, R.G. and Kim, Y.S., 1972, J. Chem. Phys., 56:3122.

Islam, M.S., Leslie, M., Tomlinson, S.M. and Catlow, C.R.A., 1988, J. Phys. C., 21:L109.

Jackson, R.A., Murray, A.S., Harding, J.H. and Catlow, C.R.A., 1986, Phil. Mag. A., 53:27.

Jackson, R.A.. and Catlow, C.R.A., 1988, Molecular Simulation, 1:207.

Jacobs, P.W.M., Corish, J. and Catlow, C.R.A., 1980, J. Phys. C, 13:1977.

Kleppmann, W.G. and Bilz, H., 1976, Commun. Phys., 1.

Leslie, M., 1985, Physica B, 131:145.

Mackrodt, W.C., Colbourn, E.A. and Kendrick, J., 1981, Report No. CL-R1/81/1637A, ICI Corporate Laboratory.

Sanders, M.J., Leslie, M. and Catlow, C.R.A., 1984, J. Chem. Soc. Chem. Comm., 1273.

Singh, R., 1982, Rev. Mod. Phys. 12:358.

Vessal, B., Leslie, M. and Catlow, C.R.A., 1989, Molecular Simulation, in press.

Vessal, B., Amini, M., Fincham, D. and Catlow, C.R.A., Phil. Mag., to be published.

Wedepohl, P.T., 1967, Proc. Phys. Soc. 92:79.

EMBEDDED ATOM METHOD:

MANY-ATOM DESCRIPTION OF METALLIC COHESION

Murray S. Daw

Theoretical Division
Sandia National Laboratories
Livermore, CA 94550

INTRODUCTION

Daw and Baskes[1,2] have proposed a new framework for calculating the energetics of metals which they call the Embedded Atom Method (EAM). In this approach, the energy of the metal is viewed as the energy to embed an atom into the local electron density provided by the remaining atoms of the system. In addition, there is an electrostatic interaction. The ansatz they used is

$$E_{coh} = \sum_i F_i \left(\sum_{j \neq i} \rho_j^a(R_{ij}) \right) + \frac{1}{2} \sum_{i,j(j \neq i)} \phi_{ij}(R_{ij}) \tag{1}$$

where F is the embedding energy, ρ^a is the spherically averaged atomic electron density, and ϕ is an electrostatic, two-atom interaction. The background density for each atom in Eq. 1 is determined by evaluating at its nucleus the superposition of atomic density tails from the other atoms. Equation 1 combines the computational simplicity needed for large-scale systems with a physical picture which includes many-atom effects and avoids the ambiguities of the pair potential scheme. This method has been applied successfully to such problems as phonons[3], liquid metals[4], defects[1,2,5], alloys[5], impurities[1,2], interdiffusion in alloys[6], fracture[1,7], surface structure[8-10], surface adsorbate ordering[11,12], surface segregation[13], surface order-disorder transitions[14], surface ordered alloys[15], and surface phonons[16]. The computational requirements of the EAM are not significantly more than that required for pair potential calculations.

Several years ago, J. Friedel[17] related the solution of hydrogen in copper to the solution of hydrogen in jellium. This work was recently extended by the independent work J. Nørskov and N. Lang (Effective Medium[18]) and of M. J. Stott and E. Zaremba (Quasiatom[19]). Nørskov and Lang showed that the heat of solution of a light, interstitial impurity (e.g., H and He) in metals could be calculated by replacing the host with a suitable

181

effective medium, which in this case was jellium. Nørskov and his co-workers have had great success in calculating from first principles the heats of solution[20] and heats of chemisorption[21] of hydrogen in metals, the primary information being the energy gained by putting a hydrogen atom into jellium (i.e., the embedding energy). The optimal density of the jellium was determined by weighting the background metallic density by the Hartree potential of the impurity. Stott and Zaremba arrived at a similar concept, based on viewing the impurity as a quasiatom in a nearly uniform electron gas.

Daw and Baskes[1] made a significant generalization with the EAM by proposing to view the cohesive energy of a metallic solid as comprised of the embedding energy plus electrostatic interactions. In this view, each atom in the metal is embedded into the electron gas created by the other atoms. Atoms near a defect such as a surface are embedded into an electron gas of different density profile than atoms in the bulk. The energy is apportioned so that the cohesive energy is manifestly symmetrical in the atomic index. They suggested the ansatz (Eq. 1) which will be discussed here. They then obtained the functions empirically by fitting to properties of the bulk metals. The generality of the functions was tested by applying them to surfaces and other defects. This generalization allowed calculations of complex metallic structures to be done within the approximate embedding energy framework. The EAM is thus a significant improvement in simplified total energy calculations for metallic systems.

Even more recently, Jacobsen, Nørskov, and Puska[22], Manninen[23], and Kress and DePristo[24] re-examined the ansatz used in the EAM with arguments based on the effective medium approach. Jacobsen, *et al.* demonstrated how the cohesive energy of a metallic system could be related to the embedding energies, with corrections accounting for the d-d hybridization in the transition metals. Their approach showed that with the neglect of the d-d hybridization (valid for simple metals and also presumably for early and late transition metals), the EAM expression is recovered. The density of the effective medium was taken to be an unweighted average of the background density over the Wigner-Seitz cell of the atom. Kress and DePristo suggested using as a weighting function the electron density of the atom itself, so that the background density is related to the overlap of charge densities. A correction to the embedding function based on a local energy functional was used in their Corrected Effective Medium Method to correct for inaccuracies in the definition of the optimal effective medium density.

Several other methods, equivalent to the EAM, have been proposed since the original work of Daw and Baskes. These methods all take the form of Eq. 1, with differences due to specific parameterizations or functional forms. Finnis and Sinclair[25] proposed that the d-d hybridization in the second moment approximation could be described by Eq. 1. Their "N-body" potential has been applied to defects in BCC metals. The Finnis-Sinclair model is formally equivalent in the case of homogeneous metals to the EAM, though the physical motivation is quite different. For alloys, however, the Finnis-Sinclair model leads to a different form from the EAM. The "local volume forces" represents an alternative parameterization of the EAM by Voter and Chen[26], and has been applied to surface relaxations and grain boundaries in binary metallic alloys. The "glue model" of Ercolessi, Tosatti, and Parrinello[27] also belongs to the EAM class of models, and has been applied mainly to surface reconstructions.

In this paper, we discuss several points of interest, from a heuristic derivation to a review of applications. The derivation gives us a feeling for the physical origin of the functions and their basic form. The theoretical basis is then used to justify functional forms used in the semi-empirical fitting. These semi-empirical functions are then applied to a large variety of important problems.

THEORETICAL BASIS

Our goal is to derive an approximate expression for the cohesive energy of a metallic system which is an explicit function of the positions of the atoms and which is simple to evaluate (i.e., Eq. 1). This has been discussed in detail in Ref. 28, and we will review the highlights here. We start with the electron density of the solid, $\rho(r)$, and the energy functional of the electron density, $E[\rho]$[29]. We will assume that the kinetic, exchange, and correlation contributions to the energy, $G[\rho]$, are semi-local:

$$G[\rho] = \int g\,(\rho(r), \nabla \rho(r), \nabla^2 \rho(r), \ldots) \tag{2}$$

where the integral is understood here and elsewhere to be over real space. The cohesive energy of a solid is then:

$$E_{coh} = G[\rho] + \frac{1}{2} \sum_{i,j}' \frac{Z_i Z_j}{R_{ij}} - \sum_i \int \frac{Z_i\,\rho(r)}{|r - R_i|} + \frac{1}{2} \int \int \frac{\rho(r_1)\rho(r_2)}{r_{12}} - E_{atoms} \tag{3}$$

where the sums over i and j are over the nuclei of the solid, the primed sum indicates omission of the i=j term, Z_i and R_i are the charge and position of the i-th nucleus, and the integrals are over r (or r_1 and r_2). E_{atoms} is the energy of the system when the atoms are removed from each other.

In order to make Eq. 3 useful, we will express the electron density ρ explicitly in terms of the positions of the nuclei. We start by assuming that the electron density of the solid can be described as a linear superposition of the densities of the individual atoms ($\rho_s(r) \equiv \sum_i \rho_i^a(r - R_i)$). (The effect of charge redistribution is treated in detail in Ref. 28.) Substituting $\rho(r) = \rho_s(r)$ into Eq. 3 gives:

$$E_{coh} = G[\sum_i \rho_i^a] - \sum_i G[\rho_i^a] + \frac{1}{2} \sum_i' U_{ij}^a \tag{4}$$

with $U_{ij}^a \equiv \int \int \dfrac{n_i^a(r_1) n_j^a(r_2)}{r_{12}}$ and $n_i^a(r) \equiv \rho_i^a(r - R_i) - Z_i\,\delta(r - R_i)$. Th

The first two terms in Eq. 4 involve the difference in kinetic, exchange, and correlation energies in going from the case of isolated atoms to the solid. The last term is the electrostatic energy of the overlapping charge distributions.

Consider the region around atom i. The electron density near atom i can be viewed as the sum of the density of atom i, ρ_i^a, and a background density due to the surrounding atoms. Let us define the background density for atom i to be $\rho_{b,i}(r) \equiv \sum_{j \neq i} \rho_j^a(r - R_j)$. In most of this region, the density ρ_i^a dominates in ρ_s; i.e., $\rho_{b,i}(r)$ is small and slowly varying compared to ρ_i^a. Thus it seems plausible to approximate $\rho_{b,i}(r)$ by a constant $\bar{\rho}_i$. Let us therefore define the embedding energy for an atom in an electron gas of some constant density $\bar{\rho}$ (neutralized by a positive background):

$$G_i(\bar{\rho}_i) \equiv G[\rho_i^a + \bar{\rho}_i] - G[\rho_i^a] - G[\bar{\rho}_i] \tag{5}$$

Using the embedding energy, we can rewrite the cohesive energy as:

$$E_{coh} = \sum_i G_i(\bar{\rho}_i) + \frac{1}{2} \sum_{i,j} U_{ij}^a + E_{err} \tag{6}$$

where the error is

$$E_{err} = G[\sum_i \rho_i^a] - \sum_i G[\rho_i^a + \bar{\rho}_i] + \sum_i G[\bar{\rho}_i]$$

$$= \int_\infty [\ g(\sum_i \rho_i^a) - \sum_i g(\rho_i^a + \bar{\rho}_i) + \sum_i g(\bar{\rho}_i)\ ] \tag{7a}$$

Setting $E_{err} = 0$ provides a definition for the optimal background densities $\bar{\rho}_i$. Let us break up the integral into a sum of integrals over the regions around each atom, and set each individual integral to zero. For the sake of discussion, let us consider only the dominant part (in Ref. 28, the full solution is considered). This dominant part can be written

$$\int_{\Omega_i} g(\rho_i^a + \rho_{b,i}) \approx \int_{\Omega_i} g(\rho_i^a + \bar{\rho}_i) \tag{7b}$$

and this expresses the relationship between $\bar{\rho}_i$ and $\rho_{b,i}$ more directly.

In the limit that the background density $\rho_{b,i}$ is very slowly varying compared to ρ_i^a, then $\rho_b(r) = \rho(0)$ and the solution to Eq. 7b is $\bar{\rho} = \rho(0)$. If we assume that the background density is largely constant with small variations, we can approximate the background density $\rho_b(r) \approx \rho(0) + \mathbf{r} \cdot \nabla \rho(0) + \frac{1}{2} \mathbf{rr} \cdot \cdot \nabla \nabla \rho(0)$, and expand in a Taylor's series involving powers of $\nabla \rho(0)$ and $\nabla \nabla \rho(0)$. We find that the lowest order corrections are

$$\bar{\rho} = \rho(0) + \alpha \nabla^2 \rho(0) + \beta |\nabla \rho(0)|^2 \tag{8}$$

with α and β being constants depending on the atom i and the function g. This relation is only approximate, due to the breakdown in the Taylor's series expansion near the edge of each atomic cell. However, Eq. 8 implies that the dominant contribution to the optimal background density is the density near the origin, with small corrections due to the gradient and curvature of the density.

In summary, the cohesive energy can be written $E_{coh} = \sum_i G_i(\bar{\rho_i}) + \frac{1}{2} \sum_{i,j} U_{ij}^a$,

where U contains the two-atom electrostatic interactions and the embedding function is given by $G_i(\bar{\rho_i}) \equiv G[\rho_i^a + \bar{\rho_i}] - G[\rho_i^a] - G[\bar{\rho_i}]$. The constant background density is given

by solving $E_{err} = 0$. This derivation is based on the assumption that the charge distribution in the solid is not very different from a superposition of atomic electron densities; thus this form is restricted to simple metals and early or late transition metals.

In Ref. 28, the procedure outlined in this section was tested on a model system. A model energy functional based on the Thomas-Fermi-Dirac-von Weizsacker kinetic energy was developed to describe FCC nickel quite well. Within this model, the embedding function and electrostatic interactions were obtained, and the equation for the optimum

electron density was solved in detail. A simple, linear relationship between $\bar{\rho}$ and $\rho(0)$ was found to hold, with some error attributed to gradient corrections as in Eq. 8. In the semi-

empirical EAM, the assumption has been to set $\bar{\rho} = \rho(0)$.

This completes the heuristic derivation of the EAM.

MANY-ATOM INTERACTIONS

A measure of the energetics of defects can best be obtained by calculating changes in the total energy due to distortions in the system. For small distortions, the change in the energy within the EAM is equivalent to the change in the sum of effective two- and three-atom interactions[4]:

$$\psi_{ij}(R) \equiv \frac{1}{2} \{ [U_{ij}(R) + 2G_i'(\bar{\rho_i}) \rho_j^a(R) + G_i''(\bar{\rho_i}) \rho_j^a(R)^2] + [(i \leftrightarrow j)] \} \tag{9a}$$

$$\chi_{ijk}(R_i, R_j, R_k) \equiv \{ G_i''(\bar{\rho_i}) \rho_j^a(R_{ij}) \rho_k^a(R_{ik}) + (jki) + (kij) \} \tag{9b}.$$

Higher-body interactions can also be obtained. The interactions in Eqs. 9 demonstrate that the effective interactions within the EAM are *environment-dependent*, in that the interaction between two atoms depends on the slopes of their embedding functions, which depend on the $\bar{\rho}$ for each atom. As one goes to the surface, the effective two-atom interaction becomes stronger and the bond-length shortens. Within the EAM, this is a result of the requirement that $G'' > 0$. In this way, the EAM describes the effect of coordination on bond strength: the less-coordinated surface atoms tend to have stronger bonds and shorter bond lengths. This in fact is the strength of the EAM, and is directly connected to the nonlinearity of the embedding function. Of course, this effect is neglected in the traditional pair potential formulation.

SEMI-EMPIRICAL IMPLEMENTATION

The embedding function and pair interaction can be obtained from first principles, as is demonstrated in Ref. 28 with a model density functional for nickel (also see Ref. 22, for a different treatment). These first principles functions do reasonably well in describing the properties of nickel. However, in practice, we have taken a semi-empirical approach, where the fundamental theory has guided the fitting of the embedding function and pair interaction to basic bulk properties. In general, we have fitted to lattice constant, cohesive energy, elastic constants, and vacancy formation energy. We have also found it useful to require that the cohesive energy as a function of lattice constant follow the "universal binding curve" of Ref. 30. In one paper, we have also fitted to the dilute heats of alloying for binary alloys. In general, the fits are overdetermined, and the fact that we can achieve any kind of fit is in some sense a check of self-consistency. Functional forms have varied: in some cases, the functions are determined by general splines, in other cases, we have found it more useful to use analytic functions.

There is one fact to be aware of in doing the fitting. The following transformation does not change the total energy:

$$F_i(\rho) \rightarrow F_i(\rho) - c_i\rho$$
$$\phi_{ij}(R) \rightarrow \phi_{ij}(R) + c_i\rho_j^a(R) + c_j\rho_i^a(R).$$

Therefore, the embedding function cannot be determined to within a linear function. This is related to the fact that some of the electrostatic interaction can be grouped either as part of the pair term or as part of the embedding energy: the choice is arbitrary. This ambiguity also confuses the comparison of functions obtained by different workers. Some workers have required that $\phi(R)$ be positive definite, which amounts to a particular choice of the constants in the transformation. Other workers have required that the embedding functions correspond to those obtained from atom-in-jellium calculations, which amounts to a different choice of the constants. However, the different choices of constants are arbitrary: very different looking sets of EAM functions may in fact give exactly the same energetics. Instead of comparing embedding functions and pair interactions directly, one should compare the effective pair and trio interactions (Eqs. 9a and b). These effective interactions are unaffected by the transformation, and therefore provide a way of comparing rather different looking sets of EAM functions.

In obtaining new functions, or in applying existing ones, there are several traps to avoid:

(1) The functions should be smooth.

(2) The embedding function should have positive curvature. This ensures that the bond strength decreases with increasing coordination (see the section on many-atom interactions). This is a basic property of embedding functions, as can be demonstrated from fundamental principles[19,20,22,28].

(3) A direct consequence of the previous point is that metals with negative Cauchydiscrepancy evidently have some basic properties which are neglected in the EAM.The approximations used in justifying the EAM break down in these cases.

(4) Several workers[2,5] have used the restriction of $\phi_{ij} = \sqrt{\phi_{ii}\phi_{jj}}$. This was usedonly to restrict the number of free parameters and is generally not a good approximation.

(5) Functions which have been fitted to the properties of two elemental metals with no regard for the alloying properties will be very lucky to describe the alloy properly. For example, Daw and Baskes[2] give functions for Ni and functions for Pd. These functions were fitted to the elemental properties but do not describe the Ni-Pd system well. On the other hand, Foiles, Baskes, and Daw[5] fitted the dilute heats of alloying and do well for Ni-Pd.

(6) Avoid covalent materials, or ones with negative Cauchy discrepancy. The EAM does not describe the basic physics related to these properties.

APPLICATIONS

We have implemented the EAM in four different types of calculations: (1) energy minimization, (2) molecular dynamics, (3) Monte Carlo, and (4) vibrational normal mode analysis. The current practical limitations of EAM calculations on a CRAY-XMP/24 allow molecular dynamics simulations of up to 35,000 atoms. Simulated times for molecular dynamics typically run up to tens of picoseconds. For typical problems, the calculational cost scales mostly like the number of atoms. Monte Carlo simulations based on the EAM are in practice capable of typically 10^5 iterations per atom. These simulations are most useful for "annealing" structures and obtaining transition temperatures or equilibrium concentration profiles. Vibrational normal mode calculations are currently limited by memory to unit cells of less than 200 atoms.

We have applied the EAM to a large variety of problems involving intrinsic defects and surfaces of metals. From the EAM we obtain information on the structure, dynamics, phase transitions, and phonons. We review briefly some of the applications in the literature.

Various properties of the bulk other than those used in the fitting have been calculated to test the range of applicability of the EAM. The lattice dynamical matrix was derived in Ref. 3 for bulk phonons in transition metals. The comparison with experiment showed that the EAM provides a reasonable description of elementary vibrational excitations in solids. The agreement near the zone center is to be expected, because the functions are fitted to the elastic constants. However, agreement over the full zone was obtained as well, and this was not guaranteed by the fitting. The linear coefficients of thermal expansion of FCC metals were computed[31]. The results were in good agreement with experiment. These results demonstrated the importance of including the equation of state in the semi-empirical fitting. Above the melting point, the static structure factor of various liquid transition metals were computed[4]. The results were in good agreement with experimental data. These calculations of basic bulk properties demonstrated that the EAM provides a good description of the interatomic interactions in metals.

A range of surface properties of fcc metals was calculated, again to test the robustness of the EAM. The energetics and structure of several fcc (110) surfaces were calculated[9,10], demonstrating that for Pt and Au this surface was stabilized by a missing row reconstruction. The atomic relaxations were computed and compared in detail to experimental analysis, showing excellent agreement. Experimentally, it is known that the missing-row structure of Au disorders at 650K. EAM/Monte Carlo was used to study the structure and order-disorder transformation of Au and Pt(110) surfaces[14]. It was found that the missing-row structure of Au disordered at 570K, in very good agreement with experiment. Pt(110) was predicted to disorder at 750K, which prompted experiments currently in progress. Relaxations and vibrations were found to be essential for the accurate calculation of the critical temperature, implying that the bulk of the literature of lattice gas calculations must be carefully interpreted. The energies and polarizations of phonons on the clean Cu(100) surface were calculated with the EAM.[16] The atomic relaxations and the dynamical matrix were obtained, from the usual bulk semi-empirical functions, in one consistent calculation. Excellent agreement with experiment was obtained. The force constants in the EAM were found to be qualitatively different than those obtained by fitting two-atom central-potential models, reflecting the many-atom nature of the EAM. Most of these surface calculations were performed on systems of relatively high symmetry. The agreement with the simpler systems gives us benchmarks and some confidence in extending the EAM to more complicated systems. The simplicity of the EAM allows calculations of the properties of much more complicated, reconstructed surfaces than can be approached with any other method.

Internal interfaces have also been investigated with the EAM. Foiles has performed careful calculations of the structure of grain boundaries in gold. The atomic positions calculated for the $\Sigma 13$ (100) twist boundary are in excellent agreement with experiment.[32] The $\Sigma 5$ (210) tilt shows a very interesting reconstruction which doubles one of the periodic lengths in the plane of the boundary.[33] These studies have demonstrated the unique capabilities of the EAM.

Metallic alloys also represents a fruitful ground for testing the EAM and for performing calculations that cannot be reliably investigated with other methods. The EAM was used to calculate phase stability, phonons, point defect properties, antiphase boundary energies, and surface energies and relaxations for Ni_3Al.[34,35] Further calculations revealed an intriguing relationship between bulk stoichiometry and structural and compositional order in grain boundaries. The problem of segregation to the surfaces of binary alloys was treated in a series of papers by Foiles[13,36,37]. The EAM/Monte Carlo simulations provided equilibrium concentration profiles. The results showed that the composition may vary nonmonotonically near the surface, in agreement with experimental studies. Calculations using the EAM showed that Au forms ordered surface layers on the low index faces of Cu[15]. These "ordered surface alloys" were demonstrated to exist in equilibrium with a bulk containing dilute amounts of Au. The results are in excellent agreement with experimental results for the structure of Au deposited on Cu.

Hydrogen in metals has also been a promising area of investigation using the EAM. The phasé diagram of hydrogen on Pd(111) was calculated[11,12], using EAM/Monte Carlo. The symmetry and critical temperatures of ordered phases were predicted in excellent agreement with experiment. A unique feature of the theoretical results was the prediction of occupation of subsurface sites by the hydrogen.

Mechanical properties of metals has been the focus of a series of calculations using the EAM. Molecular dynamics calculations of dislocation dynamics in FCC Ni were performed, showing the motion of an edge dislocation under applied stress.[7] The results were compared quite favorably with continuum elasticity solutions of the same problem. Molecular dynamics studies were also performed of crack propagation in FCC Ni.[1,7] Both brittle and ductile fracture were studied. Hydrogen atoms were added and demonstrated that hydrogen near the crack tip could reduce both the brittle fracture stress and the stress required to generate dislocations from the crack tip.

The wide range of problems to which the EAM has been applied demonstrate that the method describes ground state structural properties of fcc metals quite well. These benchmark calculations now provide us with the confidence to do more complicated systems.

CONCLUSIONS

The Embedded Atom expression for the cohesive energy of a solid can be derived from approximations to density functional theory. The energy can be divided into an embedding energy contribution with an electrostatic, two-atom correction. The procedure defines an optimal constant background density, $\bar{\rho}$, for the embedding function, in terms of the actual background density $\rho_b^i(r)$ at each site. The EAM form is expected to be most applicable to simple metals and early or late transition metals The expressions developed here for the EAM can serve as a basis for the parameterization of the functions used in the semi-empirical technique. The EAM has been tested on a large variety of situations and has proven very versatile in describing interatomic interactions in a metallic environment. The EAM should replace the use of pair potentials in simulations of metallic systems.

ACKNOWLEDGEMENTS

Much of the work on the EAM and applications have been done in collaboration with Drs. S. M. Foiles and M. I. Baskes of this laboratory. Other collaborators have included Mr. C. L. Bisson, Drs. J. B. Adams, J. S. Nelson, E. C. Sowa, and W. G. Wolfer of this laboratory and Prof. R. D. Hatcher of Queens College. This work was supported by the U. S. Department of Energy, Office of Basic Energy Sciences, Division of Materials Science.

REFERENCES

1.   M. S. Daw and M. I. Baskes, Phys. Rev. Lett. 50:1285 (1983)
2.   M. S. Daw and M. I. Baskes, Phys. Rev. B 29:6443 (1984).
3.   M. S. Daw and R. L. Hatcher, Sol. State Comm. 56:697 (1985).
4.   S. M. Foiles, Phys. Rev. B 32: 3409 (1985).
5.   S. M. Foiles, M. I. Baskes, and M. S. Daw, Phys. Rev. B 33:7983 (1986).
6.   J. B. Adams, S. M. Foiles, and W. G. Wolfer, J. Mater. Res. (in press).
7.   M. S. Daw, M. I. Baskes, C. L. Bisson, and W. G. Wolfer, in: "Modeling Environmental Effects on Crack Growth Processes", edited by R. H. Jones and W. W. Gerberich (The Metallurgical Society, Warrendale, Pa., 1986).
8.   M. S. Daw and S. M. Foiles, J. Vac. Sci. Technol. A4:1412 (1986)
9.   M. S. Daw, Surf. Sci. 166: L161 (1986)
10.   S. M. Foiles, Surf. Sci. 191:L779 (1987).
11.   M. S. Daw and S. M. Foiles, Phys. Rev. B35:2128 (1987).
12.   T. E. Felter, S. M. Foiles, M. S. Daw, and R. H. Stulen, Surf. Sci. 171:L379 (1986).
13.   S. M. Foiles, Phys. Rev. B32:7685 (1985).
14.   M. S. Daw and S. M. Foiles, Phys. Rev. Lett. 59:2756 (1987).
15.   S. M. Foiles, Surf. Sci. 191:329 (1987).
16.   J. S. Nelson, E. C. Sowa, and M. S. Daw, Phys. Rev. Lett. 61:1977 (1988).
17.   J. Friedel, Phil Mag. 43:153 (1952).
18.   J. K. Nørskov and N. D. Lang, Phys. Rev. B21:2131 (1980).
19.   M. J. Stott and E. Zaremba, Phys. Rev. B22:1564 (1980).
20.   J. K. Nørskov, Phys. Rev. B 26:2875 (1982).
21.   P. Nordlander, S. Holloway, and J. K. Nørskov, Surf. Sci. 136:59 (1984).
22.   K. W. Jacobsen, J. K. Nørskov, and M. J. Puska, Phys. Rev. B35:7423 (1987).
23.   M. Manninen, Phys. Rev. B34:8486 (1986).
24.   J. D. Kress and A. E. DePristo, J. Chem. Phys. (submitted).
25.   M. W. Finnis and J. E. Sinclair, Phil. Mag. A50:45 (1984).
26.   S.-P. Chen, A. Voter, and D. L. Srolovitz, Phys. Rev. Lett. 37:1308 (1986).
27.   F. Ercolessi, E. Tosatti, and M. Parrinello Surf. Sci. 177:314 (1986) and Phys. Rev. Lett. 57:719 (1986).
28.   M. S. Daw, Phys. Rev. B (in press).

29.   P. Hohenberg and W. Kohn, <u>Phys. Rev</u>. B136:864 (1964).
30.   J. H. Rose, J. R. Smith, F. Guinea, and J. Ferrante, <u>Phys. Rev</u>.
         B29:2963 (1984).
31.   S. M. Foiles and M. S. Daw, <u>Phys. Rev. B</u> (in press).
32.   S. M. Foiles, <u>Acta Met</u>. (in press).
33.   S. M. Foiles in: "Interfacial Structure, Properties and Design", ed. M. H. Yoo,
         W. A. T. Clark, and C. L. Briant (Mater. Res. Soc., Pittsburgh) (in press).
34.   S. M. Foiles and M. S. Daw, <u>J. Mater. Res</u>. 2:5 (1987).
35.   S. M. Foiles, "Mat. Res. Soc. Symp. Proc". Vol. 81 (1987).
36.   S. M. Foiles, "Mat. Res. Soc. Symp. Proc". Vol. 83 (1987).
37.   S. M. Foiles, <u>J. Vac. Sci. Technol</u>. A5:889 (1987).

APPLICATION OF MANY-BODY POTENTIALS TO NOBLE METAL ALLOYS

G.J.Ackland and V.Vitek

Department of Materials Science and Engineering
University of Pennsylvania
Philadelphia, PA19104-6272

INTRODUCTION

A significant recent development in the field of interatomic potentials has been the introduction of many-body terms which model the effect of the local electronic density. Different schemes of this type[1,2,3], although derived from different approaches, yield strikingly similar models in the case of pure metals. In numerous applications many-body potentials have given reasonable results, generally similar to those obtained using pair potentials, but these models also yield good results in simulating vacancies and surfaces; applications for which traditional pair potentials are inadequate.

We will show that the properties of noble metal alloys can also be reproduced by many-body potentials. This is done using a simple extension of the Finnis-Sinclair[1] (FS) model. The noble metal alloys provide an interesting variety of systems. Silver-gold forms a disordered f.c.c. based alloy; copper-silver shows only a limited solid solubility; copper-gold forms ordered alloys. These structures have recently been accounted for using an ASW based density functional theory model to calculate the ground state energies and a cluster variation method to investigate the entropy effect[4,5]. We will compare our results with this more sophisticated calculation. An important question we address during the construction of these potentials is the local relaxation in disordered alloys and near substitutional impurities. We also investigate various plausible approaches to calculating the properties of disordered f.c.c. based alloys, and find significant differences between them.

POTENTIAL MODEL

The FS potential formalism is based on a second moment approximation to the tight-binding theory[6]. The energy of an atom i in a perfect crystal of pure material is written as:

$$E_i = \sum_j V(R_{ij}) - \sqrt{\sum_j \phi(R_{ij})}$$

(1)

The pairwise $V(R)$ term represents interaction between the atoms at sites i and j which is strongly repulsive for small separation and arises primarily from core electron interactions. The square root term models the bonding. ϕ can be interpreted as a sum of squares of hopping integrals.

When extending the model to alloys, both V and ϕ are dependent on the species of the atoms at both i and j. The species dependence is inseparable. Consequently, for a binary system we need two sets of three different functions, V_{AB} and ϕ_{AB}, where A and B denote the atomic species. We assume that these functions are independent of the concentration of the alloys and identify V_{AA}, V_{BB}, ϕ_{AA} and ϕ_{BB} with those of pure metals. For noble metals they were constructed in a previous work[7]. To minimise the amount of new empirical fitting, the function ϕ_{AB} was chosen as a geometrical mean of ϕ_{AA} and ϕ_{BB}. This is consistent with its interpretation in terms of hopping integrals.

The possibility of also determining V_{AB} from pure metal properties was inviting. We examined the form $2V_{AB} = V_{AB} + V_{AB}$. However, with this model there are too many mathematical constraints: for example, none of the random alloys is stable against separation. Thus V_{AB} must be refitted empirically. The model is then adequate to describe alloying energies and equilibrium densities.

FITTING THE FUNCTION V_{AB}

Previous works[8,9] have used the energy of a single substitutional atom for empirical fitting. For each binary system this gives two experimental results to which to fit. The advantage of this property is that it is easy to calculate theoretically. There are, however, problems which are seldom discussed:

1) Experimental data for single substitutional atoms does not exist as such. The data which is usually used is extrapolated from a region of 10% alloy concentration. At this concentration 72% of impurity atoms have at least one impurity as a nearest neighbour (based on a random distribution of species in an f.c.c. lattice), and only 0.01% of impurities can be regarded as isolated on a third neighbour model.

2) The effects of relaxation of the lattice around the impurity are significant, as we shall see later. Moreover, the relaxation will be very different for a 10% alloy than for an isolated impurity.

We fitted V_{AB} to observed alloy formation energies at finite concentrations. In all cases a two point cubic spline was found to give a reasonable fit. Details of the co-efficients of the corresponding polynomials are presented elsewhere[10]. The precise nature of the fitted data varies between the three alloy systems, so we shall examine each case separately. However, since the calculation of random alloy properties is not straightforward, we shall discuss this before describing the fitting.

RANDOM ALLOY MODELS

There are numerous models to describe a random f.c.c. based alloy. None of these are exact in modelling the relaxed ground state. We shall discuss five different possible approximations. In all cases we consider an f.c.c. lattice with two types of atom, A and B, present with atomic fractions c_A and c_B.

The hydrostatic total smearing model (HTSM): Each site is regarded as occupied by a particle consisting of c_A of an A atom and c_B of a B atom[9]. This can be regarded as a smearing out of both atomic types at each site. The lattice has perfect cubic symmetry, with all sites identical, and so will not relax locally. The alloying energy of the FS model in this approximation is given by:

$$\Delta E_{al} = E^{rand} - E^{separated} \tag{2}$$

$$E^{rand} = c_A^2 \sum_j V_{AA}(R_{ij}) + c_B^2 \sum_j V_{BB}(R_{ij}) + 2c_A c_B \sum_j V_{AB}(R_{ij})$$
$$- \sqrt{c_A^2 \sum_j \phi_{AA}(R_{ij}) + c_B^2 \sum_j \phi_{BB}(R_{ij}) + 2c_A c_B \sum_j \phi_{AB}(R_{ij})} \tag{3}$$

$$E^{separated} = c_A \sum_j V_{AA}(R_{ij}) + c_B \sum_j V_{BB}(R_{ij}) - c_A \sqrt{\sum_j \phi_{AA}(R_{ij})} - c_B \sqrt{\sum_j \phi_{BB}(R_{ij})} \tag{4}$$

Environmental smearing models: The species of a central atom i is fixed, but each of its neighbours is regarded as c_A of an A atom and c_B of a B atom. In this way the central atom has a distinct species, but its surroundings consist of smeared atoms as in the HTSM. There are thus two types of site, A and B, which exist in the same environment. There are two possibilities for relaxation. In the *local environmental smearing model* (LESM) the regions surrounding each of the two sites are relaxed independently. The corresponding neighbour distances, denoted by R^A and R^B respectively, are different and cannot be fitted together without some additional strain, so the LESM is overrelaxed.

$$E^{rand} = c_A^2 \sum_j V_{AA}(R_{ij}^A) + c_B^2 \sum_j V_{BB}(R_{ij}^B) + c_A c_B \sum_j V_{AB}(R_{ij}^A) + c_A c_B \sum_j V_{AB}(R_{ij}^B)$$
$$- c_A \sqrt{c_A \sum_j \phi_{AA}(R_{ij}^A) + c_B \sum_j \phi_{AB}(R_{ij}^A)} - c_B \sqrt{c_B \sum_j \phi_{BB}(R_{ij}^B) + c_A \sum_j \phi_{AB}(R_{ij}^B)} \tag{5}$$

In the *hydrostatic environmental smearing model* (HESM) the regions surrounding

195

each type of site are constrained to have the same lattice parameter. This means that, although the energy is minimised with respect to the overall lattice expansion or contraction, no local relaxation is permitted and neighbour distances are independent of the central atom species, otherwise E^{rand} is identical to the LESM.

Configurational sampling models: Every atom is regarded as being either species A or B with probabilities c_A and c_B respectively. There is no smearing. All possible configurations of atoms around a central atom, i, are considered. Atoms are placed on perfect lattice sites, and the lattice parameter is determined by minimising the energy. No local relaxation of the configurations is allowed in the *hydrostatic configurational sampling model* (HCSM) - each configuration is constrained to have the same lattice parameter. This gives an upper bound to the alloying energy. In the *local configurational sampling model* (LCSM), each configuration is relaxed separately (while maintaining local cubic symmetry). The local lattice parameters are denoted by R^α, where α represents the local atomic configuration. This model overcompensates for the lack of relaxation in the HCSM. Although each configuration is locally relaxed, it is not possible to build a crystal without some additional strains. We expect that the LCSM will provide a lower bound to the energy of a random crystal, although it does not allow for local shear relaxation of asymmetric configurations which would further decrease the energy.

We can write the total number of configurations degenerate due to the central approximation in terms of binomial coefficients. Defining N_n^A as the number of A atoms which are nth neighbours of the central atom i, and Z_n as the total number of atoms in the nth shell, the probability P_α for any of the above configurations is

$$P_\alpha = P(N_1^A, N_2^A, N_3^A, \ldots) = \prod_n \binom{Z_n}{N_n^A} c_A^{N_n^A} c_B^{N_n^B} \qquad \text{where} \quad Z_n = N_n^A + N_n^B \tag{6}$$

The energy per atom of a random alloy in the LCSM is then given by

$$E^{rand} = \sum_\alpha P_\alpha \left[c_A \left\{ \sum_{n=1}^{3} N_n^A V_{AA}\left(R_n^{A,\alpha}\right) + N_n^B V_{AB}\left(R_n^{A,\alpha}\right) - \sqrt{\sum_{n=1}^{3} N_n^A \phi_{AA}\left(R_n^{A,\alpha}\right) + N_n^B \phi_{AB}\left(R_n^{A,\alpha}\right)} \right\} \right.$$
$$\left. + c_B \left\{ \sum_{n=1}^{3} N_n^A V_{AB}\left(R_n^{B,\alpha}\right) + N_n^B V_{BB}\left(R_n^{B,\alpha}\right) - \sqrt{\sum_{n=1}^{3} N_n^A \phi_{AB}\left(R_n^{B,\alpha}\right) + N_n^B \phi_{BB}\left(R_n^{B,\alpha}\right)} \right\} \right] \tag{7}$$

The expression for the HCSM energy is similar, except that the interatomic spacings R are now the same for species A and B and depend on the neighbour shell n only. In general this sampling of all possible configurations will be expensive on computation because of the number of configurations to be considered. This number is much reduced in the case of short-range potentials. In the present third-neighbour model 4550 possible configurations must be considered.

COPPER-SILVER ALLOYS

Ag-Cu alloys have a simple eutectic phase diagram[11]. They exhibit phase separation at 0K and limited solid solubility at higher temperatures. To find a suitable function for V_{AB} we fit to mixing enthalpies, ΔH. These are tabulated across the whole range of concentrations for the liquid phase, and for solid solutions up to the solubility limit[11]. We would like to fit zero temperature enthalpies, but as these are not available we make the approximation that ΔH is independent of temperature below the melting point. In the solid complete local relaxation is impaired by the crystal structure, and the alloying at low concentrations will be best described by the HCSM. In the liquid we can expect the fully-relaxed local configurations of the LCSM to be a better approximation. We expect that the theoretical values obtained from the HCSM to be too large, and those from the LCSM to be to small. We determined V_{AB} such that the experimental ΔH lies between these limits.

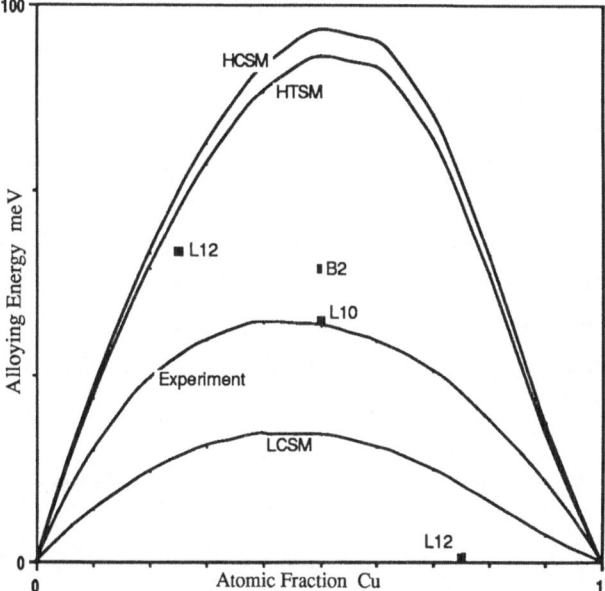

Figure 1.   Concentration dependence of the alloying energy in silver-copper alloys with the various approximations.

The alloy energies calculated for the various random alloy models and selected ordered alloys are shown in figure 1, along with the experimental values. The ordered alloys were fully relaxed. The dependence of ΔH on concentration is more symmetric in the HCSM than in the LCSM. This can be explained in terms of the size difference between the atoms. Since materials are harder in compression than expansion, a silver atom squeezed into a copper site will possess more energy than a copper atom residing in a more spacious silver site. Consequently, relaxation of the local environment will have a bigger effect in the former case than in the latter. This effect also occurs in Cu-Au, but not in Au-Ag where there is negligible size difference. This calculated departure from symmetry is consistent with experimental data[11]. Another interesting feature is that the peak of the ΔH curve is displaced away from the 50-50 mixture toward the Cu rich end for the HCSM and toward the Ag rich end for the LCSM. This is again a consequence of relaxation being more important at the Cu end due to the size effect. It corresponds to the experimental observation that ΔH is greater at the Cu end in a solid solution (HCSM) while for the liquid mixture (LCSM) the peak is displaced toward the Ag end. A third feature is that at low concentration the second derivative of the ΔH plot is positive. This suggests that presence of very small amounts of impurity make it more unfavourable for further alloying to occur, and that small groups of impurity atoms are unstable and will dissociate. The reason for this is not entirely clear, but it is in contrast to the prediction of a simple bond-breaking model in which the second impurity atom would have one less (unfavoured) AB bond if it were near the first. In that case the alloying of a di-impurity is favoured over two single impurities. The variations of volume with concentration are shown in figure 4. This curve shows very little departure from Vegard's law: a reasonable result, although there is no experimental data to confirm it.

SILVER-GOLD ALLOYS

The silver-gold phase diagram shows a complete mutual solubility in the solid alloys. No ordered alloys are known to be stable[11]. It is impossible for a pair potential model to give a stable random f.c.c. alloy at 0K[12], and it seems reasonable to suppose that this is true for the present potentials also. For the fitting of V_{AB} we used the alloying enthalpy[11] and the atomic volume[13]. In this fitting we described the random alloy using the HCSM, although the LCSM gave almost identical values. The accuracy of the fit is shown in figure 2. The curve is very symmetric, and the results obtained using different

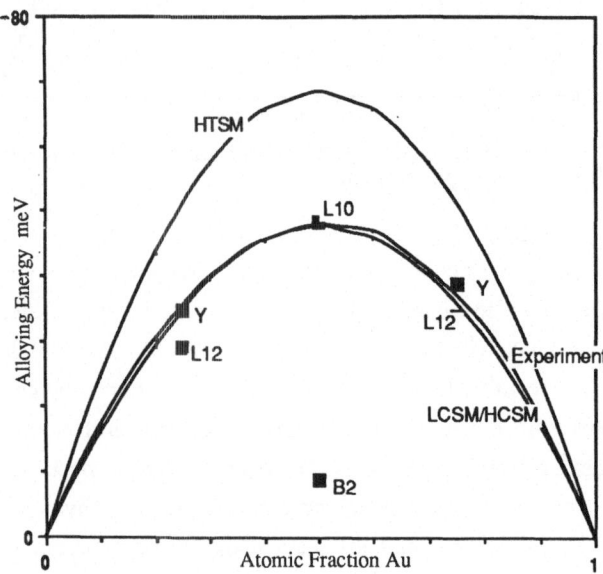

Figure 2.　Concentration dependence of the alloying energy in silver-gold alloys with the various approximations.

models are similar. Since silver and gold have similar atomic volumes, it is not surprising to find that local relaxation is of little importance in this system.

Several possible ordered structures were examined in the present work. In all cases the ordering enthalpies were very small. This is consistent with the observation of a random alloy in all experimental situations. The largest of these ordering energies corresponds to a thermal energy of about 30K. In view of the approximations made in the model we consider these ordering energies to be negligible. We note also that these are much smaller than those obtained from electronic structure calculations (68meV[5],which suggests an order-disorder transition at about 300K[4]). The lowest ordering energy (most stable structure) which we found is for a theoretically constructed crystal structure, denoted as structure Y. This can be regarded as a deformed $L1_2$ structure with repeated 1/2[011](111) stacking faults. The mathematical significance of this structure is that it maximises the number of unlike second neighbours. The elastic constants of the alloy were also calculated using the HCSM and the HTSM. In general the calculated values are slightly higher than experimental data[14], however, the experiments were done at finite temperature, and we expect the elastic constants to decrease with temperature. It is not clear how the local relaxation will affect the elastic constants; indeed the effect differs between systems.

COPPER-GOLD ALLOYS

Experimentally, copper-gold has a phase diagram consisting of various ordered alloys, which undergo an order-disorder transition at high temperature[11]. Again we fit V_{AB} to the random alloy alloying enthalpies, taking the values from high temperature data and using the approximation of temperature independence. By analogy with other estimates of transition temperatures[4], we expect to find ordered structures which are more stable than the random alloy by tens of eV. Once again, the exact value of the random alloying energy is dependent on the model used: here we fit to the HCSM because our data is for a solid solution (albeit at high temperature). A potential was found which gives stable ordered alloys $L1_2$ and $L1_0$. This agrees with the experimental situation at low temperature. These structures were also more stable than other ordered structures which were examined. The exact ordering energies are dependent upon the model chosen for the random alloys.The alloying energies for various ordered and random alloys are shown in figure 3 and the lattice parameters in figure 4. With our model the elastic constants of the $L1_2$ structure are systematically about 8% too high compared with experiment. This is very good agreement for a quantity which is determined by the curvature of the potential. We are unaware of data for the elastic constants of disordered CuAu alloys except at a 3:1 ratio[14] where the lowest temperature considered is 293K, and there is a steady increase of C_{ij} with decreasing temperature. Our model shows that alloying tends to cause a significant elastic stiffening of the material. An exception to this rule occurs at low Au concentration, where alloying appears to soften the material.

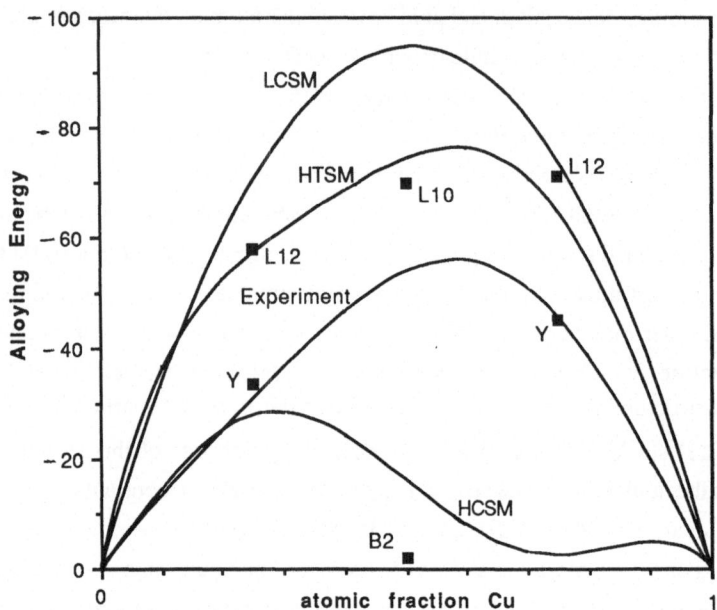

Figure 3.   Concentration dependence of the alloying energy in copper-gold alloys with the various approximations.

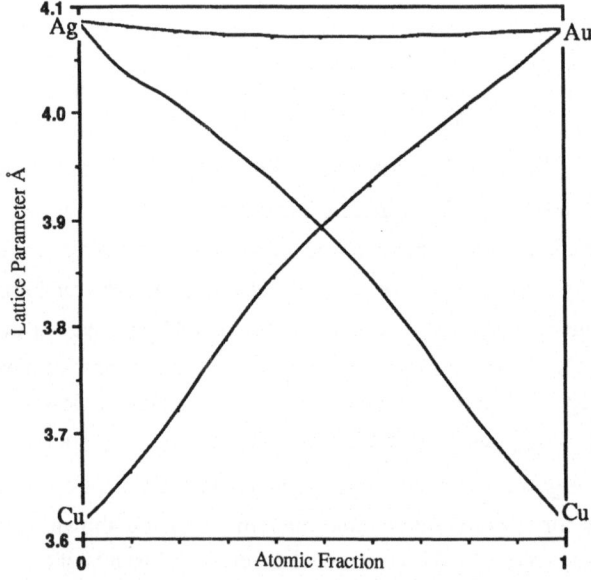

Figure 4.   Concentration dependence of the random alloy lattice parameter for Cu-Ag. Au-Ag and Cu-Au using the HCSM approximations.

SINGLE SUBSTITUTIONAL IMPURITIES

The energy of a single substitutional impurity was also calculated. We allowed relaxation of three shells of neighbours of the impurity atom. In the present model this meant incorporating interactions with neighbours up to the twelfth shell. The energies of the relaxed and unrelaxed configurations were very different, as shown.

Impurity	Host	Unrelaxed (eV)	Relaxed (eV)
Cu	Ag	0.348	0.199
Ag	Cu	0.190	0.098
Au	Ag	-0.187	-0.211
Ag	Au	-0.191	-0.199
Cu	Au	-0.188	-0.298
Au	Cu	-0.156	-0.280

DISCUSSION

The intention of this work was to show that a simple extension of the FS potential scheme can lead to reasonable potentials for noble metal alloys. Throughout, we have adhered to a philosophy of minimising the amount of empirical re-fitting and keeping the functional forms as simple as possible. We were aware that compromising this commitment to simplicity, while enabling our results to look even more impressive, would have been a pointless exercise because of the inherent simplicity of the FS-scheme. We have thus derived a model which maximises the advantages of many-body potentials over electronic structure calculations - the speed at which interatomic forces can be calculated, and the transparency of the results for interpretation. The ability of the model to describe three distinct types of alloying behaviour with only a small amount of empirical refitting is encouraging. The direct application of functions derived from pure metal properties suggests that the model has some physical significance. The simplicity of the (arbitrarily chosen) functional form of V_{AB} is further evidence of this.

In the case of random alloys we have shown that local relaxation is of considerable importance, especially when there is a large discrepancy between atomic sizes. We have also shown that the many different models of a random alloy are not only inequivalent, but also lead to significant differences in calculated values for alloy properties. The stability of the theoretically constructed structure Y for the AuAg potential is curious. That the stability should be determined by the second neighbour interaction is typical of a purely central interaction viewpoint. We believe that structures of this type are also likely to be stable in many-body models, and that in checking global stability they are more important than non-close packed structures such as B2.

The constructed potentials should prove useful in applications where calculation of electronic structure is impractical. The demonstration here of the importance of local relaxation and relative unimportance of detailed electronic structure suggests that more reliable results may be obtained from relaxed molecular statics or dynamics calculations than from single configuration electronic structure calculation. The many-body formalism will also enable studies of properties for which pair potentials are known to be inadequate.

REFERENCES

1. M.W.Finnis and J.E.Sinclair, Phil. Mag.A, **50** 45 (1984)
2. M.S.Daw and M.I.Baskes, Phys.Rev.Lett., **50** 1285 (1983); Phys.Rev.B, **29** 6443 (1984)
3. F.Ercolessi, E.Tosatti and M.Parrinello, Phys.Rev.Lett. **57** 719 (1986)
4. T.Mohri, K.Terakura, T.Oguchi and K.Watanabe, Acta Metall. **36** 547 (1988)
5. K.Terakura, T.Oguchi, T.Mohri and K.Watanabe, Phys.Rev.B, **35** 2169(1987)

6.   G.J.Ackland, M.W.Finnis and V.Vitek, TP1279; published in J.Phys.F
7.   G.J.Ackland, G.Tichy, V.Vitek and M.W.Finnis, Phil. Mag.A, **56** 735 (1987)
8.   S.M.Foiles, M.I.Baskes and M.S.Daw, Phys.Rev.B **33** 7983 (1986)
9.   K.Maeda, V.Vitek and A.P.Sutton, Acta Metall. **30** 2001 (1982)
10.   G.J.Ackland and V.Vitek, to be published.
11.   R.Hultgren, R.L.Orr, P.D.Anderson and K.K.Kelley, "Selected Values of the Thermodynamic Properties of Metals and Binary Alloys", Wiley, New York, (1963)
12.   G.J.Ackland, to be published.
13.   W.B.Pearson,"Handbook of Lattice Spacings and Structures of Metals and Alloys",Pergamon, Oxford (1967)
14.   R.O.Simmons and H.Wang, "Single Crystal Elastic Constants and Calculated Aggregate Properties: a Handbook", MIT Press, Cambridge MA (1971)

MANY-BODY POTENTIALS FOR HEXAGONAL CLOSE-PACKED METALS

Masaaki Igarashi, M. Khantha and V. Vitek

Department of Materials Science and Engineering
University of Pennsylvania
3231 Walnut Street,
Philadelphia, Pennsylvania 19104-6272, U.S.A.

INTRODUCTION

Empirical many-body potentials are currently replacing pair potentials in studies of defects in metallic materials. The two widely used forms of potentials are the embedded-atom (Daw and Baskes 1984) and the Finnis-Sinclair types (Finnis and Sinclair 1984). Although these two approaches are based on different physical models, they reduce to very similar schemes on the empirical level (Johnson 1988). Recently, both approaches have been successfully used in a number of studies of defects in cubic transition and noble metals, most notably, in studies of surfaces (Daw and Baskes 1984, Ackland, Tichy, Vitek and Finnis 1987), point defects (Finnis and Sinclair 1984, Ackland et al. 1987) and cracks (Baskes, Foiles and Daw 1988). In these schemes, the total energy of a system of atoms consists of two parts, a many-body term and a pair potential, which are both determined by empirical fitting. An important point is that the many-body term plays a role analogous to that of the density dependent term in pair-potential schemes, but is now an explicit function of atomic positions. Hence, in contrast to pair-potentials, constant pressure calculations can be carried out straightforwardly and density variations can be easily accounted for.

Many-body potentials have been constructed almost exclusively for materials which crystallize in a cubic structure. However, in this structure, pair-potentials have often been successful in revealing important structural properties of defects (Vitek and De Hosson 1986). On the other hand, pair-potentials have been much less useful in non-cubic structures. Hence, a more general applicability of the many-body potentials can be demonstrated if they are constructed for structures of lesser symmetry for which the structural stability cannot usually be attained using pair-potentials. One class of such materials is the hexagonal close-packed metals, in particular, those with non-ideal c/a ratios. For modelling of extended lattice defects, the potentials must describe a lattice which is stable not only with respect to small (elastic) strains but also large homogeneous deformations (Yamaguchi, Vitek and Pope 1981, and Vitek 1988). Furthermore, the vacancy formation energy and the stacking-fault energy calculated using the potentials must be positive. Pair-potentials which satisfy these criteria have been constructed by Bacon and Liang (1986) for hexagonal metals, but only for the case of the ideal c/a ratio. However, it is likely that differences in mechanical behavior of different hexagonal metals are linked to the significant variation of c/a ratio from material to material.

The first attempt to construct many-body potentials for hexagonal metals was made by Oh and Johnson (1988) who applied the embedded atom method to Mg, Ti and Zr. They fitted the cohesive energy, average bulk and shear moduli and vacancy formation energy, and showed that the hexagonal lattice was more stable than the corresponding f.c.c. and b.c.c. lattices of the same density. The calculated second order elastic constants were in satisfactory agreement with the experimental data. However, the equilibrium c/a ratios for Ti and Zr deviated considerably from the experimental values. Furthermore, since the stability with respect to the f.c.c. lattice was only marginal, the stacking-fault energy is expected to be low in their scheme.

We construct Finnis-Sinclair type potentials for six hexagonal metals, Zn, Mg, Co, Zr, Ti and Hf, by fitting the cohesive energy, vacancy formation energy, five second order elastic constants and the stability conditions for the hexagonal lattice with the corresponding non-ideal c/a ratio. In this paper, we present results for Zn (c/a=1.856) and Hf (c/a=1.581) for which the c/a ratios deviate significantly from the ideal one (1.633). We test the stability of the hexagonal lattice with respect to large deformations using these potentials. The calculated stacking fault energy on the basal plane has reasonable values for all the elements. As a final test, we calculate the phonon frequencies and find dispersion relations in satisfactory agreement with experiments.

THE MODEL

In the Finnis-Sinclair scheme (Finnis and Sinclair 1984), the total energy of the configuration of N atoms is written as

$$U_{tot} = \frac{1}{2}\sum_{ij} V(R_{ij}) - \sum_i f(\rho_i) , \qquad (1)$$

where the summations extend over all the atoms. $V(R_{ij})$ is a pair-potential between two atoms i and j separated by a distance R_{ij}. f is the so-called embedding function which is taken to be of a square root form in the Finnis-Sinclair scheme. This functional form is based on the second moment approximation to the tight-binding theory (Ackland, Finnis and Vitek 1988). $\rho_i = \sum_j \phi(R_{ij})$ where $\phi(R_{ij})$ is another pair-potential. We have used the square root functional form for f in the case of Mg, Co, Zr, Ti and Hf, but a slight modification is necessary for Zn as described below.

204

In analogy with the many-body potentials for noble f.c.c. metals (Ackland et al. 1987), we choose $V(r)$ and $\phi(r)$ as cubic splines:

$$V(r) = \sum_{k=1}^{6} A_k (R_{ak} - r)^3 H(R_{ak} - r), \tag{2}$$

$$\phi(r) = \sum_{k=1}^{4} B_k (R_{bk} - r)^3 H(R_{bk} - r), \tag{3}$$

where R_{ak} (R_{bk}) are chosen knot points such that $R_{a1} > R_{a2} > \text{--} > R_{a6}$ and $R_{b1} > R_{b2} > R_{b3} > R_{b4}$. $H(x) = 0$ for $x < 0$ and $= 1$ for $x > 0$. V and f are always cut-off between 6th and 9th neighbor shells so that the interactions included 38 to 56 neighbors. An additional term $A_7(R_{a7} - r)^3$ is introduced to improve the structural stability of Zn under large compression. However, R_{a7} is less than the first nearest neighbor distance so that this part has no effect on equilibrium properties.

The cohesive energy of an atom i is $E_c^i = \frac{1}{2} \sum_j V(R_{ij}) - f(\rho_i)$. To first order in f, the unrelaxed vacancy formation energy, E_v^f is related to the cohesive energy, E_c^i by

$$E_v^f = -E_c^i - f(\rho_0) + f'(\rho_0) \rho_0, \tag{4}$$

where ρ_0 is the value of ρ_i in the ideal lattice. The stresses $\sigma_{\alpha\beta}$ vanish at equilibrium and the second order elastic constants, $C_{\alpha\beta\gamma\delta}$, can be obtained by applying an infinitesimal homogeneous strain to the perfect crystal and expanding the energy to second order with respect to this strain (Born and Huang 1954). Using Eq.(1), we obtain

$$\Omega_0 \sigma_{\alpha\gamma} = \sum_{ij} \frac{\partial U_{tot}}{\partial R_{ij\alpha}} R_{ij\gamma} = \frac{1}{2} \sum_{ij} \frac{R_{ij\alpha} R_{ij\gamma}}{R_{ij}} V'(R_{ij}) - \sum_i f'(\rho_i) \sum_j \frac{R_{ij\alpha} R_{ij\gamma}}{R_{ij}} \phi'(R_{ij}), \tag{5}$$

$$\Omega_0 C_{\alpha\beta\gamma\delta} = \sum_{ij} \frac{\partial^2 U_{tot}}{\partial R_{ij\alpha} \partial R_{ij\beta}} R_{ij\gamma} R_{ij\delta}$$

$$= \frac{1}{2} \sum_{ij} \left[\frac{R_{ij\alpha} R_{ij\beta} R_{ij\gamma} R_{ij\delta}}{R_{ij}^2} \left(V''(R_{ij}) - \frac{V'(R_{ij})}{R_{ij}} \right) + \delta_{\alpha\beta} \frac{R_{ij\gamma} R_{ij\delta}}{R_{ij}} V'(R_{ij}) \right]$$

$$- \left[\sum_i f'(\rho_i) \sum_j \left\{ \frac{R_{ij\alpha} R_{ij\beta} R_{ij\gamma} R_{ij\delta}}{R_{ij}^2} \left(\phi''(R_{ij}) - \frac{\phi'(R_{ij})}{R_{ij}} \right) + \delta_{\alpha\beta} \frac{R_{ij\gamma} R_{ij\delta}}{R_{ij}} \phi'(R_{ij}) \right\} \right.$$

$$\left. + \sum_i f''(\rho_i) \sum_j \frac{R_{ij\alpha} R_{ij\beta}}{R_{ij}} \phi'(R_{ij}) \sum_j \frac{R_{ij\gamma} R_{ij\delta}}{R_{ij}} \phi'(R_{ij}) \right]. \tag{6}$$

Here $\mathbf{R}_{ij}$ is the vector connecting the atoms i and j, $R_{ij\alpha}$ is the α th component of this vector, $\delta_{\alpha\beta}$ is the Kronecker delta and Ω_0 is the equilibrium atomic volume.

The dynamical matrix needed for the calculation of phonon dispersion relations (Asker 1985) has been obtained from the force constants (Finnis and Sinclair 1984) as

$$D_{\alpha\beta}(q) = \sum_{ij} \frac{\partial^2 U_{tot}}{\partial R_{i\alpha} \partial R_{j\beta}} (\delta_{kk'} - e^{i q \cdot R_{ij}}),$$ (7)

where k and k' denote the atomic sites at the origin and the center of the unit cell, respectively.

Eqs.(1) and (4)-(6) were used to determine the potential parameters A_j and B_j in Eqs.(2) and (3) by fitting the cohesive energy, E_c^i, the vacancy formation energy, E_v^f, the five independent elastic constants, $C_{11}, C_{12}, C_{13}, C_{33}$, and C_{44} and the equilibrium conditions $\sigma_{11} = \sigma_{22} = 0$ and $\sigma_{33} = 0$. The latter conditions impose an important restriction upon the functional form of $f(\rho)$ such that $(C_{12} - C_{66}) \cdot f''(\rho_0) < 0$. When $f = \sqrt{\rho}$, as in the case of Mg, Co, Zr, Ti and Hf, this is satisfied since $f''(\rho_0) = -1/4(\rho_0)^{-3/2} < 0$ and $(C_{12} - C_{66}) > 0$. However, in the case of Zn, $(C_{12} - C_{66}) < 0$ and, therefore, the simple square root functional form of f cannot be used. For this reason, we have modified the function f for Zn to be

$$f(\rho) = \rho^{\frac{1}{2}} (1 + \frac{\rho}{2\rho_0}).$$ (8)

The coefficients A_j and B_j in Eqs.(2) and (3) found by these fitting procedures can be obtained from the author on request.

RESULTS AND DISCUSSION

Structural Stability

The mechanical stability of the hexagonal lattice with respect to large strains was tested in several different ways. First, the dependence of the energy per atom was calculated as a function of the c/a ratio and V/Vo, where V is the volume per atom of the deformed lattice and Vo the equilibrium volume. For Hf this dependence is shown in Fig.1 in the form of constant energy contours. It is seen that for 1.1 < c/a < 2.1 and density changes of ± 25% the only minimum is at the equilibrium density for the fitted c/a ratio of Hf. Several cross sections of this energy surface corresponding to different values of V/Vo < 1 are shown in Fig.2. These curves show that as the compression increases the c/a ratio increases and converges to the ideal value of 1.633. At high pressures the atoms behave like hard spheres owing to the strong hard core repulsion between the atoms at separations much smaller than the equilibrium nearest neighbor spacing, resulting in a close-packed configuration. Similar dependences of the energy on density and c/a ratio as well as the convergence of the c/a ratio towards the ideal one were found for all the other hexagonal metals considered. However, as described in the previous section, in the case of Zn, an additional repulsive term had to be added to the pair-potential for separations of atoms smaller than the nearest neighbor spacing in order to attain stability with respect to large compressions.

The structural stability of the hexagonal lattice with respect to other possible crystal structures was also tested by calculating the energy per atom of a body-centered tetragonal lattice(b.c.t.) with lattice parameters a_1 and a_2, as a function of V/Vo and a_2/a_1. For Hf this dependence is shown in Fig.3 in the form of energy contours. At certain values of a_2/a_1, the b.c.t. lattice reduces to a special symmetric configuration which may be energetically more favoured than other less symmetric ones. The two most symmetric configurations are f.c.c. (at $a_2/a_1 = \sqrt{2}$) and b.c.c. (at $a_2/a_1 = 1$). It is seen from Fig.3 that in the case of Hf the constructed potential gives at certain volumes metastable configurations corresponding to the f.c.c.and possibly b.c.c. lattices. However, the energies of these lattices are about 0.04 eV higher than the energy per atom of the fitted hexagonal structure which is, therefore, a lower energy structure than either the f.c.c. or the b.c.c. lattice. The results of this stability test are similar for all the other hexagonal metals considered.

Stacking Fault Energy

Since the hexagonal structure is favored over the f.c.c. structure the stacking-fault energies of the I_1 type (stacking: *ABABACACA*) and I_2 type (stacking: *ABABCACAC*) faults on the basal plane are expected to be at least positive. The calculated values of these energies are 58 and 111 mJ/m^2 for Hf and 28 and 49 mJ/m^2 for Zn, respectively. While these are not low stacking fault energies they are appreciably lower than the reported values of 390 mJ/m^2 for Hf (for the I_2 type; Legrand 1984) and 140 mJ/m^2 for Zn (for the I_1 type; Murr 1975), respectively, but attempts to reproduce such high stacking fault energies by suitable modification of the potentials were not successful.

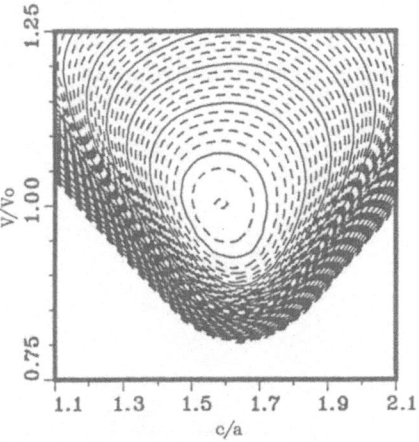

Fig.1 Contours of constant energy for different values of V/Vo and c/a. Neighboring contours corresponding to the energy difference of 0.02 eV.

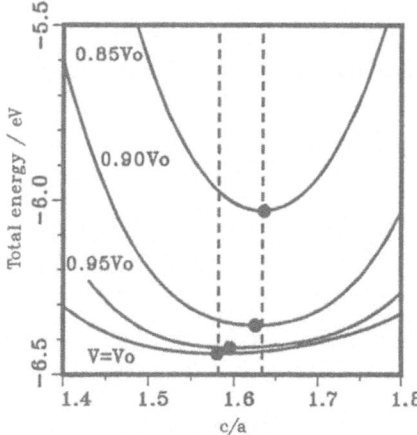

Fig.2 Cross section of the energy surface corresponding to different values of V/Vo.

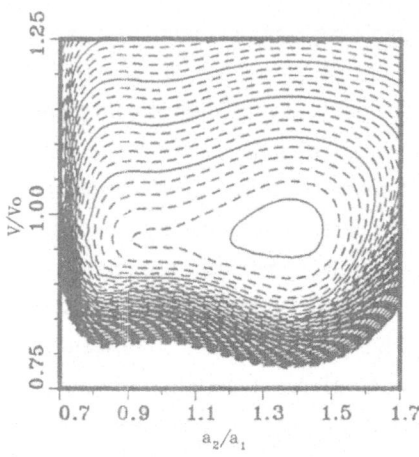

Fig.3 Contours of constant energy of the b.c.t. structure for different values of V/Vo and a_2/a_1.

Phonon Dispersion Relations

Phonon dispersion relations were calculated along three high symmetry directions in the hexagonal Brillouin zone. The results are shown as full lines in Fig.4 for Hf and Fig.5 for Zn; available experimental data are depicted by symbols. Since the potentials have been fitted to elastic constants, a good agreement with experiments is expected for acoustic branches in the long wave length limit. For Hf the agreement is very good for all the acoustic branches and for the transverse optical branches. However, the longitudinal optical branch in the [001] direction shows an anomaly typical for transition metals which is, presumably, related to directional character of bonds and cannot be expected to be reproduced in our scheme. For Zn the agreement is good for the longitudinal and one of the transverse acoustic branches. However, it is poor for the other transverse acoustic branch in the [100] and [110] directions, and the possible reason is the importance of directional bonds and long-range interactions beyond the cut-off of the potentials used (Almqvist and Stedman 1971). The optical branches are in a qualitative agreement.

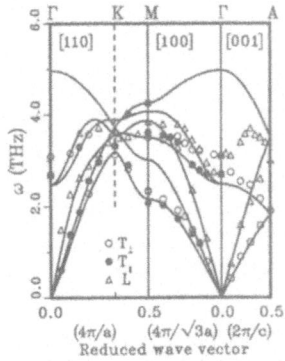

Fig.4 Phonon dispersion relations for Hf. Experimental data were taken from Stassis et al. (1981).

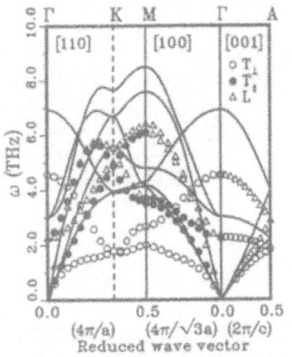

Fig.5 Phonon dispersion relations for Zn. Experimental data were taken from Almqvist and Stedman (1971).

208

CONCLUSIONS

Finnis-Sinclair type many-body potentials have been constructed for the hexagonal metals, Zn, Mg, Co, Zr, Ti and Hf. Each potential ensures that the hexagonal lattice with the correct non-ideal c/a ratio is stable with respect to both small and large deformations and also with respect to other crystal structures like f.c.c. and b.c.c. The calculated stacking fault energies of the I_1 and I_2 type faults on the basal plane are positive but smaller than the reported values. Calculated phonon dispersion relations are in reasonable agreement with experiments although the anomalies in the transition metals are not predicted as expected. It is concluded that these potentials are suitable for studies of extended lattice defects in hexagonal metals.

ACKNOWLEDGEMENTS

This research was supported by the National Science Foundation Grant No DMR85-01974 (MK and VV) and by Sumitomo Metal Industries, Ltd. (MI).

REFERENCES

Ackland, G.J., and Finnis, M.W., and Vitek, V., 1988, *J. Phys.* F, **18**, L153.

Ackland, G.J., Tichy, G., Vitek, V., and Finnis. M.W., 1987, *Phil. Mag.* A, **56**, 735.

Almqvist, L., and Stedman, R., 1971, *J. Phys.* F, **1**, 785.

Askar, A., 1985, *'Lattice Dynamical Foundations of Continuum Theories'*, World Scientific, Singapore.

Bacon, D.J., and Liang, M.H., 1986, *Phil.Mag.* A, **53**, 163.

Baskes, M.I., Foiles, S.M., and Daw, M.S., 1988, *J. Phys.*, Paris, (to be published).

Born, M., and Huang, K., 1954, *'Dynamical Theory of Crystal Lattices'*, Clarendon Press, Oxford.

Daw, M.S., and Baskes, M.I.,1984, *Phys. Rev.* B, **29**, 6443.

Finnis, M.W., and Sinclair, J.E.,1984, *Phil. Mag.* A, **50**, 45.

Johnson, R.A., 1988, *Phys. Rev.* B, **37**, 6121.

Legrand, P.B., 1984, *Phil. Mag.* B, **49**, 171.

Murr, L.E., 1975, *'Interfacial Phenomena in Metals and Alloys'*, Addison-Wesley, Reading, Mass.

Oh D.J., and Johnson, R.A., 1988, *J. Mater. Res.* **3**, 471.

Stassis, C., Arch, D., McMasters, O.D., and Harman, B.N., 1981, *Phys. Rev.* B, **24**, 730.

Vitek, V., 1988, *Phil. Mag.* A, **58**, 193.

Vitek, V., and de Hosson, J.Th.M., 1986, in *'Computer-Based Microscopic Description of the Structure and Properties of Materials'*, MRS, Pittsburgh, 137.

Yamaguchi, M., Vitek, V., and Pope, D.P., 1981, *Phil. Mag.* A, **43** 1027.

DERIVATION OF MANY BODY POTENTIALS TO REPRODUCE ELASTIC

AND VIBRATIONAL QUALITIES OF FCC AND BCC METALS

James Eridon
Condensed Matter and Radiation Sciences Division
Naval Research Laboratory
Washington, DC 20375

INTRODUCTION

One of the great advantages of the local density methods of materials simulation (Daw and Baskes, 1984; Finnis and Sinclair, 1984) over simple pair potentials is their ability to model such things as elastic properties and defect properties in regions of low symmetry with a degree of computational difficulty only approximately twice that involved in pair models. Although the model itself is conceptually simple, it can be quite difficult to implement owing to the effort involved in deriving the functions necessary to accurately describe the material of interest. This task is even more daunting when an alloy system is under investigation, in which case the functions must reproduce not only the empirically observable qualities of the pure materials, but also such things as the heat of formation and stable structure of the alloys of interest. Several authors have undertaken to investigate an assortment of alloys with a variety of functions derived specifically for that purpose. In general, the functions so derived are not useful in conjunction with any other materials. This necessitated, for example, the derivation of new nickel functions for use with nickel-aluminum alloys (Foiles and Daw, 1987) even after a set of nickel functions had already been derived which were compatible with copper, silver, gold, platinum, and palladium (Foiles et al., 1986).

The derivation of a set of "universal" functions which are applicable in any combination is probably not possible. It is not even clear what form such functions would be required to take. For example, in investigations of nickel and aluminum alloys, different authors (Voter and Chen, 1987; Foiles and Daw, 1987) have derived functions in which the attractive and repulsive forces are assumed by either the embedding function or the pair potential, with equally good results. In view of the difficulty involved in deriving these functions, and the need for a wide variety of functions for use with different alloys, it would be useful if some common features could be identified among all those functions which meet certain minimum criteria, such that the derivation could proceed in a more orderly fashion, as opposed to the often tedious trial and error techniques that are often encountered when functions are characterized by a variety of fitting constants of almost unlimited magnitude. A minimum criteria for metals requires that the model reproduce the lattice constant in the stable structure and heat of sublimation of the metal of interest. It is also generally acknowledged that the model should reproduce the vacancy and interstitial formation energies, be stable against collapse under high compression (Rebonato et al., 1987; Ackland and Thetford, 1987) and, at least for cubic metals, reproduce the elastic constants. In some cases, niobium for example, it is also necessary to match the zone boundary phonon frequencies, owing to the long range forces in this material and the many anomalies in its phonon dispersion curves.

The three functions involved in the model, which may be termed the pair potential, $V(R)$, the local density function, $\phi(R)$ (which is interpreted here as an effective electron density), and the embedding function, $f(\rho)$, must be derived in such a way as to meet the criteria specified above. In addition, they should have certain features. For example, the embedding function should not be "kinky" and should tend toward a straight line with constant slope at high values of electron density (Puska et al., 1981). The electron density should be non-negative and have a value of zero as well as vanishing first and second derivatives past some cutoff radius. The pair potential should have the same properties as ϕ at its cutoff radius, but is not required to be non-negative everywhere. It should, however, tend to positive infinity as R goes to zero. With these restrictions, it is possible to derive a wide range of functions which meed the model requirements. In the following section, a derivation of the form of the pair potential which meets the criteria on lattice constant, elastic constants, and phonon frequencies is presented. This is followed by a description of an additional restriction on $V(r)$ which guarantees the value of the unrelaxed vacancy formation energy. The next section describes a simple form for the embedding function. The electron density function is assumed to have been chosen in a form matching the requirements set forth above. The use of either actual atomic wavefunction data (Clementi and Roetti, 1974) or a simple polynomial expansion (Finnis and Sinclair, 1984) is adequate. The next section describes the implementation of the model in the cases of iron and niobium. This is followed by a brief discussion. Although the method works equally well for fcc metals, and has been applied to both copper and aluminum, space limitations preclude the presentation of these results. All of what follows, however, applies to both fcc and bcc metals.

PAIR POTENTIALS AND FORCE CONSTANTS

The Finnis-Sinclair or Embedded Atoms expression for the energy of a single atom in a crystal is:

$$U_i = \frac{1}{2} \sum_{j \neq 1} V(R_{ij}) + f(\rho_i)$$

where:

$$\rho_i = \sum_{j \neq 1} \phi(R_{ij})$$

The first term is the pair potential contribution wile the second represents the local density dependent contribution. The expression for the force constants derived from this potential (Finnis and Sinclair, 1984) has the following form for bcc and fcc metals:

$$\Phi_{ij}^{\alpha\beta} = \left[V'(R_{ij}) + 2f'(\rho) \, \phi'(R_{ij}) \right] \left[\delta^{\alpha\beta} - \frac{R_{ij}^{\alpha} \, R_{ij}^{\beta}}{R_{ij} \, R_{ij}} \right] \frac{1}{R_{ij}}$$

$$+ \left[V''(R_{ij}) + 2f'(\rho) \, \phi''(R_{ij}) \right] \frac{R_{ij}^{\alpha} \, R_{ij}^{\beta}}{R_{ij} \, R_{ij}}$$

$$+ f''(\rho) \sum_{k \neq ij} \phi'(R_{ki}) \, \phi'(R_{kj}) \frac{R_{ki}^{\alpha} \, R_{kj}^{\beta}}{R_{ki} \, R_{kj}}$$

where $\Phi_{ij}^{\alpha\beta}$ is the force exerted in direction α on atom i due to an infinitesimal movement of atom j in direction β. This expression is linear in the derivatives of the pair potential, but non-linear in the functions $f(\rho)$ and $\phi(R)$. However, for a given function $\phi(R)$, it is possible to derive expressions for f' and f'', leaving the force constants as linear functions of the first and second derivatives of the pair potential evaluated at the neighbor locations (Vi = V(Ri), Ri = atomic shell radii). This is done by invoking the condition for zero stress (R.A. Johnson, 1972; Daw and Baskes, 1984) as follows:

$$f' = \frac{A_{11}}{B_{11}}$$

where:

$$A_{11} = \frac{1}{2} \sum_{j \neq \iota} V'(R_{ij}) \frac{R_{ij}^1 \, R_{ij}^1}{R_{ij}} \qquad B_{11} = \sum_{j \neq i} \phi'(R_{ij}) \frac{R_{ij}^1 \, R_{ij}^1}{R_{ij}}$$

This leaves f' as a linear function of V_i'. The value of f'' may be found from a consideration of the Cauchy pressure (Daw and Baskes, 1984) as follows:

$$C_{12} - C_{44} = f'' \, B_{11}^2 / \Omega$$

where Ω is the atomic volume and C_{12} and C_{44} are elastic constants. Therefore, we have effectively transformed the equation for the force constants in a cubic material into a set of linear equations of the first two derivatives of the pair potential evaluated at the atomic shell radii. For both fcc and bcc metals, the elastic constants (and phonon frequencies) are linear functions of these force constants (Squires, 1963). We may therefore write equations for these measurable quantities in terms of the derivatives of $V(R)$. In general, there will be more variables (derivatives of V) than there are equations. For example, if $V(R)$ ranges out to third neighbors, we have six variables: the first two derivatives of $V(R)$ for each neighbor shell. If we are only interested in matching the elastic constants, then we have only two equations (one equation involving the elastic constants in used in fitting f" above). We may therefore select two of the variables to obtain a basic solution, and also obtain four additional fitting functions, any multiple of which added to the basic solution will also be a solution. This gives us great latitude in defining a set of pair potentials which meet the criteria on the elastic constants, while also restricting the set of possibilities to a manageable few.

Note that until now nothing has been said about the actual magnitude of the pair potential. Although it must go smoothly to zero at the cutoff radius, there is another restriction on the size of $V(R)$ which arises from the vacancy formation energy. By expanding the equation for the equilibrium energy of an atom within the local density approximation, it is a simple matter to derive the following approximation for the value of the unrelaxed vacancy formation energy (Daw and Baskes, 1984):

$$E_f^v = \overline{V} - \overline{\rho} \; f' \; (\overline{\rho}) + \frac{1}{2} \sigma f'' \; (\overline{\rho})$$

where:

$$\sigma = \sum_{j \neq i} \phi^2 \; (R_{ij})$$

$\overline{\rho}$ is the equilibrium electron density, and $\overline{V}$ is the pair energy at a lattice site (the first sum in the energy expression). Since $f'(\overline{\rho})$ and $f''(\overline{\rho})$ have already been calculated, this equation restricts the size of the pair energy at a lattice site to a value determined by the vacancy formation energy, which is typically about one third of the sublimation energy for metals. Since $f'(\overline{\rho})$, is actually a function of the derivatives of $V(R)$ through the equilibrium equation, and the $f''(\overline{\rho})$ term is relatively small, this equation effectively relates the value of the pair potential to its derivatives.

EMBEDDING FUNCTION F

After selecting a set of fitting functions to describe the derivatives of $V(R)$, and having assumed a parameterization for $\phi(R)$, the value of $f'(\overline{\rho})$ and $f''(\overline{\rho})$ can be found from the equation of equilibrium and from the Cauchy pressure, as described above. After fitting a potential function to the derivatives and conditions discussed in the previous sections, it is possible to compute $\overline{V}$, the sum of the potential energy at a lattice site. The value of $f(\bar{\rho})$ can then be found as simply the difference between $\bar{V}$ and the sublimation energy. With these three values for $f(\bar{\rho})$ and its first two derivatives, it is possible to derive an expression for $f(\rho)$ as a simple expansion in roots of ρ. That is:

$$f(\rho) = f_0 + f_1 \, \rho \; + \; f_2 \sqrt{\rho} \; + f_3 \sqrt[3]{\rho}$$

This expansion, with $f_0 = 0$, has the attractive properties that $f(0) = 0$ and $f(\rho)$ tends to a linear function at large values of ρ. Also, if $f_3 < 0$, this function has an infinite negative slope at $\rho = 0$, which is consistent with the interpretation of $f(\rho)$ as the energy of embedding an atom in an electron gas (Puska et al., 1981). This is in contrast to, for example, an expansion in powers of ρ, which would contain ρ raised to the third power and which could result in some unpleasant kinkiness at large values of ρ. By solving the simple set of equations arising from this expansion at the equilibrium value of ρ, we find that:

$$\frac{f_3}{9} \, [\bar{\rho}]^{1/3} = \bar{\rho}^2 \, f''(\bar{\rho}) + \frac{1}{2} \, [f(\bar{\rho}) - \bar{\rho} \, f' \, (\bar{\rho})]$$

Using the equations for the vacancy formation energy and the sublimation energy, and the sublimation energy, and ignoring terms in f'', the right hand side of this equation becomes simply $(E_s + E_v)/2$. Since the sublimation energy is (by our convention) negative, and the vacancy formation energy is much smaller in magnitude than the sublimation energy, the infinite negative slope at $\rho = 0$ is certain provided the f'' term is small, which is generally the case.

APPLICATION TO NIOBIUM AND IRON

The forces in iron are fairly short range as shown by the magnitude of the force constants determined from phonon dispersion data (Brockhouse, et al., 1967; Minkiewicz et al., 1967). Therefore, a fit was attempted using a pair potential with forces out to third neighbors and an electron density ranging out to second neighbors. The electron density was parameterized as $(R_{rho} - R)^3$, with $R_{rho} = 3.5$ Å. This left six derivatives of $V(R)$ to determine. The fitting conditions were chosen as the three elastic constants and two phonon frequencies: the (1 0 0) zone boundary frequency of 8.56 THz and the (1/2 1/2 1/2) critical point frequency of 7.21 THz. This left two variables free for use as fitting constants. In order to show the wide variety of potentials possible within the boundaries of the method, two different solutions were created, one of which is termed the positive potential solution, since it is repulsive at the first neighbor shell, and the negative potential solution, which is attractive at all three neighbor shells. The positive potential solution was formed using variables $V'(R_3)$ and $V''(R_2)$ as fitting parameters, while the negative potential solution was formed using $V''(R_1)$ and $V''(R_2)$ as fitting parameters. Any choice of fitting parameters will, of course, yield linearly dependent solutions, provided the equations remain non-singular. The solutions are shown in Table 1, which gives the basic solution and the two fitting solutions for each case in terms of the derivatives of $V(R_i)$ as well as the final values of all the derivatives. With these values for the derivatives, a polynomial fit to the potential was created subject to the conditions on the derivatives and the value of $V(R_{cut} = 4.6Å) = 0$ and the condition on the vacancy formation energy, which was estimated as 1.8 eV. The polynomial uses powers of $(R/R_{cut})n$ from $n = -1$ to 8. If the $n = -1$ coefficient of the expansion is positive, this is a sufficient condition to guarantee against a "black hole" instability at very high compression, although at moderate compression an instability is still possible. The resulting pair potentials and embedding functions for the two cases are shown in Figures 1 and 2. Note that the embedding functions both start with negative slope, but that the negative solution quickly turns around so as to have a positive slope at the equilibrium value of electron density, $\rho = 9.954$.

Both the positive and negative solutions have good behavior with respect to the relaxed vacancy formation energy (1.76 eV and 1.80 eV, respectively), interstitial formation energy (2.87 eV and 3.28 eV, respectively) and stability against transformation to the fcc structure (about 0.01 eV in both cases).

TABLE 1.   Derivatives of the pair potential at neighbor shell locations for bcc iron for both positive and negative potential solutions. Units are eV/Å and eV/Å^2.

		Positive Pair Potential			Negative Pair Potential		
		Neighbor Shell #			Neighbor Shell #		
		1	2	3	1	2	3
Basic Solution	V'	0.7054	0.2837	0.0000	4.3445	1.6816	-2.5379
	V''	1.7451	0.0000	0.0050	0.0000	0.0000	1.2571
Fitting Solution #1	V'	-1.4339	-0.5508	1.0000	-2.0853	-0.8011	1.4543
	V''	0.6876	0.0000	-0.4934	1.0000	0.0000	-0.7175
Fitting Solution #2	V'	-0.8172	-0.3167	0.0000	2.5323	0.9699	-2.3359
	V''	1.6062	1.0000	0.0000	0.0000	1.0000	1.1524
Final Solution	V'	-2.0240	-0.7736	0.1248	0.6999	0.2816	0.0038
	V''	6.8439	3.1210	-0.0566	1.7477	0.0000	0.0031

In the case of niobium, the forces are appreciable all the way out to the eighth neighbor shell (Nakagawa and Woods, 1963). For modeling purposes, the forces were allowed to range to fifth neighbors, providing ten variables. The fitting conditions were chosen as the elastic constants plus the phonon frequencies at six points in symmetry directions, including (1 0 0), (1/2 1/2 1/2), the transverse mode at (1/2 0 0), and the three modes at (1/2 1/2 0). This again leaves two variables available for fitting purposes, which in this case were chosen to be $V''(R_3)$ and $V''(R_4)$. With a procedure similar to that employed for iron, an embedding function, pair potential, and electron density function were produced which satisfied conditions of stability against transformation to fcc, reasonable vacancy and interstitial formation energies, and fits to all the elastic constants and selected phonon frequencies. The phonon dispersion curves for niobium show several anomalies, including the crossing of the two transverse modes in the (qq0) direction. The curves computed from these functions reproduce this anomaly, as well as several others, and provide a fit about as good as a fifth neighbor Born-von Karman analysis (Sharp, 1967).

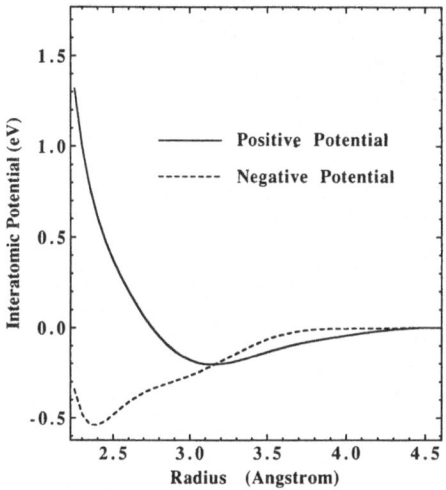

Figure 1. Pair potentials for iron.

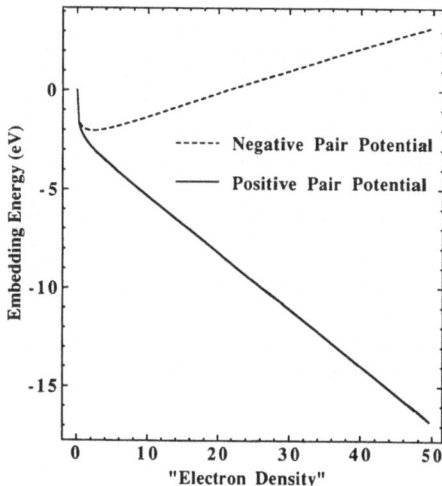

Figure 2 Embedding functions for both iron pair potentials.

SUMMARY AND DISCUSSION

A method has been developed to derive pair potentials and embedding functions for fcc and bcc metals in such a way that the resulting model exactly reproduces long-wavelength (elastic) and short wavelength (zone boundary frequency) parameters of the metal, as well as the lattice constant and heat of sublimation. Additionally, these functions are derived so as to provide stability against transformation to another cubic lattice, stability against collapse at high compression, and a reasonable value of the vacancy formation energy. For a given electron density and range of forces, there is a wide variety of pair potentials and embedding functions which satisfy the constraints mentioned above. These functions may be constructed from a set of fitting functions which are solutions to a set of linear equations derived from the equations for the force constants, elastic constants, and phonon frequencies of cubic materials. While the mathematics involved is quite simple, the derivation of the equations themselves can be tedious. A simple code has been written to facilitate the computation of the equations and their solutions for both fcc and bcc materials, and is available from the author on request. An obvious question raised by these results is to what extent this method may be applied to hexagonal metals. While the equations for the force constants, elastic constants, and phonon frequencies are similar for hcp metals, they are much more complex owing to the lower symmetry of hexagonal lattices with non-ideal c/a ratios. Nevertheless, it should be possible to extend the method to these materials as well.

REFERENCES

Ackland, G.J. and Thetford, R., 1987, Phil. Mag. A, **56**, 15.
Brockhouse, B.N., Abou-Helal, and Hallman, E.D., 1967, Solid State Comm., **5**, 211.
Clementi, E., and Roetti, C., 1974, Atomic and Nuclear Data Tables, **14**, 177.
Daw, M.S. and Baskes, M.I., 1984, Phys. Rev. B, **29**, 6443.
Foiles, S.M., Baskes, M.I., and Daw, M.S., 1986, Phys. Rev. B, **33**, 7983.
Foiles, S.M. and Daw, M.S., 1987, J. Mater. Res., **2**, 5.
Finnis, M.W. and Sinclair, J.E., 1984, Phil. Mag. A, **50**, 45.
Johnson, R.A., 1972, Phys. Rev. B, **6**, 2094.
Minkiewicz, V.J., Shirane, G., and Nathans, R., 1967, Phys. Rev., **162**, 528.
Nakagawa, Y. and Woods, A.D.B., 1963, in "Lattice Dynamics", edited by R.F. Wallis (Pergamon Press).
Puska, M.J., Nieminen, R.M., and Manninen, M., 1981, Phys. Rev. B, **24**, 3037.
Rebonato, R., Welch, D.O., Hatcher, R.D., and Bilello, J.C., 1987, Phil. Mag. A, **55**, 655.
Sharp, R.I., 1969, J. Phys. C (Solid St. Phys), **2**, 421.
Squires, G.L., 1963, Ark. Fys., **25**, 21.
Voter, A.F. and Chen, S.P., 1987, in MRS Proceedings, **82** (Materials Research Society) Pittsburgh, Pennsylvania.

EMBEDDED ATOM POTENTIAL FOR BCC IRON

Ralph J. Harrison

Materials Reliability Division
U.S. Army Materials Technology Laboratory
Watertown, MA 02172-0001

Arthur F. Voter and Shao-Ping Chen

Theoretical Division
Los Alamos National Laboratory
Los Alamos, NM 87545

We have used the embedded atom method (EAM)[1-4] to construct interatomic potentials for use with BCC iron. Our original motivation for this work was to model the grain boundaries in iron[5]. The version of the EAM we have used is essentially the same as that described in references 3 and 4, where the total energy was given as the sum of two body terms summed over pairs of atoms i,j, together with an embedding term given by the sum of embedding functions whose arguments are the total electronic charge density at the sites i. A Morse potential

$$\phi(r_{ij}) = v_1[1 - \exp[(-v_3(r_{ij} - v_2)]]^2 - v_1$$

with the three parameters v_1, v_2 and v_3, is used to represent the two-body terms. The total electronic charge densities at each individual site which are the arguments of the embedding functions, are obtained from the linear superposition of atomic 4s Slater function charge densities of form

$$r_{ij}^6[\exp[-v_4 r_{ij}] + 512 \exp[-2v_4 r_{ij}]].$$

The fifth parameter v_5 is taken as the cutoff distance for the atomic charge densities and for phi (r_{ij}), which are adjusted for smoothness at cutoff. The embedding function is determined by the fit of the cohesive energy of bcc iron as a function of interatomic distance to the "universal" cohesive energy function described by Rose et al.[6-8]. In this fit, the depth and curvature and distance scale of the cohesive energy function is matched to the experimental cohesive energy, the bulk modulus B, and equilibrium lattice constant, a_0, but introducing no additional explicit matching parameters. The parameters v_1 through v_5 are selected by fitting (using a simplex optimization procedure[9]) to selected experimental properties of iron that we shall now describe. The first are the remaining elastic constants, either c_{11} or c_{12} (since $3B = c_{11} + 2c_{12}$) and c_{44}. We next try to match the observed vacancy formation energy[19]. This is known somewhat less precisely than are the elastic constants, but it is important for matching purposes since the deviation from this value is a measure of the importance of many body forces due to the fact that the unrelaxed vacancy energy would be equal to the cohesive energy if only pair forces operated.

For a magnetic material such as iron there is a contribution to the vacancy energy arising from the altered magnetism on the neighboring atoms to the vacancy. There are also important contributions arising from magnetic effects in the energy difference between ferromagnetic bcc and nonmagnetic fcc and hcp iron. Our EAM calculation takes no explicit account of spin polarization effects, whereas quantum computations including these effects[10] indicate that they are crucial in providing phase stability of the bcc phase. However, just as the adjustment of fitting parameters to vacancy energy may account for some of the magnetic energy in an empirical way, we try to make this fit also provide phase stability. Therefore we have made it a requirement that the minimum energies of the fcc or hcp phases should always be higher than that of the bcc phase of iron. This requirement was deemed essential in order to be able to utilize the resulting potential for the computation of defect configurations by energy minimization techniques. It turned out that this constraint on ensuring the stability of the bcc phase was restrictive enough to prevent us from finding a choice of parameters which exactly fit the elastic constants and vacancy energy. We might have gotten some additional freedom in fitting if we had chosen a two body potential with additional parameters. However, we hesitated to do so since we wanted to avoid unrealistic structure in the potential which might arise from too much freedom in curve fitting.

In Table I we show some results with computed EAM potentials. We list the experimental and calculated values of elastic constants and vacancy energy for the bcc phase, and the calculated cohesive energies and lattice constants which minimize energy for the hcp and fcc phases. We also tabulate the values of the fitting parameters for the different potentials.

The column marked FEA is for the potential published in reference 5 where we tabulated interplanar spacings near the surface as well as surface and grain boundary energies obtained with its use. Cheung and Yip[11] also used this potential in preliminary computations studying thermal instability by means of molecular dynamics techniques. They found that a density discontinuity occurred for a 4x4x4 periodic cell of 128 atoms maintained at zero pressure using the Parinello-Rahman boundary conditions[12], at 1750K. This is somewhat lower than the melting temperature of iron (1809K), while one might expect an instability temperature to be higher than melting. An additional molecular dynamics "experiment" was done on the uniaxial deformation of iron with this potential, in an attempt to simulate the martensitic Bain transformation to the fcc phase[13]. A phase transformation to the hcp phase occurred rather than to the expected fcc phase. This result may be due to the fact that the hcp phase for this potential is only 0.009 eV above the bcc ground state, while the fcc phase is 0.02 eV above bcc. This closeness of another phase might even have contributed to the lower than expected instability temperature. We therefore tried to compute other potentials which would show greater separation from the ground state and possibly reverse the order of the fcc and hcp phases. As the results listed in the other columns show we have increased the separation of the energies of the hcp and fcc phases from that of the bcc phase, although we have not been able to reverse the order. There is also a somewhat better fit to the elastic constants. For the potentials FEC and FEC, the energies of the hcp and fcc are <u>identical</u>, a consequence of the fact that the cutoff distance is between second and third neighbors; since the distances and numbers of first and second neighbors are the same for hcp and fcc there will be zero stacking fault energy. It is suggestive that perhaps a short range many body potential coupled with a two body potential having a longer range and additional structure might give an optimal description.

TABLE I   Fitting parameters for EAM potentials for BCC iron

Exactly fit by all potentials:
Lattice constant $a_0 = 2.870$Å   Cohesive energy = 4.28 eV;
Bulk modulus = 1.73×10^{12} dyn/cm^2.

Elastic Constants (10^{12} dyn/cm^2)	Experimental	FEA (previous[5])	FEB	FEC (short range)	FED	FEB err
c_{11}	2.432	1.93	2.131	2.178	2.299	12.5%
c_{12}	1.381	1.63	1.534	1.507	1.447	11.1%
c_{44}	1.219 (Ref. 18)	1.05	1.165	1.175	1.179	4.4%
Vacancy Energy (eV) unrelaxed		1.93	2.623	1.953	2.207	
relaxed	2.0 ± 0.2 (Ref. 19)	1.64	2.181	1.792	2.032	
Parameters:						
v_1 x10^{-2} Hartrees		2.323	2.849	2.684	2.771	
v_2 angstroms		2.092	2.721	2.690	2.852	
v_3 1/(angstroms)		1.343	0.746	1.225	1.293	
v_4 1/(angstroms)		6.262	8.508	6.580	6.423	
v_5 (cutoff distance, A.)		4.524	4.5096	3.625	3.626	
Phase separation:						
E_{fcc} - E_{bcc} (eV)		0.02	0.057	0.074	0.108	
E_{hcp} - E_{bcc} (eV)		0.009	0.048	0.074	0.108	

Cheung and Yip[11] have also made some initial studies with FEB and FEC of the Bain transformation and find, in accord with the much greater separation from the ground state of the hcp and fcc states, that much greater uniaxial stress than previously can be applied without causing any instability. They have not yet carried these studies to the point of any transformation.

We conclude this presentation of EAM potentials by remarking that the test of empirical potentials must be their robustness in describing various types of dynamical as well as static structural phenomena. The study of surface spacings and energies[5], grain boundary energies[5] and of transformations under stress and temperature[11] are the beginnings of such a testing program. Whether any potential that does not explicitly consider magnetic effects can have the required robustness is a question for the future. We have not tried to review other related work, but we must certainly mention that of Finnis and Sinclair[15] whose iron potential has already been used to compute surface spacings and energy and energies and configurations of particular defects[16,17].

ACKNOWLEDGMENT

We thank Kin Cheung and Professor Sidney Yip for discussion of their work cited in reference 11.

REFERENCES

1.   M.S. Daw and M.I. Baskes, Phys. Rev. B29:6443 (1984).
2.   S.M. Foiles and M.S. Daw, J. Mater. Res. 2:5 (1987).
3.   S.P. Chen, A.F. Voter, D.J. Srolovitz, Phys. Rev. Letters 57:1308 (1986).
4.   A.F. Voter and S.P. Chen, Mater. Res. Soc. Symp. 82, 175 (1987).
5.   An account of some results obtained with a preliminary version of this potential was presented at the 34th Sagamore Army Materials Research Conference, Sept. 1987. (Proceedings to be published by Plenum Press, 1988).
6.   J.H. Rose, J.R. Smith, F. Guinea, J. Ferrante, Phys. Rev. B29:2963 (1984).
7.   J.H. Rose, J.R. Smith, J. Ferrante, Phys. Rev. B28:1835 (1983).
8.   J.R. Smith, J. Ferrante, J.G. Gay, R. Richter, J.H. Rose, in "Chemistry and Physics of Fracture", R.H. Jones and R.M. Latanision, eds. (Martinus Nyhoff, Hingham, MA, 1987), p. 329.
9.   J.A. Nelder and R. Mead, Comp. J. 7:308 (1965).
10.   G. Krasko and G.B. Olson, "Energetics of BCC-FCC Phase Transformation in Iron", to be published in Materials Research Society Symposium 1988, on Atomic Scale Calculations in Materials Science.
11.   K. Cheung and S. Yip, Private Communication.
12.   M. Parinello and A. Rahman, J. Appl Phys. 52:7182 (1981).
13.   R. Najafabadi and S. Yip, Scripta Met. 17:1199 (1983) carried out a simulation of the Bain transformation by the Monte Carlo technique using the Johnson I iron potential (R.A. Johnson, Phys. Rev. 134A:1332 (1964).
14.   L. Kaufman and H. Bernstein, Computer Calculations of Phase Diagrams, Academic Press, New York (1970), indicate from extrapolation to low temperatures, at high pressures, the hcp phase of iron may be more stable than the fcc phase.
15.   M.W. Finnis and J.E. Sinclair, Phil. Mag. A 50:45 (1984).
16.   C.C. Matthai and D.J. Bacon, Phil. Mag. A, 52:1(1985).
17.   J.M. Harder and D.J. Bacon, Phil. Mag. A, 58:165 (1988).
18.   J.A. Rayne and B.S. Chandresekhar, Phys. Rev. 122:1714 (1961).
19.   L. De Schepper, D. Segers, L. Dorikens-Vanpraet, M. Dorikens, G. Knuyt, L.M. Stals and P. Moser, Phys. Rev. B27:5257 (1983).

EFFECTS OF B AND S ON Ni$_3$Al GRAIN BOUNDARIES

A.F. Voter, S.P. Chen, R.C. Albert, A.M. Boring, and P.J. Hay

Los Alamos National Laboratory
Los Alamos, NM 87545

INTRODUCTION

In many materials, the mechanical behavior is controlled by the grain boundary (GB) properties. An extreme example is the ordered alloy Ni$_3$Al, which, as a single crystal, is ductile, while as a pure polycrystal, exhibits severe intergranular brittleness, making it useless as a technological material. However, it has been found [1] that doping Ni$_3$Al that is slightly Ni-rich (76% Ni) with small amounts of boron (~1-2 atomic %) restores the ductility almost to the level of the single crystal. While it is known experimentally that B segregates to grain boundaries [1], the mechanism by which ductilization occurs is not known. We present here results of our initial investigation into the effects of impurities on Ni$_3$Al grain boundaries, using interatomic potentials of the embedded atom [2,3] form, coupled with molecular statics techniques.

Theoretical studies of grain boundaries in homonuclear systems have been reported by many authors [4,5]. Both molecular statics and molecular dynamics techniques have been employed. As a result of these studies, the general structural features and, to a lesser extent, the energetics, of pure element GBs are now understood. More recently, studies of alloy GBs have appeared [6,7,8]. In the present work, we study Ni$_3$Al grain boundaries with B and S impurities. The S impurity is of interest because it is known to have the opposite effect from B on GBs: it promotes intergranular fracture.

The next two sections describe the procedure used to fit embedded atom style potentials to the pure elements Ni, Al, B and S, and the cross potentials necessary for treating alloys and segregants. Parts of this have appeared in detail elsewhere [3], but are included here because an emphasis on interatomic potentials is appropriate for the present proceedings. It should be of interest that the potentials for B and S, and their cross potentials with Ni and Al, were fit almost exclusively to results from electronic band structure calculations. The last section presents and discusses results from molecular statics simulations using these potentials.

INTERACTION POTENTIALS FOR Ni, Al, B, AND S

The interatomic potentials employed here have the same form as the embedded atom method of Daw and Baskes [2], though the method of parameterization is different [3]. In the embedded atom approach, the energy of an n-particle homonuclear system is written as

$$E = \frac{1}{2} \sum_{\substack{ij \\ (i \neq j)}}^{n} \phi(r_{ij}) + \sum_i^n F(\bar{\rho}_i) , \tag{1}$$

where r_{ij} is the distance between atoms i and j, ϕ is a pairwise interaction potential, F is the embedding function, and $\bar{\rho}_i$ is the density at atom i due to all its neighbors,

$$\bar{\rho}_i = \sum_{j(\neq i)}^{n} \rho(r_{ij}) . \tag{2}$$

The pairwise potential is taken to a Morse potential,

$$\phi(r) = D_M \{1 - \exp[-\alpha_M(r - R_M)]\}^2 - D_M \tag{3}$$

The three parameters, D_M, R_M, and α_M, define the depth, distance to the minimum, and a measure of the curvature at the minimum, respectively. The density function, $\rho(r)$, is taken as

$$\rho(r) = r^6 [e^{-\beta r} + 2^9 e^{-2\beta r}] , \tag{4}$$

where β is an adjustable parameter. This is the density (ignoring normalization) of a hydrogenic 4s orbital, with the second term added to ensure that $\rho(r)$ decreases monotonically with r over the whole range of possible interaction distance (2^9 is the relative normalization factor for a 4s orbital with a doubled exponent). This was originally chosen for describing first-row transition metals, but has been found to work well for a number of fcc metals.

Rose et al. [9] have shown that the cohesive energy of most metals can be scaled to a simple universal function, which is approximately

$$E_U(a^*) = -E_0(1 + a^*) e^{-a^*} . \tag{5}$$

where a^* is a reduced distance variable and E_0 is the depth of the function at the minimum ($a^* = 0$). Following Foiles et al. [10], $F(\bar{\rho})$ is specified by requiring that the energy of the fcc crystal obeys Eq. (5) as the lattice constant is varied. The appropriate scaling is obtained by taking E_0 as the equilibrium cohesive energy of the solid (E_{coh}), and defining a^* by

$$a^* = (a/a_0 - 1)/(E_{coh}/9B\Omega)^{1/2} , \tag{6}$$

where a is the lattice constant, a_0 is the equilibrium lattice constant, B is the bulk modulus, and Ω is the equilibrium atomic volume. Thus, knowing E_{coh}, a_0, and B, the embedding function is defined by requiring that the crystal energy from Eqs. (5) and (6) match the energy from Eq. (1) for all values of a^*. By fitting $F(\bar{\rho})$ in this way, the potential should be appropriate for a large range of densities. Note that because a^* cannot be expressed neatly as a function of $\bar{\rho}$, the construction of $F(\bar{\rho})$ is performed numerically once $\phi(r)$ and $\rho(r)$ are known.

To be suitable for use in molecular dynamics and molecular statics simulations, the interatomic potential and its first derivatives with respect to nuclear coordinates need to be continuous at all geometries of the system. This is accomplished by forcing $\phi(r)$, $\phi'(r)$, $\rho'(r)$ to go smoothly to zero at a cutoff distance by defining

$$f_{smooth}(r) = f(r) - f(r_{cut}) + \frac{r_{cut}}{m}\left[1 - (\frac{r}{r_{cut}})^m\right]\left(\frac{df}{dr}\right)_{r=r_{cut}},\qquad (7)$$

where $f(r)$ denotes $\phi(r)$ or $\rho(r)$ and m = 20. The cutoff distance, r_{cut}, is used as a parameter in the fitting procedure. So that $F(\bar{\rho})$ is properly defined, $E_U(a^*)$ is also modified to go smoothly to zero when the expanded crystal has a nearest neighbor distance equal to r_{cut}.

Having specified the functional forms for $\phi(r)$, $\rho(r)$, $F(\bar{\rho})$, we now describe the fitting procedure. Because of the way $F(\bar{\rho})$ is determined, the potential always gives a perfect fit to the experimental values of a_0, E_{coh}, and B for any choice of $\phi(r)$ and $\rho(r)$. The remaining five parameters, R_M, D_M, α_M, β, and r_{cut}, are determined by minimizing the root-mean-square deviation (χ_{rms}) between calculated and experimental thermodynamic data. This is accomplished using a simplex procedure[11].

For Ni and Al, the experimental data consists of the three cubic elastic constants (c_{11}, c_{12}, and c_{44}), the vacancy formation energy (ΔE_V^f) and the bond length (R_e) and bond energy (D_e) of the diatomic molecule. In addition, the search was constrained by the requirement that the fcc crystal structure be more stable than either bcc or hcp with ideal c/a ratio. As can be seen in Ref. 3, the fit is excellent for Ni (χ_{rms} = 0.75%) and quite good for Al (χ_{rms} = 3.85%).

For B and S, because experimental thermodynamic data are less readily available and the native crystal forms are more complex, we employed a different approach. Local density band structure calculations were performed on B and S in hypothetical crystal structures using the linearized muffin tin orbital (LMTO) method [12]. The thermodynamic properties calculated using LMTO for three different crystal structures (fcc, bcc, and fcc with a periodic vacancy structure obtained by removing one of the four cubic sublattices) were then used in the parameter searches. The only experimental information guiding the fits were the diatomic data for B_2 and S_2. These fits are shown in Table I.

CROSS POTENTIALS FOR Ni$_3$Al WITH B AND S

For a general alloy system, the energy expansion becomes

Table I. Best fit to LMTO and experimental data for B and S. The values of a_0, E_{coh} and B for the fcc phase match exactly due to the way $F(\bar{\rho})$ is determined.

Source	Property	B "Expt."	B Calc.	S "Expt."	S Calc.
LMTO	fcc a_0 (Å)	2.91		4.143	
LMTO	fcc E_{coh} (eV)	5.33		2.18	
LMTO	fcc B (10^{12}erg/cm^3)	2.02		7.55	
LMTO	bcc a_0 (Å)	2.34	2.33	2.23	3.29
LMTO	bcc E_{coh} (eV)	4.95	5.31	2.40	2.18
LMTO	bcc B (10^{12}erg/cm^3)	2.11	2.12	1.00	0.75
LMTO	3/4 fcc a_0 (Å)	2.78	2.65	3.97	3.94
LMTO	3/4 fcc E_{coh} (eV)	5.78	5.66	2.60	2.12
experiment	diatomic R_e (Å)	1.59	1.62	1.89	1.83
experiment	diatomic D_e (eV)	3.02	3.06	4.41	4.35

$$E = \frac{1}{2} \sum_{\substack{ij \\ (i \neq j)}}^{n} \phi_{t_i t_j}(r_{ij}) + \sum_{i}^{n} F_{t_i}(\bar{\rho}_i) , \qquad (8)$$

where

$$\bar{\rho}_i = \sum_{j}^{n} \rho_{t_j}(r_{ij}) . \qquad (9)$$

and the subscript t_i indicates the type of atom i. For the present work, we construct potentials for alloys with up to 3 components (we do not put B and S into Ni$_3$Al simultaneously). Most of the alloy functions required by Eq. (8) are already known from the pure element fits (ϕ_{NiNi}, ϕ_{AlAl}, ϕ_{BB}, ϕ_{SS}, ρ_{Ni}, ρ_{Al}, ρ_B, ρ_S, F_{Ni}, F_{Al}, F_B, F_S), leaving ϕ_{NiAl}, ϕ_{NiB}, ϕ_{AlB}, ϕ_{NiS}, and ϕ_{AlS} to be determined. Just as for the pure elements, these pair functions are taken as Morse potentials. In addition to optimizing the pair potentials, there are two other types of variation that can be utilized to improve the fit to alloy properties. Examination of Eq. (1) shows that the energy of a homonuclear system is invariant under two types of transformation, scaling $\rho(r)$ and adding a linear term to $F(\bar{\rho})$. For an atom of type A, the ρ-scaling transformation is

$$\rho_A(r) \rightarrow s_A \, \rho_A(r) \tag{10a}$$

$$F_A[\,\overline{\rho}\,] \rightarrow F_A\left[\frac{\overline{\rho}}{s_A}\right], \tag{10b}$$

and the linear embedding term transformation is given by

$$F_A[\overline{\rho}\,] \rightarrow F_A[\overline{\rho}\,] + g_{A\overline{\rho}} \tag{11a}$$

$$\phi_{AA}(r) \rightarrow \phi_{AA}(r) - 2g_{A\rho A}. \tag{11b}$$

While these transformations do not affect the pure element energy, they do affect the energy of an alloy system. We thus use these additional g and s parameters in optimizing the fits to alloy properties. For an n-component alloy, there are $n - 1$ independent scaling transformations [Eq. (10)] and n independent linear term transformations [Eq. (11)].

We first fit the Ni-Al cross potential [ϕ_{NiAl} $(D_M, R_M, \alpha_M, r_{cut}), s_{Al}, g_{ni}, g_{al}$] using experimental thermodynamic data on Ni$_3$Al and B$_2$ phase NiAl, as shown in Ref. 3. Special attention was required to obtain positive energies for the antiphase boundaries (APBs) and the superlattice intrinsic stacking fault energies. Since the time this fit was originally performed, new experimental values have been determined for the {100} and {111} APBs (90 mJ/m^2 and 111 mJ/m^2, respectively [13]), which actually agree better with the calculated values (83 mJ/m^2 and 142 mJ/m^2, respectively). To obtain the cross potentials with B and S, we utilized LMTO properties of the hypothetical alloys Ll$_2$ phase Ni$_3$B, Al$_3$B, Ni$_3$S, and Al$_3$S, and B$_2$ phase NiB, AlB, NiS, and AlS, as shown in Table II. The fit for B is seen to be better than the fit for S. The optimized parameters from all the fits are shown in Table III.

Despite the fact that the various crystal structures of B and S used here are not naturally occurring forms, potentials that give a reasonable description of these higher energy structures are perhaps as well suited to the task at hand, that of describing B and S as minority alloy constituents at Ni$_3$Al GBs, as potentials that are fit to the most stable structures. Two possibly more important questions are (a) how accurate are the LMTO results to which the potentials are fit, and (b) how well can the embedded atom method be expected to do in describing systems containing B and S. Regarding question (a), we note that using the same LMTO method to calculate thermodynamic data for the known crystals, Ni, Al, and Ni$_3$Al leads to deviations from experiment of at most 10% and often significantly less. We thus feel that using the LMTO results instead of experimental data is justified. As for question (b), we note that the embedded atom potential does not have any terms that can describe angular effects arising from directional covalent bonding.

Table II. Results of fit to LMTO data for Ni-B, Al-B, Ni-S, and Al-S cross potentials using Ll$_2$ phase Ni$_3$B, Al$_3$B, Ni$_3$S, and Al$_3$S and B$_2$ phase NiB, AlB, NiS and AlS. For the Ni-B potential, an estimated bond length for diatomic NiB (Re = 1.60 Å) was also included in the fit.

Property	Ni-B		Al-B		Ni-S		Al-S	
	LMTO	Calc.	LMTO	Calc.	LMTO	Calc.	LMTO	Calc.
Ll$_2$ a$_0$ (Å)	3.39	3.39	3.81	3.91	3.59	3.66	4.19	4.30
Ll$_2$ E$_{coh}$ (eV)	4.80	4.80	3.72	3.92	3.96	4.18	3.60	2.89
Ll$_2$ B (10^{12}erg/cm^3)	2.56	2.56	0.93	0.92	2.16	3.15	0.44	0.45
B2 a$_0$ (Å)	2.55	2.58	2.96	2.96	2.97	2.98	3.22	3.56
B2 E$_{coh}$ (eV)	5.30	5.30	4.24	4.27	3.50	3.78	3.90	2.59

Table III. Optimized parameters for embedded atom potentials for Ni-Al-B and Ni-Al-S.

	D$_M$ (eV)	R$_M$ (Å)	α$_M$ (Å^{-1})	β (Å^{-1})	r$_{cut}$ (Å)	s (Å)	g (eVÅ3)
Ni	1.5335	2.2053	1.7728	3.6408	4.7895	1.0000	6.5145
Al	3.7760	2.1176	1.4859	3.3232	5.5550	0.61723	-0.22050
B	0.7182	1.6517	3.1915	2.0108	4.3716	0.0034936	-0.078785
S	0.5976	0.2495	2.0445	4.7632	5.0795	11.9960	2.1058
Ni-Al	3.0322	2.0896	1.6277		5.4639		
Ni-B	0.2822	2.3149	2.4852		2.8181		
Al-B	0.1295	2.8876	1.3904		3.0000		
Ni-S	0.9479	2.4498	2.9784		8.0382		
Al-S	0.7422	2.1196	1.1299		5.2622		

Directional bonding is certainly more important for both B and S than for either Ni or Al. It is possible that when B or S are placed into the highly coordinated, metallic environment, the importance of directional bonding is reduced, though we have no direct evidence of this. The results presented in the following section should thus be viewed with an awareness of this limitation of the interatomic potentials.

GRAIN BOUNDARY SIMULATIONS

The potentials described above have been used for simulations of surfaces [14] and of grain boundaries in Ni, Al, and Ni$_3$Al, and Ni$_3$Al with either B or S segregants [6,15,16]. We summarize the salient features of the GB results here. All simulations were performed using molecular statics techniques, in which a steepest descent search was used to find the minimum energy geometry. Energies thus correspond to T = 0 K. Periodic boundary conditions were employed in the GB plane, while the simulation block was terminated with free surfaces in the direction normal to the GB.

Properties of pure Ni, Al, and Ni$_3$Al [001] symmetric tilt GBs are presented in detail in Refs. 6 and 15. As expected, the GB energy (the energy increase relative to bulk per unit area of the planar GB defect) varies with both the tilt angle and the local composition (excess Al or Ni) of the GB, with the Al rich GBs usually highest in energy. GB energies correlate strongly with the local volume expansion of the boundary. The GBs in pure Al are found to be much lower in energy than in pure Ni. The average GB energy in Ni$_3$Al is very similar to pure Ni, implying that the difference in ductility between the two probably arises from either the difference in yield stress of the surrounding matrix [17], or from a reduced mobility of dislocations in the Ni$_3$Al GB region [18].

When B is introduced into Ni$_3$Al, it is found to prefer interstitial sites, with typical binding energies as follows: (relative to B atom at infinity) interstitial binding site at GB, 5.7-6.9 eV; bulk, 6-Ni octahedral interstitial site (favored over tetrahedral) 4.6 eV; interstitial site near free surface, ~5 eV: surface adsorption site ~5.5 eV. These results are consistent with the experimental observation that B segregates to GBs [19]. In Ni$_3$Al that is slightly Ni rich (the common experimental stoichiometry is 76%Ni), excess Ni atoms exist as antisite defects. Moving an antisite Ni defect from the bulk to the (210) GB is favorable by 0.14 eV, while moving a B from an octahedral interstitial site to this GB is favorable by 1.9 eV. Moving both a Ni and B to the (210) GB is favorable by 2.5 eV, representing a favorable cosegregation effect of 0.46 eV. Because excess Ni at the boundary disrupts the Ll$_2$ ordering, it is likely that dislocation mobility is also enhanced [18]. Ni cosegregation may thus be an important component of the B ductilizing effect.

While we have not yet performed simulations of dislocations near a GB, we have investigated the effect of B on the boundary cohesion. This was done in a simple fashion, by pulling apart the optimized GB with clamps four lattice constants away on either side. At successive strain increments of 2%, the system was allowed to relax to the lowest possible energy. With this approach the system may not break at the same position or in the same way as it would in a slowly strained, thermalized system , but the results are qualitatively instructive nonetheless. The resulting stress-strain curves are shown in Fig. 1 for the Ni$_3$Al (210) GB with segregated Ni and/or B (more detail may be found in Refs. 16 and 20). (Note that both the stress and the strain are much larger than typical experimental values for a macroscopic stress-strain curve. This is because the present simulations give local GB properties). The addition of B and Ni is seen to significantly enhance both the maximum stress the GB can accommodate and the total work done on the GB (given by the area under the stress-strain curve) before failure. In this case, B alone has only a small effect on the GB cohesion; for the (310) GB, B alone has a large effect. We thus conclude that boron may play two roles in enhancing the ductility of polycrystalline Ni$_3$Al: by promoting segregation of Ni to the GB, it disrupts the local order and increases the cohesive strength of the GB.

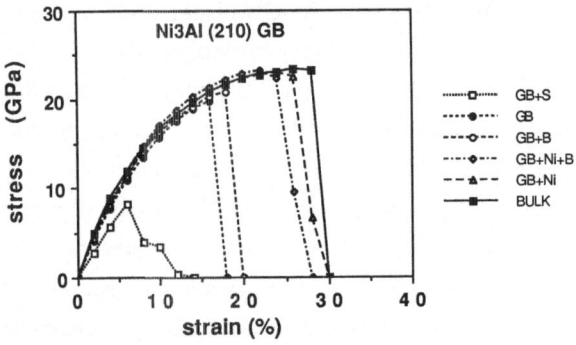

Fig. 1   Stress-strain relationship for Ni₃Al [001] symmetric tilt (210) grain boundary. The computational method is described in the text.

In contrast to B, which prefers to bind GBs, S binds most favorably to the surface (~4.5 eV), in agreement with experimental observations [19]. In the bulk or at the boundaries, S occupies substitutional sites, preferentially displacing Al (bulk Al substitution = 3.0 eV, bulk Ni substitution = 2.2 eV, GB Al substitution = ~0.5 eV), while interstitial binding is repulsive (-3.4 eV for bulk interstitial). If S is placed at the GB, it has a pronounced weakening effect, as shown in Fig. 1. This is consistent with experimental observations of sulfur embrittlement in other materials [21].

REFERENCES

1.   C.T. Liu, C.L. White and J.A. Horton, Acta Metall., 33:213 (1985).
2.   M.S. Daw and M.I. Baskes, Phys. Rev. Lett. 50:1285 (1983); Phys. Rev. B 29:6443 (1984).
3.   A.F. Voter and S.P. Chen, Mater. Res. Soc. Symp. Proc. Vol. 82:175 (1987).
4.   "Grain Boundary Structure and Kinetics", ed. by R.W. Balluffi, ASM, Metals Park, OH (1980).
5.   A.P. Sutton, Int. Metals Rev. 29:377 (1984).
6.   S.P. Chen, A.F. Voter, and D.J. Srolovitz, Scripta Met. 20:1389 (1986).
7.   S.M. Foiles, MRS Proceedings 81:51 (1987).
8.   J.J. Kruisman, V. Vitek, and J.Th.M. DeHosson, Acta. Met. 36:2729 (1988).
9.   J.H. Rose, J.R. Smith, F. Guinea, and J. Ferrante, Phys. Rev. B 29:2963 (1984).
10.   S.M. Foiles, Phys. Rev. B 32:7685 (1985); S.M. Foiles, M.I. Baskes, and M.S. Daw, Phys. Rev. B 33:7983 (1986).
11.   J.A. Nelder and R. Mead, Comp J. 7:308 (1965).
12.   O.K. Andersen, Phys. Rev. B 12:3060 (1975); H.L. Skriver, "The LMTO Method: Muffin-Tin Orbitals and Electronic Structure", Springer-Verlag, Berlin (1984).
13.   J. Douin, P. Veyssière, and P. Beauchamp, Phil. Mag. A54:375 (1986).
14.   S.P. Chen, A.F. Voter, and D.J. Srolovitz, Phys. Rev. Lett. 57:1308 (1986).
15.   S.P. Chen, D.J. Srolovitz and A.F. Voter, J. Mater. Res, 4:62 (1989).
16.   S.P. Chen, A.F. Voter, R.C. Albers, A.M. Boring, and P.J. Hay, Scripta Met., 23:217 (1989).
17.   J. Hack, D.J. Srolovitz, and S.P. Chen, Scripta Met. 20:1699 (1986).
18.   V. Vitek, S.P. Chen, A.F. Voter, J.J. Kruisman, and J.Th.M. DeHosson, in: "Grain Boundary Chemistry and Intergranular Fracture", G.S. Was, ed., Trans. Tech. Publications (1989).

19. C.L. White, R.A. Padgett, C.T. Liu, and S.M. Yalisov, <u>Scripta Met</u>. 18:1417 (1984).

20. S.P. Chen, A.F. Voter, R.C. Albers, A.M. Boring, and P.J. Hay, <u>J. Mater. Res</u>., submitted.

21. "Embrittlement of Engineering Alloys", C.L. Briant and S.K. Banerji, eds. Academic Press, New York, NY (1983).

EMBEDDED ATOM METHOD MODEL FOR CLOSE-PACKED METALS

Dirk J. Oh and Robert A. Johnson

Department of Materials Science
University of Virginia
Charlottesville, VA 22901

INTRODUCTION

A simple embedded atom method (EAM)[1,2] model for close-packed metals has recently been presented[3]. In this model, both the electron density function and the two-body potential are taken as exponentially decreasing functions, and the embedding energy was obtained from a modification of the equation of state given by Rose et al.[4] using the method developed by Foiles[5]. In this model, however, the cutoff procedures for the electron density function and the two-body potential were not specified and the embedding energy was a tabulated function of total electron density. Thus, we refined this model by introducing: (1) a smooth cutoff function for the electron density function and the two-body potential, and (2) an analytic form of the embedding energy function.

THE MODEL AND APPLICATION

The basic equations of the EAM for a monoatomic crystal are[2,3,6]

$$E_{tot} = \sum_i E_i \tag{1}$$

$$E_i = F(\rho_i) + 1/2 \sum_{j \neq 1} \phi(r_{ij}) \tag{2}$$

$$\rho_i = \sum_{j \neq 1} f(r_{ij}) \tag{3}$$

where E_{tot} is the total internal energy of the crystal, E_i is the internal energy associated with atom i, ρ_i is the total electron density at atom i due to all other atoms, $F(\rho_i)$ is the embedding energy of atom i into the electron density ρ_i, $\phi(r_{ij})$ is the two-body central potential between atom i atom j separated by the distance r_{ij}, and $f(r_{ij})$ is the contribution to the electron density at atom i due to atom j at the distance r_{ij} from atom i.

The f, ϕ, and F functions must be given for a complete EAM model. In the present model the simple exponential forms of the f and ϕ functions in Ref. 3 are modified to smooth cutoff exponentials at a cutoff distance r_c:

$$f(r) = f_{old}(r) - f_c(r) \tag{4}$$

$$\phi(r) = \phi_{old}(r) - \phi_c(r) \tag{5}$$

where

$$f_{old}(r) = f_e \, e^{-\beta(r/r_e - 1)} \tag{6}$$

$$\phi_{old}(r) = \phi_e \, e^{-\gamma(r/r_e - 1)} \tag{7}$$

$$f_c(r) = f_{old}(r_c) + g(r) \, f'_{old}(r_c)/g'(r_c) \tag{8}$$

$$\phi_c(r) = \phi_{old}(r_c) + g(r) \, \phi'_{old}(r_c)/g'(r_c) \tag{9}$$

$$g(r) = 1 - e^{\delta(r/r_e - r_c/r_e)} \tag{10}$$

Here r_e is the equilibrium interatomic separation distance. The cutoff functions $f_c(r)$ and $\phi_c(r)$ ensure $f(r_c) = f'(r_c) = 0$ and $\phi(r_c) = \phi'(r_c) = 0$, respectively. This cutoff is similar to that used by Voter et al.[7].

As a form of $F(\rho)$, we choose

$$F(\rho) = a(\rho/\rho_e)^n + b(\rho/\rho_e) \tag{11}$$

where, from Eq. (3),

$$\rho_e = \sum_m f(r_e^m) \tag{12}$$

Here r_e^m is the mth neighbor distance to a particular atom at equilibrium. If n equals 0.5, then this embedding function reduces to that of Finnis-Sinclair[8]. We find that this form of embedding function most closely fits the embedding function predicted in Ref. 3 and therefore yields an equation of state very close to the one given by Rose et al.[4].

There are nine EAM parameters to be determined for the present model: f_e, β, δ, r_c, γ, ϕ_e, a, b, and n. The parameter f_e cancels in a monatomic crystal because only ratios of

ρ's occur in Eq. (11), and therefore should be determined from alloy properties such as heat of mixing as done by Foiles[5,9] which will not be considered in the present model. Although β can be obtained from fitting the free atom densities, it can be treated as an adjustable parameter. The cutoff function parameter δ is taken as 20 in the present model. The cutoff distance r_c is arbitrarily chosen, except for general range from considering the lattice stability against other lattice structures. The remaining five EAM parameters, γ, ϕ_e, a, b, and n are determined from the following five governing equations, details of which are given in Ref. 3:

$$F(\rho_e) + 1/2 \sum_m \phi(r_e^m) = -E_c \tag{13}$$

$$1/2 \sum_m [(r_{i,e}^m)^2/r_e^m] \psi'(r_e^m) = 0, \quad i = 1,2,3 \tag{14}$$

$$1/2 \sum_m (r_e^m)^2 \psi''(r_e^m) + F''(\rho_e) (r_e \rho_e')^2 (1 - 3Q) = 15\Omega G_v \tag{15}$$

$$1/2 \sum_m (r_e^m)^2 \psi''(r_e^m) + F''(\rho_e) (r_e \rho_e')^2 = 9\Omega B_v \tag{16}$$

$$1/2 \sum_m \psi(r_e^m) - F''(\rho_e)/2 \sum_m [f(r_e^m)]^2 = -E_v^{uf} \tag{17}$$

where

$$\psi(r) = 2F'(\rho_e) f(r) + \phi(r) \tag{18}$$

$$\rho_e' = \sum_m [r_e^m / r_e] f'(r_e^m) \tag{19}$$

$$Q = q_1 q_2 + q_2 q_3 + q_3 q_1 \tag{20}$$

$$q_i = \{ \sum_m [r_{i,e}^m)^2 / r_e^m] f'(r_e^m) \}/r_e \rho_e' \tag{21}$$

Here $r_{i,e}^m$ is the ith Cartesian component of r_e^m, E_c is the cohesive energy, Ω is the atomic volume at equilibrium, B_v is the Voigt average Bulk modulus, G_v is the Voigt average shear modulus, and E_v^{uf} is the unrelaxed vacancy formation energy.

Eq. (14) is the EAM equilibrium condition. In cubic metals, due to cubic symmetry, the three conditions in Eq. (14) reduce to

$$1/2 \sum_m r_e^m \psi'(r_e^m) = 0 \tag{22}$$

However, in hcp metals, we have the following equilibrium condition in addition to Eq. (22):

$$1/2 \sum_m [(z_e^m)^2 / r_e^m] \psi' (r_e^m) = 0 \qquad (23)$$

where the z direction is along [0001] of hcp crystal. This equation is used to fit c/a ratios of hcp metals.

Using Eq. (15), Eq. (16) can be rewritten to

$$F''(\rho_e) (r_e \rho_e')^2 Q = 3\Omega B_v - 5\Omega G_v \qquad (24)$$

This equation describes the Cauchy discrepancy as shown in Ref. 3.

The Q is equal or close to 1/3 for cubic metals or hcp metals with c/a ratios near the ideal value of $\sqrt{8/3}$, respectively. Thus, Eq. (15) can be simplified to

$$1/2 \sum_m (r_e^m)^2 \psi'' (r_e^m) = 15\Omega G_v \qquad (25)$$

This equation is strictly true for cubic metals, but a very good approximation for hcp metals with near ideal c/a ratios.

Since $F''(\rho_e)$ is much smaller than $F'(\rho_e)$ in the present model, Eq. (17) can be approximated as

$$1/2 \sum_m \psi(r_e^m) = -E_v^{uf} \qquad (26)$$

The effective pair potential, ψ, defined in Eq. (18), is considered by many authors[2,6,8,10]. The ψ is reminiscent of empirical pair potentials[11,12]. In constructing an empirical pair potential, the vacancy formation energy, instead of the cohesive energy, was chosen to adjust the well depth of potential, causing the cohesive energy-vacancy formation energy dilemma[13]. Eq. (26) justifies this approach since it demonstrates that the sum of the pair energies over all neighbors is related to the vacancy formation energy, not the cohesive energy. Also, Eq. (25) shows that pair potentials yield only the shear resistance of metals (G_v), not the resistance to volume change (B_v), which is the basis of the Cauchy discrepancy. The EAM clears these two well-known paradoxes associated with pair potentials through the embedding energy $F(\rho)$ as shown in Eqs. (13) and (16).

The EAM parameters for the seven fcc metals, copper, silver, gold, nickel, palladium, platinum, and aluminum, are listed in Table I. The parameters were determined by fitting to the experimental values given in Ref. 3. The stability of the fcc structure for all seven metals was examined by calculating the energy of the bcc and hcp phases. The fcc structure was found to be most stable. Formation energies of single vacancies and self interstitials were computed in the same way as described in Ref. 3. Results were very close to those reported in Ref. 3. This was expected because, with $\delta > 10$, the cutoff functions in Eqs. (8) and (9) are perturbed from the original functions in Eqs. (6) and (7) only near to r_c.

Table I.   The EAM model parameters for the seven fcc metals. Here ϕ_e and r_c are expressed in eV and r_e, respectively, the other parameters being dimensionless.

	Cu	Ag	Au	Ni	Pd	Pt	Al
β	5	6	6	6	5.5	6	5
γ	8.5	7.5	8	9	9	9.5	10.5
δ	20	20	20	20	20	20	20
r_c	1.9	1.9	1.9	1.9	1.9	1.9	1.9
ϕ_e	.36952	.85184	.54608	.63199	.42409	.48675	.12538
a	-4.0956	-5.0929	-39.701	-3.1684	-123.85	-32.563	-4.8144
b	-1.6979	-2.9961	32.428	-5.1237	117.36	23.841	.47685
n	.44217	.65777	.92504	.12501	.9796	.86934	.39948
ρ_e	12.793	12.551	12.551	12.551	12.665	12.551	12.793

Table II.   The EAM model parameters for the four hcp metals. Here ϕ_e and r_c are expressed in eV and basal lattice constant, respectively, the other parameters being dimensionless. Here [c/a] is the c/a ratio of hcp metal.

	Mg	Ti	Zr	Co
β	6	6	6	6
γ	10.5	8	8.7	10.5
δ	20	20	20	20
r_c	1.7	1.7	1.7	1.7
ϕ_e	.14720	.86629	.65727	.33557
a	-1.1049	-4.0942	-5.7121	-3.8668
b	-1.3122	-6.1649	-4.6234	-2.5909
n	.18638	.19908	.18921	.22497
ρ_e	12.316	12.437	12.391	12.316
[c/a]	1.62307	1.61564	1.61844	1.62307

The EAM parameters for four hcp metals, magnesium, titanium, zirconium, and cobalt, are listed in Table II. The parameters were determined by fitting to the experimental values given in Ref. 3. In all four metals, the hcp structure was found to be energetically favoured over the bcc and fcc structures. Relaxation calculations of formation energies of single vacancies and various self-interstitial configurations were carried in the same way as described in Ref. 3. Results were very close to those in Ref. 3.

SUMMARY

A simple EAM model for close-packed metals has been developed. Both the electron density function f(r) and two-body potential ϕ(r) are approximated by exponentially decreasing terms with a smooth cutoff as shown in Eq. (4) and Eq. (5), respectively. The embedding function $F(\rho)$ is taken as the sum of two terms; one with the electron density raised to an adjustable power and the other linear in the electron density as given in Eq. (11). This form of embedding function is chosen because it most closely fits the tabulated embedding function predicted in Ref. 3 and therefore yields an equation of state very close to the one given by Rose et al.[4] The model parameters are fitted to atomic volume, bulk and shear moduli, cohesive energy, and vacancy formation energy (also c/a ratio for hcp metals). The model has been applied successfully to seven fcc metals and four hcp metals, yielding correct stability relative to other structures. The extension of the present model to bcc transition metals was unsuccessful due to difficulty in matching the low shear anisotropy ratio C_{44}/C' with $C' = (C_{11} - C_{12})/2$ of these metals. However, the present model can be applied to the bcc alkali metals Li, Na, and K, which have the high C_{44}/C'.

ACKNOWLEDGMENTS

Support from the U.S. Department of Energy, Office of Basic Energy Sciences, Division of Materials Science by Grant Number DE-FG05-86ER45246 is gratefully acknowledged.

REFERENCES

1.   M.S. Daw and M.I. Baskes, Semiempirical, quantum mechanical calculation of hydrogen embrittlement in metals, Phys. Rev. Lett., 50:1285 (1983).

2.   M.S. Daw and M.I. Baskes, Embedded atom method: derivation and application to impurities, surfaces, and other defects in metals, Phys. Rev. B, 29:6443 (1985).

3.   D.J. Oh and R.A. Johnson, Simple embedded atom method model for fcc and hcp metals, J. Mater. Res., 3:471 (1988).

4.   J.H. Rose, J.R. Smith, F. Guinea, and J. Ferrante, Universal features of the equation of state of metals, Phys. Rev. B, 29:2963 (1984).

5.   S.M. Foiles, Calculation of the surface segregation of Ni-Cu alloys using the embedded atom method, Phys. Rev. B, 32:7685 (1985).

6.   R.A. Johnson, Analytic nearest-neighbor model for fcc metals, Phys. Rev. B, 37:3924 (1988).

7.   A.F. Voter and S.P. Chen, Accurate interatomic potentials for Ni, Al, and Ni$_3$Al, MRS Proceeding, 82:175 (1987).

8.   M.W. Finnis and J.E. Sinclair, A simple empirical N-body potential for transition metals, Philos. Mag. A, 50:45 (1984).

9.   S.M. Foiles, M.I. Baskes, and M.S. Daw, Embedded atom method functions for the fcc metals Cu, Ag, Au, Ni, Pd, Pt, and their alloys, Phys. Rev. B, 33:7983 (1986).

10.   S.M. Foiles, Application of the embedded atom method to liquid transition metals, Phys. Rev. B, 32:3409 (1985).

11.   R.A. Johnson, Interstitials and vacancies in α iron, Phys. Rev., 134:A1329 (1964).

12.   R.A. Johnson, Point defect calculations for an fcc lattice, Phys. Rev., 145:423 (1966).

13.   R.A. Johnson, Interatomic potential development in materials science, in: "Computer Simulation in Materials Science", R.J. Arsenault, J.R. Beeler, Jr., and D.M. Esterling, ed., American Society for Metals, Metals Park (1987).

BOUNDARY CONDITIONS FOR QUANTUM CLUSTERS
EMBEDDED IN CLASSICAL IONIC CRYSTALS

John M. Vail

Department of Physics
University of Manitoba
Winnipeg, MB R3T 2N2
Canada

INTRODUCTION

In order to understand the boundary condition problem, consider a single electron trapped at a negative ion vacancy in a crystal: it is called an F center. It is bound in a potential well due to the net positive charge near the vacancy. Treating the ions as point charges produces one picture. Taking account of the electronic structure of the ion produces a physically more correct picture, in which each ion adds its core repulsion to the original effective potential. In this picture, the quantum-mechanical analysis is based on a complete set of one-electron basis functions, which may be delocalized. It is possible to transform, by unitary transformation, to a new set of basis functions which are localized on individual sites. The unitary transformation leaves the physics of the system unchanged, but in the equation that determines the one-electron functions, the repulsive potentials of surronding ions are reduced. The contribution to the effective potential, from the transformation, which produces this reduction is called a Kunz-Klein localizing potential (KKLP).[1]

In practice, for a point defect such as an F center, it is not possible to include the ion-size effect of all ions in the quantum-mechanical calculation. However, if one includes only a few near neighbors, one may introduce spurious diffuseness into the solution, in which the trapped electron can tunnel beyond the barrier of a finite set of repulsive core potentials. In this case, introducing KKLP will reduce the potential barrier, possibly eliminating the spurious diffuseness, while maintaining some of the true ion-size effect. The choice of level of approximation for KKLP with a finite, incomplete basis set will affect the manifold of one-electron states which are occupied, and therefore affects the predicted physical results. Because KKLP is rigorously part of the all-electron treatment of the crystal, approximations and subsequent improvements can be clearly defined.

We mention that it is, of course, possible that the defect is truly physically diffuse. The only way to test this is to include ion-size effect for a sufficient number of neighbors to enclose the region of diffuseness.

In our work, we simulate a point defect or a region of perfect crystal lattice by a quantum-mechanical molecular cluster of ions embedded in an infinite classical ionic lattice. The cluster is treated in unrestricted Hartree-Fock (UHF) self-consistent field (SCF) approximation. The classical lattice is described by the shell-model. Consistency between the cluster and lattice is achieved variationally, by minimizing the total energy with respect to the electronic wave function and nuclear positions of the cluster, and simultaneously with respect to shell-model ion positions and dipole moments. Cluster-lattice interaction includes Coulomb forces, and either shell-model short-range forces or KKLP. We refer to this theoretical method and the associated automated computer programs as the ICECAP methodology.[2]

EQUILIBRIUM: PERFECT-LATTICE CLUSTERS

A basic test of the ICECAP methodology is whether it can simulate a region of perfect lattice. We present an example of such a test, namely the fluorine ion F^- and its six nearest-neighbors, Na^+, in NaF.[3] In Table 1, some properties of three related clusters are given. The clusters are Na_6F, $Na_6F(K4)$, and $F(K4')$, where the notations (K4) and (K4') indicate that KKLP is applied out to and including fourth neighbors of the central F^- ion. Thus (K4) has KKLP on second, third, and fourth neighbors, while (K4') has KKLP on first through fourth neighbors. The properties calculated are: d_0, the nearest-neighbor equilibrium distance from F^- in the cluster; K', the mean effective harmonic force constant with nearest neighbors for compression and expansion of the cluster, compared to values obtained from the shell-model representation of the cluster ions; and ΔE, the mean deviation of cluster from shell-model values of the energy required for such distortions at 5% of the lattice spacing.

The results in Table 1 show that K' is much closer to the ideal value of 1.0 (since the shell model is derived from macroscopic static and dynamic properties of the crystal), when KKLP is included, than when it is omitted, and that for reasonable equilibrium spacing, the nearest-neighbor ion-size effect must be represented by including these neighbors in the UHF cluster. The discrepancies $d_0 = 0.97$ (compared with 1.0) and $\Delta E = 0.04$ eV indicate that embedded $Na_6F(K4)$ is an acceptable simulation of the perfect lattice region.

CHARGE TRANSFER: F^+ CENTER IN MgO

Since the ions in MgO are divalent, a single electron trapped by an O^{2-} vacancy has net charge $+|e|$, and is called an F^+ center. There is a question whether the trapped electron is located mostly in the vacancy or on its nearest-neighbor Mg^{2+} ions, in their unoccupied 3s or 3p states, or a bit of both. We shall refer to Mg^{2+} occupancy as charge transfer from the vacancy.

In Table 2 we give results for two calculations on the ground state of the F^+ center in MgO^3. In both cases, the UHF cluster included the six nearest-neighbor Mg^{2+} ions and the vacancy, but in one case KKLP was included to fourth neighbors, and in the other case it was omitted. The table contains the Mulliken population of the vacancy site, the orbital radius of the state which is dominated by vacancy-centered basis functions, and the nearest-neighbor Mg^{2+}-vacancy distance, in units of perfect lattice spacing. We note that without KKLP, about 0.05e per Mg^{2+} ion is held in the 3s state, for a total of about 0.3e transferred from the vacancy, whereas with KKLP the charge transfer is negligible, 2%. Clearly KKLP, that is, the ion-size effect of second through fourth neighbors, prevents the trapped electron from spreading out onto its nearest neighbor Mg^{2+} ions. We also note from Table 2 that the trapped electron is well-localized, with orbital radius $\langle r^2 \rangle^{1/2} = 0.59$, and that the net charge of the defect forces the nearest Mg^{2+} ions outward, to d = 1.03, both numbers in units of lattice spacing.

LATTICE DISTORTION: DEFECTS IN MgO

Our most extensive defect calculations to date[4] have been for the ground state properties in Hartree-Fock approximation of a set of related defects: F^+, F, $[H^-]^+$, $[H^{2-}]^0$, in MgO. The notation refers to the F^+ center of sec. 3, the two-electron F center, the substitutional H^- ion at an O^{2-} site, and the corresponding H^{2-} substitutional. The latter,

Table 1.   Fluorine-Centered Perfect-Lattice Clusters in NaF. d_0 is equilibrium cluster spacing (units: perfect lattice spacing); K' is mean effective harmonic force constant (units: shell model value); DE is mean energy deviation from shell model at 5% distortion (units: eV).

	Na6F	Na6F(K4)	F(K4'
do	0.96	0.97	0.90
K'	1.43	1.19	0.86
DE	0.09	0.04	0.03

Table 2.   F+-Center Ground State in MgO. Nearest-neighbor UHF cluster with and without KKLP. Mulliken population at vacancy site (Pop.), orbital radius $\langle r^2 \rangle^{1/2}$ (units: lattice spacing); nearest-neighbor equilibrium distance d (units: lattice spacing).

KKLP	Pop.	$\langle r^2 \rangle^{1/2}$	d
with	0.98	0.59	1.03
without	0.69	0.64	1.06

along with the F center, is an electrically neutral defect, while each of the other two has net charge +|e|. In this work, considerable effort was spent in refining basis sets, and ultimately UHF clusters were used which contained the defect, and first and second neighbors, with KKLP for third and fourth neighbors. Such clusters may be denoted $Mg_6O_{12}:D(K4)$, where D refers to the defect center. Our first requirement was to determine an accurate basis set for O^{2-}, and this was done using an $Mg_6O(K4)$ cluster.

In Table 3 we give the equilibrium nearest-neighbor Mg^{2+} and second-neighbor O^{2-} distances, d_1 and d_2 respectively, from each of the four defects. For the two positively charged defects F^+ and $[H^-]^+$, the results are the same, showing that Coulomb effects predominate. In fact, the same result is obtained for the neutral $[H^{2-}]^0$ defect. The reason is that, in our results, H^{2-} consists of a $1s^2$ core, plus a 2s-like state with orbital radius of 2.95, so that to first and second neighbors the defect appears to have net charge +|e|. For the two-electron F center, its softness compared to the O^{2-} ion that it replaces, and its localization and electrical neutrality, cause both first and second neighbors to relax inward by 2%.

Table 3.   Lattice Distortion by Ground State Defects in MgO. Equilibrium first (Mg^{2+}) and second (O^{2-}) neighbor distances for defects D (units: perfect lattice distances): d_1 and d_2 respectively.

D=	F^+	F	$[H^-]^+$	$[H^{2-}]^0$
d1	1.03	0.98	1.03	1.03
d2	0.97	0.98	0.97	0.97

A sensitive test of the entire method is to compare calculated electronic spin densities with those deduced from experiment. For the F^+ center in MgO, the experimental spin density at nearest-neighbor Mg^{2+} nuclei[5] is 0.27 A^{-3}. Our calculated value is 0.17 A^{-3}, which is respectable agreement. By contrast, if lattice relaxation is ignored, we obtain a value of 0.53 A^{-3}, a result which emphasizes the sensitivity of the quantity to ionic displacement.

By deforming the embedded defect cluster in symmetrical and asymmetrical modes one can tabulate the energy as a function of configuration, and then with a shell-model representation of the cluster, determine shell-model parameters for the defect which reproduce this energy dependence. We have done this for H⁻ in MgO, with plausible results.[4,6] Furthermore, we have applied the resulting defect shell model to calculate the activation energy for H⁻ diffusion by the vacancy mechanism in MgO, with the result 4.92 eV.[6] This may be compared with the experimental value for O^{2-} self-diffusion in MgO, of 4.5 eV.[7]

LOCALIZATION: DIFFUSENESS

In the introduction we described how introducing the ion-size effect on a small number of neighboring ions could produce spurious diffuseness for a defect. We now present an example where this occurs, and another example where the result is still in question.

We first consider the unrelaxed excited state of the F^+ center in Mg[4,8]. This is the final state of the optical excitation process of the defect. The state is 2p-like, and we evaluated the energy as a function of the range of the 2p-like orbital, where range R is qualitatively similar to orbital radius $<r^2>^{1/2}$. Four cases were considered: (1) all ions, including nearest neighbors, were represented by the shell model: (2) nearest-neighbors were included with the defect in the UHF cluster, denoted $F^+: Mg_6$; (3) same as case (2), but with KKLP added, out to fourth neighbors, denoted $F^+:Mg_6(K4)$; (4) both first and second neighbors included in the UHF cluster, with KKLP to fourth neighbors, denoted $F+:Mg_6O_{12}(K4)$. In case (1), the optimal range of the F+-center orbital was about 0.5 lattice spacings (denoted a). In case (2), with UHF nearest-neighbor ion-size effect, $R \approx$ 1.5a was optimal, with a subsidiary minimum remaining at 0.5a. In case (3), with (K4) added, the optimal R moved out to 2a, with the subsidiary at 0.5a remaining. Finally in case (4) with first and second neighbor UHF ion-size effect, and (K4), the long-range energy minimum disappeared, and the optimum was $R \approx$ 0.5a. Thus, if one had not carried the ion-size effect to the full extent of case (4), one would have been misled into thinking that the state was much more diffuse than is indicated by case (4). The use of a more flexible basis set for O^{2-} in case (4) remains to be done: it may restore some diffuseness, or it may not.

In sec. 4 we noted[4] that the $[H^{2-}]^0$ 2s-like state had orbital radius 2.95a, when ion-size effect was extended to fourth neighbors, whose distance from the defect center is 2a. In this case we cannot be sure whether the diffuseness is spurious or not. To resolve the question, ion-size effect would need to extend beyond the corresponding optimal range of the 2s state.

CONCLUSION

We have briefly described the role of Kunz-Klein localizing potentials (KKLP) in providing mathematically consistent and physically appropriate boundary conditions for embedded quantum clusters. We have given an example where KKLP strongly improves the effective nearest-neighbor force constant in a perfect lattice cluster, namely for F^- in NaF. In another example, the localization of the F+ center in the vacancy in MgO, rather than partial charge transfer to nearest Mg^{2+} ions, is produced by KKLP. Our most detailed work to date, which includes KKLP to fourth neighbors associated with second-neighbor UHF clusters for defects in MgO, provides plausible lattice distortion by the defects, gives acceptable spin density for the F^+ center, and produces impurity shell-model parameters that lead to reasonable impurity diffusion results. This work also illustrates the role of adequate ion-size description in determining the extent of diffuseness of a defect state.

ACKNOWLEDGMENTS

The results presented here are the product of collaborations with R. Pandey, A.H. Harker, and A.B. Kunz. Partial financial support was provided by NSERC Canada and by AERE Harwell.

REFERENCES

1   A.B. Kunz, Phys. Rev. B26:2056 (1982); A.B. Kunz and D.L. Klein, Phys. Rev.
        B17:4614 (1978); A.B. Kunz and J.M. Vail, Phys. Rev. B, 38:1058
        (1988); A.B. Kunz, J. Meng, and J.M. Vail, Phys. Rev. B, 38:1064 (1988).
2   J.H. Harding, A.H. Harker, P.B. Keegstra, R. Pandey, J.M. Vail and
        C. Woodward, Physica 131B:151 (1985).
3   J.M. Vail, R. Pandey and A.H. Harker, Cryst. Latt. Def. and Amorph. Mat. 15:13
        (1987).
4   R. Pandey and J.M. Vail, J. Phys : Cond. Mat., in press (1989).
5   W.P. Unruh and J.M. Culvahouse, Phys. Rev. 154:861 (1967).
6   J.M. Vail, A.B. Kunz, J. Meng and R. Pandey, in preparation (1988).
7   J. Narayan, J. Appl. Phys. 57:2703 (1985).
8   R. Pandey, Ph.D. Thesis, University of Manitoba (1988).

PHYSICAL PROPERTIES OF GRAIN-BOUNDARY MATERIALS: COMPARISON OF EAM AND CENTRAL-FORCE POTENTIALS

D. Wolf, J. Lutsko, and M. Kluge

Materials Science Division
Argonne National Laboratory
Argonne, IL 60439

Three types of grain-boundary phenomena expected to be particularly sensitive functions of a local-volume dependence in the interatomic interaction potentials employed are investigated by means of many-body (embedded-atom and Finnis-Sinclair) and pair potentials. These phenomena are the zero-temperature volume expansion localized at the grain boundaries, the *local* elastic constants of grain-boundary materials, and their high-temperature stability. The same qualitative behavior is found in all these phenomena for both types of potentials, from which it is concluded that the local-volume dependence, incorporated in the many-body potentials only, does not have a strong effect on the predicted properties of grain-boundary materials. The reasons for these similarities are thought to arise from the fact that most grain-boundary properties are governed, as are those of liquids, by atoms in very close contact; i.e., by the short-range part of the interatomic potential which is of a central-force type in both sets of potentials. However, many-body potentials are expected to represent a given material better than pair-potentials since (i) they usually permit a larger number of adjustable parameters to be fitted to real-material properties, and (ii) in contrast to equilibrium pair-potentials they do not satisfy the Cauchy relation.

INTRODUCTION

A fundamental characteristic of all interfacial systems is that they are intrinsically inhomogeneous; i.e., the physical properties in the interfacial region differ (in some cases significantly) from those of the bulk perfect crystal. In grain-boundary (GB) materials the inhomogeneity in the physical properties arises from the structural disorder at the GBs and the subsequently modified (usually lower) mass density due to the volume expansion normal to the GBs.

In monatomic metals the lower mass density gives rise to a lower conduction electron density localized at the GB. The interaction of atoms in the GB region with other atoms is therefore strongly anisotropic in that interactions with atoms on the opposite side of the interface are mediated by a lower electron density than interactions with atoms on the same side. From a conceptual point of view this anisotropy rules out the use of central-force potentials for GB simulations. Nevertheless, much atomistic simulation work on GBs employing pair potentials has been performed during the last two decades. In much of this work the volume expansion at the GB was suppressed in order to avoid the conceptual difficulties arising from the anisotropic electron density near the interface. However, this volume expansion and the related optimum translational state parallel to the interface represent the most pronounced structural features of a GB. Therefore, conclusions drawn from such constant-volume simulations are questionable. On the other hand, constant-pressure GB simulations employing equilibrium pair potentials suffer from the fact that these potentials are isotropic and do not depend on volume.

With the development in recent years of many-body embedded-atom-method (EAM) potentials[1,2] these problems associated with the use of pair potentials can now be avoided, thus enabling much more realistic atomistic simulations of interfacial systems. In contrast to central-force potentials, in the EAM the local-volume dependence of the binding energy of a metal is *explicitly* taken into account via the net electron density experienced by a given ion. This density arises from the surrounding metal atoms, and inhomogeneous electron-density distributions associated with locally inhomegeneous atom configurations are thus accounted for. Conceptually EAM potentials are therefore based on a much better physical description of metallic bonding than are pair potentials.

In this article we will investigate whether, and to which degree, the conceptual differences between these two types of potential predict different physical properties of GB materials. Instead of abolishing the use of pair potentials for GB simulations altogether, in all simulations discussed below both an EAM and a pair potential have been employed in an attempt to elucidate the importance of many-body effects in GB properties.

For this comparison we will choose properties which are expected to be particularly sensitive functions of volume changes. For example, it is well known that a volume expansion of typically about 2% between zero temperature and melting causes a decrease in the elastic constants of most materials by typically about 50%. This small volume expansion is also thought to play an important role in the process of melting. With substantially larger--yet localized--volume expansions believed to be present in many GBs, the elastic behavior and high-temperature stability of GB materials should provide a particularly fruitful testing ground for an assessment of the differences between many-body and pair potentials.

The article is organized as follows: After a more detailed comparison of EAM and pair potentials in Sec. 2, the correlation between the zero-temperature energy, structure, and volume expansion of GBs in fcc and bcc metals will be reviewed in Sec. 3. The local elastic response of an isolated GB and the anomalous elastic properties of a superlattice of GBs will be discussed in Secs. 4 and 5, respectively. Finally, in Sec. 6 molecular dynamics simulations on the high-temperature stability of GBs, and the possible existence of a "premelting" transition in the GB region, will be compared for the two types of potentials, followed by some general conclusions concerning the importance of many-body effects in GBs which are formulated in Sec. 7.

COMPARISON OF MANY-BODY AND PAIR POTENTIALS

In developing interatomic potentials, instead of solving Schroedinger's equation for a complex system of atoms and electrons, empirical concepts are frequently adopted for reasons of tractability. In fact, to postulate an interatomic potential governed by the *atom* positions alone represents such an empirical concept. The "derivation" (or better, development) of an interatomic potential thus consists of two basic steps. First, some conceptually rationalizable analytical form is chosen for the total potential energy of the system of N particles, $U(\{r_N\})$, which depends on the particle positions r_N. This basic form represents the theoretical framework; it may consist of a simple central-force expression or the EAM starting expression discussed below. Second, this basic analytical form postulated to describe the dependence of $U(\{r_N\})$ on $\{r_N\}$ is fitted to a certain number of experimentally determined quantities, such as the zero-temperature lattice parameter and some or all of the perfect-crystal elastic constants, the sublimation, melting, or vacancy formation energy, or others.

The result of this two-step procedure is an empirical interatomic potential "for a given material" although the only properties of this material correctly described by such a potential are the ones used in the fitting. Whether or not other properties are predicted more or less accurately needs to be established in every case. Obviously, within a given conceptual framework (such as a central-force potential) a potential is expected to be more realistic if more experimental quantities are used in the fitting; i.e., if the basic analytical form contains more adjustable parameters. Thus, not surprisingly, with three instead of only two adjustable parameters the Morse potential has often been found to describe a wider spectrum of properties of a given material than the Lennard-Jones potential. One might ask whether other forms of pair potentials which allow fitting to even more empirical parameters can represent an even broader spectrum of properties. Indeed, as pointed out by Johnson almost 25 years ago[3,4], potentials consisting of a number of splines in different distance ranges may be fitted, in principle, to an arbitrary number of experimental results. However, owing to the intrinsic limitations of central-force potentials certain materials properties cannot be reproduced regardless of the number of disposable paramerters. For example, all equilibrium central-force potentials for cubic crystals satisfy the Cauchy relation, $C_{12}=C_{44}$; instead of the three independent elastic constants, any pair potential for a cubic crystal with the proper equilibrium lattice parameter therefore yields only two independent elastic constants. This shortcoming of central-force potentials has been a major motivation in the development of a broader conceptual framework for the description of interatomic interactions, particularly in metals.

Central-force potentials provide a good description of the short-range repulsion between the atoms. However, in metals the attraction between atoms at larger separations is intrinsically of a many-body nature and therefore not well described by a pair potential. The attraction between ions in a metal arises from the lowering of the energy of the electron gas when embedding the ion cores in the Fermi sea. The corresponding binding energy, $F_i(\rho_i^T)$, depends on the total electron density, ρ_i^T, at the position of atom i. In both formulations of the embedded atom method,[1,2] the contributions to $U(\{\underline{r}_N\})$ are written as follows:

$$U\left(\{\underline{r}_N\}\right) = \sum_i F_i\,(\rho_i^T) + \sum_{i<j} \phi(\underline{r}_{ij})\,. \tag{1}$$

In writing Eq. (1), three basic assumptions are usually made[1,2,5]: (a) The electron densities are assumed to be centrally symmetric; i.e., the density $\rho_j(\underline{r})$ contributed by some atom j at $\underline{r}$ is assumed to depend on the distance $|\underline{r}|$ only. (b) By assuming *atomic* electron densities, all electron densities are assumed to be independent of the environment; i.e., self-consistent rearrangements in the electron gas are ignored. The total electron density in (1) may then be subdivided as follows:

$$\rho_i^T = \sum_{j \neq i} \rho_j^a\,(r_{ij})\,, \tag{2}$$

where the superscript "a" refers to the atomic density. (c) The short-range repulsion is assumed to be of a central force type, i.e., $\phi(\underline{r}_{ij}) \equiv \phi(|\underline{r}_{ij}|)$.

Up to this point the two formulations of the EAM due to Daw and Baskes[1,5] and to Finnis and Sinclair[2] are identical. The two approaches differ merely in the specific expressions for $\rho^a(r)$, $F(\rho)$, and $\phi(r)$. In the original paper of Finnis and Sinclair[2] analytical expressions, including six disposable parameters, were presented for $F(\rho)$, $\rho^a(r)$, and $\phi(r)$ in Eqs. (1) and (2). Ackland and Thetford[6] have recently added a term (and two more adjustable parameters) to $\rho^a(r)$ in order to enhance the short-range repulsion for very short distances. In this modified form the many-body potentials of Finnis and Sinclair allow for eight parameters to be fitted to empirical data, although by choosing a less complicated (for example, Born-Mayer) form for $\rho^a(r)$ from the outset this number could be reduced to six or even five with probably no loss in the physical content of the potentials. In the formulation of EAM many-body potentials due to Daw et al.[1,5] numerical tables are given for the functions in Eqs. (1) and (2). As pointed out recently,[7] this renders the procedure of fitting their six disposable parameters not very transparent.[5]

We note that rather different physical arguments were put forth by Daw and Baskes and by Finnis and Sinclair to rationalize the EAM starting expression (1). Whereas Daw and Baskes argued within the framework of density functional theory, Finnis and Sinclair drew on insights gained from the tight-binding approximation and from band-structure theory. This difference is irrelevant, however, once the basic forms (1) and (2) are chosen for the fitting to the six experimental numbers. The appearance of the total charge density at the position of each atom, defined in Eq. (2), in the EAM potential has been interpreted as an explicit dependence of the potential on local density.[1,2] For example, an atom at a free surface "knows" that it is interacting with fewer atoms than in the bulk because its total charge density is lower. However, it may equally well be said that implicitly the same is true for a pair potential because (i) the perfect-crystal cohesive energy depends on the lattice parameter (i.e., volume) and (ii) in a defected crystal the energy of an atom depends on the local environment. In that sense, it appears to us that the two types of potential are rather similar physically.

In contrast to pair potentials, the basic form (1) permits a perfect cubic crystal to be described by all three elastic constants. The six experimental numbers usually used to fit the six disposable parameters in EAM potentials for monatomic metals therefore include all three elastic constants, the zero-temperature equilibrium lattice parameter, the sublimation energy, and the vacancy formation energy. For alloys an additional parameter is introduced which is fitted to the heat of mixing. Although a Johnson-type or other spline pair potential could be fitted essentially to the same parameters (with the exception of one elastic constant), the EAM potential is expected to describe a given material better since it does not require that the Cauchy relation be satisfied.

ENERGY-FREE VOLUME CORRELATION FOR GBs IN fcc AND bcc METALS

The understanding of the relationship between the structure and energy of GBs has been a subject of considerable interest during the past two decades. For the reasons outlined in the Introduction, in most GB simulations employing pair potentials the volume expansion at the GB has been suppressed. This volume expansion can now be investigated systematically by means of EAM potentials. By its very definition the excess free volume, δV, is one-dimensional (i.e., parallel to the GB-plane normal) since, in the GB plane, the unit-cell dimensions are fixed by the surrounding bulk regions. δV is thus a volume expansion per unit area, and is usually given in units of the lattice parameter a_0.

A systematic investigation of the correlation between the GB structure, energy, and volume expansion has been performed recently by Wolf[8,9] for both many-body and pair potentials. In these studies symmetrical and asymmetrical twist and tilt boundaries were considered. For lack of space we will only review a few representative results. The GB energies, E^{GB}, and related δV values obtained for symmetrical GBs on the two densest planes in fcc [the (111) and (100) planes] are shown in Figs. 1 and 2, respectively.[8] Similar curves were obtained for symmetrical GBs on the (110) and (113) planes, representing the third- and fourth-densest planes in fcc.[8] Several features in these results are particularly noteworthy.

(a) The two potentials predict a remarkably similar qualitative behavior for both $E^{GB}(\theta)$ and $\delta V(\theta)$. Notice, however, that whereas the absolute values for the GB energy are rather similar, the volume expansion is generally larger for the Lennard-Jones (LJ) potential. This difference will be discussed later in this section.

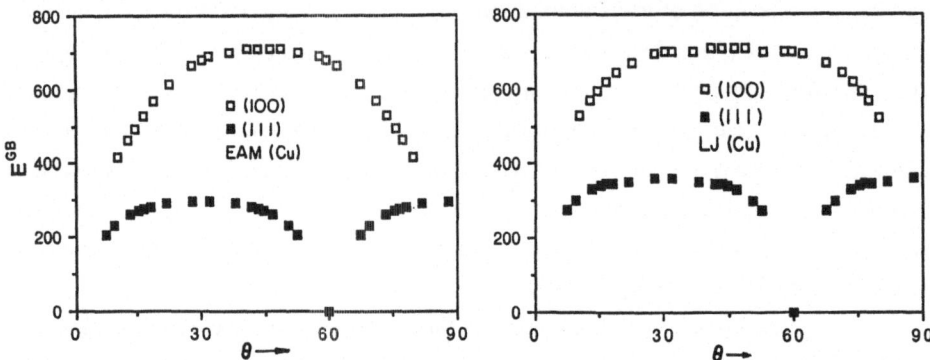

Fig. 1   Energy (in erg/cm^2) of symmetrical GBs on the (100) and (111) planes versus twist angle for an EAM and a Lennard-Jones (LJ) potential fitted to Cu (see Ref. 8).

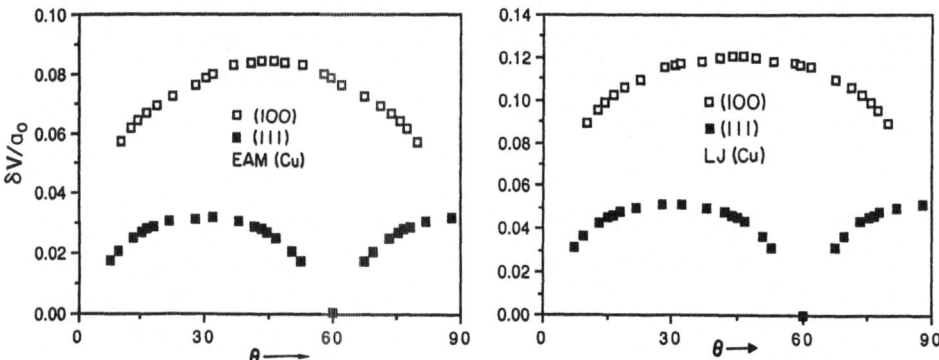

Fig. 2   Volume expansion (in units of the lattice parameter a_0) for the GBs in Fig. 1 (see Ref. 8).

(b) The $E^{GB}(\theta)$ and $\delta V(\theta)$ curves are relatively smooth and of similar shape. For the (100) plane the shape of the $E^{GB}(\theta)$ curve was investigated experimentally[11] and it was found that, with the exception of a small cusp near $\theta = 36.87°$, the curve is, indeed, smooth and levels off for larger angles.

(c) For all four lattice planes considered in Ref. 8, $E^{GB}(\theta)$ and $\delta V(\theta)$ initially increase strongly, in accordance with the Read-Shockley model for low-angle GBs.[10] For larger angles the dislocation cores overlap completely and, according to Figs. 1 and 2, both $E^{GB}(\theta)$ and $\delta V(\theta)$ level off. The GB is then considered "saturated" with elastic strain energy.

(d) Both $E^{GB}(\theta)$ and $\delta V(\theta)$ depend strongly on the GB plane. Thus the energies of the (100) boundaries are more than twice as large as those of the (111) boundaries. The energies and free volumes of the (110) and (113) boundaries are larger yet.[8]

(e) The results for both potentials are in remarkable qualitative agreement with earlier constant-volume calculations for these boundaries in which a variety of different pair potentials was employed.[12-14] Allowing for the volume expansion at the GB merely reduces the GB energy but does not alter the strong preference in the GB energy for the (111) over the (100) plane.[8]

The underlying causes for the large differences in E^{GB} and δV for the different planes were discussed in detail in Ref. 8. As illustrated there, the amount of structural disorder, as evidenced in the related radial distribution functions, increases sharply with decreasing interplanar spacing or planar mass density. Because of the direct correlation between structural disorder and energy, the GB energy subsequently increases sharply with decreasing interplanar spacing, $d(hkl)$, of the lattice planes parallel to the GB plane. We note that for the fcc lattice, $d(111)=0.577a_0$, $d(100)=0.5a_0$, $d(110)=0.354a_0$, and $d(113)=0.302a_0$.

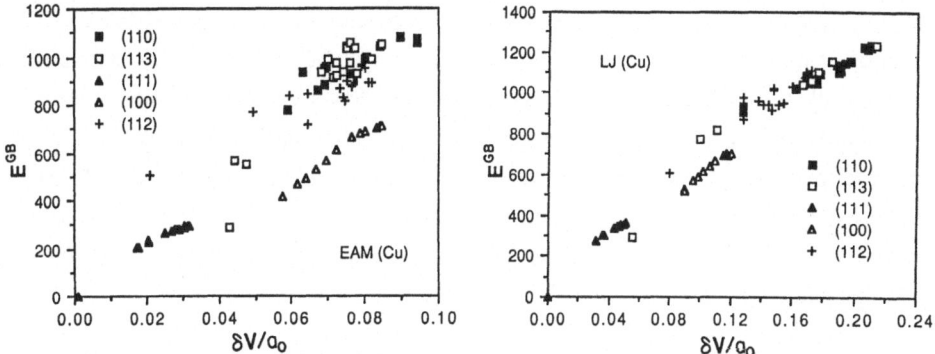

Fig. 3   GB energy vs. volume expansion for the GBs in Figs. 1 and 2 and for the symmetrical GBs on the (110), (113), and (112) planes considered in Ref. 8.

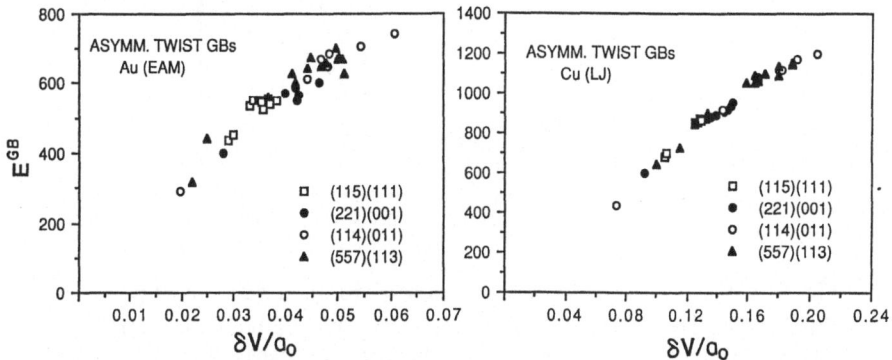

Fig. 4 GB energy vs. volume expansion for asymmetrical twist (i.e., general) GBs determined by means of an EAM potential for Au and the LJ potential for Cu employed throughout this paper. The different points, for example, for the (115) (111) combination of lattice planes differ by the twist angle about the plane normal (see Ref. 8).

The very similar shapes of the $E^{GB}(\theta)$ and $\delta V(\theta)$ plots suggest a plot of E^{GB} vs. δV. As shown in Fig. 3 both potentials show a remarkable correlation between the GB energy and volume expansion for all four lattice planes considered in Ref. 8 and for the GBs on the (112) plane. A similar correlation is observed for symmetrical tilt boundaries[8] and for asymmetrical twist (i.e., general) boundaries involving four different combinations of lattice planes in the two halves of the related bicrystals. The latter are shown in Fig. 4. Notice that the same LJ potential has been employed in Figs. 1-4; by contrast, the EAM results in Fig. 4 were obtained for an EAM potential fitted to Au, as opposed to the results in Figs. 1-3 which were obtained for an EAM potential for Cu. Since the LJ potential has a single length and energy scale, no qualitative differences arise from different parameterizations. Although the LJ potential employed here has been fitted to Cu, the qualitative results are entirely independent of this particular choice.

The significantly larger volume expansion obtained for the same GB when employing the LJ potential is readily understood. As illustrated in Fig. 5, the much deeper EAM potential (with E_{coh} = -3.43eV at the minimum) rises much more steeply for increasing lattice parameter than the shallower LJ potential (with E_{coh} = -1.03eV). The creation of new volume, for example, when generating a GB, is hence much more costly energetically for the EAM potential, with subsequently lower volume expansion at the GB. One could easily construct a different pair potential with a much deeper well and which increases more steeply for increasing values of the lattice parameter, with the likely result that volume effects such as those shown in Figs. 1-3 would be more similar for the two potentials. A similar analysis, leading to the same conclusion, can be performed for the EAM potential for Au (with E_{coh} = -3.92eV).

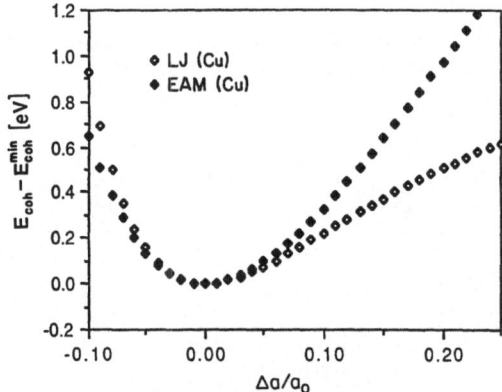

Fig. 5 Cohesive energy vs. lattice parameter for the EAM and LJ potentials fitted to Cu. The values of the cohesive energy at equilibrium, i.e., for $\Delta a = a - a_0 = 0$, were subtracted to emphasize the different shapes of the curves (see Ref. 8).

If it is, indeed, the spacing of lattice planes parallel to the GB plane that determines the GB energy and volume expansion, one would expect rather different results for GBs in the bcc structure. The densest planes in the bcc lattice are the (110) planes followed by the (100) planes (with $d(110)=0.707a_0$). As illustrated in Figs. 6 and 7, the GBs on the (110) plane show, indeed, a much lower energy and volume expansion than the GBs on the (100) plane. In these calculations[9] a many-body potential due to Finnis and Sinclair[2] (with a modified short-range repulsion[6]) was compared with a spline pair potential fitted to the bulk properties of a α-Fe by Johnson.[3] Apart from the different lattice planes involved, the general features in Figs. 6 and 7 are similar to those for the fcc boundaries outlined in (a)-(e) above. Not surprisingly, a similar correlation between E^{GB} and E^{GB} and δV is therefore exhibited in Fig. 8.

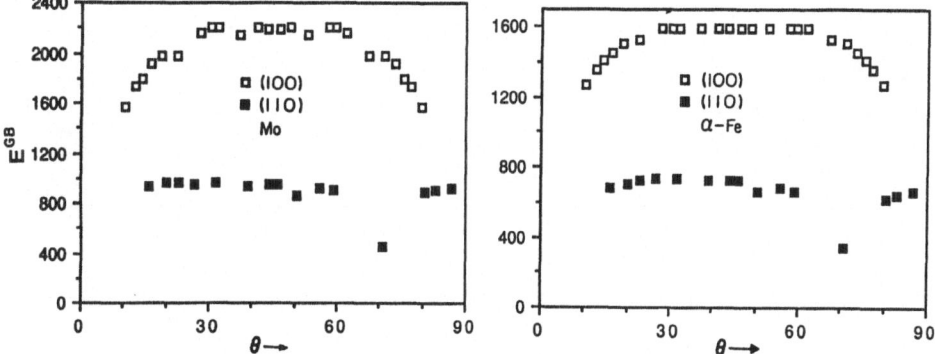

Fig. 6 Energy of symmetrical GBs on the two densest planes of the bcc lattice obtained by means of Johnson's pair potential[3] for a α-Fe and Finnis and Sinclair's many-body potential[2] for Mo as modified by Ackland and Thetford[6] (see Ref. 9).

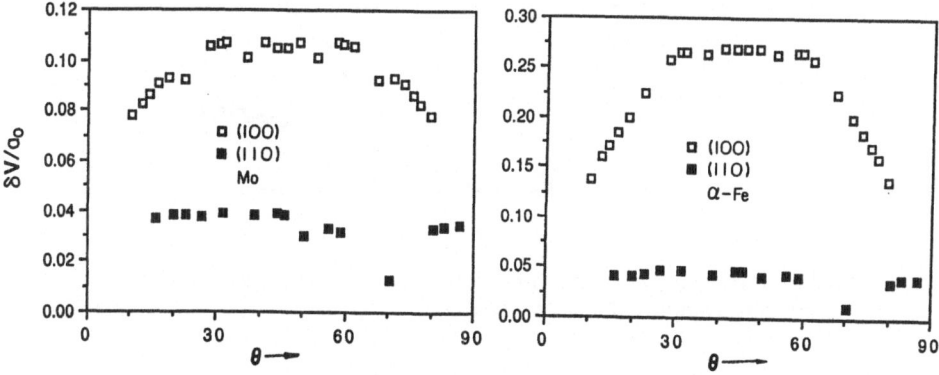

Fig. 7 Volume expansion for the GBs in Fig. 6 (see Ref. 9).

An interesting qualitative difference between the two potentials is exhibited in Fig. 8, however, which shows a positive (i.e., upward) curvature in the $E^{GB}(\delta V)$ curve for the many-body potential but a negative (i.e., downward) curvature for the pair potential. As for the fcc boundaries, this difference is readily understood in terms of the related cohesive-energy vs. lattice-parameter curves shown in Fig. 9. Notice that again (cf. Fig. 5) the many-body potential has a much larger cohesive energy (-6.82eV) at the equilibrium lattice parameter than the pair potential (-1.54eV). For a small increase in lattice parameter the many-body potential leads to a much sharper increase in E_{coh} (with positive curvature) than the pair potential (with negative curvature for only moderate expansion), with subsequently smaller volume expansion at the GB.

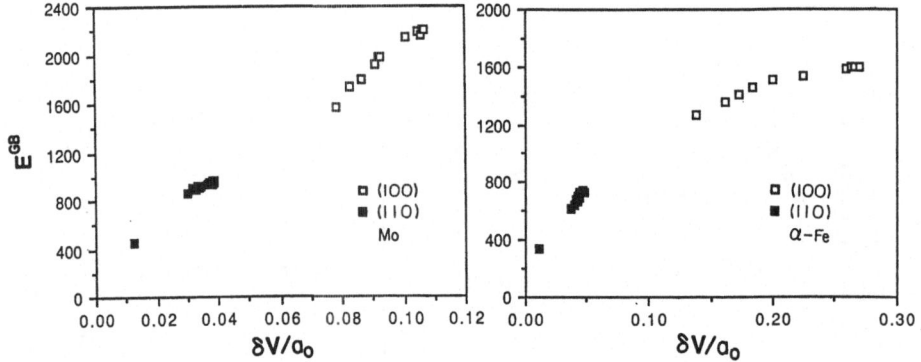

Fig. 8   GB enervy vs. volume expansion for the GBs in Figs. 6 and 7 (see Ref. 9).

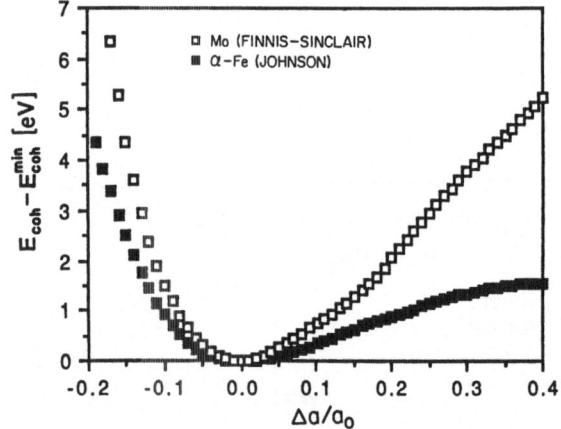

Fig. 9   Cohesive energy vs. lattice parameter, similar to Fig. 5, for the two bcc potentials employed here (see Ref. 9).

LOCAL ELASTIC CONSTANTS NEAR GBs

It has been noted in the Introduction that grain-boundary materials, by their very nature, are inhomogeneous systems.   In addition to the reduced density in the grain-boundary region which arises from the local volume expansion at the interface (see Sec. 3), these inhomogeneities are particularly noticeable in the elastic properties near interfaces.

Most phenomenological models describing interface-related properties do not take into account the local variation of the elastic constants near an interface. As examples we think of (i) segregation models in which the bulk moduli are used to describe the accomodation of an over- or undersized impurity in a GB, and (ii) fracture models based on continuum elasticity in which bulk, space-independent elastic constants and moduli are used to describe the elastic response near a crack tip. Atomistic simulations represent an ideal, and in some cases the only, method to determine such local properties in highly inhomogenous systems. In this section we will briefly summarize some very recent advances in understanding the correlation between the local elastic response near a GB and the structural disorder at the interface.

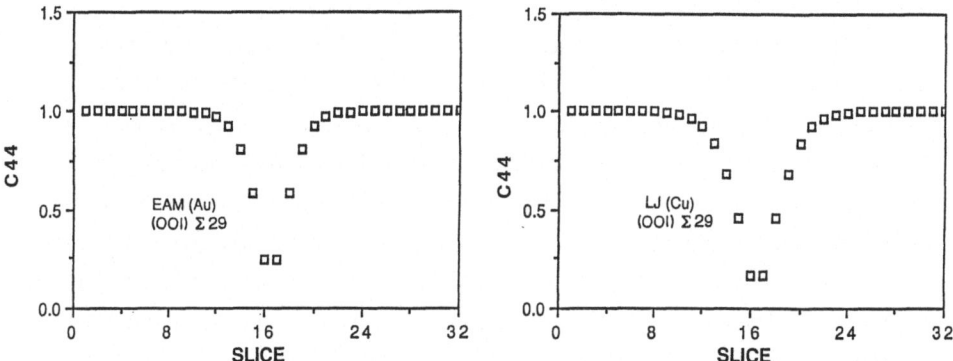

Fig. 10 Local shear elastic constant, C_{44}, normalized to its bulk value, for the (100) $\theta = 43.60°$ ($\Sigma 29$) twist boundary for an EAM potential fitted to Au and the LJ potential (see Ref. 17).

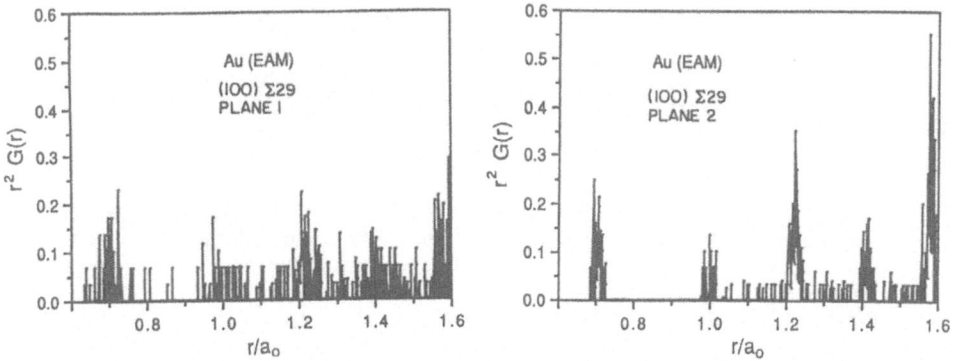

Fig. 11 Radial distribution functions, $r^2 G(r)$, for the two (100) planes nearest to the GB whose local shear elastic constants are shown in Fig. 10(a) (see Ref. 17).

Fig. 10 shows the local shear constant, C_{44}, for shear parallel to the GB plane as a function of distance from the (100) $\theta = 43.60°$ ($\Sigma 29$) twist boundary in the fcc lattice. In these zero-temperature calculations the EAM potential for Au and the LJ potential for Cu, used also in Sec. 3, were employed in a combined lattice-statics and lattice-dynamics evaluation of local-elastic constant expressions derived recently.[16] The remarkable feature in Fig. 10 is the pronounced decrease, observed for both potentials, in the shear resistance at the GB, noticeable within the four or five (100) planes nearest to the GB plane. The other elastic constants also show anomalies, although much less pronounced, in this region.[17] The interpretation of these anomalies, presented in Ref. 17, is based on the destruction of the perfect-crystal ...ABAB... stacking at the GB in the <100> direction which, together with the subsequent volume expansion, leads to the large reduction in C_{44}. The rapidly decreasing degree of structural disorder as a function of distance from the GB plane is illustrated in Fig. 11 in which the radial distribution functions, $r^2G(r)$, for the two (100) planes nearest to the GB plane are shown for the EAM potential.[8]

THE SUPERMODULUS EFFECT

The discovery of the "supermodulus effect" in the biaxial modulus of composition-modulated strained-layer superlattice structures of Au-Ni and Cu-Pd[18] and its subsequent detection in a variety of metallic superlattice materials (see, for example, Ref. 19) has drawn considerable attention to the possibility of designing interface materials with elastic properties otherwise not achievable in bulk materials. Different qualitative explanations for these anomalies, based on the different electronic properties of the constituent metals, have been proposed.[20] Remarkably, these models disregard any role the interfaces, as structural defects, might play, and hence predict the absence of such elastic-constant anomalies in GB materials.

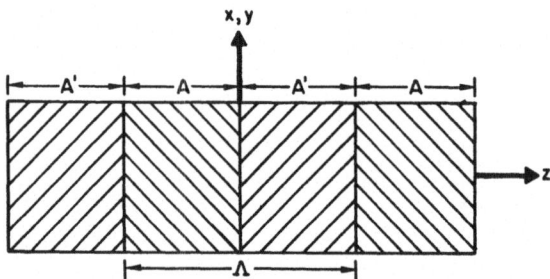

Fig. 12   Periodic arrangement of thin slabs A and A' to form a "grain-boundary superlattice". A and A' are slabs of the same material which were rotated about the plane normal (z ‖ [100]) to form a periodic array of twist boundaries (x-y plane).

To eliminate electronic-structure differences, Wolf and Lutsko[21,22] have determined the zero-temperature elastic constants in a so-called "grain-boundary superlattice." Such a (somewhat hypothetical) material consists of a periodic arrangement, ...|A|A'|A|A'|..., of thin slabs of A and A' of equal thickness, $\Lambda/2$ (see Fig. 12). In contrast to a composition-modulated alloy, however, A and A' consist of the same material and are merely rotated with respect to each other about the z axis by the twist angle θ (between A|A') or $-\theta$, respectively (between A'|A). The variation with wavelength of the elastic moduli normalized to their bulk values are shown in Fig. 13 for the LJ potential and in Fig. 14 for the Au (EAM) potential. In the right halves of these figures the elastic moduli exhibiting the largest anomalies as functions of the modulation wavelength are shown.

The main difference between the GB superlattice calculations discussed here and the consideration of an isolated GB in the preceding section is that only in the $\Lambda \rightarrow \infty$ limit are the GBs arranged in the superlattice embedded in bulk matter. The isolated GB considered in Figs. 10 and 11 therefore represents the $\Lambda \to \infty$ limit for the GB superlattice. Due to this embedding the volume expansion at the GB (see Sec. 3 above) cannot give rise to a contraction in the x-y plane as one would expect from the Poisson effect. Instead, the x-y lattice parameter in the GB region is determined by the bulk perfect crystals between which the GB region is embedded.

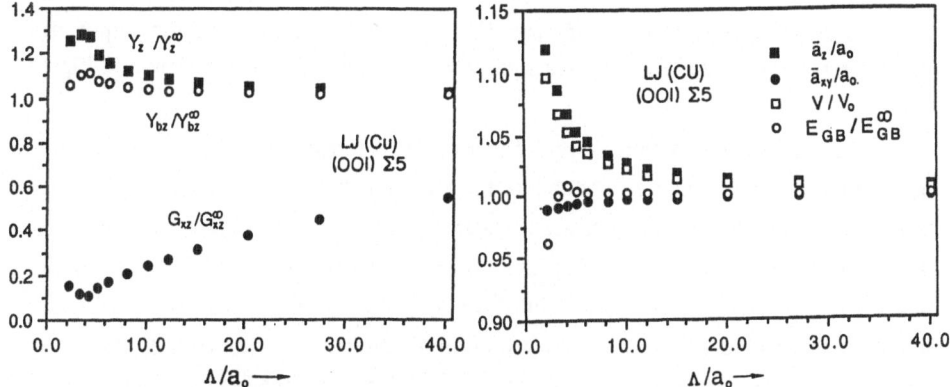

Fig. 13 Normalized elastic moduli (left) and normalized average lattice parameters,volume expansion, and GB energy (right) for a superlattice of (100) $\theta = 36.87°$ ($\Sigma 5$) twist GBs determined by means of the LJ potential for Cu[22].

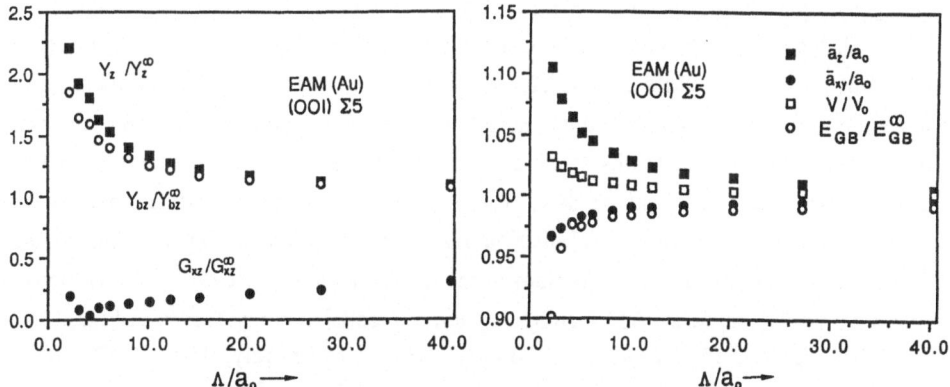

Fig. 14 Normalized elastic moduli (left) and normalized average lattice parameters,volume expansion, and GB energy (right) for a superlattice of (100) $\theta = 36.87°$ ($\Sigma 5$) twist GBs determined by means of the EAM potential fitted to properties of Au[22].

When, for decreasing values of Λ, these perfect-crystal regions are gradually removed, the Poisson effect can actually occur. The volume expansion at the GBs (which results in an increase in the average lattice parameter, $\bar{a}_z$, in the z direction) consequently gives rise to a contraction in the GB (x-y) plane, i.e., a decrease in the average lattice parameter, $\bar{a}_{xy}$. The breaking of the degeneracy in the lattice parameter is illustrated in

Figs. 13(b) and 14(b). Also shown is the steady increase in sample volume, V/V_0, and the decrease in the average GB energy as Λ decreases.

The GB considered in these superlattice calculations is the (100) $\theta = 36.87°$ (so called $\Sigma 5$) twist boundary in the fcc lattice. As shown in Ref. 17, the local elastic constants of this GB are very similar to those of the $\Sigma 29$ boundary considered in Sec. 4 above. This is not suprising if one recalls that these two GBs have practically the same energies and volume expansions (see Figs. 1 and 2). Similarly, the superlattice calculations for the $\Sigma 29$ GB yield practically the same results as those shown in Figs. 13 and 14.

Figures 13 and 14 show several remarkable features. First, the strong increase in both Young's and the biaxial modulus in the z direction, Y_z and Y_{bz}, is coupled with a large decrease in the resistance towards shear parallel to the GB plane. This increase is present in spite of the *increase* in the average z lattice parameter (see Figs. 13(b) and 14(b)). Second, a minimum in the shear modulus, G_{xz}, is observed for $\Lambda \approx 15$ Å, which is in agreement with a similar minimum observed experimentally for several dissimilar-material strained-layer superlattices.[23,24] Third, for the Lennard-Jones potential a maximum is observed in Y_z and Y_{bz} at the same value of Λ at which G_{xz} shows a minimum. By contrast, the EAM potential shows no such maximum. Finally, in qualitative agreement with experiments (see, for example, Ref. 24), a close connection exists between elastic-modulus anomalies in Figs. 13(a) and 14(a) and the accompanying changes in interatomic spacings in Figs. 13(b) and 14(b).

A detailed interpretation of these anomalies in terms of the structural disorder at the GBs was presented in Refs. 21 and 22. In particular, it was shown that the increase in the Young's modulus in the z direction, in spite of an *increase* in the mean distance between the atoms, is due to some atoms which are pushed together more closely than the perfect-crystal nearest-neighbor distance d_{NN}. These atoms give rise to a significant density for $r < d_{NN}$ in the corresponding radial distribution functions similar to Fig. 11. Lutsko and Wolf[22] also showed that the minimum in the shear modulus arises from the attractive interaction between GBs. As illustrated by the related GB-energy curves in Figs. 13(b) and 14(b), for $\Lambda \lesssim 15$ Å ($\approx 4a_0$) the GB energy decreases strongly; thus the GBs attract one another.

Figures 13 and 14 display some interesting qualitative differences between the LJ and EAM results. Whereas the LJ potential shows a maximum in Young's and the biaxial modulus, the EAM potential yields a practically steady increase in both as Λ decreases. These differences were shown[21,22] to arise from the rather different magnitudes of the Poisson contraction in the x-y plane obtained for the two potentials. For the superlattice with the smallest value of Λ, $\bar{a}_{xy} \approx 0.989a_0$ ($0.966a_0$) for the LJ (EAM) potential. When this effect is suppressed the results shown in Fig. 15 are obtained. Without the Poisson effect the EAM potential now also shows a maximum in Y_z and Y_{bz}; however, the magnitude of the anomalies in these moduli is substantially reduced and is now of the same order of magnitude as for the LJ potential (see Fig. 13). Also, the attraction between the

GBs is greatly reduced when the Poisson effect is suppressed. The comparison of Figs. 15 and 14 illustrates the role of the Poisson effect in *enhancing* the anomalies because lower GB energy means less disorder in the overall system.[8] The appearance of the maximum in Young's and the biaxial modulus therefore arises from the competition between the Poisson effect and the attraction between the interfaces. (For further details see Ref. 22.) The qualitative differences obtained for the two potentials are therefore closely connected with the different magnitudes predicted for these competing processes. As mentinoed in Sec. 3, the creation of new volume is much more costly energetically for the EAM than for the LJ potential (see Fig. 5). The Poisson effect in response to the volume expansion at the GB is therefore much larger for the EAM potential.

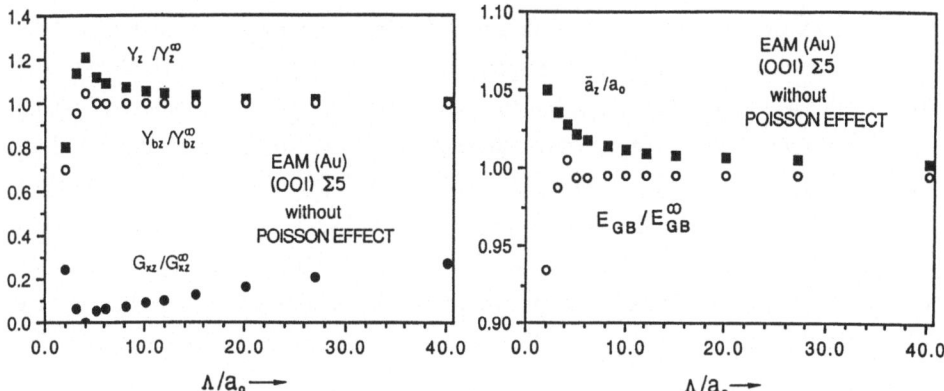

Fig. 15   Normalized elastic moduli (left) and normalized average lattice parameters, volume expansion, and GB energy (right) for a superlattice of (100) $\theta = 36.87°$ ($\Sigma 5$) twist GBs in which the Poisson contraction parallel to the (100) GB planes was prevented.[22]

HIGH-TEMPERATURE STABILITY OF GRAIN BOUNDARIES

Computer simulations of the high-temperature behavior of grain boundaries (GBs) in metals have, in the last few years, suggested that these GBs may undergo an order-disorder structural transformation at temperatures of the order of half the melting temperature of the undefected ideal crystal.[25-29] However, the results of several such simulation studies are inconsistent with one-another in that in some[25,29] a continuous disordering of the GB region was observed as the melting temperature is approached[25] whereas in others[26-28] a relatively sharp first-order phase transition seems to occur in which the solid-solid interface becomes a solid-liquid-solid "premelted" system.[27,28] Also, these simulation studies disagree with experiments in which a transition to a solid-liquid-solid interface could not be detected even very close to the actual melting point.[30]

One of the key difficulties in all these studies is the effect of the border conditions on the simulated behavior. The borders present a problem because planar interfacial systems are, by definition, inhomogeneous in the direction perpendicular to the interface. The imposition of ordinary periodic borders in three dimensions (3-d PBCs) is, thus, somewhat artificial, giving rise to two interfaces in the simulation cell. In the case of grain boundaries this results, at high temperatures, in the extreme consequence of the two grain boundaries migrating towards each-other and mutually annihilating.[25] To avoid these problems, a new border condition has been proposed recently[31] which introduces movable perfect-crystal blocks at the two ends of the simulation cell in the z-direction. This model allows the simulation of isolated interfaces embedded in bulk perfect crystal.

In the first application of this new border condition, the high-temperature stability of the (100) $\theta = 43.60°$ ($\Sigma29$) grain boundary was investigated.[32] The zero-temperature equilibrium structure and local elastic constants of this boundary were investigated in Secs. 3 and 4, respectively, The two factors motivating the choice of this geometry were the relatively large interplanar spacing of the (001) planes and the relatively large primitive planar unit cell of the $\Sigma29$ grain boundary. Because of the large interplanar spacing, the lattice planes are easily distinguished thus making clear the onset of disorder should the boundary prove unstable at elevated temperatures. The large planar unit cell allows us to consider this a "generic" high-angle boundary as opposed to boundaries with small planar unit cells for which the energy is known to be especially sensitive to translations and for which multiple primitive planar unit cells must be combined in a non-primitive simulation cell to satisfy the condition that no particle interact with its own image. Similarly, a *twist* rather than a *tilt* boundary (as has been commonly used in the study of the high-temperature stability of GBs) was chosen because the symmetrical tilt boundary on a given lattice plane has the smallest planar unit cell of all GBs on that plane and, hence, is also highly sensitive to translations.

Because of the inhomogeneity of the GB system, properties were monitored locally by dividing the system into 32 slices in the z-direction with each slice initially containing one (001) plane. Properties such as potential energy, temperature, mean square displacement in the plane and perpendicular to the plane were then tabulated for each slice. Finally the squared magnitude of the static structure factor, $S(\underline{k})$, was also monitored for each slice. The static structure factor is the Fourier transform of the density, and its squared magnitude is given by:

$$\left| S(\underline{k}) \right|^2 = \left\{ \frac{1}{N} \sum_{i=1}^{N} \cos\left(\underline{k} \cdot \underline{r}_i\right) \right\}^2 + \left\{ \frac{1}{N} \sum_{i=1}^{N} \sin\left(\underline{k} \cdot \underline{r}_i\right) \right\}^2 , \tag{3}$$

where $\underline{r}_i$ is the position of the ith atom. For the overall $S(\underline{k})$, the sums include all atoms while for the planar structure factor, $S_p(\underline{k})$, they only include the atoms in a given plane. When the wave vector is chosen as a reciprocal lattice vector in the x-y plane, $S_p(\underline{k})$ is a measure of planar order. At zero temperature, in the ideal crystal, $S_p(\underline{k})$ then equals unity while in the liquid it fluctuates close to zero. Since the lattice planes in the two halves of the bicrystal are rotated relative to one another, two wave vectors must be used (each a reciprocal lattice vector in the corresponding half of the bicrystal) to monitor the two planar symmetries present in the system. These wave vectors are simply related by the relative rotation, characterized by the twist angle $\theta = 43.60°$ for the $\Sigma29$ boundary. Thus, for each plane we monitor $S_p(\underline{k_1})$ and $S_p(\underline{k_2})$ to determine how much it deviates from the symmetry (i.e., planar order and relative rotation) of a perfect (001) plane in the corresponding half of the bicrystal.

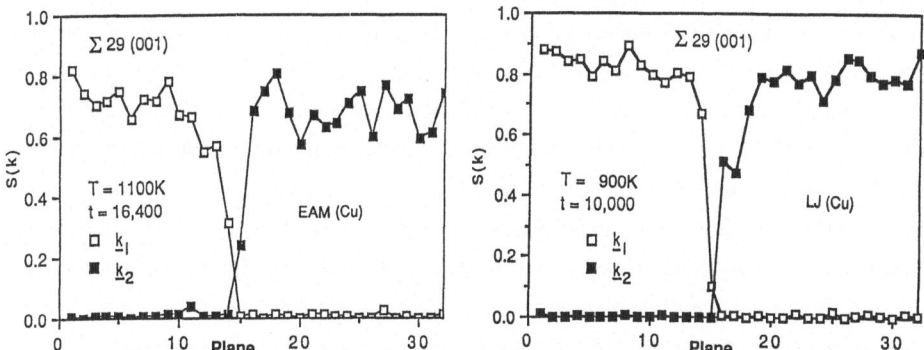

Fig. 16   Planar structure factors in the two halves of a (100) $\theta = 43.60°$ ($\Sigma 29$) bicrystal for the Cu EAM and LJ potentials. The LJ results at T = 1050K are practically identical to the ones shown here at 900K.[31]

Figure 16 shows the plane-by-plane planar static structure factor profiles obtained after full equilibration of the GB at 1100K for the EAM potential[32] and 900K for the LJ potential.[31] Results for the latter at 1050K are virtually indistinguishable from those at 900K shown in Fig. 16.[31] To assess the relevance of these results it is important to know the thermodynamic melting point, T_m, obtained for the two potentials. From a free-energy analysis they were found to be approximately 1170K and 1100K, respectively, for the Cu (EAM) and Cu (LJ) potentials.   According to Fig. 16, at these temperatures, corresponding to 94% and 95.5% of the melting point, the GB is very sharply defined with no indication of a premelting transition of any kind. Moreover, a careful and systematic investigation of bicrystals, using the EAM potential for Cu and the Stillinger-Weber potential for silicon,[33] has shown that thermodynamic melting is nucleated at the GBs with subsequent spreading of the two solid-liquid interfaces into the solid.[32,34] The propagation velocity for the kinetics of this spreading of the liquid phase was found to increase with increasing temperature above T_m.

CONCLUSIONS

In the preceding four sections we have reviewed the computer simulations of grain-boundary phenomena which are particularly sensitive functions of the volume expansion at the interface. In all cases considered a central-force potential yields qualitatively the same results as a many-body (EAM or Finnis-Sinclair) potential. The differences between the two types of potential, mostly quantitative, have been shown to arise from the differences in the related cohesive-energy vs. lattice-parameter curves for the perfect crystal.

This result is particularly relevant if one considers that GB systems are intrinsically *inhomogeneous* whereas the cohesive-energy curves are those for a *homogeneous* system (i.e., the perfect crystal). We therefore conclude that the local-volume dependence of the interatomic interactions, incorporated only in the many-body potentials, does not play an important role in the basic generic properties of grain-boundary materials. Contrary to what appears to be a widespread misconception, we therefore conclude that the basic physical properties of GB materials may equally well be studied by means of pair and many-body potentials, provided that (i) in the simulations the volume expansion at the GBs is permitted and (ii) an *equilibrium* pair potential is employed (in contrast to pair potentials not fitted to the proper zero-temperature lattice parameter, such as those derived from pseudopotential theory). Considering that for the same value of the cutoff radius simulations employing many-body potentials are at least twice as CPU time intensive, it appears that in many cases it may be more advantageous to investigate basic phenomena primarily in terms of central-force potentials and limit the use of many-body potentials to cases in which a better quantitative agreement with experimental results is desired.

The better quantitative performace of many-body potentials has two origins. First, in contrast to equilibrium pair potentials, the former do not automatically satisfy the Cauchy relation; i.e., instead of only two, all three elastic constants of a perfect cubic crystal may be used in the empirical fitting. Second, many-body potentials for monatomic systems are usually fitted to six experimental numbers instead of, for example, the two for a Lennard-Jones potential or three for a Morse potential. This larger number of adjustable parameters, combined with a more complete representation of the elastic constants, not surprisingly, leads to a better description of the overall behavior of a given material in terms of the many-body potentials.

The fact that qualitatively the same results were obtained throughout this paper for both many-body and pair potentials is probably due to the fact that the physics of grain boundaries, as that of liquids, is governed by (i) the short-range repulsion between atoms in very close contact and (ii) by the anharmonicity of the interatomic interactions. Since both types of potentials are of a central-force type for distances much shorter than the nearest-neighbor distance, it is not surprising that they yield the same qualitative behavior. Extensive electronic structure calculations are obviously needed to assess for which materials this assumption of central-force repulsion is valid.

We conclude this article by pointing out the unique capabilities of atomistic-level computer simulations in the investigation of the local physical properties of inhomogeneous systems and to provide insight into the underlying causes. Particularly in interfacial systems in which these inhomogeneities may be concentrated over very small distances (of the order of only several atomic spacings), these methods may provide the only means in some cases to investigate the local behavior near or at interfaces.

ACKNOWLEDGEMENTS
This work was jointly supported by the U. S. Department of Energy, BES Materials Sciences, under Contract W-31-109-Eng-38, and by the Office of Naval

Research, under Contract No. N00014-88-F-0019. We have benefitted from many discussions with S. Phillpot.

REFERENCES

1.   M. S. Daw and M. I. Baskes, Phys. Rev. Lett. 50:1285 (1983).
2.   M. W. Finnis and J. E. Sinclair, Phil. Mag. A 50:45 (1984).
3.   R. A. Johnson, Phys. Rev. 134, A1329 (1964).
4.   See, for example, R. A. Johnson, in "Computer Simulation in Materials Science,"
        R. J. Arsenault, J. R. Beeler, Jr. and D. M. Esterling (American
        Society for Metals, Metals Park, 1987), p. 29.
5.   S. M. Foiles, M. I. Baskes, and M. S. Daw, Phys. Rev. B33:7983 (1986).
6.   G. J. Ackland and R. Thetford, Phil. Mag. A56:15 (1987).
7.   D. J. Oh and R. A. Johnson, J. Mater. Res. 3:471 (1988).
8.   D. Wolf, Acta Metall. (to be published).
9.   D. Wolf, Phil. Mag. A (to be published).
10.   W. T. Read and W. Shockley, Phys. Rev. 78:275 (1950).
11.   Siu-Wai Chan and R. W. Balluffi, Acta Metall. 26:1113 (1985).
12.   D. Wolf, Acta Metall. 32:245 (1984) and Acta Metall. 32:735 (1984).
13.   D. Wolf, Physica 131B:53 (1985).
14.   D. Wolf in: "Computer Simulation in Materials Science," (edited by
        R. J. Arsenault, J. R. Beeler and D. M. Esterling), American
        Society for Metals, 1987, Metals Park, Ohio, p. 111.
15.   A. Seeger and G. Schottky, Acta Metall. 7:495 (1959).
16.   J. Lutsko, J. Appl. Phys. 64:1152 (1988).
17.   M. Kluge, J. Lutsko, and D. Wolf, Scripta Metall. (submitted).
18.   W. M. C. Yang, T. Tsakalakos, and J. E. Hilliard, J. Appl. Phys. 48:876 (1977).
19.   For a review see I. K. Schuller, IEEE 1985 Ultrasonics Symposium,
        B. R. McAvoy, ed. (IEEE, New York, 1985), p. 1093.
20.   T. B. Wu, J. Appl. Phys. 53:5265 (1982).
21.   D. Wolf and J. Lutsko, Phys. Rev. Lett. 60:12 (1988).
22.   J. Lutsko and D. Wolf, Phys. Rev. B. (submitted).
23.   A. Kueny, M. Grimsditch, K. Miyano, I. Banerjee, C. M. Falco, and
        I. K. Schuller, Phys. Rev. Lett. 48:166 (1982).
24.   M. Grimsditch, to be published in: "Brillouin Scattering from Metallic
        Superlattices," M. Cardona and G. Günthcrodt, eds., Springer,
        Heidelberg.
25.   M. Guillope, G. Ciccotti, and V. Pontikis, Surf. Sci. 144:67 (1984).
26.   P. Deymier, G. Kalonji, R. Najafabadi, and S. Yip, Surf. Sci. 144:77 (1984).
27.   P. S. Ho, T. Kwok, T. Nguyen, C. Nitta, and S. Yip, Scripta Metall. 19:993
        (1985); T. Nguyen, P. S. Ho, T. Kwok, C. Nitta, and S. Yip,
        Phys. Rev. Lett. 57:1919 (1986).
28.   P. Deymier, A. Taiwo, and G. Kalonji, Acta Metall. 35:2819 (1987).
29.   J. Q. Broughton and G. H. Gilmer, Phys. Rev. Lett. 56:2692 (1986).
30.   R. W. Balluffi, J. Phys. Paris, 49:C5-337 (1988).
31.   J. F. Lutsko, D. Wolf, S. Yip, S. R. Phillpot, and T. Nguyen, Phys. Rev. B
        (to be published).
32.   J. F. Lutsko, D. Wolf, S. Phillpot, and S. Yip, Phys. Rev. B (submitted).
33.   F. H. Stillinger and T. A. Weber, Phys. Rev. B31:5262 (1985).
34.   S. Phillpot, J. F. Lutsko, D. Wolf, and S. Yip, Phys. Rev. B (submitted).

TEMPERATURE DEPENDENCE OF INTERATOMIC FORCES

Adrian P. Sutton

Department of Metallurgy and Science of Materials
Oxford University
Oxford OX1 3PH, England

INTRODUCTION

In computer modelling of defects in crystalline materials it is common practice to use a pair potential or N-body potential to describe the energy of the system as a function of the atom positions. The equilibrium atomic positions at 0 K are determined by minimization of the total potential energy with respect to each atomic coordinate in the model. Having obtained an equilibrium structure at 0 K it is natural to ask how the structure changes when the temperature is increased and what are the thermodynamic properties of the defect. A popular approach is to use molecular dynamics or Monte Carlo simulations to explore the classical phase space of the system and derive thermal averages of the atom positions and thermodynamic functions. Atoms are assumed to interact via the same potential as they were at 0K and temperature enters the simulations either through the classical kinetic energies in molecular dynamics or the acceptance criterion of trial configurations in Monte Carlo. The burden of this paper is to describe an alternative approach to determining equilibrium structure and thermodynamic properties at a finite temperature. In this method the (Helmholtz) free energy of the system for a given (thermally averaged) atomic configuration and interatomic potential is expressed in the harmonic approximation. The total free energy comprises this harmonic term and the potential energy arising from the interatomic potential. The *total free energy is minimized* at a given temperature by varying the volume of the system and the positions of atoms within it. During the relaxation the frequencies of vibrational normal modes change and thus the free energy arising from them also changes. This procedure is often called the *quasiharmonic* approximation to distinguish it from the harmonic approximation in which the same lattice vibrational free energy is calculated but for fixed atom positions that remain the same as those at 0K. Thus in the quasiharmonic approximation the atom positions and the volume of the system are variational parameters with which to minimize the free energy of the system. No such minimization is attempted in the harmonic approximation. The thermally averaged force acting on an atom is given by the negative of the gradient of the free energy with respect to the atom position. We shall show that because the interatomic potential is anharmonic this force is dependent on temperature. The equilibrium structure is obtained by requiring that the thermally averaged force on each atom is zero, and since the forces are temperature dependent so is the equilibrium structure. Thus one can use a molecular static relaxation code with temperature dependent interatomic forces to study thermodynamic properties of defects. This approach to including temperature in interatomic forces is similar to recent studies of thermal expansion coefficients using quasiharmonic theory (e.g., Harding, 1985). At a given temperature the equilibrium lattice parameter is determined by requiring that the derivative of the free energy with respect to volume is zero. This derivative, which is the negative of the pressure, varies with temperature owing to the aharmonicity in the potential and thus the equilibrium lattice parameter changes with temperature. The quasiharmonic approximation cannot account for all the effects of the anharmonicity of the interatomic potential because the energy associated with thermal vibrations is expanded only to second order in the displacements of atoms from their equilibrium positions. Nevertheless, the success with which thermal expansion coefficients have been calculated in quasiharmonic theory suggests that it does account for most of the anharmonicity.

For a given set of atomic positions the free energy arising from thermal motions of atoms is expressed in harmonic theory as an integral over vibrational normal modes of the system. As the lattice expands or equilibrium atomic positions move off lattice sites the normal modes change in a complicated way and the change in the free energy is normally evaluated by numerical differentiation of the integral. An alternative, more approximate, approach is presented in the next section. This approach leads to analytic expressions for the thermally averaged force and pressure which are seen to be N-body in nature. The approximation is tested by comparing the Gruneisen constants computed for the Finnis-Sinclair transition metal potentials (Finnis and Sinclair, 1984) with those computed by full quasiharmonic theory (Marchese et al., 1988). It is found that the approximation embraces most of the physics of the full quasiharmonic theory and a simple explanation for why some of the Gruneisen constants are negative is offered. Finally the temperature dependence of the structure of a $\Sigma = 13$ (001) twist boundary is modelled with a Lennard-Jones potential augmented with temperature dependent forces. Qualitative agreement with experimental measurements of the mean square displacements (Fitzsimmons et al., 1988) in this boundary in Au is obtained and considerable insight into the thermodynamic behavior of the boundary is gained.

FORMULATION

Analytic Expressions for Thermodynamic Functions at the Atomic Level

If the total density of phonon states in the system is $N(\omega)$, where ω is the angular frequency, then, in the harmonic and quasiharmonic approximations, the (Helmholtz) free energy is given by

$$F = E_p + kT \int_0^\infty N(\omega) \ln \left[2 \sinh \frac{\hbar\omega}{2kT} \right] d\omega \qquad (1)$$

E_p is the potential energy for the atomic configuration which is expressed in terms of some interatomic potential. $N(\omega)$ is determined by the usual equations of motion of lattice dynamics:

$$m_i \omega^2 u_{i\alpha} = \sum_{j\beta} D_{i\alpha j\beta} u_{j\beta} \qquad (2)$$

where roman subscripts are site indices and greek subscripts are directions x,y,z. m_i is the mass of atom i, $u_{i\alpha}$ is the displacement of atom i the α direction and $D_{i\alpha j\beta} = \partial^2 E_p / \partial u_{i\alpha} \partial u_{j\beta}$. The normal modes of the system are determined by diagonalizing (2) and the n'th eigenvalue is $\omega^{2(n)}$. The total density of states expressed as a function of ω^2, $G(\omega^2)$, is given by

$$G(\omega^2) = \sum_n \delta(\omega^2 - \omega^{2(n)}) = N(\omega)/2\omega. \qquad (3)$$

To evaluate the force (i.e. the thermally averaged force) on an atom we require the change in $N(\omega)$ (or $G(\omega^2)$) when the atom is displaced by a small amount. This is normally done numerically (De Lorenzi and Jacucci 1986) by rediagonalizing (2). At this point we *assume* that the local atomic environment around the atom we are considering is more important in determining the change of $N(\omega)$ than more remote atoms. Just as in tight binding theory we can express $N(\omega)$ as a sum over local densities of states (LDOS's) $n_i(\omega)$ and we can evaluate the change in each $n_i(\omega)$ analytically in the second moment approximation. Thus (3) can be expressed exactly as

$$G(\omega^2) = \sum_i g_i(\omega^2) = N(\omega)/2\omega = \sum_i n_i(\omega)/2\omega$$

where

$$g_i(\omega^2) = \sum_n \left| u_i^{(n)} \right|^2 \delta(\omega^2 - \omega^{2(n)}) = n_i(\omega)/2\omega \qquad (4)$$

and $u_i^{(n)}$ is the displacement vector of atom i in the n'th normal mode. The second moment of $n_i(\omega)$ is $\mu_2^i = \int_0^\infty n_i(\omega) \omega^2 d\omega$ and it is equal to the first moment, M_1^i of $g_i(\omega^2)$:

$M_1^i = \int_0^\infty g_i(\omega^2) \omega^2 d\omega^2$. It follows from (2), and the orthonormality of the eigenvectors $u^{(n)}$, that

$$\mu_2^i = \nabla_i^2 E_p / m_i \tag{5}$$

$$= -1/m_i \sum_{j \neq i} \sum_\alpha D_{i\alpha j\alpha} \tag{6}$$

where the last equality follows from the condition of invariance of E_p under a translation of the whole system. From (5) it is clear that μ_2^i *measures the stiffness of the environment* at site i. We now assume a functional form for $g_i(\omega^2)$ and fit M_1^i. In three dimensions we know that $G(\omega^2)$ has square root singularities at the lower and upper band edges. The simplest assumption is therefore an elliptic form for $g_i(\omega^2)$.

$$g_i(\omega^2) = \frac{6}{\pi b_i^2} \left[b_i^2 - (\omega^2 - b_i)^2 \right]^{1/2} \tag{7}$$

where $b_i = M_1^i = \mu_2^i$ and the normalization constant ensures that the zeroth moment of $g_i(\omega^2)$ is 3. The LDOS $n_i(\omega)$ is then given by

$$n_i(\omega) = \frac{12 \, \omega^2}{\pi \, (\mu_2^i)^2} \left[2 \, \mu_2^i - \omega^2 \right]^{1/2} \tag{8}$$

At low frequencies $n_i(\omega)$ is proportional to ω^2, giving the T^3 dependence of the low temperature lattice specific heat and a speed of sound in the perfect crystal of approximately $0.3 \sqrt{\mu_2^i}$.

Standard expressions for thermodynamic functions in lattice dynamics are all of the form of an integral over the total density of states as in (1). Using (4) we may easily derive expressions for the projections of these thermodynamic functions onto individual atomic sites. For example (1) may be written as

$$F = \sum_i E_p^i + F_i \tag{9}$$

where

$$F_i = kT \int_0^\infty n_i(\omega) \ln \left[2 \sinh \frac{\hbar\omega}{2kT} \right] d\omega$$

$$= \frac{48kT}{\pi} \int_0^1 y^2 [1 - y^2]^{1/2} \ln \left[2 \sinh \frac{c_i y}{2} \right] dy \tag{10}$$

and

$$c_i = \frac{h\sqrt{2\mu_2^i}}{kT} \tag{11}$$

The last equality in (10) is obtained by substituting (8) for $n_i(\omega)$. E_p^i is the projection of the total potential energy onto site i. We may think of F_i loosely as the local vibrational contribution to the free energy from site i, but clearly F_i is affected by the disposition of neighboring sites. Other standard thermodynamic functions projected onto site i are as follows:

$$S_i = \frac{48k}{\pi} \int_0^1 y^2 [1 - y^2]^{1/2} \left[\frac{c_i y}{2} \coth \frac{c_i y}{2} - \ln \left[2 \sinh \frac{c_i y}{2} \right] \right] dy \tag{12}$$

$$U = \sum_i E_p^i + U_i$$

$$U_i = \frac{24kT}{\pi} c_i \int_0^1 y^3 [1 - y^2]^{1/2} \coth \frac{c_i y}{2} dy \tag{13}$$

$$C_v^i = \frac{12kc_i^2}{\pi} \int_0^1 \frac{y^4 [1 - y^2]^{1/2}}{\sinh^2 \frac{c_i y}{2}} d \tag{14}$$

$$\left\langle u_i^2 \right\rangle = \frac{24h^2}{m_i \pi kTc_i} \int_0^1 y [1 - y^2]^{1/2} \coth \frac{c_i y}{2} dy \tag{15}$$

S_i, U_i, and C_v^i are the entropy, vibrational internal energy, and specific heat at constant volume projected onto site i. $<u_i^2>$ is the mean square displacement at site i. At low temperatures, i.e. typically where $c_i > 1$, not all the modes are excited and quantum effects dominate these expressions. At high temperatures, $c_i < 1$, the crystal behaves classically. The temperature, Θ_i, at which $c_i = 1$ is analogous to a local Debye temperature and it is given by $\Theta_i = h\sqrt{2\mu_2^i}/k$. At high temperatures $U_i = 3kT$, $C_v^i = 3k$ and

$$F_i = 3kT [\ln (c_i/2) + 1/4] \tag{16}$$

$$S_i = 3k [3/4 - \ln(c_i/2) \tag{17}$$

$$\left\langle u_i^2 \right\rangle = 6kT/\nabla_i^2 E_p \tag{18}$$

Interatomic Forces

The thermal expectation value of the force, f_i, acting on atom i is equal to the negative gradient of the total free energy with respect to the position of atom i. Using (5, 9, 10, 11 and 13) we obtain

$$f_i = - \nabla_i \bar{E}_p - \sum_j \nabla_i F_j = - \nabla_i E_p - \sum_j \frac{U_j}{2} \nabla_i (\ln \nabla_j^2 E_p) \tag{19}$$

It is clear from (19) that in addition to the usual force, $-\nabla_i E_p$, arising from the potential energy, there is an additional, temperature dependent force. Even for a pair potential description of E_p the temperature dependent contribution is N-body in nature; *it is non-zero for all anharmonic interaction potentials* since it depends on third derivatives of E_p. Minimization of the free energy of the system, using the forces given in (19), will result in equilibrium atomic configurations for defects in crystals which change with temperature. This free energy minimization can be done with an ordinary molecular statics code. After the relaxation the other thermodynamic functions (12-15) also change in a consistent fashion.

Thermal Expansion

The central quantity in determining the thermal expansion coefficient and other anharmonic properties of a crystal is the Gruneisen constant averaged over all vibrational modes. It may be defined by (see Ashcroft and Mermin 1976, eqn. 25.19):

$$\gamma = - \frac{\int_0^\infty d\omega \, N(\omega) \frac{\partial \ln \omega}{\partial \ln V} \hbar\omega \frac{\partial}{\partial T} \left[\frac{1}{\exp(\hbar\omega/kT - 1)} \right]}{\int_0^\infty d\omega \, N(\omega) \hbar\omega \frac{\partial}{\partial T} \left[\frac{1}{\exp(\hbar/kT) - 1} \right]} \tag{20}$$

By replacing $N(\omega)$ by $n_i(\omega)$ we may define a local Gruneisen constant, γ_i, at site i. ($\gamma = \gamma_i$ in a monoatomic perfect crystal). In reality γ and γ_i are dependent on temperature at low temperatures because $\partial \ln \omega/\partial \ln V$ is dependent on frequency. In our model (as in the Debye model) $\partial \ln \omega/\partial \ln V$ is the same for all modes. Thus,

$$\gamma_i = - 1/2 \, \partial \ln \nabla_i^2 E_p/\partial \ln V. \tag{21}$$

γ_i *is seen from (21) to be a measure of the anharmonicity of the potential energy at site* i.

The total pressure P in the system is given by

$$P = - \left[\frac{\partial F}{\partial V} \right]_T = - \frac{\partial E_p}{\partial V} - \sum_j \frac{U_j}{2} \frac{\partial \ln \nabla_j^2 E_p}{\partial V} = - \frac{\partial E_p}{\partial V} + \sum_j \frac{U_j \gamma_j}{V} \tag{22}$$

It is the balance between these two terms in the pressure that determines the thermal expansion of the solid. The linear thermal expansion coefficient α is defined as

$$\alpha = \frac{1}{3B} \left[\frac{\partial P}{\partial T} \right]_V \tag{23}$$

270

where B is the bulk modulus $-V(\partial P/\partial V)_T$. Using (13,14,22) we find that

$$B = \frac{V\partial^2 E_p}{\partial V^2} + \sum_j \frac{V}{4} (U_j - C_v^j T) \left[\frac{\partial \ln \nabla_j^2 E_p}{\partial V}\right]^2 + \frac{U_j V}{2} \frac{\partial^2 \ln \nabla_j^2 E_p}{\partial V^2} \tag{24}$$

and

$$\left[\frac{\partial P}{\partial T}\right]_V = \sum_j \frac{C_v^j \gamma_j}{V} \tag{25}$$

In these expressions V is the volume of the whole solid. The thermal expansion coefficient is obtained by substituting (24,25) into (23). The summation term in (24) is a correction to the bulk modulus arising from the volume dependence of the phonon pressure. This correction is normally negative and small in comparison with the term arising from the potential energy. The main reason why the bulk moduli of most materials decrease with increasing temperature is that the thermal expansion causes a decrease in the bulk modulus arising from the potential energy. To a good approximation the summation term in (24) may be neglected in comparison with the change in the bulk modulus due to thermal expansion.

In order to use this formulation to derive thermodynamic functions, interatomic forces and Gruneisen constants we need an explicit form for E_p so that $\nabla_i^2 E_p$, $\nabla_j \nabla_i^2 E_p$ and $\partial \nabla_i^2 E_p/\partial \ln V$ may be evaluated. Expressions for these quantities have been derived for pair potential and N-body ("embedded atom") potential descriptions of E_p.

Pair Potentials

The potential energy is assumed to be given by

$$E_p = \frac{1}{2} \sum_{i \neq j} \sum \phi (R_{ij}) \tag{26}$$

where $\phi(R_{ij})$ is the pair potential energy for an interatomic separation R_{ij}. We obtain the following expressions:

$$\nabla_i^2 E_p = \sum_{m \neq i} \phi'' (R_{im}) + 2 \phi' (R_{im})/R_{im} \tag{27}$$

$$\nabla_j \nabla_i^2 E_p = \Gamma(R_{ij}) (R_j - R_i)/R_{ij}$$

provided $j \neq i$ where

$$\Gamma(R_{ij}) = \left[\phi'''(R_{ij}) + 2\phi''(R_{ij})/R_{ij} - 2\phi'(R_{ij})/R_{ij}^2\right] \tag{28}$$

and

271

$$\nabla_i \nabla_i^2 E_p = - \sum_{j \neq i} \nabla_j \nabla_i^2 E_p \qquad (29)$$

The force f_i given by (19) becomes

$$f_i = \sum_{j \neq i} \phi'(R_{ij})(R_j - R_i)/R_{ij}$$

$$+ \frac{1}{2} \sum_{j \neq i} \left[\frac{U_i}{\nabla_i^2 E_p} + \frac{U_j}{\nabla_j^2 E_p} \right] \Gamma(R_{ij})(R_j - R_i)/R_{ij} . \qquad (30)$$

The second term in (30) is the contribution arising from the vibrational free energy. Finally, to evaluate local Gruneisen constants (21) we need the following expression

$$\frac{\partial \nabla_i^2 E_p}{\partial \ln V} = \frac{1}{3} \sum_{j \neq i} R_{ij} \Gamma(R_{ij}) . \qquad (31)$$

N-body potentials

The potential energy is assumed (Finnis and Sinclair 1984, Daw and Baskes 1984) to be given by

$$E_p = E_p^P + E_p^N \qquad \text{where}$$

$$E_p^P = \frac{1}{2} \sum_{i \neq j} \sum \phi(R_{ij}) \quad \text{and} \quad E_p^N = \sum_i f(\rho_i) \qquad (32)$$

$f(\rho_i)$ is a function which has been interpreted as the energy of embedding an atom in an electron gas of local density ρ_i (Daw and Baskes 1984) or as the covalent bond energy at site i in the second moment approximation of tight binding theory (Finnis and Sinclair, 1984). ρ_i is expressed empirically as a sum over functions extending to neighboring sites j:

$$\rho_i = \sum_{j \neq i} \Psi(R_{ij}) \qquad (33)$$

In the previous section we derived the relevant expressions for E_p^P. Let $f_i = f(\rho_i)$. Then we obtain the following expressions involving E_p^N.

$$\nabla_i^2 E_p^N = f_i'' \sum_{j \neq i} \sum_{k \neq i} \Psi'(R_{ij}) \ \Psi'(R_{ik}) \cos\theta_{jik}$$

$$+ \sum_{j \neq i} f_j'' \ [\Psi' \ R_{ij})]^2$$

$$+ \sum_{j \neq i} (f_i' + f_j') \ [\Psi''(R_{ij}) + 2\Psi'(R_{ij})/R_{ij}] \tag{34}$$

θ_{jik} is the angle between the bonds between atoms i and j and between atoms i and k. The first term in (34) shows that there will be a contribution to the force in (19) which depends on angles between bonds, even though there is no directionality in the N-body potential. This term is zero in a cubic environment but not in the vicinity of a symmetry breaking defect. In a perfect cubic crystal we obtain the following expression for $\partial \nabla^2 E_p^N / \partial \ln V$ which is required to compute the Gruneisen constant:

$$\frac{\partial \nabla^2 E_p^N}{\partial \ln V} = f_i''' \chi_i \sum_{j \neq i} [\Psi'(R_{ij})]^2$$

$$+ 2f_i'' \sum_{j \neq i} \chi_i \left[\Psi''(R_{ij}) + \frac{2\Psi'(R_{ij})}{R_{ij}} \right] + \Psi''(R_{ij})\Psi'(R_{ij})R_{ij}$$

$$+ 2f_i' \sum_{j \neq i} \left[R_{ij} \Psi'''(R_{ij}) + 2\Psi''(R_{ij}) - \frac{2\Psi'(R_{ij})}{R_{ij}} \right] \tag{35}$$

where

$$\chi_i = \sum_{j \neq i} \Psi'(R_{ij})R_{ij} \tag{36}$$

GRUNEISEN CONSTANTS FOR FINNIS-SINCLAIR TRANSITION METAL POTENTIALS

Using quasiharmonic theory Marchese *et al.* (1988) have computed Gruneisen constants for the Finnis-Sinclair potentials for transition metals (Finnis and Sinclair 1984). For Nb, V and Cr they obtained negative Gruneisen constants which is an unsatisfactory conclusion since it indicates that the thermal expansion coefficients for those potentials are negative. The Gruneisen constants for the other potentials were too small compared with experimental values. We have computed Gruneisen constants for the Finnis-Sinclair potentials using (21, 31 and 35) for comparison with the results of Marchese *et al*. The results are given in Table 1. The Gruneisen constants obtained by Marchese *et al* are in the column labelled γ [MJF]. It is seen that qualitative agreement has been obtained, although the magnitude for Nb is not well reproduced. Since $\nabla_i^2 E_p$ must be positive in order for the crystal to be at least metastable the Gruneisen constant is positive provided $\partial \nabla_i^2 E_p / \partial \ln V$ is negative. We see that the Gruneisen constants for V, Nb and Cr are negative because $\partial \nabla_i^2 E_p^P / \partial \ln V$ is too large and positive. We conclude that the pair potential energy E_p^P, rather than the N-body potential energy E_p^P, is responsible for the negative Gruneisen constants. Similarly, the positive Gruneisen constants are too small because $|\partial \nabla_i^2 E_p / \partial \ln V|$ is too small. Rebonato *et al*. (1987) have modified the Finnis Sinclair potentials for V, Nb, Ta and Mo by adding a term $K(R_1 - R_{ij})^3$, for $R_{ij} \leq R_1$, to the pair potential $\varphi(R_{ij})$ in (32), where R_1 is the first neighbor separation. This correction makes $\partial \nabla_i^2 E_p^P / \partial \ln V$ more negative while not affecting $\nabla_i^2 E_p$, and hence the Gruneisen constants become larger and positive: 2.06, 3.22, 1.09 and 0.66, respectively. However, these potentials (Rebonato *et al*. 1987) should not be used in the manner described here to optimize the atomic structures of defects at elevated temperatures because the third derivative of the pair potential changes discontinuously at R_1.

TABLE 1

Potential	$\nabla_i^2 E_p^N$	$\partial \nabla_i^2 E_p^N / \partial \ln V$	$\nabla_i^2 E_p^P$	$\partial \nabla_i^2 E_p^P / \partial \ln V$	γ	γ[MJF]
V	-5.301	- 9.920	21.268	13.583	-0.12	-0.42
Nb	-9.637	-16.062	23.977	24.362	-0.29	-1.02
Ta	-5.177	-10.704	31.927	-1.748	0.23	0.37
Cr	11.313	-3.989	25.290	7.873	-0.05	-0.09
Mo	-1.361	-5.985	44.912	3.435	0.03	0.06
W	0.588	-4.608	55.942	-13.908	0.16	0.09
Fe	13.709	-6.894	5.705	-4.518	0.29	0.60

$\Sigma = 13$ (001) TWIST BOUNDARY RELAXED AT FINITE TEMPERATURE

The thermal properties of the $\Sigma = 13$ (001) twist boundary in Au have been studied recently by Fitzsimmons *et al*. (1988) using X-ray diffraction. Those authors concluded that at 298K the mean square displacement of atoms in the grain boundary is $0.012 \pm 0.002 \text{Å}^2$ compared with $0.008 \pm 0.0005 \text{Å}^2$ in the bulk. Thus, on average, atoms at the grain boundary have 50% larger mean square displacements than atoms in the bulk. The thermal expansion coefficients of the bicrystal parallel, $\alpha_{//}$, and perpendicular, $\alpha_{\perp}$, to the boundary were also measured: at 255K they are $1.2 \pm 0.2 \times 10^{-5} \text{ K}^{-1}$ and $4.3 \pm 0.4 \times 10^{-5} \text{ K}^{-1}$, respectively. Since the Debye temperature in Au is 170K the measurements were made at temperatures where the lattice vibrational properties behave classically. We note that the assignment of a thermal expansion tensor to the grain boundary region implies a division of the bicrystal into a grain boundary region and a bulk crystal region. In an atomistic model such a division is very arbitrary and any calculated "grain boundary thermal expansion coefficient" will depend on where the division is made. By contrast the rigid body translation t of one half of the bicrystal with respect to the other does not involve such a division and it can be readily calculated as a function of temperature.

We have calculated the equilibrium atomic structure and thermal properties of a $\Sigma=13$ (001) twist boundary at 3 temperatures using a Lennard-Jones pair potential augmented with temperature dependent interactions as outlined in equations (26-31). We have evaluated mean square displacements and local Gruneisen constants at sites in the boundary and the change in the equilibrium structure, which includes t, at each temperature. Excess grain boundary free energy, entropy and specific heats have also been evaluated at each temperature.

Lennard-Jones potentials of the form

$$\varphi(R) = A/R^m - B/R^n \qquad (37)$$

where $20 \geq m > n \geq 4$ were fitted to the lattice parameter of Au at 0K (which was taken to be 4.0786 Å), and the bulk modulus of Au at 0K (which was taken to be 171 GPa). Using the atomic mass for Au, 196.97, the equilibrium lattice parameter and thermodynamic properties of the perfect crystal, were determined at each temperature between 0 and 2000K in 20K intervals. A wide range of (only positive) Gruneisen constants was obtained by varying the exponents m and n in the potential. As is intuitively obvious the larger the difference in m and n the larger the Gruneisen constant becomes. A number of potentials emulated the thermal expansion properties of Au reasonably well. The potential that was used in the simulation had the following parameters:

$$m = 12; n = 6; A = 123516.729284 \text{ eVÅ}^{12}; B = 366.603947222 \text{ eVÅ}^6$$

and it was truncated at $R = 2$ lattice parameters, incorporating 140 neighbors. At 0K the Gruneisen constant is 3.03 and the temperature Θ at which $c = 1$ is 490K. Therefore, at 600K the vibrational properties are in the classical regime. At 600K (1200K) the equilibrium lattice parameter is 4.1193044266 Å (4.1721596376 Å)., $c = 0.745421$ (0.330214), $\gamma = 3.11$ (3.23), $C_v^i = 2.932k$ (2.986k), $<u^2> = 8.97 \times 10^{-3} \text{Å}^2$ ($2.26 \times 10^{-2} \text{Å}^2$) and $\alpha = 1.82 \times 10^{-5} \text{ K}^{-2}$ ($2.74 \times 10^{-5} \text{ K}^{-1}$).

The $\Sigma = 13$ (001) was relaxed using the molecular statics program developed by Sutton and Finnis in 1985 (unpublished). Periodic boundary conditions were applied to the square repeat cell bounded by $1/2<510>$. vectors in the boundary plane. Normal to the boundary plane the explicitly relaxed region, which contained 195 atoms, was embedded within "floating" perfect crystals with lattice parameters appropriate for the simulation temperature. Thus the boundary was free to expand normal, but not parallel, to the boundary plane. The free energy of the boundary was minimized using the forces given in (30) and a variable metric minimization routine. At each temperature only one metastable state of the boundary was found and it had zero translation component parallel to the boundary away from coincidence. The layer group is $p42_12'$ at each temperature, and the boundary structure cannot be distinguished from that shown in Fig. 8 of Schwartz *et al.* (1985). At 0K the boundary energy is 1091 mJ/m^2 and the boundary expansion, e, is 0.1096a_0, where a_T is the lattice parameter at temperature T. At 600K (1200K) the excess free energy of the boundary is 1028 mJ/m^2 (948 mJ/m^2) and the boundary expansion is 0.1148a_{600} (0.1231a_{1200}). The fact that e is an increasing function of a_T indicates that the boundary expands more with temperature than the adjoining crystals. This is in qualitative agreement with Fitzsimmons *et al.* (1988). During the relaxation at 600K and 1200K the vibrational contribution to the excess free energy became more negative at the expense of a more positive contribution from the pair potential, and the overall excess free energy decreased by much less than the changes in either of these contributions. At 600K (1200K) the grain boundary excess entropy and specific heat are 0.06637 (0.1555) mJ/K/m^2 and 2.556 x 10^{-3} (1.163 x 10^{-3}) mJ/K/m^2.

By far the most significant effect of temperature on the boundary structure was the increase in expansion. Atomic sepatations in the plane of the boundary changed by an order of magnitude less than components normal to the boundary. As the temperature increased the anharmonicity of the atomic interactions, measured by the local Gruneisen constants, also increased. For this reason, and the fact that the local internal energies, U_i, in (30) are larger at a higher temperature, the boundary structure changes more at a higher temperature. This shows that our model is behaving sensibly.

It is useful to define normalized excess mean square displacements and excess local Gruneisen constants at temperature T by,

$$<u_i^2>_T^{xs} = \frac{<u_i^2>_T - <u_o^2>_T}{<u_o^2>_T} \quad ; \quad \gamma_{iT}^{xs} = \frac{\gamma_{iT} - \gamma_{oT}}{\gamma_{oT}} \tag{38}$$

where i denotes a grain boundary site and o denotes a perfect crystal site. $<u_i^2>_T$ is given by (15), which simplifies to (18) at high T; γ_{iT}, which is given by (21), is implicitly dependent on temperature because the local atomic environment changes with temperature. Table 2 summarizes the numbers of atoms in each layer with values of $<u_i^2>_T^{xs}$, expressed as a percentage, larger than 1% at 600K and 1200K. Sites corresponding to the larger values $<u_i^2>_T^{xs}$ also correspond to the most significant values of γ_{iT}^{xs}; indeed there is a 93% correlation between $<u_i^2>_T^{xs}$ and γ_{iT}^{xs}. The geometrical boundary plane passes between the first layers of either grain. Data for only one grain is given since the data is identical for the other grain. It is seen from Table 2 that the most significant values of $<u_i^2>_T^{xs}$ and γ_{iT}^{xs} occur in the first planes on either side of the boundary. If we average $<u_i^2>_T^{xs}$ over these planes we obtain 27% at 600K and 33% at 1200K. Note that there are also negative values of $<u_i^2>_T^{xs}$; they correspond to sites that are compressed. Since the γ_{iT}^{xs} are predominantly positive the boundary region expands more than the adjoining grains when the temperature is increased: interactions within the boundary are more anharmonic than those outside and become even more so as the temperature increases.

It follows from (10-15) that as c_i (given by (11)) increases F_i and U_i increase whereas S_i, $<u_i^2>$ and C_V^i decrease. There is a remarkably strong correlation (91% at 600K and 94% at 1200K) between c_i and the local pressure, P_φ^i, at site i arising from the potential:

$$P_\varphi^i = -\frac{1}{6\Omega} \sum_{j \neq i} R_{ij} \varphi' (R_{ij}) \qquad (39)$$

where Ω is the average atomic volume. Thus, as is intuitively obvious, compressed sites are stiffer environments and they are associated with higher local internal and free energies and lower local entropies, specific heats and mean square displacements. There is also a strong negative correlation (-97% and -91%) between P_φ^i and the local Gruneisen constant γ_i. Thus the more tensile the site the greater the degree of local anharmonicity. This is consistent with the strong negative correlation found between c_i and γ_i (-93% and -91%). The local pressure arising from the vibrational free energy has a strong negative correlation (-93% and -89%) with c_i and a 100% positive correlation with γ_i, which is consistent with it being negatively correlated with P_φ^i (-97%). There is a relatively weak correlation (-46%) between c_i and the local Von Mises shear stress, τ_i, and no significant correlation between γ_i and τ_i.

TABLE 2

Layer	$<u_i^2>^{xs}$ 600	$<u_i^2>^{xs}$ 1200	γ_i^{xs} 600	γ_i^{xs} 1200	No. of Atoms
1	19.1%	22.8%	0.64%	1.24%	4
	-5.0%	-4.1%	-3.30%	-3.57%	4
	51.7%	61.7%	5.40%	7.59%	4
	90.6%	109.4%	12.21%	16.53%	
2	1.2%	2.4%	-0.19%	-0.02%	4
	1.4%	2.9%	-0.34%	-0.17%	4
	4.4%	6.4%	0.30%	0.65%	4
	25.7%	31.6%	3.53%	4.89%	1
3	2.5%	2.7%	0.36%	0.44%	4
	-2.0%	-1.3%	-0.35%	-0.29%	1
4	1.1%	1.1%	0.16%	0.19%	4
	-1.9%	-1.7%	-0.31%	-0.33%	1

These correlations suggest a simple criterion for selecting atomic sites that will most influence the thermodynamic properties of a defect knowing only the hydrostatic stress at each atomic site. In general tensile sites are those with relatively large local volumes and/or relatively low coordination numbers. These sites should be associated with the largest contributions to the defect excess entropy, specific heat and thermal expansion. A similar conclusion was reached by Hashimoto *et al.* (1981) in their study of thermodynamic properties of grain boundaries in the harmonic approximation using the recursion method to compute phonon LDOS's. Structural rearrangements involving *local shears*, such as those that may be involved in a grain boundary structural transformation, are not well described in our second moment approximation. As seen in (5) the second moment μ_2^i contains information only about the local curvature of the potential, and not the local resistance to shear. The fourth moment, μ_4^i, is given by

$$\mu_4^i = 1/m_i^2 \sum_j \sum_\alpha \sum_\beta D_{i\alpha j\beta} \, D_{j\beta i\alpha} \qquad (40)$$

It is seen that μ_4^i contains information about the local resistance to shear. Our formulation can be readily extended to include μ_4, but at the expense of considerably greater complexity.

ACKNOWLEDGEMENTS

I have benefited from useful discussions with Brian Cantor, Mike Finnis, Peter Flynn, Mike Gillan, Volker Heine, David Pettifor and Ricardo Rebonato. Steve Sass kindly sent me a preprint of Fitzsimmons *et al.* (1988). The continuing support of the Royal Society is gratefully acknowledged.

REFERENCES

Ashcroft N.W. and Mermin N.D., "Solid State Physics", Holt, Reinhart and Winston, New York (1976).
Daw M.S. and Baskes M.I., Phys. Rev. B, 29, 6443 (1984).
De Lorenzi G. and Jacucci G., Phys. Rev. B, 33, 1993 (1986).
Finnis M.W. and Sinclair J.E., Phil. Mag., A50, 45 (1984).
Fitzsimmons M.R., Burkel E. and Sass S.L., Phys. Rev. Letts. 61, 2237 (1988).
Harding J.H., Phys. Rev. B, 32, 6861 (1985).
Hashimoto M., Ishida Y., Yanamoto R. and Doyama M., Acta Metall. 29, 617 (1981).
Marchese M., Jacucci G., and Flynn C.P., Phil. Mag. Letts., 57, 25 (1988).
Rebonato R., Welch D.O., Hatcher R.D. and Bilello J.C., Phil. Mag., A55, 655 (1987).
Schwartz D., Vitek V. and Sutton A.P., Phil. Mag., A51, 499 (1985).

NEW, SIMPLE APPROACH TO DEFECT ENERGIES IN SOLIDS

VIA EQUIVALENT CRYSTALS

John R. Smith and Tom Perry

Physics Department
General Motors Research Laboratories
Warren, MI 48090-9055

Amitava Banerjea

NASA Lewis Research Center
Cleveland, OH 44135

INTRODUCTION

It is evident from many of the presentations of this symposium that it is now reasonable to consider the computation and simulation of real materials phenomena. It would be desirable, for example, to have an atomistic model of crack initiation and growth or of adhesion and friction processes. This would require going beyond pair potentials with a method that is quantum mechanical in origin, accurate, and yet simple enough to allow one to deal with the low symmetries found in these phenomena.

In the following, we will describe a new method[1,2] which was created with these characteristics in mind. The total energy of a solid with a defect is written as a perturbation expansion, where the unperturbed system is a single crystal. This is an exact relationship, and can be simplified by varying the lattice constant of the single crystal until the perturbation series sums to zero. Then the energy of that crystal is equal to the energy of the solid containing a defect. This is an important simplification, because the energy of a single crystal as a function of its lattice constant is of a universal form, as will be shown below, which greatly facilitates the energy computation. In fact, this can be done for each atom, so that one can picture the local energetics of an atom in a defect as if it were in a crystal of appropriate lattice constant. Because the formulation is exact, one can identify the important contributors to the total energy and arbitrarily improve the accuracy. The method can also be approximated, yielding very simple computations of total energies which we will see in the following to be accurate. For example, we will see that predicted surface energies are found on average to be within 10% of first principle values, and predicted changes in interlayer spacings due to surface relaxation are typically within experimental error bars, often within 0.01 Å.

We begin by discussing universal features of energy relations in solids and molecules. Next, we will introduce the equivalent crystal theory, and finally we will make applications of this new method.

UNIVERSAL ENERGY RELATION

A universal relationship between total energies and distances between atoms has been discovered[3,4] for adhesion, cohesion, and chemisorption of metals and covalent semiconductors, certain diatomic molecules, and even nuclear matter. This universal form can be obtained via a simple scaling of the total energy $E(a)$:

$$E^*(a^*) = E(a) / \Delta E \tag{1}$$

where

$$a^* = (a - a_m) / l, \tag{2}$$

a is the interatomic spacing, a_m is the equilibrium spacing, and ΔE is the minimum value of the total energy or the equilibrium binding energy. The scaling length l is defined for convenience so that $[d^2 E^*(a^*) / da^{*2}]_0 = 1$ for all scaled curves:

$$l = \sqrt{\frac{\Delta E}{(d^2 E(Aa) / da^2)_{a_m}}} . \tag{3}$$

Figure 1 shows representative values of $E^*(a)^*$ for cohesion, chemisorption, adhesion, and a diatomic molecule. The energetics of these seemingly diverse phenomena fall accurately on a single curve.

This accuracy is typical[4] of what one finds for other available data in chemisorption, adhesion and cohesion. The solid line in Fig. 1 is the Rydberg function:

$$E^*(a^*) = - (1 + a^*) e^{-a^*} . \tag{4}$$

It is an accurate approximation to the universal energy relation. Some progress has been made in the understanding of why there is a universal energy relation. This is discussed in Ref. 4.

One might look at Fig. 1 and conclude that an energy relation that general should make it possible to determine total energies of defects in solids. While we will see that the universal energy relation is important for that task, one must go beyond it for general defect studies. This because the paths that the atoms were allowed to follow in each of the four processes represented in Fig. 1 were constrained. For example, in cohesion the crystal structure is fixed as the atoms move apart or together.

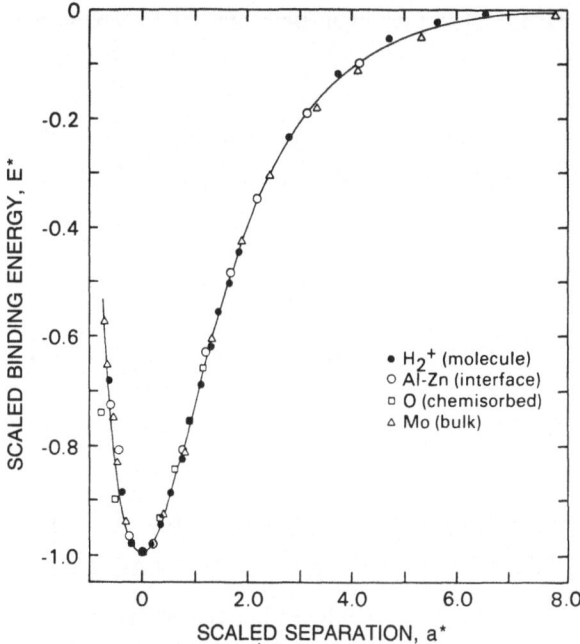

Figure 1. Scaled binding E^* plotted against scaled separation a^* for representative cases of cohesion, chemisorption, adhesion, and a diatomic molecule. The solid line is the Rydberg function, Eq. (4). See Ref. 4.

Thus, for total energies of defects, we must develop a new method. As the general defect has low symmetry, our goal is to find a simple method which is nevertheless accurate. In recent years, there has been progress toward that goal. The embedded atom method[5] EAM and related methods[6,7], which are discussed elsewhere in this symposium, have been very useful in modeling defects. Here the energy of an atom U_i is written as the sum of an embedding term which depends on a local density ρ_i and a pairwise term: $U_i = -f(\rho_i) + 1/2\Sigma_j V_{ij}$. Parameters in both the embedding and pair terms are determined empirically. The effective medium[8] and quasiatom[9] approaches are similar to each other and are based on a first principles solution of a simplified model. The energy of an impurity in a solid is computed as if the impurity were in a uniform electron gas whose density is a local electron density in the actual solid.

EQUIVALENT CRYSTAL THEORY OF DEFECT ENERGIES

Here we take a different approach. Since we are interested in treating semiconductors and transition metals from the outset, we will not employ a uniform electron gas model. It is also desirable to have a method which can be made exact. In that way, accuracies can be systematically improved when applying the method to new systems. It also allows systematic identification of the physical phenomena which are important to the particular application.

Consider first a single crystal of an elemental solid. Next introduce a defect or an array of lattice defects into the crystal. The total energy E *(defect)* of the crystal containing the defects is equal to the energy of a single crystal E *(crystal)* plus a perturbation series, where the perturbing potential is the difference between the array of ion core potentials of the crystal containing defects and those of the single crystal:

$$E(defect) = E(crystal) + Perturbation\ Series\ . \tag{5a}$$

As shown in Fig. 1, the energy of the single crystal is of universal form:

$$E(crystal) = \Delta E\ E^*(a^*)\ . \tag{5b}$$

Here ΔE is the cohesive energy, $a^* = (r_{WS} - r_{WSE}) / l$ (see Eq. (2)), r_{WS} is the Wigner-Seitz radius of equilibrium value r_{WSE} ($3/4\pi r^3_{WSE} \equiv$ bulk atom density), $l = \sqrt{\Delta E /(12\pi B r_{WSE})}$, and B is the bulk modulus of the ground state single crystal.

It is important that the perturbation be small so that the perturbation series converges rapidly. We will be interested in treating problems such as a surface, which is a large perturbation on a ground state single crystal. Actually, the perturbation series can be made to sum to zero, even in the case of a surface:

$$Perturbation\ Series = 0\ , \tag{6}$$

by varying the lattice constant of the single crystal so that

$$E(defect) = E(crystal) = \Delta E\ E^*(a_0^*)\ , \tag{7}$$

where a_0^* is the value of a^* which satisfied Eq. (7). Thus the energy of the single crystal is equal or *equivalent* to the energy of the solid which contains defects. You see, the solid with defects is the real system and the single crystal is just a model system which facilitates the computation of the energy of the solid with defects. Thus we can adjust the single crystal lattice constant to our advantage. The energy of the defect solid is always greater than the energy of the ground state single crystal (for the latter, a* = 0, Fig. 1). One can see from Fig. 1 that the single crystal energies for $a^* \neq 0$ are also greater than the ground state single crystal energy. Thus, we can in fact adjust a^* until the single crystal and defect energies are equal. The problem of determining the defect energies then becomes the formally simple problem of solving one equation in one unknown, i.e., determining a_0^* such that Eq. (6) is satisfied.

Now let us take a closer look at Eqs. (5):

$$E = \Delta E\, E^*(a^*) + \frac{Z^2}{2N} \left\{ \sum_{\substack{l,m \\ l \neq m}} \frac{1}{|\vec{R}_l' - \vec{R}_m'|} - \frac{1}{|\vec{R}_l - \vec{R}_m|} \right\} + E_1 + E_2 + \ldots \; .$$

(8)

where

$$E_1 = \frac{1}{N} \sum_l \int_{V_l} d\vec{r}\, n(\vec{r})\, \delta v(r) \; ,$$

(9)

$$\delta v(\vec{r}) = \sum_m \left(v(\vec{r} - \vec{R}_m') - v(\vec{r} - \vec{R}_m) \right) \; ,$$

(10)

$v(\vec{r})$ is the ion core potential,

$$E_2 = \frac{1}{2N} \sum_l \int_{V_l} d\vec{r}\, \delta n(\vec{r})\, \delta v(\vec{r}) \; ,$$

(11)

$$\delta n(\vec{r}) = \int d\vec{r}\, L(\vec{r}, \vec{r}')\, \delta v(\vec{r}') \; ,$$

(12)

$L(\vec{r}, \vec{r}')$ is the linear response function of the equivalent crystal, Z is the valence, $\vec{R}_l$ is the set of position vectors of the single crystal and $\vec{R}_l'$ that of the solid with lattice defects, N is the number of atoms, V_l is the unit-cell volume enclosing the lth ion core of the equivalent crystal and $n(\vec{r})$ is the valence-electron density distribution in the equivalent crystal.

We see that electron screening enters first through E_2, with $\delta n(\vec{r})$ having the effect of screening $\delta v(\vec{r})$. Screening is an important effect which we will want to include in our calculations.

In terms of Eq. (8), Eq. (6) becomes:

$$\frac{1}{N} \sum_l \int_{V_l} d\vec{r}\, \delta v(\vec{r})\, [n(\vec{r}) - Z\delta(\vec{r} - \vec{R}_l')]$$

$$- \frac{Z}{2N} \sum_l \int_{V_l} d\vec{r}\, \delta v(\vec{r})\, [\delta(\vec{r} - \vec{R}_l') - \delta(\vec{r} - \vec{R}_l)] + E_2 + \ldots = 0 \; .$$

(13)

Here, the terms in the perturbation series of Eq. (8) have been combined so that each quantity in the brackets derives from a neutral charge distribution. This means that only a few neighbors of the lth cell are required for contributions to $\delta v(\vec{r})$ in evaluating these integrals. In fact, it will be seen in the applications given in the next section that second neighbors are more than adequate for most calculations of interest. One can write without approximation $n(\vec{r}) = \Sigma_l \, n_A(\vec{r} - \vec{R}_l)$ where $n_A(\vec{r} - \vec{R}_l)$ is an electron distribution localized about R_l, since $n(\vec{r})$ is the electron density of a single crystal. Integrals in Eq. (13) including E_2 are of the form

$$\int_{V_l} d\vec{r} \left(n_A \left(\vec{r} - \vec{R}_l \right) - Z \delta \left(\vec{r} - \vec{R}_l \right) + \delta n(\vec{r}) / 2 \right) v \left(\vec{r} - \vec{R}_m' \right) = g(\vec{R}_m' - \vec{R}_l, \, a_0^*) \, . \qquad (14)$$

While Eq. (13) is formally simple in that it is one equation for the one unknown a_0^*, its terms are complicated. We will need then to make some approximations. Eq. (14) involves a sum of integrals involving neutral charge distributions overlapping ion-core potentials. We approximate these integrals as

$$g \propto R^p \, e^{-\alpha R} \qquad (15)$$

for nearest neighbors and

$$g \propto R^p \, e^{-\alpha R} \, e^{-R/\lambda} \qquad (16)$$

for next nearest neighbors and beyond, where $p = 2n - 2$, n = principal quantum number, and λ = electronic screening length. This derives from taking the form of $n_A(\vec{r} - \vec{R}_l)$ to be that of the highest partially occupied s orbital, which has the largest range to overlap neighboring ion-core potentials $v(\vec{r} - \vec{R}_m')$. Screening enters naturally through the $\delta n(\vec{r})$ term in Eq. (14). In the approximate forms, it enters both through the value of α and through the additional screening function $e^{-R/\lambda}$. Perturbing potentials which are at next nearest neighbor distances and beyond from site l are more strongly screened than those which are only at nearest neighbor distances because of nearest neighbors being between the perturbation and the site, and because the larger distances allow the screening cloud to have more of an effect. We also must take note of the fact that in many cases defects are localized, so that only a relatively small fraction of the sites have a different environment from that of a site in a single crystal. Thus, the total energy change associated with defect formation becomes a small difference between large numbers. This can be error prone, and so it is preferable to choose an equivalent crystal lattice constant a_{0l}^* for each atomic site l. a_{0l}^* is determined so that the integral of the energy density of the crystal in the vicinity of site l is the same as the integral of the energy density of the actual solid in the vicinity of the corresponding site. When this procedure is repeated for all defect sites, it is equivalent to Eq. (7) but now one can take differences between small numbers to find the excitation energy δE associated defect formation:

$$\delta E = \sum_{l=1}^{P} \left[E_l \, (defect) - E_l \, (crystal) \right] \, , \qquad (17)$$

where E_l *(defect)* and E_l *(crystal)* are the energies associated with site l for the defect crystal and the ground state single crystal respectively, and P is typically a modest number of sites. For example, for the surface relaxation calculation discussed below, $P = 6$ is more than sufficient, where P is the number of surface layers included and in the case chosen only one atom in each layer need be considered because of periodicity within each layer. Including terms in the perturbation series of Eq. (13) through E_2 and combining Eqs. (13-16), a_{0l}^* for each site are determined from

$$b_1 R_1^P(l)\, e^{-\alpha R_1(l)} + b_2 R_2^P(l)\, e^{-\alpha R_2(l)}\, e^{-R_2(l)/\lambda}$$

$$= \sum_{defect\ n} R_i'^P(l)\, e^{-\alpha R_i'(l)} + \sum_{defect\ nn} R_i'^P(l)\, e^{-\alpha R_i'(l)}\, e^{-R_i'(l)/\lambda} \quad , \tag{18}$$

where on the left side we have site l equivalent crystal values for b_1 and b_2, the number of nearest and next neighbors respectively; $R_1(l)$ and $R_2(l)$, the respective nearest and next nearest neighbor distances; $a_{0l}^* = (R_1(l)\,/\,c_1 - r_{WSE})\,/\,l$; and c_1, the ratio of the equilibrium or ground state nearest neighbor distance to r_{WSE}. The first sum over defect crystal sites on the right side is over nearest neighbors to site l, while the second is over next nearest neighbors. Again, we will see that one does not need to go beyond second neighbors for the applications presented in the next section. The expression for the surface energy σ which is consistent with Eqs. (15-18) is

$$\sigma = (\Delta E/A) \sum_{l=1}^{P} \left[F^*(a_{0l}^*) + \sum_{m=1}^{M} \theta_{ml} F\,(a_{ml}^*)\,/\,L_{ml} \right] , \tag{19}$$

where

$$F^*(a^*) = 1 + E^*(a^*) = 1 - (1 + a^*)\,e^{-a^*} , \tag{20}$$

where A is the surface area, P = number of layers with energy different from bulk (≤ 6), M = number of nearest neighbors for layer l, $\theta_{ml} = 1$ if $a_{ml}^* \leq 0$ and $\theta_{ml} = 0$; otherwise, $a_{ml}^* = (R_{ml}/c_1 - r_{WSE})/l$, R_{ml} is the distance between atoms m and l, and L_{ml} is the number of nearest neighbors of atom m or l, whichever is smaller. The first term in Eq. (19) depends on average neighbor distances (Eq. (18)). It is zero for an equilibrium crystal, and ≥ 0 for average neighbor distances which are either larger or smaller than the equilibrium crystal value (see Fig. 1). The second term is a bond compression term which is zero for the equilibrium crystal and in general zero for bonds which are not compressed. We will see that relaxed atomic configurations arise from a competition between these two terms.

There are some similarities between Eq. (19) and the expression for U_i in the embedded atom method[5] written below Eq. (4). They are both written as functions of basically two terms, and are of similar simplicity in application. There are basic differences in formulation and in results, however. The first term in Eq. (19) is not an embedding function as it is in the EAM. It does depend on electronic screening effects through Eq. (18), which we find to be important. The second term in the EAM is a pair potential, while our second term is not. Our second term is zero for isotropic deformation or when bonds are not compressed, while a pair potential is nonzero in both cases. Finally, one can arbitrarily improve the accuracy of Eq. (19) through the systematic perturbation theory framework of Eq. (8). This also helps to clarify the nature of the approximations made.

APPLICATIONS AND RESULTS

We are now ready to apply Eqs. (18-19). Note that the calculations truly have become trivial. We must solve only one equation, Eq. (18), for each value of a_{0l}^*. For a typical surface calculation, we need find only three values of a_{0l}^*, one each for three atomic layers nearest the surface. The energy is then obtained immediately from Eqs. (19-20). The three input parameters ΔE, r_{WSE}, and l are properties of the equilibrium bulk crystal which we obtain from experiment. They are listed in Tables I and II of Ref. 10 for much of the periodic table. The remaining input parameter, α, is determined so that the vacancy formation energy is given accurately by the method. The values of α in Å^{-1} used in these applications are[11] Cu, 2.944; Ag, 3.336; Ni, 3.015; Al, 2.105; Fe, 3.124; W, 4.232; and Si, 2.138.

Surface Energy Anisotropies

The first quantity that we will compute is the surface energy as a function of surface plane or surface grain orientation. Although such surface energy anisotropies are much discussed in materials science research, there is little quantitative information on them in the literature. Thus results would be expected to be useful. Also, recently, some values for particular crystal planes have been obtained from first principles, providing a test of our predictions. Finally, a solid surface is perhaps the most severe defect that a crystal can have, providing a significant test of our method which is based on perturbation theory applied to a single crystal. For now we will rigidly fracture the solids, keeping surface atoms in the same positions relative to their remaining neighbors that they had in the bulk. Later on, we will allow the atoms to relax to compute their new locations with corresponding small decreases in surface energies. As discussed in Ref. 1, additivity of bond energies does not apply to metals, with electronic reorganization in remaining bonds being significant. Such effects are of course included in the exact approach of Eq. (8), and are also in the approximate Eqs. (18-20). This is perhaps most apparent in Eq. (20), where the universal energy relation contains all electronic effects for lattice expansion and contraction. Surface energies computed with and without second neighbors are nearly equal, the second neighbor contribution being typically less than 1% to the surface energy.

This implies that the sum over neighbors which was truncated at second neighbors in Eq. (18) is rapidly convergent. Electronic screening plays a significant roll in this convergence.

Table I. Surface energies, in erg/cm^2, for some selected metals.

Element	Crystal Face	Theory (this work)	Theory (LDA)	Experiment[a]
FCC Metals				
Cu	(111)	1830	2100[c]	
	poly			2016
	(100)	2380	2300[b]	
	(110)	2270		
Ag	(111)	1270		
	poly			1543
	(100)	1630	1650[b]	
	(110)	1540		
Ni	(111)	2400		
	poly			2664
	(100)	3120	3050[b]	
	(110)	2980		
Al	(111)	920		
	poly			1169
	(100)	1280		
	(110)	1310	1090[d]	
BCC Metals				
Fe	(110)	1810		
	poly			2452
	(100)	3490	3100[b]	
W	(110)	3330		
	poly			4435
	(100)	5880	5100[e]	

[a] Ref. 12   [c] Ref. 14   [e] Ref. 16
[b] Ref. 13   [d] Ref. 15

Results are found in Table I. The third column contains the values computed from Eqs. (18-20). Experimental values are typically available only for polycrystalline surfaces, and these are listed[12] for all the metals. For well-annealed polycrystalline surfaces, one should expect that the lowest surface energy plane would predominate, with some admixture of the next lowest surface energy plane. Thus the experimental polycrystalline value should fall in between the lowest and next lowest surface energies for each element. We see that is satisfied for all elements in Table I. The next test is against first principles values, listed as Theory (LDA) in Table I. Here LDA refers to the commonly used local density approximation. One can see that the agreement between the first principles values and the predictions from Eqs. (18-20) is very good, the average difference being less than 10%. This difference is within the expected error range of the first principles values. The

EAM is roughly equivalent in ease of application, so perhaps it is appropriate to inquire about accuracy of predicted surface energies from that method. Surface energy predictions[5,17] from the EAM are substantially smaller than the values in Table I. For example, the fcc (111) predictions of Ref. 5 are 35% to 50% lower than the experimental polycrystalline values.

Oscillatory Surface Relaxation

Now let's move on to a test which in some ways is more difficult. When a surface is created, the atoms in the surface region will tend to relax away from their rigid (bulk) interatomic distances. In the case of Ni, Cu, Al, and Ag, it has been found experimentally that a planar relaxation occurs. In fact, in recent years low energy electron diffraction (LEED) and ion scattering results for relaxation distances have come into relatively good agreement with each other. We can calculate this relaxation by looking for minima in the surface energy (Eq. (19)) as a function of interplanar spacings. Of course, the planes are coupled by the energy minimization, so one has to vary all the interplanar spacings (d_{12}, d_{23}, etc. - see Fig. 2) simultaneously to find the correct minimum.

The results can be found in Table II. One should note first the size of the changes in interplanar spacings. They are typically less than 0.1Å. The energy changes per surface atom are a few hundredths of an eV. These are very small changes and represent a difficult test for any predictive theory.

Secondly, note the good agreement between theory and experiment. As mentioned earlier, the theory is typically within the experimental error bars, and often within 0.01Å of an experimental result.

Third, note that $\Delta d_{12} \leq 0$ for all predicted values and for all experimental values except for Al (111). Even in that case, the experimental Δd_{12} is small. The equivalent crystal approach provides for a simple understanding of why $\Delta d_{12} \leq 0$. Atoms in the surface layer are missing neighbors and so the average neighbor distance is larger than it is in the bulk. That is, Eq. (18) will yield $a_{01}^* > 0$ where a_{01}^* is the equivalent crystal lattice constant for a surface atom. Surface atoms will then be represented by a point to the right of the minimum in Fig. 1. Thus the surface layer will move to decrease the average neighbor distance in order to lower the energy through the first term of Eq. (19). This will amount to moving toward the minimum in Fig. 1. This can be accomplished by decreasing d_{12}, i.e., by decreasing the distances to substrate neighbors. However, the movement toward the energy minimum by decreasing d_{12} is a highly anisotropic process. While $a_{01}^* > 0$, so that the equivalent crystal lattice constant for atoms in the surface layer is *larger* than that of the equilibrium bulk value, some bond lengths become *smaller* than the bulk value due to bond compression accompanying the decrease in d_{12}. The energy rises rapidly in compression, as can be seen from Fig. 1. This enters the surface energy σ through the second term in Eq. (19). It is the competition between this term and the first term which yields a minimum in σ (d_{12}).

Table II. Percentage changes in interlayer spacings due to relaxation. Calculated absolute changes are also given in Å for reference.

Element	$\Delta d_{n,n+1}$	Theory	Experiment	Technique (reference)
Cu(110)	Δd_{12}	$-6.5\%(-0.083\text{Å})$	$-8.5 \pm 0.6\%$ $-7.5 \pm 1.5\%$	LEED (18) Ion scattering (19)
	Δd_{23}	$+2.7\%(+0.034\text{Å})$	$+2.3 \pm 0.8\%$ $+2.5 \pm 1.5\%$	LEED (18) Ion scattering (19)
Cu(100)	Δd_{12}	$-3.5\%(-0.063\text{Å})$	$-2.1 \pm 1.7\%$ $-1.1 \pm 0.4\%$	LEED (20) LEED (21)
	Δd_{23}	$+1.6\%(+0.029\text{Å})$	$+0.45 \pm 1.7\%$ $+1.7 \pm 0.6\%$	LEED (20) LEED (21)
Cu(111)	Δd_{12}	$-2.8\%(-0.058\text{Å})$	$-0.7 \pm 0.5\%$	LEED (22)
	Δd_{23}	$+1.4\%(+0.029\text{Å})$		
Ag(110)	Δd_{12}	$-5.4\%(-0.078\text{Å})$	-5.7% $-7.8 \pm 2.5\%$	LEED (21) Ion scattering (23)
	Δd_{23}	$+2.4\%(+0.035\text{Å})$	$+2.2\%$ $+4.3 \pm 2.5\%$	LEED (21) Ion Scattering (23)
Ag(100)	Δd_{12}	$-2.9\%(-0.059\text{Å})$		
	Δd_{23}	$+1.4\%(+0.029\text{Å})$		
Ag(111)	Δd_{12}	$-2.3\%(-0.054\text{Å})$		
	Δd_{23}	$+1.1\%(+0.026\text{Å})$		
Ni(110)	Δd_{12}	$-6.8\%(-0.084\text{Å})$	$-8.7 \pm 0.5\%$ $-9.0 \pm 1.0\%$	LEED (24) Ion scattering (25)
	Δd_{23}	$+3.2\%(+0.040\text{Å})$	$+3.0 \pm 0.6\%$ $+3.5 \pm 1.5\%$	LEED (24) Ion scattering (25)
Ni(100)	Δd_{12}	$-3.7\%(-0.065\text{Å})$	$-3.2 \pm 0.5\%$	Ion scattering (26)
	Δd_{23}	$+1.7\%(+0.030\text{Å})$		
Ni(111)	Δd_{12}	$-2.9\%(-0.059\text{Å})$	$-1.2 \pm 1.2\%$	LEED (27)
	Δd_{23}	$+1.4\%(+0.028\text{Å})$		
Al(110)	Δd_{12}	$-10.1\%(-0.14\text{Å})$	$-8.6 \pm 0.8\%$ $-8.5 \pm 1.0\%$	LEED (28) LEED (29)
	Δd_{23}	$+4.8\%(+0.068\text{Å})$	$+5.0 \pm 1.1\%$ $+5.5 \pm 1.1\%$	LEED (28) LEED (29)
	Δd_{34}	$-0.4\%(-0.006\text{Å})$	$-1.6 \pm 1.2\%$ $+2.2 \pm 1.3\%$	LEED (28) LEED (29)
	Δd_{45}	$0.0\%(0.0\text{Å})$	$+0.1 \pm 1.3\%$ $+1.6 \pm 1.6\%$	LEED (28) LEED (29)
Al(100)	Δd_{12}	$-5.0\%(-0.10\text{Å})$		
	Δd_{23}	$+2.1\%(+0.042\text{Å})$		
Al(111)	Δd_{12}	$-3.6\%(-0.084\text{Å})$	$+0.9 \pm 0.7\%$	LEED (30)
	Δd_{23}	$+1.8\%(+0.042\text{Å})$		

Because plane 1 has moved closer to plane 2, the average neighbor distance from an atom in the second layer, as computed from Eq. (18), has decreased. That is, a_{02}^* has become negative and the representative point on the universal energy curve of Fig. 1 lies to the left of the minimum for the second layer. Thus in order to decrease the total energy, it will be necessary for that point to move to the right, toward the minimum. This in effect means that atoms in the second layer move to increase the average neighbor distance. This can be accomplished by increasing d_{23} without sacrificing the energy lowering due to the decrease in d_{12}.

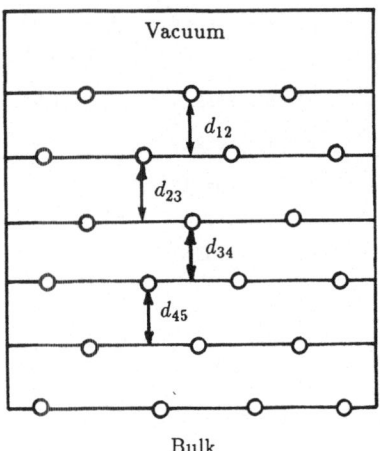

Fig. 2. Schematic showing interplanar spacings near the surface.

Of course when d_{23} is increased, this decreases the average neighbor density around plane 3, so that $a_{03}^* > 0$ and the representative point on the universal curve of Fig. 1 for plane 3 lies to the right of the minimum. The energy can then be lowered by $\Delta d_{34} < 0$.

This oscillatory relaxation continues on into the bulk. Such oscillatory relaxation effects had been detected experimentally as can be seen in Table II. Alternation with depth of the sign of the change in interlayer spacing has also been seen in surface calculations on Al[31,32], Fe[33], and Al-Ni alloys[6]. Here we have the first quantitative agreement with experiment for a series of transition metals. More importantly, the simplicity of the equivalent crystal approach allows one to understand the principal physical mechanisms leading to oscillatory relaxation.

It should be noted that oscillatory relaxation will not be found for all surfaces because it depends on a delicate balance between the two terms of Eq. (19). For example, if layer 2 has an a_{02}^* which is sufficiently large and positive prior to relaxation, then one might expect $\Delta d_{23} < 0$. Also there are a number of trends in Table II which bear further discussion, and the interested reader is referred to Ref. 2.

Equivalent Crystal Theory of Semiconductor Defects

There is considerable interest in a simple method for computing energies of defects in semiconductors (see, e.g., Refs. (34-35)). Semiconductors represent a different challenge from the metals which we have applied the equivalent crystal approach to thus far. Semiconductors have a more open structure and more directed bonds than metals. absolute changes are also given in Å for reference.

Model potentials[34-35] for semiconductors employ bond angle expressions which were not involved with potentials for metals[5].

There is evidence[4], however, that the universal energy relation applies as well to covalent semiconductors as to metals. Also, the equivalent crystal approach (Eq. (5a)) is in principle exact if the perturbation series is carried to all orders. The question, then, is whether the approximate relations, Eqs. (18-20), can give accurate results for semiconductors.

We decided to test these equations as they are - unchanged - on calculations of Si surface properties. No additional parameters are required for semiconductors than for metals. As noted above, ΔE, r_{WSE}, and l are listed in Ref. 10 and α is 2.138 as determined from the vacancy formation energy. Because of additivity of bond energies in semiconductors, $L_{ml} = 4$ for Si. The results are shown in Table III. Semiconductor surfaces usually reconstruct in addition to planar relaxation, because of their more open structure. However, for Si (100) there are first principles theoretical results[37] for the planar relaxed (unreconstructed (1x1)) surface.

Thus we can make a direct comparison with the type of accuracy we obtained for metals. We see in Table III that the total surface energies and relaxation energies are in rather good agreement. The Δd_{12} values are also in good agreement. The agreement for Δd_{23} is not quite as good, but the values are small.

The Si (111) - (1x1) surface has been obtained[36] experimentally via a pulsed ruby laser, and the planar relaxations have been determined from LEED experiments. The agreement with the equivalent crystal predictions is very good. The question of whether a good, well-ordered Si (111) - (1x1) surface can be obtained via laser annealing is somewhat controversial. See Ref. 38 and references therein for a summary of more recent experimental results. Fortunately, there are first principles calculations of relaxation of the Si (111) - (1x1) surface to compare with[39]. Northrup and Cohen[39] found $\Delta d_{12} = -25.1\%$ and $\Delta d_{23} = +0.4\%$, in reasonable agreement with both our calculations and with experiment[36], especially for Δd_{12}.

The results of these preliminary tests suggest that the equivalent crystal approach as approximated in Eqs. (18-19) has accuracy for Si surface properties which is comparable to that exhibited in Table I and II for metals. This suggests that higher order perturbation terms or, perhaps, bond angle energy terms are not required for these planar relaxations. Bond angle energy terms may play a more important role in reconstruction, however. Reconstruction calculations are currently being carried out for semiconductor surfaces, and

Table III. Results for Si.

I. Si(111)—(1x1)

	Theory	Experiment (Ref. 36)
Δd_{12}	−25.6%(−0.20Å)	−25.5 ± 2.5%
Δd_{23}	+2.3%(+0.054Å)	+3.2 ± 1.5%

II. Si(100)—(1x1)

	Theory	Theory (Ref. 37)
Surface Energy	2.6eV	2.5eV
Relaxation Energy	0.04eV	0.03eV
Δd_{12}	−7.2%(−0.098Å)	−5.0%
Δd_{23}	+0.8%(+0.011Å)	−2.1%

this should tell us more about the accuracy of the equivalent crystal method for semiconductors and, in particular, about the importance of higher order terms.

SUMMARY

We have developed a new method for computing energies and atom locations in solid defects which is both simple and accurate. Solid defects are formulated in terms of an exact perturbation expansion in which the unperturbed system is a single crystal. The defect formation energy can be simply obtained from the cohesive energy of a single crystal whose lattice constant is chosen so that the perturbation expansion sums to zero. A universal energy relation is used for the cohesive energy as a function of lattice constant.

This simple method was tested against first principles and experimental surface energies for the low index surface planes of a series of transition metals. Good agreement was obtained in all cases. Planar relaxations in the surface region were computed for the low index surfaces of Al, Cu, Ag, and Ni and compared with LEED and ion scattering data. Agreement was typically within experimental error bars, often within 0.01Å for the interplanar spacings. The transparency of the method allows one to simply understand the origin of the oscillatory relaxation that is observed both experimentally and theoretically.

The same method was tested against theoretical results for the surface energy and planar relaxation distances for Si (100) and against experimental results for planar relaxation of Si (111). The accuracy was comparable to what we found for metals, suggesting that directed bonds, more open lattices, and other differences between semiconductors and metals are well described by the method, at least for these applications.

The success of these initial tests bodes well for the future of this method. Further testing is under way. Currently, reconstruction and impurity effects are being treated.

REFERENCES

1. J.R. Smith and A. Banerjea, Phys. Rev. Letters 59:2451 (1987).
2. J.R. Smith and A. Banerjea, Rapid communications, Phys. Rev. B, 37:10411 (1988).
3. J.H. Rose, J. Ferrante, and J.R. Smith, Phys. Rev. Letters, 47:675 (1981).
4. A. Banerjea and J.R. Smith, Phys. Rev. B, 37:6632 (1988), and references therein.
5. S.M. Foiles, M.I. Baskes and M.S. Daw, Phys. Rev. B, 33:7983 (1896) and references therein.
6. S.P. Chen, A.F. Voter and D.J. Srolovitz, Phys. Rev. Letters, 57:1308 (1986).
7. M.W. Finnis and J.E. Sinclair, Phil. Mag. A, 50:45 (1984).
8. J.K. Norskov, Phys. Rev. B, 20:446 (1979). See also J.K. Norskov and N.D. Lang, Phys. Rev. B, 21:2131 (1980) and M. Manninen, Phys. Rev. B, 34:8486 (1986) and references therein.
9. M.J. Stott and E. Zaremba, Phys. Rev. B, 22:1564 (1980).
10. J.H. Rose, J.R. Smith, F. Guinea, and J. Ferrante, Phys. Rev. B, 29:2963 (1984).
11. Unfortunately, the ordering of the α values as listed in Ref. 2 is incorrect for Ag, Ni, and Al.
12. H. Wawra, Z. Metallk. 66:395, 492 (1975).
13. J.G. Gay, J.R. Smith, R. Richter, F.J. Arlinghaus, and R.H. Wagoner, J. Vac. Sci. Technol. A, 2:931 (1984); J.R. Smith, J. Ferrante, P. Vinet, J.G. Gay, R. Richter, and J.H. Rose, in: "Chemistry and Physics of Fracture", R.H. Jones and R.M. Latanision, eds., p. 329 (Martinus Nijoff, Hingham, MA, 1987).
14. J.A. Appelbaum and D.R. Hamann, Solid State Commun., 27:881 (1978).
15. K.M. Ho and K.P. Bohnen, Phys. Rev. B, 32:3446 (1985).
16. C.L. Fu, S. Ohnishi, H.J.F. Jansen, and A.J. Freeman, Phys. Rev. B, 31:1168 (1985).
17. G.J. Ackland and M.W. Finnis, Philos. Mag. A, 54:301 (1986).
18. D.L. Adams, H.B. Nielsen, J.N. Andersen, I. Stensgaard, R. Feidenhansl and J. E. Sorense, Phys. Rev. Letters, 49:669 (1982).
19. M. Copel, T. Gustafsson, W.R. Graham and S.M. Yalisove, Phys. Rev. B, 33:8110 (1986).
20. R. Mayer, C. Zhang, K.G. Lynn, W.E. Frieze, F. Jona and P.M. Marcus, Phys. Rev. B, 35:3102 (1987).
21. H.L. Davis and J.R. Noonan, Surf. Sci. 126:245 (1983).
22. S.A. Lindgren, L. Wallden, J. Rundgren and P. Westrin, Phys. Rev. B, 29:576 (1984).
23. Y. Kuk and L.C. Feldman, Phys. Rev B, 30:5811 (1984).
24. D.L. Adams, L.E. Petersen and C.S. Sorensen, J. Phys. C. : Solid State Phys. 18:1753 (1985).
25. S.M. Yalisove, W.R. Graham, E.D. Adams, M. Copel and T. Gustafsson, Surface Sci., 171:400 (1986).
26. J.W.M. Frenken, J.F. van der Veen and G. Allan, Phys. Rev. Letters, 51:1876 (1983).
27. J.E. Demuth, P.M. Marcus, and D.W. Jepsen, Phys. Rev. B, 11:1460 (1975).
28. J.N. Andersen, H.B. Nielsen, L. Petersen and D.L. Adams, J. Phys. C. : Solid State Phys., 17:173 (1984).
29. J.R. Noonan and H.L. Davis, Phys. Rev. B, 29:4349 (1984).
30. D.L. Adams, H.B. Nielsen and J.N. Andersen, Physica Scripta, T4:22 (1983).
31. R.N. Barnett, U. Landman and C.L. Cleveland, Phys. Rev. Letters, 51:1359 (1983) and references therein.
32. K.M. Ho and K.P. Bohnen, Phys. Rev. B, 32:3446 (1985).
33. P. Jiang, F. Jona and P.M. Marcus, Phys. Rev. B, 35:7952 (1987).
34. M.I. Baskes, Phys. Rev. Letters, 59:2666 (1987).
35. B.W. Dodson, Phys. Rev. B, 35:2795 (1987).
36. D.M. Zehner, J.R. Noonan, H.L. Davis, and C.W. White, J. Vac. Sci. Technol., 18:852 (1981).
37. M.T. Yin and M.L. Cohen, Phys. Rev. B, 24:2303 (1981).

38. F. Jona, P.M. Marcus, H.L. Davis, and J.R. Noonan, <u>Phys. Rev. B</u>, 33:4005 (1986).
39. J.E. Northrup, J. Ihm, and M.L. Cohen, <u>Phys. Rev. Letters</u>, 47:1910 (1981). See also J.E. Northrup and M.L. Cohen, <u>Phys. Rev. B</u>, 29:5944 (1984).

GRAIN-BOUNDARY AND FREE-SURFACE INDUCED THERMODYNAMIC

MELTING: A MOLECULAR DYNAMICS STUDY IN SILICON

S.R. Phillpot, J.F. Lutsko, and D. Wolf

Materials Science Division
Argonne National Laboratory
Argonne, IL 60439

S. Yip

Department of Nuclear Engineering
Massachusetts Institute of Technology
Cambridge, MA 02139

INTRODUCTION

It is well known that melting of a solid generally proceeds from the surface. For example, it was observed in measurements on silica[1] and phosphorous pentoxide[2] that melting was not a homogeneous process; invariably it occurred at free surfaces and grain boundaries. A variety of experimental data now exists which points to the controlling role of an extrinsic surface.[3] Small atomic clusters, with a significant fraction of the particles on or close to the surface, have been observed to exhibit quite different melting behavior from that of the bulk substance; for example, melting-point depression of up to 30% has been measured in metal clusters of diameter 20-30Å[4], as has substantial superheating of argon bubbles of similar size formed in an aluminum lattice,[5] and of hydrogen bubbles in amorphous silicon.[6] Superheating has also been observed recently in small single crystals of silver coated with gold, the latter with a higher melting point.[7] The implication of these results is that while melting is a thermodynamic transition, in general it is initiated at either an external surface or an internal interface, such as a grain boundary or a dislocation.

In this paper we report on some recent calculations[8] of the melting of silicon in the presence of a grain boundary (GB) and a free surface. The question of disordering or "premelting" of the GB below the thermodynamic melting temperature has been discussed elsewhere[9] and will not be addressed here. Silicon is particularly suitable for this study as it is known to have a low GB mobility,[10] thereby avoiding confusion between the disordering associated with grain-boundary migration and the disordering associated with melting. Also, the Stillinger-Weber (SW) potential[11] for silicon, used here, has proved successful in recent studies of the bulk crystal,[11,12] the liquid,[12] the crystal-liquid interface[13,14] and grain boundaries.[15]

The GB chosen for our investigation is the so-called $\Sigma 11$ (110) high-angle twist boundary. This boundary on the (110) plane is obtained by rotation about [110], by an angle of $\theta = 50.48°$, of two perfect semi-crystals with (110) faces. The area of the rectangular unit cell of this coherent interface is $\Sigma = 11$ times greater than that of the corresponding primitive planar unit cell ($\Sigma = 1$) on the (110) plane; each lattice plane thus contains 22 atoms. Our computational cell was chosen to contain 32 (110) planes, i.e. 704 atoms. Two-dimensional periodic border conditions, described in detail elsewhere,[16] were applied parallel to the boundary (x-y) plane. The lattice parameter in the x-y plane was fixed to that of the ideal crystal at the temperature simulated. In the z direction (z ‖ [110]) the simulation cell was embedded between two semi-infinite perfect-crystal blocks which, during the simulation, were allowed to move as rigid units thus allowing both GB sliding and volume expansion at the interface.[16] The zero-temperature structure and energy of this boundary were investigated in Ref. 14. For the surface simulations, two free surfaces were generated at the ends of an ideal crystal of the same unit-cell dimensions, area and number of atoms as the bicrystal.

To monitor the inhomogeneity of the bicrystal in the z direction, the system was subdivided into 32 slices; initially every slice thus contained one (110) plane. In every slice the instantaneous values of the potential energy, temperature and mean square displacement (msd) were monitored. The total number of defected atoms, N_{def}, in the system was also calculated. An atom is considered defected if its number of nearest neighbors is different from that in an ideal crystal. As a merely semi-quantitative measure we define the nearest-neighbor shell to end half way between the ideal-crystal first and second nearest-neighbor distances. Thus, any disordering or melting of the system, and subsequent coordination change will be apparent as an increase in N_{def}.

To investigate the breakdown of the crystalline order upon melting we also monitored the square of the magnitude of the static structure factor, $S(\underline{k})$ (which for brevity we simply denote $S^2(\underline{k})$). For the overall $S^2(\underline{k})$ all atoms in the simulation cell are included, whereas for a planar structure factor $S_P^2(\underline{k})$ only atoms in a given lattice plane are considered. For an ideal crystal at zero temperature, $S_P^2(\underline{k})$ then equals unity for any wave vector, $\underline{k}$, which is a reciprocal lattice vector in the plane. By contrast, in the liquid state (without long-range order in the plane), $S_P^2(\underline{k})$ fluctuates near zero. As the two halves of the bicrystal are rotated with respect to each other about the GB-plane normal two different wave vectors, $\underline{k}_1$ and $\underline{k}_2$, are required, each corresponding to a principal direction in the related half.

GRAIN-BOUNDARY INDUCED MELTING

Prior to the GB simulation a perfect 3-d periodic crystal containing 216 particles was superheated. This system was found to melt at between 2400K and 2600K, which is consistent with previous results[12,13] and is considerably higher than the thermodynamic melting temperature, $T_m = 1691 \pm 20K$, determined from free-energy analysis.[12] This well-known discrepancy shows that, on this time scale, even at temperatures significantly higher than the thermodynamic melting temperature, a defect-free perfect crystal is stable against any lattice instability.

In all the simulations the time step was $1.15 \times 10^{-15}s$, for which energy was found to be conserved to six significant figures for simulations of several thousand time steps. At the beginning of each simulation run, particles were given random velocities corresponding to a temperature of 1200K. To reach the desired simulation temperature the system was then heated by 100K every 200 time steps. As we are primarily interested in phase transitions which involve a latent heat, when the desired temperature was reached a thermostat was applied by rescaling the particle velocities.

Our bicrystal simulations were performed at six different temperatures: 1650K, 1750K, 1800K, 1900K, 2000K and 2100K. After 9000 time steps at 1650K the system was found to be completely crystalline all the way up to the grain boundary, with neither an increase in the width of the GB region, nor a change in the location of the GB plane.[9]

At all the higher temperatures the system was unstable with respect to melting induced at the grain boundary. Figure 1, snapshots of the planar structure factor after 7600 and 18600 time steps at 1900K, demonstrates the spreading of a disordered or liquid phase through the bicrystal, starting from the grain boundary.

A similar analysis of the mean square displacement as a function of time shows a time-independent and small msd for the perfect-crystal regions but a linearly increasing msd for the disordered planes. The diffusion constant at 1900K of $\approx 0.7 \times 10^{-8}$ m^2/s was extracted from the time dependence of the msd in the disordered region. This value is close to that found for the liquid at this temperature.[12]

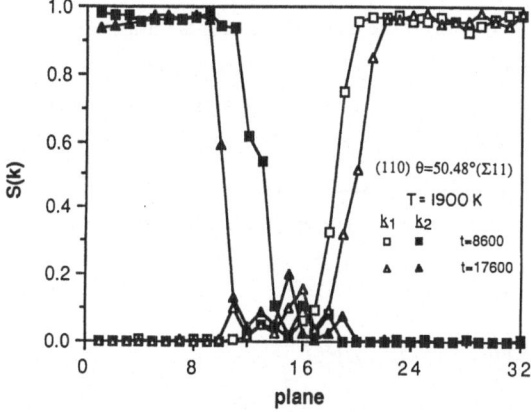

Fig. 1   Instantaneous values of S_P^2 ($\underline{k}_1$) and S_P^2 ($\underline{k}_2$) for the 32 slices parallel to the (110) $\Sigma 11$ GB after 7600 and 18600 time steps at 1900K.

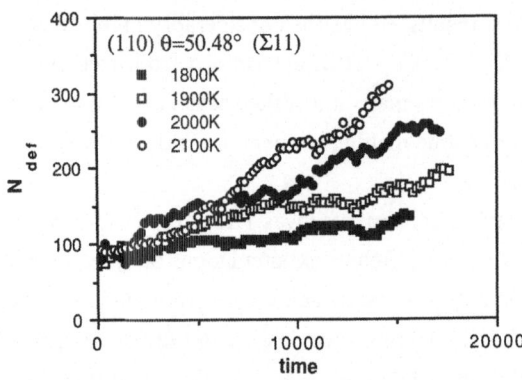

Fig. 2   Number of defected atoms in the bicrystal as a function of time for four temperatures.

The linear increase in the size of the disordered region also gives rise to a linear decrease in time of the length of the computational cell. The number of defected atoms in the system was also found to increase linearly in time, as is shown in Figure 2.   The interface propagation velocities extracted from the slopes of the data in Figure 2 are plotted as a function of temperature in Figure 3. We note that although at 1750K it is clear that the system is melting, the nucleation time is sufficiently long ($\approx$10000 time steps at 1750K vs $\approx$3000 time steps at 1900K) and that the kinetics of melting are sufficiently slow, that no reliable propagation velocity can be extracted. At 2100K the interface velocity was 76 ± 9 m/s, which is to be compared with a propagation velocity of 191 m/s measured experimentally in the propagation of a (100) interface at a temperature of less than 2870K.[17]

The evident decrease in the rate of growth of the liquid phase with decreasing temperature implies the existence of a temperature T_0 at which the growth rate vanishes. From a quadratic fit to the data, shown in Figure 3, we estimate T_0 to be $1665 \pm 50K$. We cannot, of course, directly verify the exact value of this temperature, since the time scale of the disordering becomes prohibitively large as T_0 is approached. The value obtained for T_0 is in remarkable agreement with the thermodynamic melting temperature, $T_m = 1691 \pm 20K$, determined from free-energy analysis[12] for the same potential. From this agreement we conclude that the propagation of the solid-liquid interface seen in all but the 1650K simulation is due to thermodynamic melting of the bicrystal.

FREE-SURFACE INDUCED MELTING

Surface-induced melting was simulated at four temperatures: 1900K, 2000K, 2100K and 2200K. At 1900K nucleation of melting took approximately 10,000 time steps, as compared with only ≈3000 time steps in the GB at the same temperature. Because of this long nucleation time, no attempt was made to simulate surface-induced melting at temperatures below 1900K. As the temperature is increased this difference between nucleation times at the free surface and at the GB decreases such that it is negligible at 2100K.

Despite the longer nucleation time, our simulations were long enough to observe significant melting at all four temperatures. The behavior of the structure factor, msd and N_{def} were similar over this temperature range. As evidence by the time dependence of both the planar structure factor and the msd, it is clear that a disordered region spreads from the free surfaces into the bulk. Again this disordered region may be identified as liquid on the basis of the diffusion constant of about 1.2×10^{-8} m^2/s at 2000K, a value comparable to that of the bulk liquid.[12]

Interface propagation velocities were extracted from the time rates of increase of the number of defected atoms. These velocities, together with those calculated for the GB induced melting are shown in Table 1. For the three temperatures at which simulations were performed for both the free surface and the GB, the interface velocity is identical within statistical errors. The implication is that after the initial nucleation, the melting behaviors are the same. This is consistent with our interpretation in the previous section that the melting observed is ordinary thermodynamic melting.

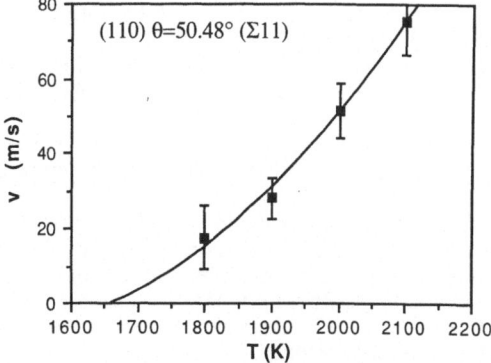

Fig. 3 Propagation velocity in m/s as function of temperature for grain-boundary nucleated melting. The curve is a quadratic fit to the data.

Table I. Propagation velocity of the solid-liquid interface in m/s when nucleated from a grain boundary and from a free surface.

	1800K	1900K	2000K	2100K	2200K
Grain Boundary	18±9	28±5	52±7	76±9	-------
Free Surface	------	22±3	56±6	77±4	95±2

CONCLUSIONS

In this paper we have presented results of molecular dynamics simulations on the characteristics of melting in Si initiated at a high-angle twist grain boundary and at a (110) free surface at temperatures above the thermodynamic melting point, T_m, associated with the Stillinger-Weber potential.

We have seen that a defect-free ideal crystal may be considerably superheated above the thermodynamic melting temperature. At temperatures above T_m, but below the maximum superheating temperature, melting is nucleated at a GB or a free surface and spreads throughout the system on a time scale that is significantly faster than that of any lattice instability that the system may have. These observations are entirely consistent with the experiments discussed in the Introduction, which show the dominant role that extrinsic defects play in the melting of real materials.

ACKNOWLEDGMENT

This work was supported by U.S. Department of Energy, BES-Materials Science, under Contract W-31-109-Eng-38. We are indebted to Prof. H. Gleiter for several helpful suggestions and discussions. The authors wish to acknowledge a grant of computer time at the Energy Research CRAY XMP at the Magnetic Fusion Computational Center at Livermore.

REFERENCES

1. N.G. Ainslie, J.D. Mackenzie and D. Turnbull, Melting Kinetics of Quartz and Cristobalite, J. Phys. Chem. 65:1718 (1961).
2. R.L. Cormia, J.D. Mackenzie, and D. Turnbull, Kinetics of Melting and Crystallization of Phosphorous Pentoxide, J. Appl. Phys. 34:2239 (1963).
3. R.W. Cahn, Melting and the Surface, Nature 323:668 (1986).
4. Ph. Buffat and J.-P. Borel, Size Effect of the Melting Temperature of Gold, Phys. Rev. A13:2287 (1976).
5. J.B. Boyce and M. Stutzmann, Orientational Ordering and Melting of Molecular H_2 in an a-Si Matrix: NMR Studies, Phys. Rev. Lett. 54:562 (1985).
6. C.J. Rossouw and S.E. Donnelly, Superheating of Small Solid-Argon Bubbles in Aluminum, Phys. Rev. Lett. 55:2960 (1985).
7. J. Daeges, H. Gleiter, and J .H. Perepezko, Superheating of Metal Crystals, Phys. Lett. A119:79 (1986).
8. S.R. Phillpot, J.F. Lutsko, D. Wolf and S. Yip, Molecular Dynamics Study of Lattice-Defect Nucleated Melting in Silicon, submitted to Phys. Rev. B.
9. S.R. Phillpot, J.F. Lutsko and D. Wolf, Nucleation and Kinetics of Thermo-dynamic Melting: A Molecular Dynamics Study of Grain-Boundary Induced Melting in Silicon, to be published.
10. D.A. Smith and T.Y. Tan, Effects of Doping and Oxidation on Grain Growth in Polysilicon in: "Grain Boundaries in Semiconductors", H.J. Leamy, G.E. Pike, and C.H. Seager, ed. (North Holland, 1982).
11. F.H. Stillinger and T.A. Weber,, Computer Simulation of Local Order in Condensed Phases of Silicon, Phys. Rev. B31:5262 (1985).
12. J.Q. Broughton and X.P. Li, Phase Diagram of Silicon by Molecular Dynamics, Phys. Rev. B35:9120 (1987).
13. U. Landman, W.D. Leudtke, R.N. Barnett, C.L. Cleveland, M.W. Ribarsky,

E. Arnold, S. Ramesh, H. Baumgart, A. Martinez and B. Kahn, Faceting at the Silicon (100) Crystal-Melt Interface: Theory and Experiment, Phys. Rev. Lett. 56:155 (1986).

14. U. Landman, W.D. Luedtke, M.W. Ribarsky, R.N. Barnett, and C.L. Cleveland, Molecular-Dynamics Simulations of Epitaxial Growth from the Melt. I--Si (100), Phys. Rev. B37:4637 (1988); W.D. Luedtke, U. Landman, M.W. Ribarsky, R.N. Barnett, and C.L. Cleveland, Molecular-Dynamics Simulations of Epitaxial Growth from the Melt. II--Si (111), Phys. Rev. B37:4647 (1988).

15. S.R. Phillpot and D. Wolf, Atomistic Simulation of Silicon Grain Boundaries, Proc. MRS Symp. on "Interfacial Structure, Properties, and Design in Solids", Reno, NV 1988; S.R. Phillpot and D. Wolf, Structure-Energy Correlation for Grain Boundaries in Silicon, submitted to Phil. Mag. A.

16. J.F. Lutsko, D. Wolf, S. Yip, S.R. Phillpot and T. Nguyen, A Molecular Dynamics Method for the Simulation of Bulk Interfaces at High Temperatures, Phys. Rev. B., 38:11572 (1988).

17. J.Y. Tsao, M.J. Azia, M.O. Thompson and P.S. Peercy, Asymmetric Melting and Freezing Kinetics in Silicon, Phys. Rev. Lett. 56:2712 (1986).

INTERATOMIC POTENTIALS AND THE BONDING ENERGETICS OF

POLYTETRAHEDRAL PACKING IN TRANSITION METAL ALLOYS

R.B. Phillips and A.E. Carlsson

Department of Physics
Washington University
St. Louis, Missouri 63130

INTRODUCTION

The discovery of quasicrystals[1], recent work on the theory of metallic glasses,[2,3] and continued interest in the Frank-Kasper phases all provide impetus for understanding the bonding energetics associated with polytetrahedral atomic configurations as building blocks for complex structures. Previous theoretical work has successfully explained the observed stability regions for a few specific PTP phases[4,5] (we define PTP phases to be those close packed phases in which all atoms sit at the vertices of tetrahedra, such as the Frank-Kasper phases and possibly some quasicrystals) within the framework of tight binding theory. In addition, Watson and Bennett[6] have made a general empirical study of the trends in the relative stability of the majority of the Frank-Kasper phases, using atomic size concepts and d-band hole counts. In this work we provide a more general theoretical analysis, based on total energy calculations, of the factors favoring PTP over fcc packing. This analysis is directly complementary to the empirical analysis of Watson and Bennett,[6] and explains empirically observed trends in the relative stability of PTP on the basis of angle-dependent effective interatomic potentials.

To calculate the bonding energies we use a tight binding d-band model Hamiltonian of the form

$$H = \sum_i \sum_{\alpha=1}^{5} \varepsilon_i \, |i,\alpha\rangle \langle i,\alpha| + \sum_{\substack{ij \\ i \neq j}} \sum_{\alpha=1} \sum_{\beta=1} h_{\alpha\beta}^{ij} \, |i,\alpha\rangle \langle j,\beta| \tag{1}$$

where i,j are site indices, α, β are d-orbital indices, the ε_i are the d-orbital single site energies, and the $h_{\alpha\beta}^{ij}$ are hopping integrals of the Slater-Koster form[7]. This d-band model is relevant for binary transition metal compounds such as many of the Frank-Kasper phases. While the neglect of the s-p electrons and the electron-electron interactions is a severe approximation, the approach[4,5,8-10] used here often provides a good description of chemical trends in structural energies. Since we aim to understand the dependence of the bonding energy upon the local atomic configurations, we characterize the bonding energy by the low order moments of the local density of states,[8] which are rigorously determined by the local environment. The moments on site i are defined by[8-10] $\mu_n^i \equiv \Sigma_{\alpha=1}^5 <i,\alpha |H^n| i,\alpha>$, where n labels the moment of interest. Once the moments are determined,[11] we reconstruct the local density of states (DOS) on the site of interest using the "maximum entropy" method.[10,12] Given the DOS $\rho(E)$, the bonding energy is approximated as the one electron energy sum (the factor of two accounts for spin degeneracy): $E_{coh} = 2 \int_{-\infty}^{\varepsilon_F} E\rho(E)dE$. We have also developed E_{coh} in a cluster potential expansion, which yields an r-space picture of the structural energy differences. The potentials are given in terms of the moments by[13]

$$V_n^{eff} (\vec{R}_1, ..., \vec{R}_n) = \frac{\partial E_{coh}}{\partial \mu_n} \mu_n^{1...n} . \qquad (2)$$

Here $\mu_n^{1...}$ is the contribution of sites $\vec{R}_1, ..., \vec{R}_n$ to the n'th moment, and the derivative term is obtained numerically from the approximate DOS. For $n \geq 3$ these potentials correspond to angle-dependent forces.

We consider the bonding energetics associated with the following PTP phases: 1) A15 structure, 2) $MgCu_2$ Laves phase, 3) σ-phase and 4) the Ti_2Ni structure. We also consider the central icosahedral site in two other hypothetical structures having PTP: an extension of the Mackay icosahedron with 147 atoms, and an icosahedral site in polytope {3,3,5}, a hypothetical packing of 120 atoms on the surface of a four dimensional sphere. The fcc and bcc structures are considered as examples of simpler packing schemes.

We first consider the elemental case (ε_i = constant). The Slater-Koster[7] couplings are chosen to reproduce a typical transition metal d-band width of 8.5 eV: $h_{dd\sigma}$ = -1.230 eV, $h_{dd\pi}$ = 0.664 eV, and $h_{dd\delta}$ = -0.049 eV. To compute the Hamiltonian matrix for the polytope in curved space we use a generalization of earlier work by Widom,[3] which is described elsewhere.[14] The results of calculations using six moments are shown in Figure 1. We show only the icosahedral sites due to space limitation. The graph shows the difference in bonding energy between the icosahedral sites and an fcc site as a function of the number of d-electrons per atom (N_d). We believe that the results are fairly well converged as a function of the number of moments included because the calculated energies show only a small change on going from four to six moments, and in several test cases the trends are in close correspondence with much more accurate calculations[4,15] we consider couplings only between nearest neighbors and assume a r^5 radial dependence. (We define the near neighbor shell to be those atoms at a distance less than 1.2 r_d, where r_d is the distance to the closest neighbor. We see that for all of the calculated icosahedral configurations the most favorable band fillings are within about 5% of 3.5 d-electrons per atom. We also observed that the energy differences of the various icosahedral configurations, relative to fcc, are within roughly 10% of each other. It is also found that the higher fold PTP sites are energetically favored for the same region of d-band filling.

Figure 1 shows that for between 2 and 5 d-electrons per atom, we would expect the icosahedral sites to be energetically stable relative to the fcc structure if only the electronic band energy contributed. Although our model does not have the quantitative accuracy required to make unambiguous predictions for specific systems, our results are consistent with the observation[16] of W, Nb and Cr in metastable A15 structures, which contain 25% icosahedral sites. Pair potential cluster calculations[17-19] would suggest that icosahedral sites are *always* preferred locally, but are not always observed because global packing constraints cannot be satisfied. Our calculations show that, on the contrary, even if the global packing constraints are artificially ignored, as in polytope (3,3,5), the icosahedral sites are not universally favored.

To simulate the effect of alloying, we introduce a single site energy difference $\Delta\varepsilon$ on the icosahedral sites. The region of preferability for the icosahedral sites moves to between roughly 5 and 9 d-electrons per atom for values of $\Delta\varepsilon$ from -2 eV to -4 eV, typical of many transition metal alloys, while the region of preferability for the higher fold coordinated sites moves to lower N_d. The shift in the region of energetic preferability resulting from the inclusion of alloying effects is due to changes in the third moment of the DOS[20].

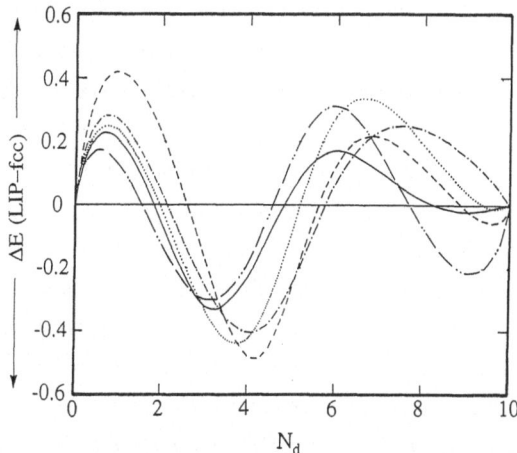

Fig. 1 Structural energy difference (ΔE) between LIP site and fcc sites, in eV. Solid curve: polytope-fcc. Dashed curve: A15-fcc. Dotted curve: Ti_2Ni-fcc. Dash-dotted curve: Mackay-fcc. Dash-double dotted curve: $MgCu_2$-fcc.

The similarity in the calculated values of the PTP-fcc energy difference for all of the structures we have considered suggests some common energetic factors distinguishing the PTP phases from the fcc structure. We now present an intuitive explanation for the calculated behavior on the basis of explicit interatomic potentials. The four moment results for the structural energy differences are within 20% of the six moment results. We therefore analyze the former for simplicity. By Eq. (2), these are determined by contributions from clusters with up to four atoms. These four-atom clusters constitute the dominant contributions to the structural energy differences considered here, although both lower and higher-order potentials are necessary to treat more subtle structural distinctions. Figure 2 defines the variables used to describe the four-atom geometry. The effective potentials given by Eq. (2) are shown in Figure 3, with $\Delta\varepsilon = -2$ eV, for $\phi = 0$ and $\phi = 90$. For simplicity, we display results only for clusters with four equal bond lengths. The potentials are d-band filling dependent, and are shown for a band containing 6.7 d-electrons per atom, for which local icosahedral packing is favored. Above the potential we show the effective number of 4-atom clusters $N_{eff}(\theta)/N_{max}$ at each θ for the σ, A15, bcc, and fcc structures. N_{eff} is defined by counting the number of paths at a given θ for $\phi < 45$, properly weighted by the number of times the site appears in the unit cell. These are usually the most important contributions, since $V_4^{eff}(\theta,\phi)$ drops off rapidly with increasing ϕ (cf. Fig. 3). Each four atom contribution is scaled by the product of the $h_{\alpha\beta}^{ij}$ around the path. We replace the r^{-5} radial dependence by $(r_0 - r)^\alpha$, where α is determined by insuring that the logarithmic derivative of this scheme equals that of the r^{-5} form at the near neighbor distance. This form insures that the couplings go to zero beyond r_0, which we take to be the second neighbor distance in the fcc structure, while still preserving the r^{-5} form at the near neighbor distance. The two dominant contributions lowering the energy of the PTP schemes relative to the fcc packing are readily seen here: First, in going from fcc packing to PTP, a large repulsive contribution from V_4^{eff} at $\theta = 90°$ disappears in all cases. Second, clusters in the fcc structure having $\theta = 60°$ (120°) are shifted to angles closer to the minima in V_4^{eff}, again lowering the energy. These effects are present in *all* of the structures we have considered. The phase cancellation resulting from the angular dependence of the couplings of the d-orbitals is crucial to these effects. For example, in an s-band model, $V_4^{eff}(\theta,\phi)$ is a constant, and as a result, contributions from different clusters exhibit no cancellation effects. In this case, the PTP phases would in general have a *higher* four-atom contribution to the energy since they have more four atom clusters per site.[21]

In conclusion, we have seen that electronic structure calculations for several hypothetical and observed PTP structures, within the framework of approximate tight-binding theory, reveal a close similarity in bonding trends as the d-band filling varies. The stability of polytetrahedral packing relative to fcc packings may be attributed to the presence of common geometric motifs in PTP which correspond to minima in the four-atom potential, which are not present in the fcc structure. Via analysis of a cluster expansion for the bonding energy, we find that the four-atom angle dependent potentials yield the dominant contribution to the difference in bonding energies between PTP and fcc packing. The energy penalty paid for *square* four-atom clusters in the fcc structure is seen to be an important part of this difference, and appears to be one of the major factors, in addition to atomic size, driving local icosahedral packing in transition metal systems.

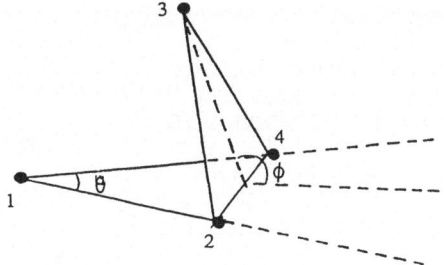

Fig. 2   Four-atom cluster geometry for defining potential V_4.

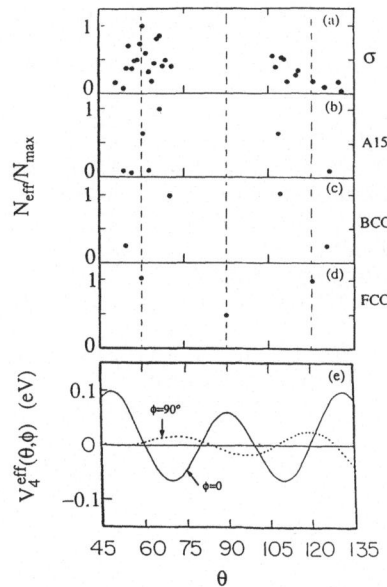

Fig. 3. (a)-(d): Number of four-atom clusters with opening angle θ. (a) σ-phase structure; (b) A15 structure; (c) bcc structure, and (d) fcc structure; (e) V_4 vs. θ (eV). Solid curve: $V_4(\theta,\phi)$ with $\phi = 0$. Dotted curve: $V_4(\theta,\phi)$ with $\phi = 90°$.

ACKNOWLEDGMENTS

This work was supported by the Department of Energy under Grant Number DE-FG02-84ER45130.

REFERENCES

1.   D. Shechtman, I. Blech,. D. Gratias, and J.W. Cahn, Phys. Rev. Lett. 53:1951 (1984).
2.   D.R. Nelson and M. Widom, Nucl. Phys. B240:113 (1984).
3.   M. Widom, Phys. Rev. B, 31:6456 (1985).

4. P. Turchi, G. Treglia, and F. Ducastelle, <u>J. Phys</u>. F13:2543 (1983).
5. R. Haydock and R.L. Johannes, <u>J. Phys</u>. F5:2055 (1975).
6. R.E. Watson and L.H. Bennett, <u>Acta Met</u>. 32:477 (1984).
7. J.C. Slater and G.F. Koster, <u>Phys. Rev</u>. 94:1493 (1954).
8. F. Cyrot-Lackmann, <u>J. Phys. Chem. Solids</u> 29:1235 (1968); F. Ducastelle and
    F. Cyrot Lackmann, <u>J. Phys. Chem. Solids</u> 32:285 (1971).
9. P. Turchi and F. Ducastelle in: "The Recursion Method and Its Applications",
    D.G. Pettifor and D.L. Weaire, eds. (Springer-Verlag, 1985), p. 104.
10. R.H. Brown and A.E. Carlsson, <u>Phys. Rev. B</u> 32:6125 (1985).
11. The moments are computed using either path counting techniques or the recursion
    method. Our calculations are done with the same second moment for all
    geometries considered. For justification of this approximation see
    D.G. Pettifor and R. Podloucky, <u>Phys. Rev. Lett</u>. 53:1080 (1984).
12. E.T. Jaynes, <u>Phys. Rev</u>. 106:620 (1957); 108:171 (1957) L.R. Mead and
    N. Papanicolaou, <u>J. Math. Phys</u>. 25:2404 (1984).
13. A.E. Carlsson, <u>Phys. Rev. B</u> 32:4866 (1985). For n atoms there can also be
    contributions from moments of order higher than n. Our main interest
    lies in four-atom interactions in a four-moment calculations, in which
    case these contributions are absent.
14. R.B. Phillips and A.E. Carlsson, <u>Phys. Rev. B</u> 37:10880 (1988).
15. R.B. Phillips and A.E. Carlsson, unpublished.
16. W.R. Morcoh, W.L. Worrell, H.G. Sell, and H.I. Kaplan, <u>Metall Trans</u>.
    5:155 (1974); G.R. Stewart, L.R. Newkirk, and F.A. Valencia,
    <u>Phys. Rev. B</u> 21:5055 (1980); C.G. Granquist, G.J. Milanowski,
    and R.A. Huhrman, <u>Phys. Lett</u>. 54:245 (1975).
17. F.C. Frank, <u>Proc. Roy. Soc</u>. A215:43 (1952).
18. M. Widom, K.J. Strandburg, and R.H. Swendsen, <u>Phys. Rev. Lett</u>. 58:706
    (1987).
19. P.J. Steinhardt, D.R. Nelson, and M. Ronchetti, <u>Phys. Rev. B</u> 28:784 (1983).
20. R.H. Brown, Ph.D. Thesis, unpublished.
21. R.B. Phillips and A.E. Carlsson, unpublished.

TRANSFERABILITY OF TIGHT-BINDING MATRIX ELEMENTS

D.J. Chadi

Xerox Palo Alto Research Center
3333 Coyote Hill Road
Palo Alto, CA 94304

INTRODUCTION

The empirical tight-binding method was initially introduced by Slater and Koster[1] as an interpolation scheme for band structure calculations. Hamiltonian matrix elements determined by fitting the electronic energies at a few high symmetry points of the Brillouin zone could be used to calculate the entire energy band structure. Reasonably accurate fits to the electronic structure of many crystalline solids, *e.g.*, for Si, have been obtained in this way.[2-6] In the last decade the method has been used less frequently as an interpolation scheme and more as a convenient computational tool in studies of surface atomic and electronic structure,[7-11] band lineups at interfaces,[12] grain boundaries,[13,14] phonons,[15,16] and structural stabilities of atomic clusters[17] and crystalline phases.[18] The method is especially suited for studying systems with a large number of atoms. The feasibility of performing realistic total-energy calculations within the tight-binding scheme which makes it possible to do relative comparisons is an important feature of many of these studies. The tight-binding approach to structural studies of semiconductor surfaces, particularly of the prototypical group IV and III-V semiconductors Si and GaAs has been discussed in detail elsewhere.[7-11] The tight-binding theory of cohesion in transition metals and their alloys is also well-developed but will not be discussed here.[19]

A major question in the application of the tight-binding method to a comparative study of different structures for a given material is the transferability of the parameters.[6] For the case of group IV, III-V, and II-VI semiconductors, a minimal basis set derived from the atomic s, p_x, p_y, p_z orbitals has been generally found to be sufficient in most studies. The tight-binding parameters are generally determined by fitting the electronic structure of a reference crystalline phase, *e.g.*, the diamond or zincblende phases for Si and GaAs. An obvious question is whether this procedure is sufficient to provide all the information that is needed for its application to different structures. The problem of transferability of parameters is analyzed in the following section where it is shown that the electronic structure of a single reference system in its ideal state does not provide sufficient information for determining the values of even all the two-center terms which are needed for applying the method to a different atomic structure. This results from the fact that most parameters are "compound" in nature., *i.e.*, they have contributions coming from one, two, and three-center terms which are difficult to isolate from each other. In some cases a parameter may vanish because of the high symmetry of the crystal and its determination requires the consideration of symmetry reducing perturbations of the atomic coordinates. The various type of matrix elements which appear in the tight-binding approach and the problem of the transferability of parameters are discussed below.

TIGHT-BINDING MATRIX ELEMENTS

Let $f_{\alpha_i}(\mathbf{r} - \mathbf{R}_i)$ be a localized orbital centered on atom i at position $\mathbf{R}_i$ with α denoting the s,p, or d symmetry of the orbital. In the two center approximation the only Hamiltonian matrix elements considered are of the type

$$H_{\alpha_i\beta_i} = <f_{\alpha_i}(\mathbf{r} - \mathbf{R}_i)|T + V_i(\mathbf{r} - \mathbf{R}_i) + V_j(\mathbf{r} - \mathbf{R}_j)| f_{\beta_j}(\mathbf{r} - \mathbf{R}_j)> \tag{1}$$

where T is the kinetic energy operator and $V_i(\mathbf{r} - \mathbf{R}_i)$ is the ionic potential at the atomic site i. The matrix elements in Eq. (1) can be separated into three groups. The first group of terms gives the atomic like orbital energies E_{α_i}:

$$E_{\alpha_i} = <f_{\alpha_i}(\mathbf{r} - \mathbf{R}_i)| T + V_i(\mathbf{r} - \mathbf{R}_i) |f_{\alpha_i}(\mathbf{r} - \mathbf{R}_i)> \tag{2}$$

The second group of matrix elements corresponds to the "crystal field" terms. These *intra-atomic* matrix elements depend on the local environment of an atom at site i and the potentials of distant ions at site j

$$I_{\alpha_i\beta_i}(\mathbf{R}_i, \mathbf{R}_j) = <f_{\alpha_i}(\mathbf{r} - \mathbf{R}_i) | V_j(\mathbf{r} - \mathbf{R}_j) | f_{\beta_i}(\mathbf{r} - \mathbf{R}_i)> \tag{3}$$

where $i \neq j$. The diagonal matrix elements of the Hamiltonian given by

$$H_{\alpha_i\alpha_i} = E_{\alpha_i} + \sum_j I_{\alpha_i\alpha_i}(\mathbf{R}_i, \mathbf{R}_j) \tag{4}$$

are, therefore, generally structural dependent. For example, for the case of an s-state on an atom which is surrounded by Z other equivalent atms at equal distances we have

$$E_s \equiv H_{ss} = E_{s_0} + ZI_{ss\sigma} \tag{5}$$

where E_{s_0} is the atomic like term from Eq. (2) involving only the intra-atomic potential and $I_{ss\sigma}$ represents the crystal field term. For p-states the later term depends not only on the atomic coordination but also on the relative orientation of the neighboring atoms as shown below.

310

The third group of terms are the usual interatomic matrix elements $V_{\alpha_i \beta_j}$ with $i \neq j$:

$$V_{\alpha_i \beta_j} = <f_{\alpha_i}(\mathbf{r} - \mathbf{R}_i) \mid T + V_i(\mathbf{r} - \mathbf{R}_i) + V_j(\mathbf{r} - \mathbf{R}_j) \mid f_{\alpha_j}(\mathbf{r} - \mathbf{R}_j)>. \tag{6}$$

Table I of the Slater-Koster paper[1] specifies the way in which these interatomic matrix elements vary as a function of the relative orientation of the atoms at $\mathbf{R}_i$ and $\mathbf{R}_j$. The same symmetry derived arguments can be used to show that the intra-atomic terms $I_{\alpha_i \beta_i}(\mathbf{R}_i, \mathbf{R}_j)$ in Eq. (3) have the same orientational dependence on $\mathbf{R}_i$ and $\mathbf{R}_j$ as the corresponding interatomic integrals $V_{\alpha_i \beta_j}$. In special cases the intra-atomic terms can be completely incorporated into the orbital energies E_{α_i}. As shown below, this happens, for example, when the system has tetrahedral or cubic point group symmetry about each atom and a limited s, p_x, p_y, p_z basis set is used. The parameters E_{α_i} in this case are not, however, strictly transferable from one system to another. The nature of the corrections is examined below.

In the two-center approximation the orientational dependence of the tight-binding parameters is related to the direction cosines

$$(l,m,n)_{ij} = (\mathbf{R}_i - \mathbf{R}_j)/|\mathbf{R}_i - \mathbf{R}_j| \tag{7}$$

between a reference atom at $\mathbf{R}_i$ and a neighbor at $\mathbf{R}_j$. To simplify the notation in following discussion we restrict ourselves to systems consisting of a single kind of atom and to nearest-neighbor interactions only. The results are easily extended to more complex systems. The Hamiltonian matrix elements are given by

$$\left\langle s_i \mid H \mid s_i \right\rangle = E_{s_0} + \sum_j I_{ss\sigma_i} \tag{8}$$

which is equal to $E_{s0} + ZI_{ss\sigma}$ when all the Z nearest neighbors are at equal distances. Similarly,

$$\left\langle p_{x_i} \mid H \mid p_{x_i} \right\rangle = E_{p_0} + \sum_j [l_{ij}^2 (I_{pp\sigma}) + (1 - l_{ij}^2) (I_{pp\pi})_{ij}] \tag{9}$$

and,

$$<p_{x_i} \mid H \mid p_{y_i}> = \sum_j l_{ij} m_{ij} (I_{pp\sigma} - I_{pp\pi})_{ij} \tag{10}$$

It can be seen that four new matrix elements denoted in the above equations by $I_{ss\sigma}$, $I_{sp\sigma}$, $I_{pp\sigma}$, and $I_{pp\pi}$ are generally needed in addition to the usual interatomic counterparts $V_{ss\sigma}$, $V_{sp\sigma}$, $V_{pp\sigma}$, and $V_{pp\pi}$ for a proper description of the interactions. In most applications of the empirical tight-binding method the diagonal matrix elements E_s and E_p are implicitly assumed to contain the contributions arising from the $I_{ss\sigma}$, $I_{pp\sigma}$, and $I_{pp\pi}$

terms. The intra-atomic matrix elements between s and p states or between p_x and p_y orbitals sum up to zero when the local nearest neighbor environment of an atom has a high symmetry (e.g., cubic, tetrahedral, or hexagonal symmetry) so in many applications on the bulk electronic structure of ideal and undistorted lattices these parameters do not show up and their explicit inclusion in the calculations has no effect on the electronic energy band structure. The coordination and environment dependence of the parameters could be important, however, in total-energy calculations when looking at vacancies, interstitials, surfaces, clusters, lattice distortions, or phase transitions involving a change in atomic coordination. It is not possible, unfortunately, to determine these additional matrix elements from an examination of the electronic structure of only the ideal crystalline structure. By using information on deformation potentials it should be possible to determine these additional parameters and at the same time remove the problems of the simple theory with deformation potentials.

TOTAL-ENERGY EXPRESSION

The total energy can be expressed as the sum of three terms:

$$E_{tot} = E_{el\text{-}ion} + E_{el\text{-}el} + E_{ion\text{-}ion} \tag{12}$$

where the first term represents the energy associated with the electron-ion interaction, the second term gives the electron-electron interaction energy and the last term is the repulsive Coulomb energy of the ion-ion system. The total energy can be expressed in the more convenient form

$$E_{tot} = \sum_{n,k} E_n(\mathbf{k}) + (E_{ion\text{-}ion} - E_{el\text{-}el}) + \delta E_{xc} \tag{13}$$

where the first sum is over all the occupied single particle electronic states labeled by the band index n and eigenstate k, and where the second term makes a correction for the double counting of electron-electron Coulombic interactions occurring in the first term, and, the last term is a correction for the overcounting of electronic exchange-correlation energies. The term $(E_{ion\text{-}ion} - E_{el\text{-}el})$ is assumed to be short range in nature. The argument for doing this is that for two ions which are separated by a distance larger than a Thomas-Fermi screening length this term is effectively zero because the ions together with their surrounding electron cloud are neutral and there is no interaction between them. Since the exchange-correlation term is also short-ranged Eq. (13) suggests the following relatively simple expression for the total-energy which we have used in most of our work:

$$E_{tot} = \sum_{n,k} E_n(\mathbf{k}) + \sum_{i>j} (U_0 + U_1 \varepsilon_{ij} + U_2 \varepsilon_{ij}^2) . \tag{14}$$

The second sum in this equation has the form of a radially dependent pair potential between nearest-neighbor atoms i and j. The fractional change in bond length from an ideal reference structure is denoted by ε_{ij}, and the U's represent empirical terms which need to be specified for each type of bond in the system. For Si and GaAs, the values of U_0, U_1 and U_2 can be determined by fitting the bulk cohesive energy, the lattice constant and the bulk modulus. In our work we have assumed that the interatomic matrix elements have a d^{-2} dependence on nearest-neighbor distance d. The validity of Eq. (14) has been recently analyzed and the pair potential component of the total-energy has been shown to be derivable starting from density functional theory.[20-21]

The electronic energies in Eq. (14) are the principal reason why the total-energy expression is not simply reducible to pair potentials. The Taylor-like expansion in Eq. (14) was originally designed[7] to deal with systems with $\varepsilon_{ij} \ll 1$. We have found it to work well in many surface structural studies when the maximal ε_{ij} is less than 10%. The asymmetry of dimers on Si (100) surfaces as well the stability and structure of single and double layer steps on this surface are correctly predicted by this prescription for the total energy.[8] The only caveat is when the calculations predict a large charge transfer ($\Delta q \simeq$ 0.5-1 e) between two atoms. In this situation the imposition of self-consistency through the "charge neutrality condition" in which the atomic s and p energy E_s and E_p of the atom with extra electronic charge are increased while those on the other atom are decreased has been found to lead to reliable results when the charge transfer between the atoms is reduced to essentially zero.[12,14] For situations where comparisons of different structures leads to $\varepsilon_{ij} > 0.1$, such as, for example, in comparisons of molecular carbon with carbon in the diamond structure, where ε can be as large as 22%, a small modification of Eq. (1) in which the terms in the second sum are multiplied by $(d_0/d_{ij})^3$ greatly improves the convergence of the series.[22] For $\varepsilon_{ij} < 0.1$ both types of expansions lead to essentially the same results.

The main problem with the simple tight-binding approach is the limited size of the basis set. For Si, a comparison of the relative stability of the diamond structure with other closed pack configurations does lead to the correct prediction that the diamond structure is the most stable one. However, the size of the energy differences are not well reproduced when the results are compared to those from *ab initio* self-consistent pseudopotential calculations.[23] The major reason for this is the absence of d-states in the basis set which leads to unfavorable energies when the bond angles deviate significantly (by $\simeq \pm 35°$) from the ideal tetrahedral value of 109.5°. The absence of d states also affects some states in the more open diamond structure. The indirect X conduction band minimum in Si or C is not generally correctly given by the minimal basis set. The inclusion of d states, or even a single excited s* state[5] which mimics the effects of higher lying d-states, completely corrects this situation and leads to a more accurate description of the conduction bands.[24] For total-energy studies, we are not interested in any particular state but are generally more interested in averages of the electronic energies, *e.g*, over the Brillouin zone, and, in total energy *variations* not in its absolute value. This makes the failings of the method for specific electronic states much less of a severe problem. The inclusion of d states in the basis set, although straightforward, would make the empirical tight-binding approach less desirable because of the large increase in the number of independent parameters.

Another problem which becomes more important in the close packed structures is that of the orthogonality of the basis functions.[6] The inclusion of orthogonality leads to other problems, however, when it is done in an approximate fashion. For example, if only the orthogonality terms between nearest-neighbor orbitals are assumed to be nonzero, then unless the overlap integral is smaller than the inverse of the atomic coordination, situations can develop where the energy can become undefined or divergent. It is, therefore, necessary to include the orthogonality terms of more distant nearest-neighbors.[6] Such an approach has been found to give satisfactory results for the electronic structure of Si in the diamond structure as well as in the more close packed fcc, bcc, and sc structures.[6]

CONCLUSIONS

The empirical tight-bind approach has proven to be an extremely successful method in diverse applications. In this paper the problem of the transferability of the Slater-Koster parameters between systems differing in structure and, particularly, atomic coordination is addressed. It is pointed out that in nearly all such applications of the method to semiconductors up to now the "crystal field" terms given by Eq. (3) which depend on the local atomic environment have been neglected. For a reference zincblende or diamond structure crystal, these terms are implicitly included in the other parameters which are used to describe the electronic structure. However, these parameters are not then strictly transferable from one system to another. A knowledge of the deformation potentials would be very useful in the determination of the intra-atomic matrix elements $I_{\alpha_i \beta_i}$ in Eq. (3).

ACKNOWLEDGEMENTS

This work is supported in part by the Office of Naval Research through Contract N00014-82-C-0244.

REFERENCES

1. J.C. Slater and G.F. Koster, Phys. Rev. 94, 1498 (1954).
2. D.J. Chadi and M.L. Cohen, Phys. Stat. Sol. (b), 68, 405 (1975).
3. K.C. Pandey and J.C. Phillips, Phys. Rev. B, 13, 750 (1976).
4. D.A. Papaconstantopoulos, Phys. Rev. B, 22, 2903 (1980); J. D. Shore and D.A. Papaconstantopoulos, Phys. Rev. B, 35: 1122 (1987); and D.A. Papaconstantopoulos in: "Handbook of the Band Structure of Elemental Solids", Plenum Press, New York (1987).
5. P. Vogel, H.P. Hjalmarson, and J. Dow, J. Phys. Chem. Solids, 44, 365 (1983).
6. P.B. Allen, J.Q. Broughton, and A.K. McMahan, Phys. Rev. B, 34, 859 (1986).
7. D.J. Chadi, Phys. Rev. Lett. 41, 1062 (1978); Phys. Rev. B, 19, 2074 (1979); Phys. Rev. B, 29, 7845 (1984). In these references the tight-bonding total-energy method is discussed and applied to the (110) surfaces of III-V and II-VI semiconductors.
8. D.J. Chadi, Phys. Rev. Lett. 43, 43 (1979); J. Vac. Sci. Technol. 16, 1290 (1979); J. Appl. Opt. 19, 3974 (1980); Phys. Rev. Lett. 59, 1691 (1987). These references apply to Si(100) surfaces with the most

recent paper discussing stepped (100) surfaces.

9. D.J. Chadi, Phys. Rev. Lett. 52, 1911 (1984); J. Vac. Sci. Technol., B 3, 1167 (1985); Phys. Rev. Lett. 57, 102 (1986). These papers are on the polar (111) surfaces of GaAs and ZnSe.

10. G.X. Xian and D.J. Chadi, Phys. Rev. B, 35, 1288 (1987). This paper discusses the tight-binding based optimization of the Si (111)-7x7 surface atomic structure.

11. D.J. Chadi, J. Vac. Sci. Technol. A 5, 834 (1987). In this paper the atomic structure of As-stabilized GaAs (100) surface is analyzed.

12. C. Priester, G. Allan, and M. Lannoo, Phys. Rev. B, 33, 7386 (1986); Phys. Rev. Lett. 58, 1989 (1987).

13. R.E. Thomson and D.J. Chadi, Phys. Rev. B, 29, 889 (1984).

14. A.T. Paxton and A.P. Sutton, J. Phys. C, 21, L481 (1988).

15. D.J. Chadi and R.M. Martin, Solid State Commun., 19, 643 (1976).

16. D.C. Allan and E.J. Mele, Phys. Rev. Lett., 53, 826 (1984); O.L. Ollerhand, D.C. Allan, and E.J. Mele, Phys. Rev. Lett. 55, 2700 (1985).

17. D. Tománek and M.A. Schlüter, Phys. Rev. Lett, 56, 1055 (1986).

18. A.T. Paxton, A.P. Sutton, and C.M.M. Nex, J. Phys. C, 20, L263 (1987).

19. D.G. Pettifor and R. Podloucky, J. Phys. C, 19, 315 (1986).

20. W.M.C. Foulkes, PhD Thesis, University of Cambridge 1987; M.W. Finnis, in "Proceedings of this Conference"; and D.G. Pettifor, in "Proceedings of the Conference".

21. A.P. Sutton, M.W. Finnis, D.G. Pettifor, and Y. Ohta, J. Phys. C, 21, 35 (1988).

22. D.J. Chadi, J. Vac. Sci. Technol. A2, 948 (1984).

23. M.T. Yin and M.L. Cohen, Phys. Rev. Lett. 45, 1004 (1980); Phys. Rev. B, 26, 5668 (1982).

24. D.J. Chadi, Phys. Rev. B, 16, 3572 (1977).

THE TIGHT BINDING BOND MODEL

D. G. Pettifor, A. J. Skinner and R. A. Davies

Department of Mathematics
Imperial College of Science
Technology and Medicine
London SW7 2BZ

INTRODUCTION

The semi-empirical Tight Binding model[1,2] is the simplest scheme for describing the energetics of semi-conductors and transition metals within a quantum mechanical framework. Assuming only one type of valence orbital per site, the total binding energy may be written in the form[3,4]

$$U = U_{rep} + U_{bond} \tag{1}$$

where U_{rep} is a semi-empirical pairwise repulsive contribution and U_{bond} is the quantum mechanical bond energy which results from evaluating the bond density of states $n_i(E)$ at site i within the two-centre, orthogonal Tight Binding approximation. That is,

$$U_{bond} = \sum_i \int^{E_F} \left(E - E_1 \right) n_i (E) \, dE \tag{2}$$

where E_i is the effective atomic energy level at site i and E_F is the Fermi energy. The pairwise nature of the repulsive term has recently been justified[5,6] from first principles using the Harris[7]-Foulkes[5] approximation to density functional theory.[8] The Hückel-type two centre orthogonal form of the matrix elements can be justified in principle within Anderson's chemical pseudopotential theory.[9]

In this paper we wish to make three points. Firstly, within semi-empirical schemes it is essential to work with the bond energy rather than the band energy and to adjust the site diagonal energies self-consistently to guarantee local charge neutrality. This is the reason why Pettifor and Podloucky[10] and Sutton et al.[6] refer explicitly to the Tight Binding Bond (TBB) model. Secondly, we analyze in detail the nature of the pairwise contribution for hydrogen, and comment on the transferability of the pair potential and bond integrals from one atomic environment to another.[11] Finally, we show that damping down the very long range Friedel oscillations by introducing a temperature dependent Fermi factor is essential for obtaining real-space convergence of the energies and forces in metallic systems.[12]

BOND ENERGY VERSUS BAND ENERGY

The total energy within density functional theory[8] can be written

$$U = U_{nn} + U_{dc} + U_{bond} \tag{3}$$

where the first two terms are the nuclear-nuclear and double-counting contributions respectively. The last term is the band energy which is obtained by summing over all the occupied states, namely

$$U_{band} = \sum_k E_k \tag{4}$$

The E_k are the eigenvalues of the one-electron Schrödinger equation

$$\left[-\nabla^2 + V_H(r) + V_{xc}(r) \right] \psi_k(r) = E_k \Psi_k(r) \tag{5}$$

where V_H and V_{xc} are the Hartree and exchange-correlation potentials corresponding to the ground state charge density

$$\rho_{gs}(r) = \sum_k \left| \Psi_k(r) \right|^2 . \tag{6}$$

Assuming only one type of valence orbital per site for ease of discussion, the band energy may be written as

$$U_{band} = \sum_i \int^{E_F} E \, n_i(E) \, dE$$

$$= \sum_i N_i E_i + U_{bond} \tag{7}$$

Thus, unlike the bond energy, the band energy contains the shift in the site diagonal energy due to changes in the local environment. However, the so-called force theorem[13,14,15] showed that these shifts are cancelled to first order by the double-counting term. If, therefore, the double-counting term is parametrized by an empirical pair potential, these shifts in the site diagonal energy will not be correctly cancelled. It is not possible to circumvent this difficulty by neglecting these shifts in site diagonal energy, because this leads to inconsistencies to second order.[6]

As an illustration of the importance of using the bond rather than the band energies, we consider the case of evaluating the heat of formation of transition metal alloys within the simple rectangular band model[3,16] which considers only the bonding between the valence d electrons. Fig. 1 presents a schematic representation of the ionic rigid-band and metallic common-band models of bonding in AB alloys. E_A^0 and E_B^0 give the free atom d energy levels, whereas the positions of E_A and E_B in the metallic bond reflect the small shift which takes place on alloy formation in order to maintain local charge neutrality. Unlike the ionic rigid-band model, the common-band model assumes that the local density of states on a given atom changes when the atom is taken from its elemental environment to that of the alloy. The band in the alloy is formed by the quantum-mechanical overlap of neighboring atomic wave functions so that the local densities of states on the A and B sites have a common band width. The band width W_{AB} is determined by the second moment of the alloy's average density of states per atom, which may be obtained by summing over all the TB hopping paths of length two.[17] It has the simple form

$$W_{AB} = \left[1 + 3 \, (\Delta E/W)^2 \right]^{\frac{1}{2}} W \qquad (8)$$

where $\Delta E = E_B - E_A$. The bond integrals have been taken in Fig. 1 to be site independent so that $W_A = W_B = W$. The fact that in reality $W_A \neq W_B$ is considered explicitly in ref. 16.

The condition for local charge neutrality may be obtained directly from the skewed local densities of states in Fig. 1. If the skew-rectangular density of states $n_A(E)$ takes the values $10(1 + \alpha)/W_{AB}$ and $10(1 - \alpha)/W_{AB}$ at the bottom and top of the band respectively, then

$$\alpha = 3 \, \Delta E/W_{AB} \qquad (9)$$

in order that the centre of gravity coincides with the atomic energy level E_A. The charge transfer qe which accompanies an atomic energy level separation ΔE may, therefore, be calculated by filling up the bands to the Fermi energy. It is given by

$$q = \frac{1}{2} \Delta N + \frac{3}{10} \left(10 - \overline{N} \right) \Delta E/W_{AB} \qquad (10)$$

where $\overline{N} = 1/2(N_A + N_B)$ and $\Delta N = N_B - N_A$. The positions and labelling of the bands in Fig. 1 correspond to $\Delta E > 0$ and $\Delta N < 0$. Thus, the first term drives charge from the high-valence A atom to the low-valence B atom, whereas the second term takes charge in the opposite direction from the higher to lower atomic energy level. It follows from eqs. (10) and (8) that local charge neutrality is obtained within the rectangular band model for an atomic energy level separation $\Delta E_{AB} = E_B - E_A$ given by

$$\Delta E_{AB} = -W\Delta N \left\{ \frac{9}{25} \left[\overline{N} \left(10 - \overline{N} \right) \right]^2 - 3 \, (\Delta N)^2 \right\}^{\frac{1}{2}} . \qquad (11)$$

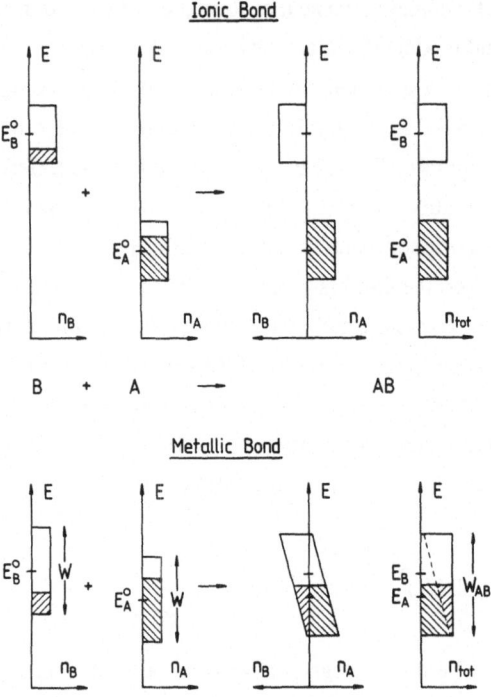

Figure 1. The rectangular band model for ionic and metallic bonding in AB compounds.

The above expressions, eqs. (9)—(11), are valid provided $\alpha \leq 1$. This corresponds to the constraint $|\Delta N| \leq 1/5\, \overline{N}(10 - \overline{N})$, so that the above simple model is applicable to all transition metal AB alloys with $\Delta N \leq 4$.

The average bond energy per atom may be written from eq. (7) as

$$U_{bond}^{AB} = \int^{E_F} En_{AB}\ (E)\ dE - \frac{1}{2}\left(N_A\,E_A + N_B\,E_B\right) \tag{12}$$

where the average <u>total</u> density of states per atom $n_{AB}(E)=10/W_{AB}$ from Fig. 1. Therefore,

$$U_{bond}^{AB}\ (\Delta E) = -W_{AB}\,\overline{N}\left(10 - \overline{N}\right)/20 - \frac{1}{4}\Delta N\ \Delta E \tag{13}$$

This expression is stationary with respect to small variations in ΔE for just that value of $\Delta E = \Delta E_{AB}$ which results from filling up the skew-rectangular <u>partial</u> densities of states and requiring local charge neutrality. Thus, this simple bond model with local charge neutrality is internally consistent up to second order in ΔN. The band model which excludes the second contribution in eq. (12) would not be internally consistent. A similar conclusion was reached[6] when discussing the second order changes involved in evaluating the metallic bulk modulus using either a homogeneous volume deformation or the long wave length phonon limit preserving total volume.

In concluding this section we should note that when eq. (11) is substituted into eq. (13) the bond energy per atom takes the transparent form

$$U_{bond}^{AB} = f\left(\overline{N}, \Delta N\right) U_{bond}\left(\overline{N}\right) \tag{14}$$

where

$$f\left(\overline{N}, \Delta N\right) = \left\{1 - \frac{25}{3}(\Delta N)^2 / \left[\overline{N}\left(10 - \overline{N}\right)\right]^2\right\}^{\frac{1}{2}} \tag{15}$$

and

$$U_{bond}\left(\overline{N}\right) = -W\,\overline{N}\left(10 - \overline{N}\right)/20 . \tag{16}$$

$U_{bond}(\overline{N})$ is just the bond energy within the virtual crystal approximation (VCA) in which all the atoms are assumed to be identical and described by the average properties of the A and B constituents. The prefactor $f(\overline{N}, \Delta N)$ represents the loss of bond order with respect to this average VCA state due to the actual mismatch in the atomic energy levels on the A and B sites ΔE_{AB}.

THE TBB MODEL FOR HYDROGEN

The Tight Binding Bond expression for the binding energy, eq. (1), may be derived[11] from first principles for the simplest case of a system of hydrogen atoms by using the Harris[7]-Foulkes[5] approximation to density functional theory and Anderson's chemical pseudopotentials.[9] Using the stationary property of the energy with respect to the true ground state charge density ρ_{gs}, Harris[7] and Foulkes[5] replaced ρ_{gs} by an approximate charge density, say

$$\hat{\rho}_{gs}\left(\underline{r}\right) = \sum_{\underline{R}} \rho_{atom}\left(\underline{r} - \underline{R}\right), \tag{17}$$

so that the density functional energy could be written exactly to first order in $(\rho_{gs} - \hat{\rho}_{gs})$ as

$$U = U_{nn} + \hat{U}_{dc} + \hat{U}_{band} \tag{18}$$

where the last two terms are the double-counting and band contributions evaluated using the approximate charge density $\hat{\rho}_{gs}$ as input. It follows from eq. (17) that the Hartree part of the double-counting term is now pairwise in nature because there is no bond charge present in $\hat{\rho}_{gs}$. The exchange-correlation part is well approximated by a pairwise form,[5,6,11] so that $U_{nn} + \hat{U}_{dc}$ is describable by a <u>pair potential</u> as assumed by the TB models.[1-4]

The Hückel-type <u>two-centre</u> orthogonal form of the TBB model may be obtained in principle within Anderson's chemical pseudopotential theory.[9] Defining localized orbitals $|\phi_A\rangle$ by

$$\left\{ \left(-\nabla^2 + V_A \right) + \sum_{B \neq A} \left[1 - |\phi_B\rangle\langle\phi_B| \right] V_B \right\} |\phi_A\rangle = E_A |\phi_A\rangle,$$ (19)

the secular equation takes the Hückel form

$$\left| \left(E_A - E \right) \delta_{AB} + \langle\phi_B| V_B |\phi_A\rangle \left(1 - \delta_{AB} \right) \right| = 0$$ (20)

where

$$E_A = \langle\phi_A| - \nabla^2 + V_A |\phi_A\rangle + \sum_{B \neq A} \langle\phi_A| V_B |\phi_A\rangle - \sum_{B \neq A} \langle\phi_A| \phi_B\rangle \langle\phi_B| V_B |\phi_A\rangle.$$ (21)

Thus, E_A comprises the free atom and crystal field terms as expected, plus the shift due to the non-orthogonality contribution.

Grouping the crystal-field term with the double-counting and nuclear-nuclear contribution in eq. (18), allows the total energy to be decomposed explicitly in terms of physically intuitive contributions as

$$U = U_{atom} + U_{es} + U_{xc} + U_{no} + U_{bond}$$ (22)

As the change in the exchange-correlation energy on bringing the atoms together from infinity is well approximated by a pairwise form,[5,6,11] the sum of the electrostatic, exchange-correlation, and non-orthogonality contributions is pairwise, so that eq. (22) provides a justification for the TBB expression eq. (1).

Fig. 2 shows the binding energy curves per atom of hydrogen as a dimer and on the infinite simple cubic (sc) and face centred cubic (fcc) lattices.[11] The left-hand panel gives the *ab initio* results of Kolos[18] for H_2 and Freeman *et al.*[19] for the bulk. The central panel shows the results of evaluating eq. (22) <u>variationally</u> assuming the Localized Orbital has variational form $e^{-\lambda r}$ and the approximate charge density $\hat{\rho}_{gs}$ is the sum of atomic densities $e^{-2\lambda r}$. In order that all the integrals are kept analytic we took the $X\alpha$ form[20] for the exchange and correlation with α chosen to give the correct binding energy for the simple cubic lattice at equilibrium, namely $\alpha = 0.871$. As is well-known the non-spin-polarized Local Density Functional calculations go off to the wrong limit as $R \rightarrow \infty$. This would be circumvented by allowing a self-consistent spin-polarized solution to develop for $R > 3.2$ a.u. (see ref. 21).

The right-hand panel shows the results of evaluating eq. (22) assuming a <u>transferable</u> pair potential and bond integral, which are chosen by freezing the Localized

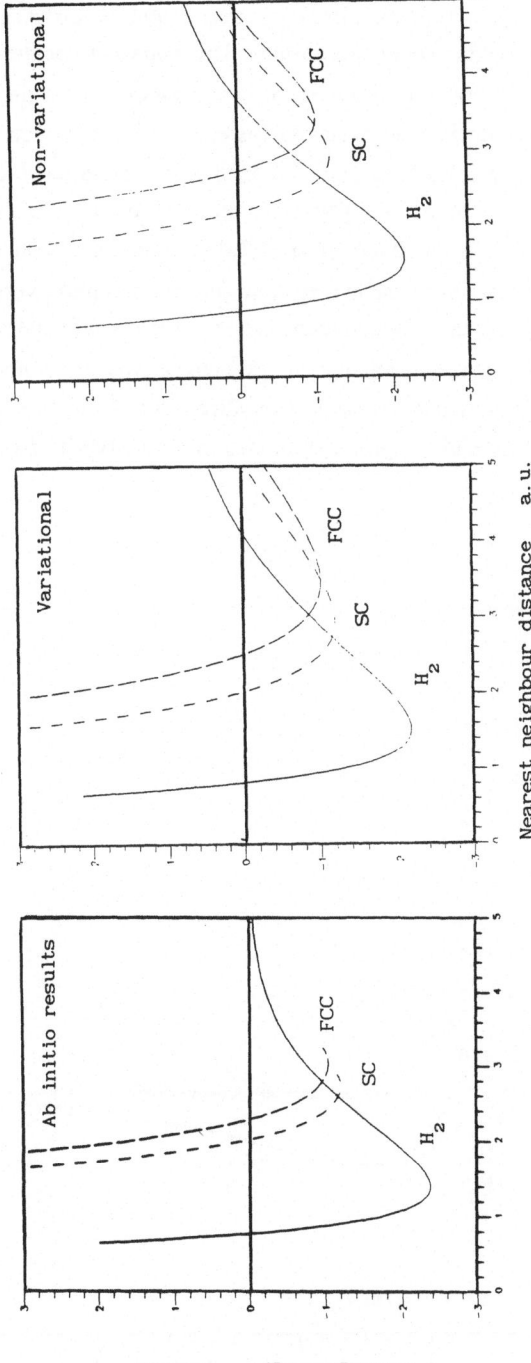

Figure 2. The binding energy curve of hydrogen as a dimer on the simple cubic (sc) and face centered cubic (fcc) lattices.

Orbital $e^{-\lambda r}$ and atomic charge density $e^{-2\lambda r}$ with their values obtained from the variational calculation for the simple cubic lattice at equilibrium, namely $\lambda_0 = 1.043$. We see that because of the variational nature of the central pannel's results, the fixed λ calculations reproduce the cohesive energy and equilibrium nearest neighbour distance to first order in $(\lambda_0 - \lambda_{eq})$ but not the curvature about equilibrium. This appears to be the price which has to be paid for using transferable parameters: it is not possible to fit the bulk moduli of different structure types with a universal set of transferable TB parameters.[11,22]

Fig.3 shows the electrostatic (es), the exchange-correlation (xc) and the non-orthogonality (no) contributions to the pair potential in addition to the ssσ bond integral curve for the case where λ has been fixed at 1.043. We see that at the equilibrium separation of the hydrogen dimer, the electrostatic contribution plays very little role. The repulsion is provided by the non-orthogonality term, the attraction by the bond integral ssσ and the change in the exchange-correlation energy as the atoms are brought together. This theory is at present being extended to the case of silicon in order to study the influence of the pseudo-potential core and the presence of two different types of valence orbitals s and p.

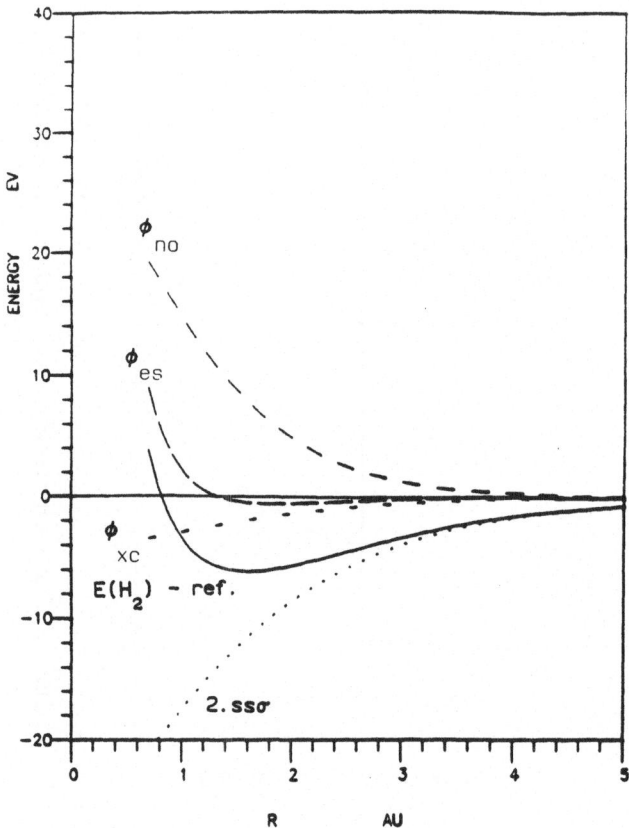

Figure 3. The electrostatic (es), exchange-correlation (xc), non-orthogonality (no), and bonds integral (sss) contribution for 1 = 1.043.

REAL-SPACE CONVERGENCE

The simple expression for the Hellman-Feynman forces within the TB model, which has been used by several contributors to this volume, is only valid provided the exact bond orders or density matrices are computed. This has posed problems for obtaining consistency between the analytic expression for the Hellman-Feynman force and the numerical derivative of the total energy when the real-space recursion method is used in metals.[24]

The underlying problem is illustrated by the top pannel of Fig. 4 in which the difference in d bond energy between the central site in a fcc and hcp cluster are plotted as a function of valence N_d for different cluster sizes.[12] (The cluster size is specified by all those atoms which can be reached in M hops or levels in the recursion algorithm. For an fcc lattice 3, 4, 5, and 6 levels correspond to 147, 309, 561, and 923 atom clusters respectively.) We see that convergence has not been obtained even within the largest cluster in Fig. 4, which is not unexpected given the very long range nature of the Friedel oscillations in metals.[25]

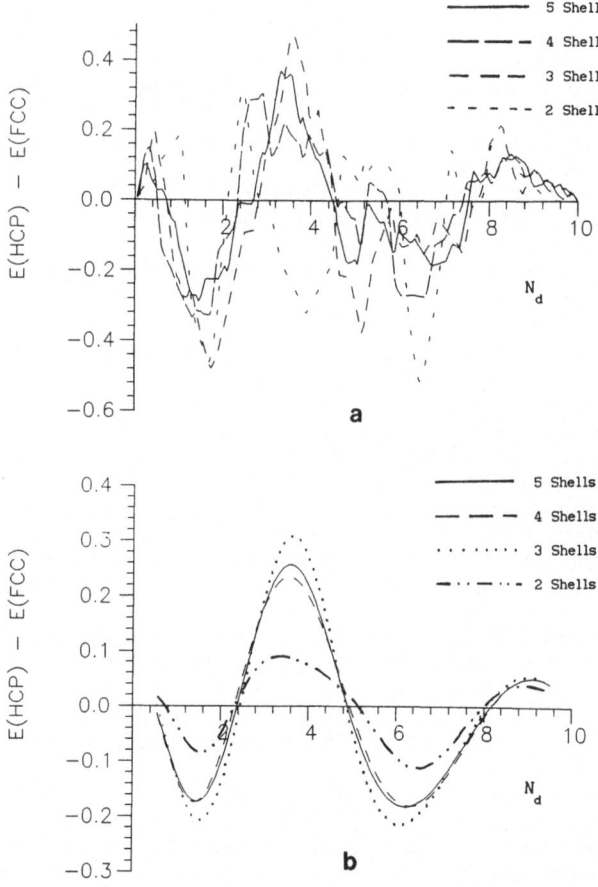

Figure 4.   Structural energy difference for HCP and FCC clusters as a function of d-band occupancy for 2,3,4 and 5 shells about a central cell (a) RT/W = 0, (b) hT/W = 1/8.

In order to have a tractable real-space relaxation algorithm, we have damped out these long-range oscillations by smearing out the Fermi surface by working at finite temperature.[12] The lower pannel of Fig. 4 shows that for kT/W=1/8 good convergence is obtained for a 4 level cluster, which is a manageable size for the atomic relaxation of defects such as surfaces or grain boundaries. With the underpinning of the TBB model by first principles theory, we can look forward to a much more quantitative model which will take us beyond pair potentials into the '90s.

ACKNOWLEDGEMENTS

D. G. Pettifor would like to thank the U. S. Department of Energy, Energy Conversion and Utilization Technologies (ECUT) Materials Program for support. A. J. Skinner would like to thank the UK Science and Engineering Research Council and the UK Atomic Energy Authority for support. R. A. Davies acknowledges the support of the UK Atomic Energy Authority.

REFERENCES

1.   W. A. Harrison, "Electronic Structure and the Properties of Solids," Freeman, San Francisco, (1980); and references therein.
2.   V. Heine, R. Haydock, D. Bullett and M. J. Kelly, Sol. St. Phys. 35 (1980); and references therein
3.   J. Friedel, in "The Physics of Metals," Cambridge University Press, London, p. 494 (1969).
4.   F. Ducastelle, J. de Physique 31:1055 (1970).
5.   W. M. C. Foulkes, Pre Ph. D. thesis (1985); Ph. D. thesis, University of Cambridge (1987).
6.   A. P. Sutton, M. W. Finnis, D. G. Pettifor and Y. Ohta, J. Phys. C 21:35 (1988).
7.   J. Harris, Phys. Rev. B31:1770 (1985).
8.   P. Hohenberg and W. Kohn, Phys. Rev. 136:864 (1964); W. Kohn and L. J. Sham, Phys. Rev. 140:1133 (1965).
9.   P. W. Anderson, Phys. Rev. Lett. 21:13 (1968).
10. D. G. Pettifor and R. Podloucky, J Phys. C19:315 (1986).
11. A. J. Skinner and D. G. Pettifor (to be published).
12. R. A. Davies and D. G. Pettifor (to be published).
13. D. G. Pettifor, Commun. Phys 1:141 (1976); J. Chem Phys. 69:2930 (1978).
14. D. G. Pettifor and C. M. Varma, J. Phys. C12:2253 (1979).
15. O. K. Andersen, §5.3 of A. R. Mackintosh and O. K. Andersen in: "Electrons at the Fermi Surface," ed. M. Springford, Cambridge University Press, (1980).
16. D. G. Pettifor, Sol. St. Phys. 40:43 (1987); and references therein.
17. M. Cyrot and F. Cyrot-Lackmann, J. Phys. F6:2257 (1976).
18. W. Kolos and L. Wolniewicz, J. Chem. Phys. 41:12 (1964).
19. B. I. Min, H. J. F. Jansen and A. J. Freeman; Phys. Rev. B30:5076 (1984).
20. J. C. Slater, "Quantum Theory of Molecules and Solids," Vol. 4, McGraw-Hill (1974).
21. O. Gunnarson and B. I. Lundquist, Phys. Rev. B13:10 (1976).
22. L. Goodwin, A. J. Skinner and D. G. Pettifor (to be published).
23. See, for example, D. J. Chadi, M. W. Finnis, A. P. Sutton and A.T. Paxton in this volume.
24. Y. Ohta, M. W. Finnis, D. G. Pettifor and A. P. Sutton (unpublished).
25. See, for example M. W. Finnis, K. L. Kear and D. G. Pettifor, Phys. Rev. Lett. 52:291 (1984).

INTERATOMIC FORCES AND BOND ENERGIES IN THE

TIGHT BINDING APPROXIMATION

A.T. Paxton

Max-Planck-Institut für Festkörperforschung
Postfach 80-06-65
7000 Stuttgart - 80
West Germany

INTRODUCTION

One might well take the view, that an interatomic potential for atomistic modeling of materials will only be useful if it can correctly reproduce the phase stability of the particular material it is supposed to describe. Yet a glance at the literature will show that prescriptions used for cohesive energies and interatomic forces are rarely tested to destruction in respect of crystal structure prediction. This is particularly regretable, if the potential in question is applied to a study of crystal defects, because of the close connection between competing crystal structures and the energies of defects. A good example of this is the relationship that exists in fcc materials between the intrinsic stacking fault energy, and the difference in energy between hcp and fcc phases. In favorable cases, these are both measurable quantities, and may therefore provide excellent guidance in the choice and testing of interatomic potentials. The present article concentrates on the semi-empirical tight-binding method, which may be thought of as a half-way stage between classical models such as pair potentials or Keating models, and first-principles total energy methods. In the next section, the development of the theory is described, emphasizing its inception as a way of understanding the crystal structures adopted by transition metals. The following two sections focus on the tight-binding theory of Si, describing recent applications of the method to the calculation of bond energies and interatomic forces.

TRANSITION METALS, MOMENTS AND CANONICAL BANDS

Even early theories of cohesion acknowledged the important role of the valence electrons in determining the crystal structures of metals and alloys. However, these theories were either based entirely on the s-p electrons, which were considered to form almost free-electron-like bands and Fermi surfaces which were spherical except where they made contact with the Brillouin zone boundaries[1]; or concentrated on the hybridization of the s and d electrons in a chemical bonding picture[2]. In order to explain the properties of transition metals, Friedel[3] proposed the theory that in these, a narrow band of d-electrons was responsible for (i) the observed parabolic dependence of the melting points, and by inference cohesive energies, as a function of number of valence electrons; and (ii) the close-packed fcc and hcp structures found at the ends of the transition series and the bcc structure of those elements in the middle of the series. As usual in tight-binding theory, consider the solid as formed by bringing atoms together from infinity. In Friedel's picture, the tightly bound d-electrons have orbital energies ε_d (with values in the solid a little smaller than in the free atoms), and hopping of d-electrons between nearest neighbors is permitted via bond integrals β whose magnitude determines the width of the narrow d-band. In a more precise theory, these bond integrals become the Slater-Koster[4] parameters $dd\sigma$, $dd\pi$ and $dd\delta$. The orbital energies and bond integrals are matrix elements of a simple model hamiltonian $\hat{H}$ for which the density of states (DOS) may be formally defined:

$$n(E) = \mathrm{Tr}\delta(E - \hat{H}).$$

This tight-binding picture gave a proper description of the structural stability of the transition metal elements[3]: at the extremes of the series where the d-band was either almost empty or almost full, the structures are close-packed; in the middle, there is a large bonding energy arising from filled bonding states and unfilled antibonding states producing the covalently bonded bcc structure with its characteristic minimum in the density at half bandfilling. Indeed, the cohesive energy is well approximated by (minus) the covalent bond energy,

$$E_{cov} = \int^{E_f} (E - \varepsilon_d)\, n_d\,(E)\mathrm{d}E$$

(E_f, the Fermi energy), and Friedel pointed out[3] that any reasonable form for the d-electron density of states (for example, $n_d(E) = $ constant) would reproduce trends in the observed sublimation energies and melting points of the transition metals which show a large maximum in the middle of the series.

The fact that the cohesive energy trends are insensitive to the detailed structure of the DOS led to the proposal by Cyrot-Lackmann[5] that it was necessary to concentrate only on the first few power moments of the distribution function $n(E)$. This, in turn, led to the development of the method of moments. The moments of the DOS are

$$\mu_r \equiv \mathrm{Tr}\, \hat{H}^r = \int_{-\infty}^{\infty} E^r\, \mathrm{b}n\,(E)\, \mathrm{d}E.$$

If the hamiltonian is represented in a basis of localized orbitals (*e.g.*, the atomic d-orbitals), then the appearance of $\hat{H}^r$ shows that the rth moment of the d-DOS is obtained by r successive operations of the hamiltonian on a d-electron. Each such operation allows the electron to hop to a nearest neighbor with a quantum mechanical amplitude determined by bond angle and bond lengths and the bond integrals β. Taking the trace of $\hat{H}^r$ implies that to obtain the first r moments of the DOS involves counting all the closed paths in the lattice of length r. By using simple model values, namely $-2dd\pi = dd\sigma \equiv \beta$ and $dd\delta = 0$, Ducastelle and Cyrot-Lackmann[6] constructed the first few moments of the bcc, fcc and hcp d-bands analytically and showed how these could be used to reconstruct the density of states and hence cohesive energy to give a qualitative description of the hcp-fcc energy difference across the transition metal series. The success of the method of moments implies, from the connection between the moments and the lattice topology, that the local electronic structure, embodied in the *local density of states* (LDOS), which is the projection of the DOS onto a local basis function, is largely determined by the local atomic structure. It is this "local invariance theorem"[7] which is exploited in the calculations to be described in this article.

The moments method of Cyrot-Lackmann and co-workers was dogged with serious problems of numerical instability[8]. Haydock *et al.*[9] in introducing the *recursion method* showed that Paige's modification of the Lanczos algorithm[10] could be used to provide information contained in the moments automatically for a given lattice topology in a computationally stable way. Nex[11] showed how exact bounds on the integrated DOS could be computed, which became narrower as further moments were found; and equal, in the limit of an exact result. This *gaussian quadrature* also did not require that the last (and most expensive) moment was thrown away, as in previous methods[8]. Recently, the maximum entropy method was introduced[12], which may be used as a replacement for gaussian quadrature when only a very small number of moments is known[13]. The success of this method will be illustrated in figure 1. The recursion method as it is usually presented in the literature[14] must be modified when used to calculate inter-atomic forces. This will be described in the next section. For the remainder of this section, and the beginning of the next, we concentrate on the question of how to find actual values for the bond integrals β. This question is still a thorn in the side for simulations in semiconductors (see next section), but has been addressed from first-principles in Pettifor's exact transformation of the KKR equations to tight-binding form[15] and in Andersen's canonical band theory for transition metals[16].

An almost exact description of transition metals is provided by the KKR or multiple scattering equations[17], which provide a good starting point for constructing simpler models. In Andersen's *atomic spheres approximation* (ASA)[16,18], the KKR equations become linear secular equations, $\det|P(E) - S(k)| = 0$, in which the scattering properties of the atom are separated, into the diagonal potential function matrix P, from the structure constant S which is determined by the structure of the lattice only. Both matrices are exressed in a basis of site and angular-momentum dependent basis functions. Simple *canonical bands* result if matrix elements of the structure constant matrix are taken as zero between different orbital angular momentum quantum numbers *l*; this amounts to neglecting hybridization between s,p,d...-electrons, and uncouples the corresponding blocks of **S** which can then be diagonalized separately. This provides matrix elements for a model d-band hamiltonian directly from first-principles theory. They are[19],

$$dd\{\sigma,\pi,\delta\} = \frac{2}{5} W \left(\frac{s}{d}\right)^5 \{-6, +4, -1\}.$$

The bond integrals are functions of the bond length *d*, and the bandwidth $W \approx 25/(m_d s^2)$, where *s* is the Wigner-Seitz radius of the lattice (in bohrs) and m_d is the effective mass of the d-band (a number usually between 1 and 10). An important point to note about this model is that both the bond length *and* volume dependence of the bond integrals goes as the inverse fifth power of *d*. The point about the canonical d-band model is that the structure constants provide pure *l* bands which are independent of the scale of the lattice and depend only on its crystal structure; for a given transition metal element, the energy bands $E_{n,\mathbf{k}}$ are then solutions of $P_l E) = S_{n,l}(\mathbf{k})$, which to first order in $(m_d s^2)^{-1}$ gives rise to the above energy and lattice constant independent bond integrals *scaled* with the bandwith *W*.

It is a feature of the transition metal series that the crystal structures of the elements show a fairly consistent trend: hcp $\rightarrow$ bcc $\rightarrow$ fcc across the series. The sequence is only broken by La, Mn and the ferromagnetic 3d elements Fe and Co. It is interesting to see how well this trend is reproduced by the canonical d-band model in the first-few-moments approximation. Figure 1 shows curves of bcc-fcc and hcp-fcc energy differences as a function of the filling of the d-band or number of valence electrons (group number in the periodic table). The curves in figure 1a are calculated[13] using canonical d-bond integrals with W = 8.8 eV extending to first neighbors in fcc and hcp, and to second in bcc. The densities of states were calculated from their first eight moments (μ_0 to μ_7) using the maximum entropy recursion method. On the one hand, it is gratifying that only eight moments are needed to reproduce all the features of the exact result for this hamiltonian[18,20]. Much of this success must be attributed to the excellence of the maximum entropy method when only a few moments are available. On the other hand, on a cautionary note, the hcp and fcc structures are in principle distinguishable from their first five moments; but the uncertainties in the method conspire to make the fourth-moment approximation unreliable. Another way to say this is that separation between the quadrature bounds on the integrated density is too wide and the resolution too poor in a three-level recursion calculation.

In figure 1, the eight-moment canonical result is compared with two more accurate calculations. Figure 1b shows the results of Pettifor's calculations[21] using a hybrid nearly-free electron tight-binding hamiltonian. In this model, the d-electron block of the hamiltonian was represented in a basis of resonant orbitals reminiscent of KKR theory,

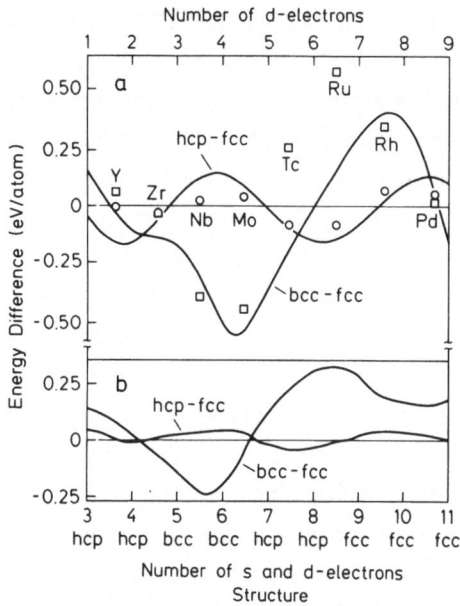

Figure 1. hcp-fcc and bcc-fcc structural energy differences in transition metals. (a) Canonical d-band model, first moments[13], circles (hcp-fcc) and squares (bcc-fcc) are ASA results[20]. (b) Results from the hybrid NFE-TB hamiltonian.[21]

while a pseudopotential-like formalism is used for the s-block. The model includes explicit hybridization between the two blocks, and was transformed into a form in which the matrix elements of the resonant block are short-ranged energy-independent integrals of Slater-Koster[4] form. Therefore the differences displayed in the curves of figures 1a and 1b can be attributed almost entirely to the neglect of the conduction s electrons and their hybridization with the d-band in the former. There is not a well-defined correspondence between the abscissae of figures 1a and 1b because if hybridization with s-electrons is included, there will be a non-integral number of d-electrons. Therefore when using the canonical model in atomistic simulation, the d-bandfilling has to be treated as an unknown parameter[22]. The squares and circles in figure 1a are taken from Skriver's ASA calculations[20] in the local density approximation which may be used to determine this parameter in the 4d metals. Skiver's calculations exploited the force theorem[23,24] by making a self-consistent density functional total energy calculation for the fcc structures, and transferring the resulting frozen potentials into the bcc and hcp structures whereupon only a single band calculation was made. The energy differences were then computed from just the sum of the occupied eigenvalues in two phases; this is in analogy with Friedel's association of the cohesive energy with the covalent bond energy alone, but is rigorously exact to first order in the atomic displacements. Both Skriver's local density calculations and Pettifor's hybrid NFE-TB model correctly reproduce the structural trends, although

Skriver's results and the canonical model systematically overestimate energy differences by a factor of about 2 compared with estimates obtained from experimental phase diagrams[25].

It is clear that the d-electrons are making the dominant contribution to cohesion in transition metals[3], and the s-electrons are playing only a minor role. Note, though, that as well as incorrectly predicting the structures at the end of the series, the canonical model incorrectly gives the bcc structure for Tc and Re which are hcp metals. Inclusion of an s-basis funcion in the empirical tight-binding basis should cure this. In fact, in recent tight-binding studies of transition metal alloys, Pasturel and co-workers[26] have included an extra s-electron using matrix elements appropriate for free-electron like bands (i.e., scaling like d^{-2}) and taking hybridization elements to be the geometric mean of d- and s-matrix elements. This improved model was able to correctly reproduce the degree of short-range ordering observed in amorphous Cu-Zr and Ni-Zr alloys[27].

A tight-binding d-band model, using less steeply scaling bond integrals than inverse fifth power (to account for the longer range of the d-electrons at small bandfillings) compensated by a repulsive Born-Mayer pair potential, has been used by Legrand in molecular dynamics studies in Ti, W and Ta. The model was able to correctly account for the non-basal slip observed in Ti on the basis of relaxed structures of dislocation cores[28]; and for experimental observations of the reconstructed (001) surface in W[22]. This work definitively vindicates the tight-binding model, in that Legrand was able to show that the model was not only sufficiently accurate to give a good description of transition metal defects, but also that it is sufficiently fast to compute inter-atomic forces in molecular dynamics algorithms and obtain relaxed defect structures.

SILICON, RECURSION AND INTERATOMIC FORCES

Friedel's covalent bond energy E_{cov}, while providing an excellent approximation to the cohesive energy differences as we have seen, becomes increasingly negative as the volume decreases until it diverges in the limit of zero bond length. This, of course, is disastrous for an atomistic simulation, and is also not physically realistic. The repulsive pair-wise energy that is invariably added to balance the covalent bond energy in empirical tight-binding has been briefly mentioned in the last paragraph, and must be elaborated upon further. We will also give a description of the recursion method in order to show how the moment theorems of the last section can be applied to the calculation of interatomic forces the the relaxation of defects. But first, the tight-binding hamiltonian for Si must be discussed.

Tight-binding in Silicon

Unfortunately, there is no canonical model in Si to draw on. This is principally due to the fact that the first-principles muffin-tin based methods such as KKR are not appropriate in covalent solids with large interstitial volumes, while the pseudopotential methods which are applicable have yet to provide simple models except in the normal metals[29,30]. An important step in developing a model tight-binding hamiltonian for Si was made by Chadi and Cohen[31] who showed that Slater-Koster[4] parameters extending only to first neighbors in Si, Ge and C could be empirically found which reproduced the valence bandstructure adequately. The tight-binding basis here is a minimal set of one s- and three p-electrons which give rise to the following four bond integrals[32] (in eV).

$$ss\sigma = -1.9375; \quad sp\sigma = 1.745; \quad pp\sigma = 3.05; \quad pp\pi = -1.075$$

Unlike the canonical d-band model in which all five d-orbitals can be assigned the same energy ε_d (usually used as the zero of energy), the model for Si includes the s-p splitting energy, $(\varepsilon_p - \varepsilon_s) = 7.45$ eV. (Harrison[33] has generalized this model to include all covalently bonded semiconductors by treating the bond integrals as universal constants, while allowing the nature of the elements involved to enter *via* the s-p splitting energy). The hamiltonian is constructed using just the above five parameters, and is conceived as having matrix elements between a set of orthogonal atomic-like s- and p-basis functions. In the transition metal canonical model, the matrix elements can be shown[18,19] to be exactly orthogonal (and two-centre) to the same first order accuracy that went into the scaling of the canonical structure constants. There is no such assurance in the model for Si, and in fact if the matrix elements are to be short ranged, there will be a large error due to neglected non-orthogonality of the basis. The final ingredient in the model is the distance scaling of the bond integrals. This is, in Chadi's model, an inverse square dependence d^{-2}, which mimics the free-electron like nature of the Si bandstructure. An inverse cube power is also often used[34]; this has the advantage that the energy does not diverge if sums over all neighbors to infinity are made. These are not likely to be encountered since one truncates the bond integrals beyond first neighbors. But it does present problems for atomistic simulation since the energy surface acquires sizable discontinuities. There does not seem to be an acceptable way around this dilemma. If the bond integrals are made to decay very steeply, then one finds that Si is stable in a close-packed structure[35,36]. Whether a canonical model for Si can be developed from first-principles ASA tight-binding theory[37] has yet to be discovered. The nature of the problem can be seen by a comparison with the canonical d-band model. Here, the tight-binding parameters are used with W taken as a constant, nominal bandwidth. Then if the crystal is dilated, the bandwidth will have the correct volume dependence because of the d^{-5}-scaling of the canonical structure constants. For free-electron-like bands, as in Si, the bandwidth must scale with a d^{-2} dependence. So if the bond integrals are to have similar form (*viz.*, $ll'm \propto \eta_{ll'm} d^{-q}$) then q must equal 2. In Harrison's expression for the bond integrals in Si, one should really associate the d^{-q} term with W in the canonical model; while the $\eta_{ll'm}$ must be identified as structure constants.

The cohesive energy

The binding energy of an assembly of atoms is defined as the energy difference between those atoms in the solid state at a given average atomic volume, and the same atoms separated at infinity *in vacuo*; it is the negative of the cohesive energy. In simple tight-binding theory, it is given by

$$E_B = E_{cov} + E_{site} + E_{rep} + \text{a constant.}$$

The first two terms make up the sum of occupied eigenvalues of the tight-binding hamiltonian, they will be discussed in detail below. The origin of the repulsive energy E_{rep} in Si has been the subject of much discussion. Chadi[38] described it as comprising the electrostatic pair potential between the ions plus the double-counting correction to the eigenvalue sum. For the purposes of atomistic simulation, the pair potential was expanded to second order in small displacements from equilibrium bond length in a Taylor series, and the three resulting disposable constants adjusted to reproduce the experimental cohesive energy, bond length and bulk modulus[39]. Harrison[40], on the other hand, argued that by far the largest contribution to E_{rep} would be the repulsive energy arising from the non-orthogonality of the basis functions. In an appeal to Hückel theory, Harrison deduced a pair potential having inverse fourth power d-dependence and without further disposable parameters. This model is able to give reasonable predictions of the lattice constants of group IV semiconductors, but severely underestimates the bulk modulus. In a comprehensive discussion of this question, van Schilfgaarde and Sher[41] have emphasized that it is not really possible to obtain a good description of Si using a pair potential with simple A/d^r form. They have pointed out that previous models have neglected the important repulsion due to non-orthogonality between the valence electrons on one atom and the core electrons on its neighbors. They demonstrated the validity of their assertion very convincingly with approximate but first-principles calculations of bulk moduli, lattice constants and cohesive energies of twelve alkali halide compounds. By including both valence-valence and valence-core terms in a pair potential for Si, van Schilfgaarde (in a private communication) arrives at a pair potential of the form $A/d^3 + B/d^8$, where the first term incorporates the $1/d$-dependence of the valence-valence overlap integrals which is more realistic than Harrison's assumed d^{-2}, or Majewski and Vogl's[42] d^{-3} forms. There is absolutely no justification for treating the non-orthogonality correction as a pair-wise energy, although double-counting terms can be shown to a very good approximation to be so[43]. This is a weakness in the tight-binding model for Si; for this reason it is very necessary to make comprehensive structural stability tests of the model. Calculating E_B in a number of crystal structures with different coordination numbers also tests the transferability of the parameterized bond integrals.

The binding energy with Chadi's bond integral parameters and Harrison's pair potential A/d^4, in which the parameter A is fitted to the lattice constant, gives energy-volume curves[36] for Si in diamond cubic, hexagonal diamond, β-tin and the close-packed fcc and bcc phases shown in figure 2a. These are compared with the best available local-density-functional calculations[44] on the same scale in figure 2b. The model is well vindicated, at least in the description of the four-fold coordinated diamond and β-tin structures; while the close-packed phases are high enough in energy not to cause problems in atomistic modeling. The reason for the poor description of non-tetracoordinated structures in the tight-binding model for Si lies in the assumption that the bond integrals do not depend on crystal structure, *i.e.*, that the $\eta_{ll'm}$ are constants[33]. Important features in these diagrams are the pressure-induced phase transformation to β-tin, and the

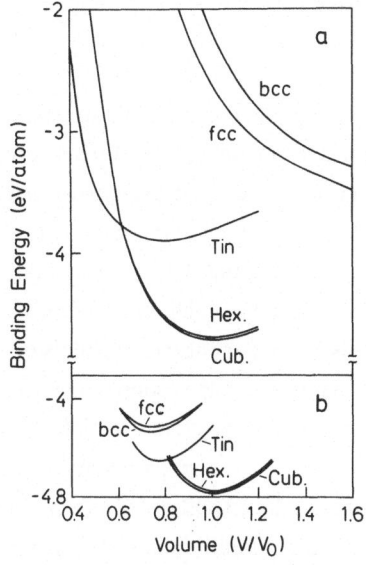

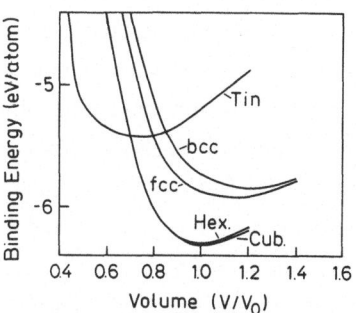

Figure 2. Energy-Volume curve for five phases in Si. (a) Empirical tight-binding[36]: $E_{rep} \propto d^{-4}$ (b)Local Density Approximation[44].

Figure 3. Tight-Binding energy-volume curves: $E_{rep} \propto d^{-5}$

non-possibility of transformation to hexagonal diamond under a purely hydrostatic stress. Both these well-established facts are properly described by the tight-binding model. Of course, it is essential that cubic diamond (rather than hexagonal) is the stable phase. The curvature of the energy-volume curve for cubic-diamond, gives a bulk modulus in this model of 0.34 Mbar which compares very poorly with the experimental value, 0.99 Mbar. To attempt to improve this situation, a pair-potential scaling like the inverse fifth power of *d* has been used. The resulting energy-volume curves are displayed in figure 3. It is gratifying that the essential features of the model are preserved under a large perturbation of one of the parameters. The only significant effect is to shift the close-packed structures below β-tin. The bulk modulus is now 0.65 Mbar, which is a useful improvement. Experience shows that the model is not sensitive to sensible changes in the bond integral parameters. For example, very much the same picture results using parameters from Harrison's Solid State Table[33]. The one critical parameter is the exponent in the scaling of the bond integrals with bond length. This should be two, at first-neighbor separation, or at most three[34]. Finally, it is worth mentioning here that the tight-binding model for Si has now been used to make a systematic study of grain boundary structure and energy, independently by two different groups: one[45] using the recursion method and an A/d^4 pair potential fitting only the lattice constant; the other[46,47] using Chadi's matrix diagonalization method and pair potential fitted to bulk modulus and cohesive energy as well. The fact that both energies and structures were in quantitative agreement argues that the form of the pair potential is not critical in the model and justifies the rather heavy concentration on calculating E_{cov} and E_{site} and the rather cavalier approach to E_{rep} that is usually adopted[48].

The Inter-Site Recursion Method

To begin with, let us state that the problem is to determine the electronic structure of an assembly of atoms in arbitrary geometry, and proceed to calculate the forces acting between these atoms. We claim that forces between atoms will be principally *local* quantities; that is, largely determined by the atomic geometry in the neighborhood of the atoms between which the force acts. In the spirit of tight-binding theory, each atom is supposed to have associated with it atomic s,p,d...-like orbitals which will be used as a basis in which to represent the hamiltonian. Nothing is known about these orbitals other than that hopping of an electron from one orbital to another is due to the action of the matrix elements of $\hat{H}$. The first 2L moments of the LDOS of an orbital α on atom i are determined by the (real) bond integrals $ll'm$ and the topology of the first L-1 neighbor shells about i. The usual moments or recursion methods then reconstruct the LDOS $n^{i\alpha}(E)$ from these moments. This is the *on-site* density of states. But for interatomic forces, it is better to concentrate on the electrons in a *bond* between atoms i and j, say, rather than the electrons on a site i. In fact, these are equivalent pictures. Moments may be calculated for the linear combination of orbitals $2^{-1/2}|i\alpha + j\beta\rangle$ and for $2^{-1/2}|i\alpha - j\beta\rangle$. These are, respectively, bonding and antibonding (or *vice versa*) combinations of the orbitals $|i\alpha\rangle$ and $|j\beta\rangle$. Because of the linearity of the moments, they may be subtracted to obtain twice the rth moment of the *inter-site* density of states $n^{i\alpha,j\beta}(E)$. The equivalence of these two pictures follows from the identity

$$\mu_{r+1}^{i\alpha} = \sum_{j\beta} H_{i\alpha,j\beta}\mu_r^{i\alpha,j\beta},$$

where $H_{i\alpha,j\beta}$ is the matrix element of $\hat{H}$ between orbital α on site i and orbital β on site j. Densities of states are also additive, so that the integral of $n^{i\alpha,j\beta}(E)$ to the Fermi energy gives the difference in occupation numbers of the bonding and antibonding orbital combinations. This is twice Coulson's partial bond order[49]. The total order of the i-j bond (bond order, for short) is half this quantity summer over the orbitals α and β:

$$\frac{1}{2}\sum_{\alpha,\beta}\int^{E_f} n^{i\alpha,j\beta}(E)\ dE.$$

The bond orders themselves provide a wealth of information about the model system; they are also the key to interatomic forces. The first two terms in the binding energy E_{cov} and

E_{site} may be written in terms of on-site and inter-site densities as follows. Summing over all the atoms in the assembly,

$$E_{cov} = \sum_{\substack{i\alpha,j\beta \\ i\alpha \neq j\beta}} H_{i\alpha,j\beta} \int^{E_f} n^{i\alpha,j\beta}(E)\,dE$$

and

$$E_{site} = \sum_{i\alpha} H_{i\alpha,i\alpha} \int^{E_f} n^{i\alpha}(E)\,dE.$$

E_{site} is the second term that was subtracted off in the first equation for Friedel's covalent bond energy. In the simple canonical d-band model it may be made zero by choosing $\varepsilon_d = 0$, but in the s-p tight-binding model it corresponds to a rehybridization energy, since it depends on the ratio of p to s electrons on each site i. In the recursion method (or indeed by matrix diagonalization) the first two terms in E_B (viz., $E_{band} \equiv E_{cov} + E_{site}$) can be seen to be calculated from two equivalent expressions:

$$E_{band} = E_{cov} + \sum_{i\alpha} \varepsilon_{i\alpha} N_{i\alpha}$$

or

$$E_{band} = \sum_{i\alpha} \int^{E_f} E n^{i\alpha}(E)\,dE.$$

The first defines $N_{i\alpha}$ as the number of electrons occupying the α-orbital on site i and $\varepsilon_{i\alpha}$ as a possibly site-dependent orbital energy; while the second is the well-known expression for the *band* or *structural* energy of the assembly of atoms. This illustrates again the equivalence of the on-site and inter-site pictures. In the former only on-site densities of states are calculated, in the latter both inter-site and on-site densities are needed but the resulting information is more detailed since the band energy is decomposed into partial bond and hybridization parts. This useful breakdown of the energy will be illustrated with specific examples in the next section.

The original recursion method solution to calculating inter-site densities was to subtract bonding and anti-bonding densities calculated separately[14]. But there is a much better method[50,51] that stems from the observation emphasized here that the moments have diagonal and non-diagonal terms in the same way as does the hamiltonian matrix. The moments and densities can therefore be calculated directly in their matrix representations. In fact, matrix moments are nothing new[52], and the classical moment problem is known to be rather generally applicable to matrix moments. In other words, the matrix densities have a unique expansion in matrix moments[52] (at least in cases of interest in the recursion method). Nex[54] has recently shown how to extend gaussian quadrature to the problem of integrating matrix densities, and the same bounding theorems as in the scalar case can be proved. The matrix or *simultaneous* recursion method has now been extensively applied[45,35] to calculations of inter-atomic forces in static relaxations and proved itself a stable algorithm. Another recent advance[55] has been to improve the maximum entropy method so that it is no longer ill-conditioned for more than about ten moments. It will be very interesting to extend the maximum entropy method to the matrix moment problem.

PHASE STABILITY AND DEFECT ENERGIES

In this concluding section, the foregoing rather formal treatment will be put on a firmer basis by citing a number of examples. First, the energy-volume curves in Si are analyzed in the inter-site method; and secondly we confirm directly the power of the local invariance theorem by examining the relationship between HD and CD crystal structures and the structure of the intrinsic stacking fault in CD crystals. Finally, the intricate reconstruction observed at the $[111]60°(11\bar{2})$ grain boundary in Si is discussed in the context of the tight-binding interatomic force model which has been used to study this reconstruction.

Phase Stability in Silicon Re-visited

The energy-volume curves shown in figures 2 and 3 gave a vindication of the tight-binding model, but no insight into why Si condenses in the cubic diamond structure, or why it transforms to β-tin under pressure. It would be a further vindication of the simple model if it were able to come up with some explanation for these phenomena. It simplifies matters to examine only the first two terms in E_B. Indeed in the energy difference between cubic diamond (CD) and hexagonal diamond (HD), these are the only non-zero terms if E_{rep} is taken to be a pair potential truncated beyond first neighbors. We have already indicated that the difference in band energies provides the major contribution to structural energy differences, at least in a qualitative picture. The way to analyze the band energy differences is to look at the partial bond energies and *s-p mixing* in the different crystal structures. In the s-p tight-binding model there are four partial bond energies[56] which are simply $B_{ss\sigma} = ss\sigma \int^{Ef} n^{is,js} (E)dE$ etc... when axes are chosen so that the positive lobe of the p_z orbital points along the i-j bond. The s-p mixing[36] on atom i is defined as N_{ip}/N_{is}, that is, the ratio of the number of p to the number of s-electrons. This can take a value between 1 (s^2p^2 configuration) and 3 (sp^3 - fully hybridized). As long as the bond is not predominantly antibonding, the partial bond energies will be negative and their magnitude will indicate the strength of the bond when summed over the four types $ss\sigma$, $sp\sigma$, $pp\sigma$ and $pp\pi$. In the structures discussed here, all atoms are equivalent, but it is important in the inter-site method to consider any non-equivalent bonds separately and sum their contributions to E_{cov}. In CD in the first-neighbor approximation, all four bonds are symmetry related. HD has three bonds related by six-fold rotation about the c-axis, while the bond point along this axis is not equivalent to these even when the axial ratio is ideal

$\sqrt{\frac{8}{3}}$ as used here and as determined experimentally[57]. In Table 1, these bonds are labelled "A" and "B" respectively. In β-tin and bcc, first and second neighbor bonds are included in the calculations and denoted "1" and "2". Table 1 shows the partial bond energies and E_{cov} in eV/bond and the site energy in eV/atom. The atomic volume relative to the equilibrium volume of CD (V/V_0) is shown , as is the number of moments $2L$ used in the calculations.

TABLE I

Phase	Bond	$B_{ss\sigma}$	$B_{sp\sigma}$	$B_{pp\sigma}$	$B_{pp\pi}$	E_{cov}	E_{site}	N_p/N_s	V/V_0	$2L$
Cubic Dia.		-0.812	-2.479	-4.245	-0.603	-8.139	-4.716	1.71	1	30
Hex. Dia.	A	-0.827	-2.485	-4.261	-0.607	-8.180				
	B	-0.821	-2.507	-4.211	-0.561	-8.100	-4.659	1.73	1	30
β-tin	1	-0.623	-1.690	-3.037	-0.505	-5.855				
	2	-0.379	-1.075	-2.338	-0.243	-4.035	-5.165	1.59	0.77	30
fcc		-0.009	-0.253	-1.439	-0.260	-5.887	-7.377	1.12	1	10
bcc	1	+0.001	-0.205	-1.608	-0.273	-4.172				
	2	-0.011	-0.053	-0.878	-0.186	-1.692	-7.764	1.07	1	10

Although CD cannot transform to HD simply by application of pressure (because their energy-volume curves do not cross), the latter phase, since its initial discovery, has recently been more and more widely reported as a consequence of both non-hydrostatic applied stresses[57,58], and ion and radiation damage[59]. Certain defects previously thought to be stacking faults have now been shown in fact to be ribbons of HD phase[59]. The study of this phase is therefore of increasing interest, so it is worth looking at the information given by the tight-binding model. The following picture emerges. In HD the three "A" bonds have more negative covalent bond energy than the bonds in CD, while the bond along the c-axis is "weaker". Overall, E_{cov} is more negative in HD; in other words the bonding is stronger, and on this basis alone one would expect HD to be the stable phase. However, in forming stronger bonds more s-electrons have been promoted to p-symmetry in order to form sp^3-hybrids at the expense of the s-p splitting energy which is large (≈ 6 eV/electron). This extra energy is reflected in the s-p mixing N_p/N_s and in the site energy E_{site}. The total result is that the band energy is higher in HD than CD and the latter structure is stable. In the close-packed structures fcc and bcc the s-p mixing is close to one; the electrons haven't bothered to form hybrids and little energy has been gained by bond formation. In β-tin the second neighbors are just 6% more distant than the first. But notice that their bond energies, especially $B_{ss\sigma}$, are considerably less negative than one would expect from the inverse square decay of the bond integrals. In all the covalent bond energy is 30% less negative rather than the expected 11% between first and second neighbors. This may be used as justification for treating β-tin as four-fold coordinated in calculating the energy-volume curves. CD can be deformed into the β-tin structure by a continuous reduction of the tetragonal axial ratio, and the second neighbors in β-tin derive from the fourth neighbors in CD. But although these atoms come very close in β-tin, they refuse to bond (or so it seems). Furthermore, the s-p mixing is not that different in β-tin at its minimum-energy volume ($V/V_0 = 0.77$ in the tight-binding model as well as in the local density approximation) as it is in diamond at V_0. The tight-binding model reveals that diamond transforms to β-tin under pressure because it can increase its atomic density while at the same time retaining the tetrahedral sp³-like nature of its bonding. These structural stability arguments are in fact supported by the charge density plots of these phases in local density calculations[44].

<u>Connection between Cubic, Hexagonal and Intrinsic Stacking Fault - Local Invariance</u>

The (111) stacking sequence in CD crystals has the well-known

$$\cdots AaBbCcAaBbCc\cdots$$

representation. If a plane of atoms is removed, the resulting intrinsic stacking fault has four atomic planes of HD stacking:

$$\cdots AaBbAaBbAaBb\cdots,$$

so that the stacking sequence of the intrinsic stacking fault is[60]

$$\cdots AaBbCcAaBbAaBbCcAaBbCc\cdots.$$

The intrinsic stacking fault energy in Si is known to be about 60 erg/cm²; arguing that this would correspond to the energy of four atomic planes of HD phase, Tan *et al.*[61] estimated that the HD-CD energy difference should be about 400 cal/mole which is close to the value 240 cal/mole (0.01 eV/atom) found both in the tight-binding model and in the local density approximation. This remarkably close parallel between the HD-CD energy difference and the intrinsic stacking fault energy is a striking example of the local invariance theorem and can be illustrated further in the following way. In figure 4, I have sketched the local densities of states (s + p in the tight-binding model) on three atoms: the left panel shows the DOS in cubic diamond, the middle panel that in hexagonal diamond and on the right, the LDOS on an atom in an intrinsic stacking fault bounding the fault plane. This atom is bounded on either side by only two planes of HD phase, yet its LDOS resembles much more closely the HD density than that of the CD phase in which the fault is embedded. It

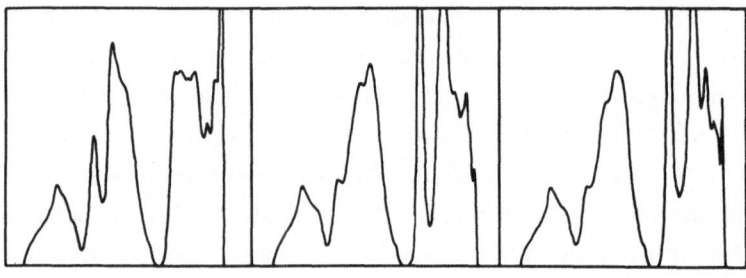

Figure 4. Illustrating Local Invariance. Local Densities of States in Si: left in CD; center in HD; right at an intrinsic Stacking Fault.

seems that the local electronic structure really is largely determined by the local atomic structure; and in fact other examples of this transferability of the LDOS can be seen in the tight-binding model densities of states at grain boundaries[35].

The energy differences in figures 2 and 3 were calculated from only the first ten moments of the DOS. The HD-CD energy difference is in fact negative if fewer than eight moments are used[56]. This is also the case for the intrinsic stacking fault energy which is only slowly convergent with the number of moments[35], and is found to be 96 ± 20 erg/cm[2]. This is in not much worse agreement with experiment than the local density approximation, which gives[62] 40 erg/cm[2]. The [111]60°(111) grain boundary, which is the interface found between deformation twins in Si[63] has

$$\cdots AaBbCcAaBbAaCcBbAa\cdots$$

stacking, and therefore just one HD structural "twin" unit per period. One would expect its energy to be half that of the intrinsic stacking fault; and this is indeed found in the tight-binding model in which the twin interface energy is 45 ± 15 erg/cm[2].

The [111] 60° (11$\bar{2}$) Grain Boundary in Si

The study of grain boundary (g.b.) structure, both in metals and semiconductors is of both academic and technological interest. Particularly in Si, one is interested both in the atomic and the electronic structure. To calculate the former, one needs a means of calculating interatomic forces, which if based in electron theory, should also yield information on the latter. As yet, the most complete studies in Si have used either the classical Keating model for two- and three-body forces[64]; or the *tight-binding bond model*[43,45] in which the interatomic force acting on atom i takes the form[43]

$$-\nabla_i E_B = -2 \sum_{\substack{j\alpha\beta \\ j\neq i}} \nabla_i H_{j\beta,i\alpha} \int^{E_f} n^{i\alpha,j\beta}(E)\, dE - \nabla_i E_{rep}$$

(The absence of the derivative E_{site} is a consequence of the *local charge neutrality condition*[43] in which the matrix elements $\varepsilon_{i\alpha}$ are incremented by amounts Δ_i in order to ensure neutral atoms). Both these models have been used in the process of elucidating the reconstruction of the [111]60°(11$\bar{2}$), g.b. in Si.

After the [111]60°(111) g.b., the [111]60°(11$\bar{2}$) g.b. is the most commonly observed in Czochralski grown Si[65]. They both separate crystals related by the same orientation relationship (*viz.*, a 60° rotation about [111]) and g.b.'s separating such crystals are usually found[65] to be faceted into segments parallel to (111) and segments parallel to (11$\bar{2}$). Unravelling the atomic structure of the [111]60°(11$\bar{2}$) g.b. was as much a question of ingenious model building as atomistic simulation. This is because of the large phase-space of atomic structures that arises from the very "open" diamond cubic structure, most of which is inaccessible in a molecular static or simulated quenching algorithm. The development is described in stages as it occurred, below; followed by the results of molecular statics calculations[45] of the various structures that were proposed in the literature.

§1. Earliest models[66,67] for the [111]60°(112̄) g.b. assumed that three- and five-fold coordinated atoms would necessarily exist in the atomic structure, a rough calculation of the energy of such an "unreconstructed" g.b. showed it to be large[68].

§2. It became clear from early electron microscope observations that the [111]60°(112̄) g.b. showed a large in-plane translation in the [111] direction[69,70]. Pond[71] and Fontaine and Smith[69] independently proposed a reconstruction of the [111]60°(112̄) g.b. that was fully four-fold coordinated and reproduced the [111] tranlation.

§3. Pond et al.[72] showed that by allowing a doubling of the period along the coincidence site lattice (c.s.l.) unit cell vector 1/2[11̄0] (tilt axis) and introducing "pinched together" dimer bonds along this direction, one could make a reconstruction that did not display a [111] translation but which had lower energy in a Keating model than the "translation twin" (§2. above).

§4. This remained an unresolved paradox until further experimental observations were made by Bourret and co-workers. They showed, (i) that the translation along [111] was less than measured previously[65]; (ii) that the [111]60°(112̄) g.b. had planar cm symmetry and the two-dimensional unit cell had twice the period of the c.s.l. along both the 1/2[11̄0] and the [111] directions in the g.b. plane[73]. In other words, the [111]60°(112̄) g.b. displays a (2x2) reconstruction. A doubling along 1/2 [11̄0] as well as a translation along [111] was achieved in a (1x2) reconstruction proposed by Papon and Petit[74].

§5. Bourret and Bacmann[75] finally showed that the Papon and Petit model could be given a doubling of the period along [111] by reconstructing the g.b. in such a way that units of the Papon and Petit structure were repeated alternately displaced along the tilt axis by 1/2 [11̄0]. They were able to confirm that this is the structure observed by careful high-resolution lattice imaging. Theoretical calculations reveal that the energy difference between the (1x2) and (2x2) Papon and Petit structures are extremely small, and although the latter has marginally lower energy in both Keating[47] and tight-binding[45] calculations, the difference is within the error bars imposed by the method. The regular, alternate arrangement of the Papon and Petit units in the [111]60°(112̄) g.b. invariably observed, has yet to be satisfactorily explained.

Table 2 displays the results of calculations of the atomic structure of the reconstructions of the [111]60°(112̄) g.b. described in §§1-5 above, using the tight-binding bond model in calculations of interatomic forces and bond and site energies. Shown in parentheses are the results of calculations in the Keating model[64]. Also shown is the experimental position. Both models correctly account for the observed atomic structure of the [111]60°(112̄) g.b. There are two areas of disagreement: (i) the volume expansions at the g.b., for which the tight-binding calculations give somewhat better agreement with experiment; (ii) the energy difference between structures §3 and §5 is not significantly different in the Keating model; whereas the tight-binding energies clearly favor the observed structure. The electronic structure is only available in the tight-binding model.

TABLE 2

	Periodicity Units of [111]x1/2[1$\bar{1}$0]	Relaxed Shubnikov Layer Group[76]	Translation State [111]	[11$\bar{2}$]	[1$\bar{1}$0]	Energy erg/cm^2
		RESULTS FROM THEORETICAL CALCULATIONS				
§1	(1x1)	p2'mm'	0	0.018	0	1200±80
§2	(1x1)	p'2mn'	1/6 (1/6)	0.050 (0.017)	1/4	670±20 (600)
§3	(1x2)	p2'mm'	0 (0)	0.023 (0.01)	0	450±20 (300)
§4	(1x2)	pm	0.092	0.032	1/4	350±30
§5	(2x2)	cm	0.092 (0.091)	0.032 (0.02)	1/4	340±30 (290)
		PRESENT EXPERIMENTAL POSITION				
Ref. 64 (Ge)			0.08-0.10	0.03-0.06	0 or 1/4	
Ref. 73 (Ge)		c m				
Ref. 77 (Si)			0.07-0.09	0.05-0.07	0 or 1/4	

This is of considerable interest following the observation that while [111]60°(111) g.b. segments are electrically inactive, [111]60°(11$\bar{2}$) g.b. segments have been seen to give a signal in Electron-Beam Induced Currents (EBIC) experiments[78,79]. The question that arises is whether this activity is due to the intrinsic structure of the g.b.; the presence of defects in the atomic structure; or the presence of impurities. The structure is fully four-fold coordinated and has a maximum bond-length strain of +5% (in the dimer bonds). The maximum bond angle strains are +14% and -17%. These have recently been shown[47] to give rise to localized states in the two-dimensional bandstructure, but these lie within the limits of the band continua; the density of states shows no structure within the gap[45,80]. This eliminates the intrinsic atomic structure as being responsible for the electrical activity.

REFERENCES

1. H.J. Jones, Proc. Phys. Soc. London 49:250 (1937).
2. S.L. Altmann, C.A. Coulson and W. Hume-Rothery, Proc. Roy. Soc. London, A240:145 (1957).
3. J. Friedel, Trans. AIME, 230:616 (1964).
4. J.C. Slater and G.F. Koster, Phys. Rev. 94:1498 (1954).
5. F. Cyrot-Lackman, J. Phys. Chem. Solids, 29:1235 (1968).
6. F. Ducastelle and F. Cyrot-Lackmann, J. Phys. Chem. Solids, 32:285 (1971).
7. V. Heine, Solid State Physics, 35:1 (1980).
8. J.P. Gaspard and F. Cyrot-Lackmann, J. Phys. C, 6:3077 (1973).
9. R. Haydock, V. Heine and M.J. Kelly, J. Phys. C, 5:2845 (1972).
10. C.C. Paige, J. Inst. Maths Applics 10:373 (1972).
11. C.M.M. Nex, J. Phys. A, 11:653 (1978).

12. L.R. Mead and N. Papanicolaou, J. Math. Phys. 25:2404 (1984).
13. S. Glanville, A.T. Paxton and M.W. Finnis, J. Phys. F 18:693 (1988).
14. R. Haydock, Solid State Phys 35:215 (1980).
15. D.G. Pettifor, J. Phys. C 2:1051 (1969).
16. O.K. Andersen, Solid State Commun 13:133 (1973).
17. J. Callaway, "Energy Band Theory", Academic Press, New York (1964).
18. O.K. Andersen, O. Jepsen and D. Glötzel, in: "Highlights of Condensed Matter Theory", F. Bassani et al. eds., North Holland, Amsterdam (1985).
19. D.G. Pettifor, J. Phys. F 7:613 (1977).
20. H.L. Skriver, Phys. Rev B 31:1909 (1985).
21. D.G. Pettifor,, J. Phys. C 3:367 (1970).
22. B. Legrand, G. Treglia, M.C. Desjonquères and D. Spanjaard, J. Phys. C 19:4463 (1986).
23. D.G. Pettifor, Commun. Phys. 1:141 (1976); J. Chem. Phys. 69:2930 (1978).
24. O.K. Andersen, H.L. Skriver, H. Nohl and B. Johansson, Pure Appl. Chem. 52:93 (1979).
25. L. Kaufman, in "Phase Stability in Metals and Alloys", P.S. Rudman et al. eds. McGraw-Hill, New York (1967); A.R. Miedema and A.K. Niessen, CALPHAD 7:27 (1983).
26. A. Pasturel, D. Nguyen Manh and D. Mayou, J. Phys. Chem. Solids 47:325 (1986).
27. D. Nguyen Manh, D. Mayou, F. Cyrot-Lackmann and A. Pasturel, J. Phys. F 17:1309 (1987).
28. B. Legrand, Phil. Mag. A, 52:83 (1985).
29. M.W. Finnis, J. Phys. F 4:1645 (1974).
30. D.G. Pettifor and M.A. Ward, Solid State Commun. 49:291 (1984).
31. D.J. Chadi and M.L. Cohen, Phys. Stat. Sol (b) 68:405 (1975).
32. Q. Gou-Xin and D.J. Chadi, Phys. Rev. B 35:1288 (1987).
33. W.A. Harrison, "Electronic Structure", W.H. Freeman, San Francisco (1980).
34. G. Allan and M. Lannoo, J. Physique 44:1355 (1983).
35. A.T. Paxton, DPhil Thesis, University of Oxford (1987).
36. A.T. Paxton, A.P. Sutton and C.M.M. Nex, J. Phys. C 20:L263 (1987).
37. O.K. Andersen and O. Jepsen, Phys. Rev. Lett. 53:2571 (1984).
38. D.J. Chadi, Phys. Rev. Lett 41:1062 (1978).
39. D.J. Chadi, Phys. Rev. B 29:785 (1984).
40. W.A. Harrison, Phys. Rev. B 27:3592 (1983).
41. M. van Schilfgaarde and A. Sher, Phys. Rev. B 36:4375 (1987).
42. J.A. Majewski and P. Vogl, Phys. Rev. Lett. 57:1366 (1986).
43. A.P. Sutton, M.W. Finnis, D.G. Pettifor and Y. Ohta, J. Phys. C 21:35 (1988).
44. M.T. Yin and M.L. Cohen, Phys. Rev. B 26:5668 (1982).
45. A.T. Paxton and A.P. Sutton, J. Phys. C 21:L481 (1988).
46. M. Kohyama, R. Yamamoto, Y. Ebata and M. Kinoshita, J. Phys. C 21:3205 (1988).
47. M. Kohyama, R. Yamamoto, Y. Watanabe, Y. Ebata and M. Kinoshita, J. Phys. C 21:L695 (1988).
48. A.P. Sutton, J. Physique Coll. 46:C4-347 (1985).
49. C.A. Coulson, Proc. Roy. Soc. London A169:413 (1939).
50. R. Jones and M.W. Lewis, Phil. Mag. B 49:95 (1984).
51. J. Inoue and Y. Ohta, J. Phys. C 20:1947 (1987).
52. F.V. Atkinson, "Discrete and Continuous Boundary Problems", Academic Press, New York (1966).
53. M.G. Krein,, Dokl. Akad. Nauk SSSR 69:125 (1949).
54. C.M.M. Nex, paper presented at MSI workshop, "Practical Iterative Methods for Large-Scale Computations" (1988).
55. I. Turek, J. Phys. C 21:3251 (1988).
56. A.T. Paxton, Phil. Mag. B 58:603 (1988).
57. P. Pirouz, R. Chaim and J. Samuels, Proc. 5th Int. Symp. "Structure and Properties of Dislocations in Semiconductors", Moscow (1986). To appear in Izv. Akad. Nauk Fiz. Ser.
58. J.L. Demenet, J. Rabier and H. Garem, "Inst. Phys. Conf. Ser. No. 87", Adam Hilger, Bristol (1987).

59. A. Bourret, "Inst. Phys. Conf. Ser. No. 87:, Adam Hilger, Bristol (1987).
60. J.P. Hirth and H. Lothe, "Theory of Dislocations", McGraw-Hill, New York (1982).
61. T.Y. Tan, H. Föll and S.M. Hu, Phil. Mag. A 44:127 (1981).
62. S.G. Louie, J. Physique Coll 46:C4-335 (1985).
63. V.G. Eremenko and V.I. Nikitenko, Phys. Stat. Sol (a) 14:317 (1972).
64. A. Bourret, L. Billard and M. Petit "Inst. Phys. Conf. Ser. No. 76", Adam Hilger, Bristol (1985).
65. F. Komminou, Th. Karakostas, G.L. Bleris and N.A. Economou, J. Physique Coll. 43:C1-9 (1982).
66. J.A. Kohn, Amer. Mineral. 43:263 (1958).
67. J. Hornstra, Physica 25:409 (1959).
68. H.J. Möller, Phil. Mag. A 43:1045 (1981).
69. C. Fontaine and D.A. Smith, Appl. Phys. Lett 40:153 (1982).
70. D.S. Vlachavas and R.C. Pond, "Inst. Phys. Conf. Ser. No. 60", Adam Hilger Bristol (1981).
71. R.C. Pond, J. Physique Coll. 43:C1-51 (1982).
72. R.C. Pond, D.J. Bacon and A.M. Bastaweesy, "Inst. Phys. Conf. Ser. No. 67", Adam Hilger, Bristol (1983).
73. A. Bourret and J.J. Bacmann, "Grain Boundary Structure and Related Phenomena", Proc. JIMIS-4 (1986).
74. A.M. Papon and M. Petit, Scripta Metall. 19:391 (1985).
75. A. Bourret and J.J. Bacmann, Surf. Sci. 162:495 (1985).
76. R.C. Pond and D.S. Vlachavas, Proc. Roy. Soc. London A386:95 (1983).
77. A. Rocher and M. Labidi, Revue Phys. Appl. 21:201 (1986).
78. P. Ruterana, A. Bary and G. Nouet, J. Physique Coll. 43:C1-27 (1982).
79. R. Sharko, A. Gervais and C. Texier-Hervo, J. Physique Coll. 43:C1-129 (1982).
80. A. Mauger, J.C. Bourgouin, G. Allan, M. Lannoo, A. Bourret and L. Billard, Phys. Rev. B 35:1267 (1986).

A NEW INTERATOMIC POTENTIAL FOR NON-METALS

M. Heggie

Department of Physics
Exeter University
United Kingdom

INTRODUCTION

Many applications of atomistic computer modelling require clusters of order a thousand atoms or more. Dislocations and cracks are good examples, and in their cases an accurate description of the highly defective core region is not possible without comparable accuracy in the surrounding elastic region (which may also be faulted). At this scale of simulation, *ab initio* calculations themselves are not feasible. Recently there have been several attempts to store the results of *ab initio* calculations on small molecules or supercells in the form of a classical interatomic potential (for example, the potentials due to Tersoff[1] and Biswas/Hamann[2]). Both initial attempts by these authors foundered on several structures. They have since been modified[3,4], but unfortunately in the case of Tersoff's potential without removing all the shortcomings[5]. For modelling line defects in diamond and graphite structures we have developed a potential in the spirit of Tersoff's potential, in which a geometrical construction is invoked for more bond information and elastic anisotropy and stacking fault energies are introduced.

This new potential starts from the assumption that the energy of an atom depends only on its local environment. This is consistent with the chemical idea of covalent bonding; it being often observed that the addition of a group to a molecule, such as >CH2 to a hydrocarbon, leads to a fixed increment in formation energy (the "group increment"). In the potential longer range interactions are shielded out smoothly, not by a radial cut-off but by a many-body function. The potential is based on the following approximate assertions:

1. Covalent bonds are mutually exclusive in space.
2. The space available for each bond is given by a simple geometrical construction for dividing all space.
3. For elemental solids this construction is the "proximity cell", which is constructed in the same way as a Wigner-Seitz cell (the minimum

polyhedron formed by planes that perpendicularly bisect vectors joining a given atom to all other atoms).

4. Each face of the cell represents a non-zero bonding interaction. Other interactions are assumed to be zero (screened out).

5. Bonding is a balance between shielding the ion-ion interaction (by interposing an electron cloud) and minimising electronic kinetic energy (which is increased by localising electrons). Therefore, the smaller the face, the less space for bonding electron density, the greater the electronic kinetic energy and the smaller is the bonding interaction. Distant faces with small areas could even feel the ion-ion repulsion alone and make a negative contribution to the bonding (for example, at stacking faults).

Figure 1 illustrates the two new parameters, r_σ and r_π, that describe a bond face in addition to the bond length r. The first is the radius of the minimum inscribed circle centered on the internuclear intercept (tangent to the edge closest to the intercept - see Appendix A). The second is a compound parameter that describes the elongation of the face parallel to the closest edge (detailed in Appendix A), which derives from two further parameters, Δ_1 and Δ_2.

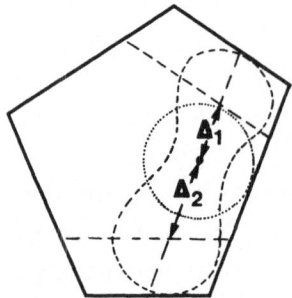

Figure 1.   Definition of parameters Δ_1, Δ_2, r_σ and r_π.

According to these definitions the main bonding features of the diamond structure are identified by four sides of

$$r_\sigma = \sqrt{2}\ r/2\ , \qquad r_\pi = r_\sigma$$

and of the graphite structure by three sides of

$$r_\sigma = \sqrt{3}\ r/2, \qquad r_\pi = c/4$$

c/2 being the interlayer separation. The second neighbours in the diamond structure have the internuclear intercept outside their respective faces and r_σ is set equal to zero for them.

Given this ability to differentiate sp^2 and sp^3 bonds, a radial potential of Tersoff's form (equation 2) for each bond is fitted to give equilibrium bond lengths, cohesive energy and bulk modulus, for diamond and graphite (for which 1/2 ($C_{11} + C_{12}$) was taken instead of the full bulk modulus). A cut-off function, f, can be used to make interactions across the ideal vacancy negligible and give an ideal vacancy formation energy equal to the cohesive energy which is approximately the case for the diamond and graphite structures. It is intended to increase this cut-off later to reproduce the breathing and tetragonal distortions of the neutral vacancy.

For a given type of bond, defined by r_π, the bond strength is diminished smoothly to zero as r_σ goes to zero by a multiplicative factor, $f(\theta, \theta_a, \theta_b)$, where $\theta = 2r_\sigma/r$ and

$$f = 0 \qquad\qquad\qquad\qquad \text{if } \theta \le \theta_a$$

$$f = \frac{1}{2}\left(1 - \cos\pi\left(\frac{\theta - \theta_a}{\theta_b - \theta_a}\right)\right) \qquad \text{if } \theta_a \le \theta \le \theta_b \qquad (1)$$

$$f = 1 \qquad\qquad\qquad\qquad \text{if } \theta_b \le \theta$$

There is a slight difference between the prefactors for the pre-exponential radial parameters A and B, each being shielded at different rates, A by $f(\theta, 0, \theta_1)$ and B by $f(\theta, 0, \theta_2)$. The same function is also used to switch between sp^3 and sp^2 bonding. So, for example, the pre-exponential A is given by

$$A = A_d + f(r_\pi, r_{\pi 1}, r_{\pi 2})(A_g - A_d)$$

$r_{\pi 1}$ and $r_{\pi 2}$ being fitting constants and g and d denoting graphite and diamond values, respectively. Likewise for the parameters $\lambda_1, \lambda_2, B, \theta_1$ and θ_2 in the bond energy, V

$$V = (1 - f(r, r_1, r_2))(Af(\theta, o, \theta_1)\exp(-\lambda_1 r) - Bf(\theta, o, \theta_2)\exp(-\lambda_2 r)) \qquad (2)$$

A measure of bond strength, C_i of a bond i is related to the attractive term

$$C_i = (1 - f(r, r_1, r_2))\beta Bf(\theta, o, \theta_2) f(\theta, \theta_1, \theta_2)\exp(-\lambda_2 r))_i \qquad (3)$$

β is a constant of proportionality to be adjusted later.

The interaction energy, E_i between a bond, i, of strength C_i and others, j, of strength, C_j is taken to be of the form of the angular part of the Keating potential[6], modified by the product $C_i C_j$ and corrected to remove contributions to the bulk modulus (second term, equation 4).

$$E_i = \sum_{i \ne j} C_i C_j (\mathbf{r}_i \bullet \mathbf{r}_j - r_{oi} r_{oj}\cos\varphi_o)^2 - \frac{1}{2} n_o C_i^2 (\mathbf{r}_i \bullet \mathbf{r}_i - r_{oi}^2)^2 \cos^2\varphi_o \qquad (4)$$

where n_o is an interpolation between the number of angles subtended by i in the diamond structure (3) and the graphite structure (2): $n_o = (\cos\varphi_o)^{-1}$ is used. φ_o depends on r_π of the two bonds in the following way

$$f_{mean} = \frac{1}{2}\left(f\left(r_\pi^i, r_{\pi 1}, r_{\pi 2}\right) + f\left(r_\pi^j, r_{\pi 1}, r_{\pi 2}\right)\right)$$

$$\cos\varphi_o = -\frac{1}{3} + f_{mean}\left(\frac{1}{3} - \frac{1}{2}\right)$$

$$r_{oi} = r_{od} + f\left(r_\pi^i, r_{\pi1}, r_{\pi2}\right)\left(r_{og} - r_{od}\right)$$

r_{og} and r_{od} are the equilibrium bond lengths in graphite and diamond respectively. The reason for using the Keating potential is that it is highly efficient and may be used alone in the harmonic regions, while matching smoothly to the anharmonic regions via the bond strength dependence. The value of β is interpolated between β_d and β_g, which are adjusted to give C_{44} in the diamond structure and C_{66} in the graphite structure, respectively.

$$\beta^i = \beta_d + f(r_\pi^i, r_{\pi1}, r_{\pi2})\,(\beta_g - \beta_d)$$

So far, then, by including more information about a bond, viz., r, r_σ and r_π we can identify the local environment of an atom in a rotationally invariant manner, interpolate between the two important covalent structures (diamond and graphite) and obtain the correct anisotropic harmonic response around these structures. (If necessary, sp bonding could be included, corresponding to $r_\pi = r_\sigma \gg \sqrt{2r}\,/\,2$).

Two final refinements allow stacking faults and graphite's C_{33}, C_{13} and C_{44} elastic constants to be reproduced. Firstly an intrinsic stacking fault gives rise to a small, distant face, for which, in the case of diamond silicon, $8r_\sigma/r^3$ is 0.074 A^{-2}. In silicon graphite[7] one type of atom has two sides with $8r_\sigma/r^3 = 0.060$ A^{-2} and the other six sides of the former. A repulsive function, E_r, which gives the diamond intrinsic stacking fault energy (60 mJ m^{-2}) and which makes AA stacking reasonably unstable with respect to AB graphite is

$$E_r = E_{ro}\, f(r, r_3, r_4)\, f(p, p_1, p_2)\, (1 - f(p, p_3, p_4))$$

where $p = 8r_\sigma/r^3$.

The function only cuts in at large distances ($r_3 = 3.2$ A) and its maximum value is small ($E_{ro} = 6$ meV), so it should only be important in nearly perfect tetrahedral or graphite environments. It contributes to C_{44} in graphite, but also causes a slight force that tends to contract the basal plane (through the r_σ dependence). To correct this, the parameters A_g and B_g should be adjusted, but this has not yet been done. The second refinement allows the crude switch representing the flipping from sp^2 to sp^3 bonds, $f(r_\pi, r_{\pi1}, r_{\pi2})$, to be augmented to give first and second differentials w.r.t. r_π at its graphite value (C/4). The new function,

$$g\left(r_\pi, r_{\pi1}, r_{\pi2}, \left(\frac{dg}{dr_\pi}\right)_{\frac{c}{4}}, \left(\frac{d^2g}{dr_\pi^2}\right)_{\frac{c}{4}}\right)$$

then replaces f multiplying B in equation 2. The new terms (first one positive, second negative) give rise to a repulsion between layers, as does the term E_r. The repulsive force is counteracted by an attractive potential E_a

$$E_a = E_{ao} \, f \, (r_\pi, r_{\pi 1}, r_{\pi 2}) \, (r_\pi - \frac{c}{4}) \frac{r}{2} \, f \, (\theta, o, \theta_2)$$

The origin of E_a might be ascribed to the exchange interaction between π bonding charge clouds.

Thus,

$$E_{ao}, \; \left(\frac{dg}{dr_\pi} \right)_{\frac{c}{4}}, \; \left(\frac{d^2 g}{dr_\pi^2} \right)_{\frac{c}{4}}, \; E_{ro}, P_1, P_2, P_3, P_4$$

can all be loosely fitted to C_{13}, C_{33}, C_{44} and to stacking fault energies and to make the force in the c-direction in graphite zero. These refinements have not been framed with the possibility of $r_\pi \gg C/4$ in mind, but it would be easy to regularize their behavior in this case (and to fit the silicon graphite interlayer bonding energy, if one were known).

Other crystal structures such as sc, fcc and bcc, are higher in energy per atom than diamond silicon. Only ideal defects have been tested so far: ideal vacancy 4.7 eV, split vacancy, 6.2 eV, hexagonal interstitial 6.8 eV, tetrahedral interstitial 13.4 eV and ideal concerted exchange saddle point 7.6 eV.

It can be seen that the behavior of the potential is promising, but its parameters at the moment are under-determined. More *ab initio* calculations are required for more detailed fitting. The table below gives the current set of parameters, which are quoted as indicators of the performance of the potential, rather than a final set. It should be noted that the potential has been framed to completely describe diamond and graphite structures. Since the values of elastic constants of silicon graphite are not known, some silicon graphite parameters have been either equated to the diamond silicon values or to zero.

APPENDIX A

The face elongation, r_π, can be gauged by constructing lines parallel to the edges displaced inwards by r_σ. The line parallel to the closest edge then includes the internuclear intercept and is cut by the other lines. The shortest distances from the internuclear intercept to these cuts in both directions are defined as Δ_1 and Δ_2. r_π will normally be dominated by the smaller and is defined by

$$r_\pi = \left(\frac{1}{2} \Delta_1^{-5} + \frac{1}{2} \Delta_2^{-5} \right)^{-1/5} + r_\sigma$$

There are some situations where this definition leads to abrupt changes in r_π, for example when there are two almost (but not quite) parallel edges about r_σ from the internuclear intercept. Even though the edges might be very long, $r_\pi = r_\sigma$. A smoother variation in r_π can be had by displacing the non-closest edge lines by $3/2\ r_\sigma - r_\sigma\ |\cos\phi|$ instead of r_σ, where ϕ is the angle between the edge and the closest edge (for $|\phi| = 0$ to $60°$).

Finally, when more than one edge is about the same distance from the internuclear intercept, then a length-weighted mean of all such edges is taken. If this is not done, the elastic properties of graphite cannot be reproduced.

Table 1.   Silicon Parameters

	A(eV)	B(eV)	$\lambda_1(\text{Å}^{-1})$	$\lambda_2(\text{Å}^{-1})$	θ_1	θ_2	β
Diamond	3254.69	87.975	3.2394	1.32345	0.8	1.0	0.2477
Graphite	3254.69	138.791	3.2394	1.49184	0.8	1.0	0.2477

	r_0 (Å)	Ec(eV)
Diamond	2.3518	-4.63
Graphite	2.249	-3.92

r_1(Å)	r_2(Å)	r_3(Å)	r_4(Å)	$p_1(\text{Å}^{-2})$	$p_2(\text{Å}^{-2})$	$p_3(\text{Å}^{-2})$	$p_4(\text{Å}^{-2})$
2.8	3.2	3.2	4.0	0.005	0.04	0.2	0.3

$r_{\pi 1}$ (Å)	$r_{\pi 2}$ (Å)	$\left(\dfrac{dg}{dr_\pi}\right)_{\frac{c}{4}} (\text{Å}^{-1})$	$\left(\dfrac{d^2g}{dr_\pi^2}\right)_{\frac{c}{4}} (\text{Å}^{-2})$	E_{ao} (eVÅ^{-2})	E_{ro}(eV)
1.7	1.9	0.	0.	0.002757	0.006

REFERENCES

1.   J. Tersoff, Phys. Rev. Lett., 1986, 56:632.
2.   R. Biswas and D. Hamann, Phys. Rev. Lett., 1985, 55:2001.
3.   J. Tersoff, Phys. Rev. B, 1987, 37:6991.
4.   R. Biswas and D. Hamann, Phys. Rev. B, 1987, 36:6434.
5.   M. Heggie, Phil. Mag., 1988, 58:75.
6.   P.N. Keating, Phys. Rev., 1966, 145:637.
7.   M.T. Yin and M.L. Cohen, Phys. Rev. B,, 1984, 29:6996.
8.   J.S. Kasper and S.M. Richards, Acta Cryst., 1964, 17:752.

TRANSFERABLE TIGHT-BINDING MODELS FROM DENSITY FUNCTIONAL THEORY

W. Matthew C. Foulkes[*]

TCM, Cavendish Laboratory
Madingley Road
Cambridge CB3 0HE
United Kingdom

INTRODUCTION

Simple two center tight-binding models have been used to describe the quantum mechanics of real solids ever since the pioneering work of Slater and Koster[1] in 1954. They viewed the tight-binding method as an empirical fitting and interpolation scheme and used it to infer electronic energy levels throughout k-space from the values at those few points where they could be calculated properly. Nowadays, there is no need to interpolate in such a manner, but the tight-binding method has outlived its original purpose and is now mainly used to calculate interatomic forces and energy differences.[2] Despite this change of emphasis, however, most tight-binding calculations are still empirical, with parameters fitted to bandstructures, elastic constants, vacancy formation energies, etc.

Although empirical tight-binding models often work very well, they are far from reliable and sets of parameters which look sensible can sometimes give results which are just plain wrong.[3,4] Combine this with the advent of density functional theory[5,6] (which has enabled accurate parameter free calculations of interatomic forces in many materials) and it becomes clear that there is a need to put the tight-binding approach on a firmer theoretical footing. In particular, we would like to find ways of calculating all the parameters and so avoid having to fit them to experiment. Once the relatonship between the tight-binding approach and, say, the full self-consistent density functional approach has been elucidated, we will have a better idea of when to trust tight-binding calculations and of the errors involved in them.

The past two or three years have seen several groups address this problem (Harris,[7] Foulkes, [8,9] Foulkes and Haydock,[10] Sutton et al.[11]; see also Andersen[12] for a different

---

[*] Current Address: AT&T Bell Laboratories, 600 Mountain Avenue, Murray Hill, NJ 07974-2070, USA.

approach to the same questions) and considerable progress has been made. Here I summarize the advances and some of the difficulties still to be overcome and discuss some encouraging preliminary results.

BASICS AND QUESTIONS

In a typical empirical tight-binding calculation, the total energy of a collection of atoms at positions $\vec{R}_\alpha$ is given by,

$$E = \sum_{\substack{i \text{ occupied}}} \varepsilon_i + \sum_{\substack{\text{pairs} \\ \alpha\beta}} U_{\alpha\beta} \, (\, |\vec{R}_\alpha - \vec{R}_\beta| \,), \tag{1}$$

where $U_{\alpha\beta}$ is a repulsive classical pair potential and the quantities, ε_i, are the eigenvalues of an independent electron Schrödinger-like equation of the form,

$$(-\frac{1}{2} \nabla^2 + V(r)) \, \psi_i \, (r) = \varepsilon_i \, \psi_i(r) \tag{2}$$

This is imagined as being solved variationally in a basis of localized atomic-like (or Wannier) orbitals, ϕ_μ, and so becomes a secular equation,

$$\det (H_{\mu\nu} - \varepsilon S_{\mu\nu}) = 0 \tag{3}$$

where

$$H_{\mu\nu} = <\phi_\mu \, | - \frac{1}{2} \nabla^2 + V(r) \, | \, \phi_\nu > \quad \text{and} \quad S_{\mu\nu} = <\phi_\mu \, | \, \phi_\nu> \tag{4}$$

The matrix elements $H_{\mu\nu}$ and $S_{\mu\nu}$, and the pair potential, $U_{\alpha\beta}$, are usually treated as empirical parameters and are fitted to experiment or to the results of other calculations.

This is to be compared with the procedure followed in a fully self-consistent density functional calculation.[5,6] There the ground state density is found by solving an independent particle Schrödinger-like equation, the Kohn-Sham equation,

$$(-\frac{1}{2} \nabla^2 + v_{KS} \, [n(r), r]) \, \psi_i(r) = \varepsilon_i \psi_i \, (r) \tag{5}$$

where the density, $n(r)$, is related to the eigenfunctions via,

$$n(r) = \sum_{\substack{i \text{ occupied}}} \psi_i^* \, (r) \, \psi_i \, (r), \tag{5}$$

and the Kohn-Sham potential, $v_{KS}[n(r), r]$, has the form,

$$v_{KS}[n(r), r] = V_{nuc}(r) + \int \frac{n(r')}{|r-r'|} \, d^3r' + \mu_{xc} \, [n(r), r]. \tag{7}$$

In this equation, $V_{nuc}(r)$ is the Coulomb potential due to the nuclear charges and μ_{xc} is the first functional derivative of the exchange and correlation (XC) energy functional, $E_{xc}[n(r)]$. This is not known exactly but is usually well represented by a local density approximation[13] (LDA). Since the Kohn-Sham potential depends on the density it generates, equation (5) is a "self-consistent" equation and must be solved by iteration from some starting density.

Once the Kohn-Sham equations have been iterated to self-consistency and the ground state density, $n_o(r)$, has been found, the density functional prescription for the ground state energy is,

$$E_o = \sum_{i\ occupied} \varepsilon_i - \frac{1}{2} \iint \frac{n_o(r)\ n_o(r')}{|r - r'|} + E_{xc}[n_o(r)]$$

$$- \int \mu_{xc}[n_o(r), r]\ n_o(r) + E_{nn}, \tag{8}$$

where E_{nn} is the nuclear-nuclear Coulomb repulsion energy. We will call the first term on the right hand side the electronic eigenvalue sum and the rest of the terms the double counting energy from now on.

We are now in a position to compare empirical tight-binding (ETB) and density functional theory (DFT). Equations (1) and (2) clearly resemble equations (5) and (8) but the details of the relationship are not clear. Why is equation (2) not self-consistent? What is the correct potential, $V(r)$, for use in equation (2)? And why can one sensibly represent the complicated double counting terms in equation (8) by the simple pair potential in equation (1)? Once these questions have been answered, we can go on to consider the reduction of equation (2) to the secular equation (3). What are the appropriate basis functions to use? Can all the matrix elements be assumed short-ranged, transferable and pairwise? What about the complicated three center integrals involved in evaluating the Hamiltonian matrix elements? And can one sensibly choose orthonormal basis functions so that the overlap matrix, S, reduces to the identity as is usually assumed?

Although I will briefly address the second set of questions later on, this article will concentrate for the most part on the questions concerning the forms of equations (1) and (2). It will turn out that they can best be understood as giving a stationary approximation to the ground state energy as calculated using DFT. With this in mind, I now introduce a rather unusual[14,7,8,9] stationary principle in density functional theory.

A STATIONARY PRINCIPLE IN DENSITY FUNCTIONAL THEORY

The standard variational principle in DFT can be formulated as follows:

Given a guess, $v_{in}(r)$, at the self-consistent Kohn-Sham potential, and a single (i.e. non-self-consistent) solution of equation (5) producing an output density, $n_{out}(r)$, and eigenvalues, ε_i, then a variational expression for the total energy is,

$$E[n_{out}] = \sum_{i \, occ} \varepsilon_i - \int v_{in} \, n_{out} + E_{nn}$$

$$+ \int V_{nuc} \, n_{out} + \frac{1}{2} \iint \frac{n_{out}(r) \, n_{out}(r')}{|r - r'|} + E_{xc}[n_{out}]. \tag{9}$$

This expression is variational in the difference between n_{out} and n_0 and so,

$$E[n_{out}] = E_o + O((n_{out} - n_o)^2), \tag{10}$$

where the second order term is greater than or equal to zero.

It was noticed, originally by Wendel and Martin,[14] and later independently by Harris[7] and Foulkes,[8,9] that in the special case where $v_{in}(r)$ is constructed from a trial density, $n_{in}(r)$, according to,

$$v_{in}(r) = v_{KS}[n_{in}(r), r],$$

then equation (9) (after some algebra, and remembering that $\mu_{xc} = \delta E_{xc}/\delta n$) reduces to

$$E[n_{out}] = \sum \varepsilon_i - \frac{1}{2} \iint \frac{n_{in}(r) \, n_{in}(r')}{|r - r'|} - \int \mu_{xc}[n_{in}] \, n_{in} + E_{xc}[n_{in}]$$

$$+ E_{nn} + O[(n_{out} - n_{in})^2]. \tag{11}$$

Hence, if we define a new functional, $\tilde{E}[n_{in}]$, via,

$$\tilde{E}[n_{in}] = \sum \varepsilon_i - \frac{1}{2} \iint \frac{n_{in}(r) \, n_{in}(r')}{|r - r'|} - \int \mu_{xc}[n_{in}] \, n_{in}$$

$$+ E_{xc}[n_{in}] + E_{nn}, \tag{12}$$

then it too is stationary about the self-consistent density ($n_{in} = n_{out} = n_0$). Because of the extra second order terms in equation (11), the new functional is not strictly variational about $n_0(r)$ and so may be either greater than or less than E_0; nevertheless, it is still stationary and all the errors are second order in small (we hope) quantities.

In fact, a more sophisticated derivation[10] shows that,

$$\tilde{E}[n_{in}] = E_o + \frac{1}{2} \iint \frac{\delta^2 E}{\delta n^2}\Big|_{n_o} (n_{in} - n_o)(n_{out} - n_o) + \begin{bmatrix} \text{higher order} \\ \text{terms} \end{bmatrix}, \tag{13}$$

where $\delta^2 E/\delta n^2|_{n_o}$ is the second functional derivative of the orginal Hohenberg-Kohn energy functional evaluated at the ground state density. The mixed nature of the second order term (involving both n_{in} and n_{out}) shows clearly why the new functional, although stationary, is not strictly variational.

Note that an evaluation of a stationary estimate of the energy using equation (12) requires neither the output wavefunctions, nor the output density, nor any self-consistent looping. All these features are shared by ETB calculations and so the relationship between ETB and DFT is already becoming clearer. Note also that both the one electron potential used and the double counting terms which must be evaluated depend only on n_{in}; if n_{in} is simple, so will they be.

It has now been realized[9,10] that the functional presented above, although it is the one which will be used here, is only a special case of the more general stationary functional,

$$E_{stat}[n_{in}, v_{in}] = \sum_i \epsilon_i - \int v_{in} \, n_{in}$$

$$+ \int V_{nuc} n_{in} + \frac{1}{2} \int\int \frac{n_{in}(r) \, n_{in}(r')}{|r - r'|} + E_{xc}[n_{in}] + E_{nn}. \qquad (14)$$

Here the eigenvalues are the solutions of

$$(-\frac{1}{2} \nabla^2 + v_{in}) \, \psi_i = \epsilon_i \, \psi_i \,, \qquad (15)$$

but there is not necessarily any relation between v_{in} and n_{in}. The second order error terms now involve both $n_{in} - n_o$ and $v_{in} - v_o$ and these must both be "small" if the energy estimate obtained is to be a good one. If we insist that v_{in} be equal to $v_{KS}[n_{in}(r), r]$, then equation (14) reduces to equation (12).

PARAMETER FREE TIGHT-BINDING MODELS

To make full use of the stationary functional, equation (12), we must use an input density of a particularly simple form, and from now on we will always choose a superposition of spherical atomic-like densities. It will turn out that the stationary principle acts efficiently and that this will be adequate in the cases we will consider here. However, one can imagine cases (highly ionic solids, for example, although see the work of Polatoglou and Methfessel[15] on NaCl) where superposing atomic densities would not be sensible and some other physically motivated choice should be made. Whether it will always prove possible to choose reasonably simple input densities and yet still obtain reliable energy estimates remains to be seen, but the results of work done so far have been encouraging.

For superposed spherical atomic-like densities, the double counting terms in equation (12) are particularly simple: the Coulomb (Hartree and nuclear-nuclear) parts are exactly a pair potential in form; and the XC contributions, although not strictly pairwise, are close enough to a pair potential that the non-pairwise bits can almost always be neglected. So the double counting terms can all be replaced by a simple pair potential which can be calculated for a dimer (a matter of evaluating a few integrals) and then used in any situation where superposing atoms gives an adequate input density.

The choice of the superposed atomic input density also simplifies the one electron potential, of course, but the exact forms of the Hamiltonian matrix elements also depend on the basis set. The choice of basis is independent of any approximations made so far and anything from plane waves to LMTO's could be used. However, since our interest lies in understanding ETB, we will choose localized atomic-like orbitals here. The evaluation of the Hamiltonian and overlap matrix elements (the tight-binding parameters) is again a matter of doing some integrals but not all of these are easy. There are complicated three center integrals needed in the evaluation of the Hamiltonian matrix and questions about when one can ignore them and make the usual two center approximation are thorny and will not be addressed further here. At any rate, all the matrix elements are "local" (in that any given matrix element depends on the positions of only a few atoms) and transferable by construction in all cases where superposing atomic-like densities is adequate. As originally intended, therefore, we now have prescriptions for calculating all the elements of a parameter free tight-binding model which gives a stationary estimate of the ground state energy.

PRELIMINARY APPLICATIONS AND CONCLUSIONS

Several groups have tested the use of the stationary functional. Harris[7] and Foulkes[8,9] used atomic-like orbitals but avoided the three center integral problem by looking only at dimers. Polatoglou and Methfessel[15] used the more sophisticated LMTO bandstructure method to study some bulk solids. All the results so far have been encouraging: interatomic spacings, cohesive energies and vibrational frequencies all differing from the accurate LDA values by only a few percent when superposed atomic input densities are used. In addition, the dimer calculations suggest that very simple basis sets, consisting of just a few atomic-like functions on each atom, may well prove sufficient. The most extreme tests so far have been H_2[9,10] and solid NaCl.[15] In the first case, a superposition of atomic densities is roughly 40% too low in the center of the bond; and since NaCl is usually thought of as ionic, one would not expect that superposing neutral atomic densities would give sensible answers. The stationary functional works well in both cases, however.

In conclusion, the stationary functional has been tested using superposed atomic densities and has performed surprisingly well in a variety of different cases. Simple basis sets proved adequate in dimers and so, if they perform similarly well in solids, it may be possible to generate reliable parameter free tight-binding models for many solids. The evaluation (or approximation) of the three center integrals may prove troublesome, but this is a technical problem which should be overcome in time.

REFERENCES

1. J.C. Slater and G.F. Koster, Phys. Rev. 94, 1498 (1954).
2. See, for example, Solid State Physics 35, H. Ehrenreich and D. Turnbull, eds., Academic, New York (1980).
3. D.J. Chadi, Phys. Rev. Lett. 43, 43 (1979).
4. K.C. Pandey, in "Proceedings of the 17th International Conference on the Physics of Semiconductors", Springer, New York (1985).
5. P. Hohenberg and W. Kohn, Phys. Rev. 136, 864 (1964). W. Kohn and L.J. Sham, Phys. Rev. 140, 1133 (1965).
6. C.J. Callaway and N.H. March in Solid State Physics 38, H. Ehrenreich and D. Turnbull eds., Academic, New York (1984).
7. J. Harris, Phys. Rev. B 31, 1770 (1985).
8. W.M.C. Foulkes, Fellowship Dissertation (unpublished, 1985).
9. W.M.C. Foulkes, Ph.D. Thesis, "Interatmic Forces in Solids", University of Cambridge (1988).
10. W.M.C. Foulkes and R. Haydock, submitted to Phys. Rev. B (1988).
11. A.P. Sutton, M.W. Finnis, D.G. Pettifor and Y. Ohta, J. Phys. C 21, 35 (1988).
12. O.K. Andersen, O. Jepsen and M. Sob in "Electronic Band Structure and Its Applications", M. Yussouf, ed., Springer (1987).
13. J. Perdew and A. Zunger, Phys. Rev. 23, 5048 (1981).
14. H. Wendel and R.M. Martin, Phys. Rev. B 19, 5251 (1979).
15. H.M. Polatoglou and M. Methfessel, Phys. Rev. B 37, 0403 (1988).

STABILITY OF THE (110) FACE IN NOBLE METALS ANALYZED WITHIN

A TIGHT-BINDING SCHEME

Bernard Legrand and Michel Guillopé

Centre d'Etudes Nucléaires de Saclay
Section de Recherches de Métallurgie Physique
91191 Gif sur Yvette, Cedex
France

INTRODUCTION

The structure of the (110) surface in f.c.c. transition and noble metals has given rise to a huge amount of studies through the last ten years, both from the experimental [1-13] and theoretical points of view [14-23]. The 5d f.c.c. metals (Ir, Pt, Au) show a (1x2) structure while the 3d (Ni, Cu) and 4d (Rh, Pd, Ag) metals keep the (1x1) structure. There is now general agreement from a great variety of experiments that the (1x2) structure is a missing-row geometry according to which every second of the [1-10] rows are missing. Moreover, relaxations in deeper layer (contraction of the first interlayer distance, pairing in the second layer and buckling in the third layer) have been exhibited experimentally [2-4]. If the structure at low temperature is well established for all these metals, the evolution of the structure with the temperature remains questionable. An order-disorder transition is observed in LEED experiments in Au [1,24-26] at $T_c \sim 660K$ and less clearly in Pt [27] between the (1x2) low temperature phase and a (1x1) disordered phase for $T > T_c$. Large amounts of disorder has been reported for Cu (110) [28,29] Pd (110) [30,31] and Ag (110) [32] and interpreted as due to the creation of thermal defects [28], uncorrelated displacements [30,31] or roughening transition [29,32].

In this work we present a theoretical study of the stability of f.c.c. (110) surfaces in noble metals. Based on a simple tight-binding scheme recently developed [33], we predict the reconstruction to the (1x2) missing-row structure and the occurrence of an order-disorder transition on Au (110), and the non-reconstruction for Cu (110) and Ag (110). In these last cases, however, we show the existence of an order-disorder transition between the (1x1) ideal structure and a (1x1) disordered structure, similar to the high temperature phase of Au (110). More details on the parameters used in this work are reported in ref. [34].

THEORETICAL METHOD

In the simple tight-binding scheme used for noble metals, the total cohesive energy is written for an atom i

$$E_c^i = -(E_R^i + E_b^i) \tag{1}$$

where E_R^i is a repulsive energy given by

$$E_R^i = \sum_j A \exp\left[-p\left(\frac{d_{ij}}{d_o} - 1\right)\right] \tag{2}$$

d_{ij} is the distance between atoms i and j, and d_o is the first neighbor distance, E_b^i is a band energy which is expressed as:

$$E_b^i = -\left[\sum_j \xi^2 \exp - 2q\left(\frac{d_{ij}}{d_o} - 1\right)\right]^{1/2} \tag{3}$$

In noble metals the band energy is not directly due to the formation of the d-band, which is full but is related to the shift of this d-band. This shift comes from the on-site term of the tight-binding Hamiltonian

$$H = \sum_{i,\lambda} |i,\lambda\rangle \varepsilon_i^\lambda \langle i,\lambda| + \sum_{i\neq j,\lambda,\mu} |i,\lambda\rangle \beta_{ij}^{\lambda\mu} \langle j,\mu| \tag{4}$$

where i and j are lattice sites, λ and μ are orbital labels.

Andersen [35] has shown that ε_i^λ depends on the local environment and his work suggests the following dependence:

$$\varepsilon_i^\lambda = \alpha \left(\sum_{j\neq i,\mu} \left(\beta_{ij}^{\lambda\mu}\right)^2\right)^{1/2} \tag{5}$$

where

$$\beta_{ij}^{\lambda\mu} = \beta_o^{\lambda\mu} \exp\left[-q\left(\frac{d_{ij}}{d_o} - 1\right)\right]$$

are the hopping integrals, and α is a constant.

When the d-band is full, and taking into account the local charge neutrality which is achieved by a correction of the on-site term $\delta\varepsilon_i$, it is easy to show [36] that the band energy:

$$E_b^i = \int\limits^{E_F} E\, n_i\, (E,\, \varepsilon_i,\, \delta\varepsilon_i)\, dE - N_d\, \delta\varepsilon_i \qquad (6)$$

can be written as (3), where ξ is proportional to $\alpha N_d \beta_0$. E_F is the Fermi level and N_d, the number of d-electrons:

$$N_d = \int\limits^{E_F} n_i\, (E,\, \varepsilon_i,\, \delta\varepsilon_i)\, dE = 10 \qquad (7)$$

Notice that the formulation of the total energy (1), (2), (3) is identical to the Embedded-Atom-Method [37] (E.A.M.) and to the effective medium theory [38] (E.M.T.).

When the d-band is not full, we have to take into account the details of the local density of states $n_i(E)$, for instance, with the recursion method [39]. If we use the second moment approximation [40], we obtain once more again equation (3) for E_b^i and we can conclude that E.A.M. is nothing but a more parametrized version of the second moment approximation. Using a better description of $n_i(E)$ (with more moments), we obtain the same result as the E.M.T. (eq. (1) of Ref. [21]). From a practical point of view the parameters (A, p, q, ξ) are determined [34] by fitting experimental cohesive energy, lattice parameter and elastic constants. The summations over j in relations (2) and (3) are extended up to the third neighbors.

RESULTS

In order to obtain the relaxed configurations, we use a quasi-dynamical technique which consists of a Molecular Dynamics calculation including a damping force [41]; this leads to a minimization of the potential energy at $T = 0$ K. The system is a slab of twenty (110) layers with periodic boundary conditions along the [001] and [1-10] directions parallel to the surface.

The reconstruction energy, E_R, defined as the difference of energy betweeen the unreconstructed (1x1) surface and the (1x2) missing row structure is presented in Table I both for unrelaxed and relaxed geometry.

Table I. Unrelaxed (unrel.) and relaxed (rel.) reconstruction energy E_R, in erg cm^2.

	Cu	Ag	Au
Unrel.	35	0.3	-35
Rel.	33	0.7	-23

Our model predicts the observed trend down the column of periodic table: Au should reconstruct, whereas Cu and Ag should not. This can be understood, using the following analytical expression of E_R for the unrelaxed case (which is not greatly different from the relaxed case, see Table I).

$$E_R = -A (r_2 + 2r_3) + \frac{\xi}{\sqrt{\alpha_1}} (-\alpha_1 b_1 + \alpha_2 b_2 + \alpha_3 b_3) \qquad (8)$$

with

$$b_1 = \sqrt{9} - \frac{1}{2}(\sqrt{7} + \sqrt{11}), \ b_2 = -\frac{2}{\sqrt{9}} + \frac{3}{2\sqrt{7}} + \frac{1}{\sqrt{11}}, \ b_3 = \frac{8}{\sqrt{9}} + \frac{9}{2\sqrt{7}} + \frac{4}{\sqrt{11}} + \frac{1}{4\sqrt{3}}$$

$$\alpha_i = \exp\left[-2q\left(\frac{d_i}{d_o} - 1\right)\right], \ r_i = \exp\left[-p\left(\frac{d_i}{d_o} - 1\right)\right]$$

and

d_i: distance to the ith neighbors.

The sign of E_R comes from a balance between terms which favor the reconstruction (the repulsive term and the first neighbors band term) and terms which work against (the second and third neighbors band term). This allows us to define a critical value of $q_c = 3.1$ (for which $E_R = 0$), almost independent of p, for reasonable values (p > 9); this q dependence of E_R explains the relative stability of the (1x1) and (1x2) structure for Cu (q = 2.43), Ag (q = 3.05) and Au (q = 4.30) [34].

Relaxed geometry [34] for the missing row structure is very similar to E.A.M. results [18] and compares fairly well with experiments.

To study the high temperature phase, we define _effective_ multiplet interactions between atoms on the surface. These interactions have been estimated by calculating the difference of energy between relaxed configurations involving either the considered multiplet or the equivalent number of isolated adatoms (an analytical expression can be obtained in the unrelaxed case [34]). We find that three pair interactions (V_1, V_2, V_3) and two triplet interactions (T_1, T_2) are sufficient to recover the exact energy of any configurations (see Fig. 1 and Table II).

In particular, E_R can be expressed as:

$$E_R = -\frac{1}{2} (V_2 + 2V_3 + 4T_2) \qquad (9)$$

This decomposition of E_R clearly indicates that the stability of the (1x2) structure is closely related to the repulsion between second and third neighbors (V_2 and $V_3 > 0$). As for E_R, the sign of the _effective_ pair interactions $V_2(V_3)$ comes from a delicate balance between a direct band attraction (contribution of the second and third neighbors to the band term,

364

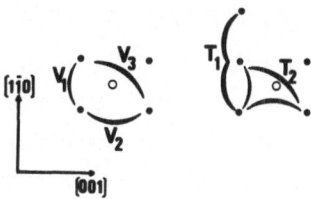

Figure 1. Effective multiplet interactions V_1, V_2, V_3, T_1, T_2.

Table II. Effective multiplet interactions V_i, T_i in relaxed configurations in eV.

	V_1	V_2	V_3	T_1	T_2
Cu	- 0.260	- 0.037	- 0.001	0.014	0.000
Ag	- 0.217	- 0.010	0.007	0.014	- 0.001
Au	- 0.176	0.019	0.017	0.016	- 0.005

proportional to α_2 and α_3), and a repulsion which involves a direct term (contribution of the second (third) neighbors to the repulsive term, proportional to r_2 (r_3)) and an indirect term via the substrate (contribution mainly due to the first neighbors band term, proportional to α_1). As for E_R, we can obtain critical q values for V_2 and V_3. Notice that Ag corresponds to a singular case, where:

$$q_c (V_3) < q_{Ag} < q_c (E_R) < q_c (V_2) \tag{10}$$

Our various multiplet interactions defined above forbid an analytical solution of the associated 2D-Ising model. We have performed Monte-Carlo (M.C.) simulations in the grand canonical ensemble on a 2D lattice gas with all the multiplet interactions V_1, V_2, V_3, T_1, and T_2 (the chosen value of the chemical potential corresponds to a zero field in the equivalent Ising spin model). The values of the critical temperatures determined from these M.C. simulations are reported in Table III.

Table III. Critical temperature in Kelvin for the order-disorder transition.

	Cu	Ag	Au
T_c	725	245	540

For Au, our calculated value Tc = 540K agrees fairly well with the experimental value Tc = 660K [24-26] and is comparable with the E.A.M. result Tc = 570 K [19]. For Cu and Ag we predict an order-disorder transition between the (1x1) ideal unreconstructed structure and a (1x1) disordered structure which is similar to the Au (110) disordered phase. This prediction can rationalize experimental indications for the existence of a large structural disorder for Cu (110) at T = 573 K [28] or T = 873 K [29], for Ag (110) at T = 723 K [32] and for Pd (110) at T = 240 K [30,31] (for Pd, a preliminary theoretical study indicates a behavior similar to Ag).

The competition between the order-disorder transition (T_c) and the roughening transition (T_R) has been studied recently within a statistical approach [42-43]. It has been concluded that T_R/T_c is between 1.10 [42] and 1.14 [43]. Then we suggest that the experimental observations on Cu and Ag (and more generally on all the other unreconstructed transition metals Ni, Rh and Pd) should be reinterpreted taking into account the existence of this order-disorder transition.

CONCLUSION

This study shows that the tight-binding scheme is a very valuable tool to study various properties in noble metals. Its analytical character is a great advantage over the E.A.M. Moreover, it allows us to study defects such as dislocation or surfaces in all the transition metals with the recursion method to reconstruct the electronic density of states [44, 45].

ACKNOWLEDGMENTS

We are grateful to A. Finel, G. Tréglia and F. Ducastelle for useful discussions. We are particularly indebted to A. Finel for performing the Monte-Carlo simulations.

REFERENCES

1.   D. Wolf, H. Jagodzinski and W. Moritz, Surf. Sci. 77:2654 (1978).
2.   W. Moritz and D. Wolf, Surf. Sci. 163: L655 (1985).
3.   C.M. Chan and M.A. Van Hove, Surf. Sci. 171:226 (1986).
4.   T. Gustafsson, M. Coppel and P. Fenter, in: "The Structure of Surfaces II", J.F. Van der Veen and M.A. Van Hove, eds., Springer-Verlag, 1988 p. 110.
5.   T. Hasegawa, N. Ikarashi, K. Kobayashi, K. Takayanagi and K. Yagi, in: "The Structure of Surfaces", J.F. Van der Veen and M.A. Van Hove, Springer Verlag, 1988, p. 43.
6.   L.D. Marks, Phys. Rev. Lett. 51:1000 (1983).
7.   I.K. Robinson, Phys. Rev. Lett. 50:1145 (1983).
8.   I.K. Robinson, Y. Kuk and L.C. Feldman, Phys. Rev. B, 29:4726 (1984).
9.   G. Binnig, H. Rohrer, Ch. Gerber and E. Weibel, Surf. Sci. 131:L379 (1983).
10.   G.L. Kellog, Phys. Rev. Lett. 55:2168 (1985).
11.   Q. Gao and T.T. Tsong, Phys. Rev. Lett. 57:452 (1986).
12.   K. Müller, J. Witt and O. Schütz, J. Vac. Sci. Tech. A5:757 (1987).
13.   H. Derks, J. Möller and W. Heiland, Surf. Sci. 188:L685 (1987).
14.   D. Tomanek, H.J. Brocksch and H. Bennemann, Surf. Sci., 138:L129 (1984).
15.   D. Tomanek, Phys. Lett. 113A:445 (1986).
16.   T. Halicioglu, T. Takai and w.A. Tiller, in: "The Structure of Surfaces",

M.A. Van Hove and S.Y. Tong, eds., Springer Verlag, 1985, p. 231.

17. M.S. Daw, Surf. Sci. 166:L161 (1986).

18. S.M. Foiles, Surf. Sci. 191:L779 (1987).

19. M.S. Daw and S.M. Foiles, Phys. Rev. Lett. 59:2756 (1987)

20. M. Garofalo, E. Tosatti and F. Ercolessi, Surf. Sci. 188:321 (1987).

21. K.W. Jacobsen and J.K. Norskov, in: "The Structure of Surfaces II", J.F. Van der Veen and M.A. Van Hove, eds., Springer Verlag, 1988, p. 118.

22. K.M. Ho and K.P. Bohnen, Europhys. Lett. 4:345 (1987) and K.M. Ho and K.P. Bohnen, Phys. Rev. Lett. 59:1833 (1987).

23. J.W. Davenport and M. Weinert, Phys. Rev. Lett. 58:1382 (1987).

24. J.C. Campuzano, M.S. Foster, G. Jennings, R.F. Willis and W. Unertl, Phys. Rev. Lett. 54:2684 (1985).

25. D.E. Clark, W. Unertl and P.H. Kleban, Phys. Rev. B, 34:4379 (1986).

26. P. Kleban, R. Hentschke and J.C. Campuzano, Phys. Rev. B, 37:5738 (1988).

27. M. Salmeron and G. Somorjai, Surf. Sci. 91:373 (1980).

28. D. Gorse and J. Lapujoulade, Surf. Sci. 162:847 (1985).

29. S.G.J. Mochrie, Phys. Rev. Lett. 59:304 (1987).

30. S.M. Francis and N.V. Richardson, Phys. Rev. B, 33:662 (1986).

31. M. Wolf, A. Goschnick, J. Loboda-Cackovic, M. Grunze, W.N. Unertl, and J.H. Block, Surf. Sci. 182:489 (1987).

32. G.A. Held, J.L. Jordan-Sweet, P.M. Horn, A. Mak and R.J. Birjeneau, Phys. Rev. Lett. 59:2075 (1987).

33. V. Rosato, M. Guillopé and B. Legrand, Phil. Mag. (in press).

34. M. Guillopé and B. Legrand, Surf. Sci. (in press).

35. O.K. Andersen, O. Jepsen and D. Glötzel, in: "Proceedings of the Enrico Fermi International School for Physics" Course LXXXIX, F. Bassani, F. Fermi, and M.P. Tosi, eds., North-Holland, Amsterdam, 1985.

36. B. Legrand, (in preparation).

37. S.M. Foiles, M.I. Baskes and M.S. Daw, Phys. Rev. B, 33:7983 (1986).

38. K.W. Jacobsen, J.K. Norskov and M.J. Puska, Phys. Rev. B, 35:7423 (1987).

39. R. Haydock, in: "Solid State Physics", H. Ehrenreich, F. Seitz and D. Turnbull, eds., Academic Press, New York, 1980, Vol.. 35, p. 215.

40. F. Ducastelle, J. de Phys. 31:1055 (1970).

41. C.H. Bennett, in: "Diffusion in Solids, Recent Developments", A.S. Nowick and J.J. Burton, eds., Academic Press, New York, 1975, p. 73.

42. F. Rys, Phys. Rev. Lett. 56:624 (1986).

43. J. Villain and I. Vilfan, Surf. Sci. 199:165 (1988).

44. B. Legrand, Phil. Mag. B, 49:171 (1984) and B. Legrand, Phil. Mag. A, 52:83 (1985).

45. B. Legrand, G. Tréglia, M.C. Desjonquères and D. Spanjaard, J. Phys. C, 19:4463 (1986).

APPLICATION OF THE TIGHT-BINDING BOND MODEL

M.W. Finnis

Theoretical Physics Division
Harwell Laboratory
Oxfordshire OX11 ORA
England

INTRODUCTION

The tight-binding model has received a lot of attention recently as a model for the total energy of metals and semiconductors which can be used for the calculation of minimum energy structures. Chadi (1979a, 1979b, 1984) pioneered the use of k-space methods for atomistic relaxation in semi-conductors with the tight-binding model. The purpose of the present paper is to summarize the present state of the art in real-space, with emphasis on the d-bonded transition metals, the tight-binding description of which was originally introduced and developed by Friedel and coworkers.

I summarize the forms of the model which have been used for atomistic simulation with their advantages and disadvantages, without giving detailed derivations, for which the reader is referred to the literature. Experience in the simulation of defect structures has proved that the recursion method, a real-space approach to calculations, is practicable for calculating interatomic forces in semiconductors and transition metals and gives new insight into for example grain boundary and surface structures. I will discuss the case of a vacancy in some detail, because it is an instructive prototype defect for testing the concepts.

I discuss the second-moment approximation of tight binding models. On the assumption that the atoms are neutral, the second moment tight-binding bond model is very similar to the effective medium or embedded atom models and essentially no more expensive to compute than simple pairwise potentials. These models are state of the art for large scale atomistic simulations which with powerful computers can handle over a million atoms.

Atomic position in silicon $\Sigma = 3$ and $\Sigma = 5$ grain boundaries have been satisfactorily relaxed with the real space TBB model (see below) based on s and p orbitals by Paxton and Sutton (1988). At the same time the $\Sigma = 5$ boundary was relaxed by Kohyama, Yamamoto, Ebata and Kinoshita (1988) using Chadi's k-space method, with remarkably similar results: the translation state of the boundary, which is not symmetry dictated, agreed to within $0.001a$. However, our experience in the relaxation of atomic positions in transition metals has revealed some subtle difficulties which beset the application of tight-binding, two of which I will discuss, concerning the rotational invariance of the energy and the consistency of force and energy expressions. These difficulties have stimulated some new techniques in the calculation of forces, e.g., the application of matrix recursion, which greatly improve the outlook.

I conclude by describing how tight-binding suggests a simple empirical approach to the modelling of metal-ionic crystal interfaces. Preliminary work described here makes plausible a description of the metal as a discrete, classical perfect conductor in which image charges are induced by the ions.

TIGHT-BINDING MODELS FOR ATOMISTIC RELAXATION

Let us recall an equation for the total energy of a material which has been the starting point for a number of calculations, e.g., Allan (1970). The total binding energy in what we refer to as the Tight Binding Bond (TBB) model takes its simplest form for transition metals, namely

$$E_{tot} = E_{cov} + E_{rep}. \tag{1}$$

where the first term is the total covalent bonding energy given by

$$E_{cov} = \sum_i E_{cov}^i , \tag{2}$$

$$E_{cov}^i = \int^{\varepsilon_F} n^i(\varepsilon)(\varepsilon - \varepsilon_{id})d\varepsilon , \tag{3}$$

and the second is an empirical repulsive potential, commonly assumed to be of Born-Mayer form:

$$E_{rep} = \frac{1}{2} \sum_{ij} V(R_{ij}) , \tag{4}$$

where

$$V(R_{ij}) = A\exp(-pR_{ij}). \tag{5}$$

The notation is as follows. Atomic positions are denoted by i and j. ε_F is the Fermi energy. $n^i(\varepsilon)$ is the local density of states on atom i. ε_d is an atomic d-level. A number of assumptions have been made in deriving (1) from the expression for total energy of local density functional theory, and as these have been documented and referenced by Sutton, Finnis, Pettifor and Ohta (1988), I will not repeat them. Suffice it to say here that since the work of Harris (1985) and Foulkes (1987) this form of tight-binding energy is now rather better justified than was thought a few years ago. Nevertheless it still has empirical ingredients such as the Born-Mayer parameters. The subtraction of the self-energy term ε_{id}

in the bond energy is remarkable in its effect. It is not, as might be thought, simply the double counting correction to the total electronic band energy - that is entirely taken care of in the repulsive pairwise energy. It represents rather an additional double counting correction term associated with the variation of the charge density at a site, going beyond the rigid atom model of Harris (1985), and it subtracts the crystal field term at a site which is supposed to be included already in the pairwise term. Its remarkable effect is to enable the simple force formula below to be derived. It also makes the model internally consistent, as discussed by Pettifor (1987), and able to account for structural trends in alloys (Pettifor and Podloucky 1986, Pettifor 1987).

The total covalent *bonding* energy has been so called because it can be written in the alternative form:

$$E_{cov} = \frac{1}{2} \sum_{i \neq j} E_{bond}^{ij} , \qquad (6)$$

where the individual bond energy is given by

$$E_{bond}^{ij} = 2 \sum_{pq} H_{pq}^{ij} \rho_{pq}^{ij} . \qquad (7)$$

H_{pq}^{ij} is the matrix element of the Hamiltonian which is assumed to have exponential of R^{-5} distance dependence according to taste. The d-orbitals are labelled p,q, where p and q range from 1 to 5. Models to date have simply cut off the Hamiltonian at second or third neighbors, although if the atomic rearrangements were large during a simulation this would cause trouble when an atom enters or leaves the range of the Hamiltonian. The matrix ρ_{pq}^{ij} is the bond order and is given by

$$\rho_{pq}^{ij} = -\frac{2}{\pi} \text{Im} \int^{\varepsilon_F} G^{ipjq} (\varepsilon) \, d\varepsilon. \qquad (8)$$

The matrix element G^{ipjq} is a matrix element of the Greenian defined by

$$G = \lim_{\delta \to 0} (\varepsilon + i\delta - H)^{-1}. \qquad (9)$$

Much research has been done to find accurate and efficient methods of calculating these Green function matrix elements, which is where the significant computation time is spent in applications of the model. The TBB model in the shape of eqn. (3) requires the diagonal elements of the Green function in order to evaluate the local density of states, which is given by

$$n^i(\varepsilon) = -\frac{2}{\pi} \text{Im} G^{ipip} (\varepsilon). \qquad (10)$$

An especially attractive feature of the bond energy representation (7) is that the forces on the atoms are simply obtained by taking derivatives of the Hamiltonian matrix elements. For example it follows from (7) that

$$\frac{\partial E_{bond}^{ij}}{\partial x_i} = 2 \sum_{pq} \frac{\partial H_{pq}^{ij}}{\partial x_i} \rho_{pq}^{ij} \ . \tag{11}$$

Equation (11) is very convenient because by virtue of the variational principle and atomic charge neutrality no derivative of the bond orders or of the site energies enters the force. A formula of this type was originally derived by Coulson (1939) and similar expressions have been rederived several times since for solid state applications, e.g., Moraitis and Gautier (1979), Sutton *et al.* (1988).

On the other hand the direct use of (3) for atomistic simulation requires forces to be calculated by some form of numerical differencing. This has been done by several authors in situations where there are not too many degrees of freedom, e.g., to study surface reconstruction (Tréglia, Ducastelle and Spanjaard 1980, Terakura, Terakura and Hamada 1981, Tréglia, Desjonquères and Spanjaard 983, Masuda-Jindo, Hamada and Terakura 1984). The results are satisfactory when compared to LEED data. Legrand (1985) applied a tight-binding model to study the relaxation of the 1/3 <11$\bar{2}$0> dislocation in titanium, using an ingenious scheme for approximating the forces. The basic difficulty is that to calculate the force on atom i by numerical differencing requires in principle a recalculation of the local density of states on all the atoms when atom i is shifted by a small amount. Legrand's scheme was to use a six moment model for the density of states on the site being moved, but to recalculate only the second moment on the other sites.

THE TERMINATION PROBLEM

In the real space approach to calculating energies, the diagonal Green function matrix elements are obtained in the form of a finite continued fraction with the recursion method, and off-diagonal elements have (in the past at least) been obtained as the difference of two continued fractions (Haydock, Heine and Kelly 1972, Heine, Haydock, Bullett and Kelly 1980). These continued fractions were either terminated by an appropriate square root, or used to generate nodes and weights for gaussian quadrature (Nex 1978). The maximum entropy method can also be used to reconstruct a 'best' density from its moments, or equivalently from the continued fraction coefficients, and all three methods have been compared by Glanville, Paxton and Finnis (1988) with the conclusion that on balance the Nex method is to be preferred. However, the termination question has taken on a new significance since we started calculating interatomic forces because it is related to two new problems in particular which are worth reporting.

The first problem came to our attention in the course of calculating the relaxed structure of a (310) grain boundary. The potential had been adjusted to ensure that the perfect lattice was in equilibrium at zero pressure with the observed lattice constant. Yet when the perfect lattice was reoriented to set up the (310) grain boundary, the perfect part of the structure was no longer in equilibrium! Our forces were not transforming properly as vectors under rotations. In fact, neither was the energy rotationally invariant. The cause of the problem was that although the exact Green function would transform as a tensor, the approximate Green function matrix elements obtained by termination or quadrature methods do not transform exactly into their symmetry related combinations under lattice rotations. The solution to this problem was realized by Inoue and Ohta (1987), which is to obtain all the symmetry related Green function matrix elements relating to a bond in one calculation by matrix recursion. This also gives off-diagonal Green function matrix elements directly as elements of a singly matrix continued fraction rather than by the difference of two continued fractions. The matrix elements obtained transform into each other correctly if the matrix continued fraction is terminated by the identity matrix. Matrix recursion had previously been applied to the calculation of off-diagonal Green function matrix elements by Jones and Lewis (1984) and it has recently been developed further by Nex (1988).

The second problem has still not been completely solved in practice, although there are proposed solutions. It is likewise due to the fact that the Green function matrix elements are not calculated exactly. In this case it is the relation between energies and forces which is not exactly preserved under approximations to the Green function matrix elements. In other words, the expression for the forces eqn. (11) is not equivalent to the derivative of the energy in eqn. (7) when *approximate* Green function matrix elements are inserted. This manifested itself dramatically in a similar grain boundary relaxation using three recursion levels, for which recursion is done on three shell clusters. When the forces had been relaxed to zero (in practice to around 0.01eV per Ångstrom), the final energy was higher than some intermediate energies by an amount comparable to the energy differences of interest between different grain boundary structures. Of course, this problem could be solved in principle simply by extending the size of the clusters used in the recursion method for each bond, thereby improving the accuracy of the Green functions. However, recent work by Davies (to be published) suggests that is may be more efficient and more consistent to use the same three shell clusters but to recur to many more levels. It is a question of squeezing as much information as possible out of the local environments of the bonds.

SECOND MOMENT MODELS

It is very simple to calculate the first and second moments of a local density of states within the tight-binding model. With this information a crude model of the bond-energy can be constructed.

We define moments of the local density of states referred to its center of gravity, thus

$$\mu_{mip} = \int\limits_{-\infty}^{+\infty} (\varepsilon - \varepsilon_{ip})^m \, n^{ip} (\varepsilon) \, d\varepsilon. \tag{12}$$

The site energies ε_{ip} must be adjusted on each site to ensure local charge neutrality, but we assume that they are not dependent on the orbital p, so that $\varepsilon_{ip} = \varepsilon_{id}$ for all p. It is easy to show that the second moments are given by

$$\mu_{2i} = \sum_{j\neq i;pq} H^2_{ipjq} , \tag{13}$$

which is a pairwise summation over neighbors. The Hamiltonian matrix can be diagonalized for each bond separately, which for the d-band metals brings eqn. (13) into the form (Allan 1970)

$$\mu_{2i} = \sum_{j\neq i} (dd\sigma(R_{ij})^2 + 2dd\pi(R_{ij})^2 + 2dd\delta \, (R_{ij})^2), \tag{14}$$

where $dd\sigma$, $dd\delta$ and $dd\pi$ are the usual Slater-Koster overlap integrals.

A common approach has been to introduce a gaussian approximation to the local density of states, from which the following expression follows for the covalent bond energy per atom

$$E^i_{cov} = -10 \left(\frac{\mu_{2i}}{2\pi}\right)^{1/2} \exp\left(-\frac{(\varepsilon_F - \varepsilon_{id})^2}{2\mu_{2i}} \right). \tag{15}$$

Allan and Lannoo (1976a,b) and Masuda and Sato (1981) have used this model. The latter authors studied with it the Peierls stress of a screw dislocation. They did not impose the charge neutrality condition, but assumed the site energies to be the same on all sites. This actually makes the problem slightly more complicated in the second moment model, except in the case of a half-filled band, for which there is no charge transfer (Carlsson and Ashcroft 1983). As Ackland, Finnis and Vitek (1988) pointed out, with the condition of atomic charge neutrality the quantity $\varepsilon_F - \varepsilon_{id}$ scales as $\sqrt{\mu_{2i}}$ from site to site, so that the exponential factor in eqn. (15) is simply a constant for all sites. The bond energy then simply scales as $\sqrt{\mu_{2i}}$. This justifies writing the energy of eqn. (1) in the form

$$E_{tot} = - \sum_j \left(\sum_j \phi(R_{ij}) \right)^{1/2} + E_{rep} . \tag{16}$$

A number of authors have used simple parameterizations of the functions $\phi(R)$ and $E_{rep} = 1/2 \, \Sigma_{i\neq j} \, V(R_{ij})$ entering (16). Finnis and Sinclair (1984) proposed a set of

374

parameters for the bcc transition metals, which were fitted to the lattice parameters, cohesive energies and elastic constants, with further adjustment to give reasonable vacancy formation energies. They were applied to study surface relaxation, and the relationship between it and surface tensions and energies (Ackland and Finnis 1986). Information could be calculated such as the inward surface relaxations and realistic surface energies which encouraged the wider application of the model, since these quantities are completely wrongly predicted with pair potentials alone. However, the results it must be emphasized, only have qualitative significance. The second moment approximation has been shown explicitly by Masuda et al. (1983) to overestimate greatly the contraction of the Mo (100) surface layer, predicting 20% compared to the 5% predicted when fourth or higher moments are included. The latter value is closer to the experimental estimates.

Cores were subsequently fitted to the Finnis-Sinclair potentials by Ackland and Thetford (1987) and Rebonato, Welch, Hatcher and Bilello (1987) and they were used to study point defect properties by for example Matthai and Bacon (1985), Maysenhölder (1986) and Harder and Bacon (1988).

Marchese, Jacucci and Flynn (1988) made a parametric variation of the α-iron potential which showed that a potential of this functional form could provide a double barrier to vacancy diffusion and thereby explain the isotope effect in α-iron. The latter authors, and Rebonato and Broughton (1987), also revealed inadequacies in the model for the bcc metals, showing that the Grüneisen constants are completely wrongly predicted, and thus the phonon spectra are poor.

A similar empirical approach was adopted by Ercolessi, Tosatti and Parinello (1986) in constructing an empirical model for the energy of gold. Fitted only to bulk properties, it was able to explain the STM data on the (100) surface, which appears roughly speaking as a rearrangement of the surface layer to become close packed, with stripes or corrugation associated with the attempt of the surface layer to achieve a commensurate match with the first sublayer. In this case, the fitting procedure also allowed variation of the function of ϕ, which in eqn. (16) is a square root. Thus, the connection to the second moment approximation is abandoned, and the procedure is best described as an empirical effective medium theory, otherwise known as the embedded atom model (Nørskov and Lang 1980, Nørskov 1982, Daw and Baskes 1983). An evident limitation of the model of Ercolessi et al. is its zero stacking fault energy.

Models of the form (16) are sufficiently rapid to compute for very large scale molecular dynamics. They must not be expected to yield quantitative information where for example structural energy differences are involved, but they are very useful for obtaining qualitative insights and suggesting the plausibility or otherwise of structures and processes. Current applications include the simulation of the events in a radiation damage cascade, and the simulation of grain boundary structures at elevated temperatures, and one is interested in possible structures and processes rather than in predicting energies. A molecular dynamics code MOLDY6 using these potentials has been adapted to run efficiently on the CRAY-2 at Harwell, with which several thousand atoms can be readily simulated. In principle the CRAY-2 memory of 256 Mwords is sufficient for up to eighteen million atoms to be simulated with this code, but that would require about an hour of CPU time per timestep, so it is not a practical possibility. The required CPU time for running the molecular dynamics code MOLDY6 scales linearly with the number of atoms.

THE MONOVACANCY IN BCC TRANSITION METALS

This is a useful test problem for models of defect energetics, since the space of plausible structures for a single isolated vacancy is much smaller than in the grain boundary situation, and the experimental energies of formation are available for comparison.

Second moment tight-binding models have been applied to the problem by Allan and Lannoo (1976), Masuda-Jindo (1982) and Matthai and Bacon (1985). Qualitatively, it is clear how this level of approximation is adequate to count for the formation energies E_v which experimentally are about a third of the cohesive energy. The argument is that the bond energy per atom varies as $\sqrt{z}$, where z is the atomic coordination. There are z neighbors of a vacancy, each of whose energy has increased from its bulk value $-\sqrt{z}$ to $-\sqrt{(z-1)}$ (omitting constant factors). This amounts to a total increase in energy of about $1/2 \sqrt{z}$, or about half of the cohesive energy. Of course, this argument is modified by the presence of a repulsive pairwise term in the energy. In terms of the bond picture, this is a simple qualitative description of the strengthening of the bonds around a vacancy compared to their bulk energies.

Ohta, Finnis, Pettifor and Sutton calculated the bond energies and forces around the vacancy in the 4d transition metals Nb, Mo, Zr and T_c, assuming all to have the bcc structure, and using matrix recursion to three levels (fifth moment). The overlap integrals $dd\sigma$, $dd\pi$ and $dd\delta$ were assumed to decay exponentially with the bond length and to be in the ratio -6 : 4 : -1. Exponents and magnitudes were chosen to reproduce the d-band width and its volume dependence obtained from self-consistent band structure calculations. The Born-Mayer parameters were fitted to the experimental equilibrium volume and bulk modulus for each metal. All the functions were truncated between second and third neighbors. The calculated force fields around the unrelaxed vacancy are shown in Fig. 1 and are different from the results of previous models of the purely pairwise force or of the second moment type. In Nb and Mo, for example, there are only very small forces on the nearest neighbors to the vacancy, whereas the forces acting on the second neighbors (200) and fifth neighbors (222) are large and inwards. These large inward forces can be understood in terms of the strengthening of bonds neighboring the vacancy.

FUTURE DIRECTIONS

It is clear that the application of tight-binding models to the simulation of defects is still in its infancy, not least because the existing force algorithms are still being improved to enable satisfactorily converged minimum energy structures in the transition metals to be

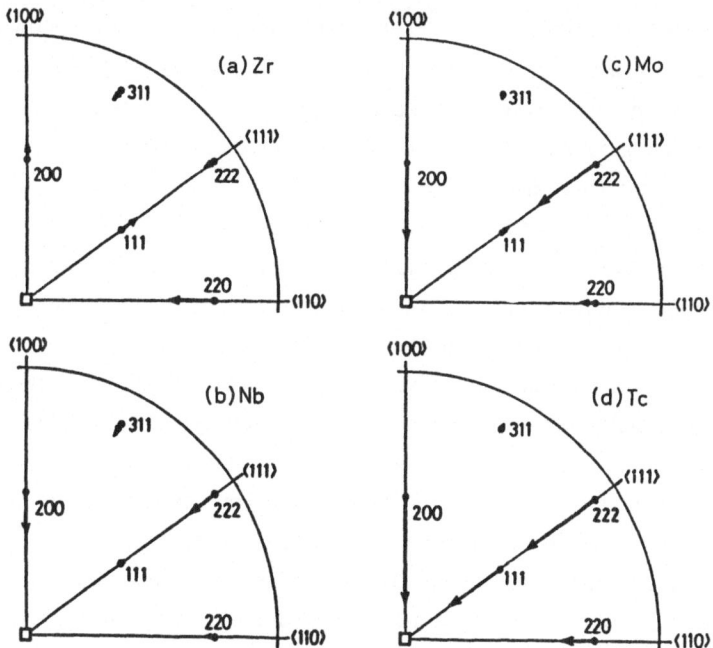

Figure 1. The force fields around the unrelaxed vacancy, after Ohta *et al.*, (1987). Arrows indicate the direction and relative magnitude of the forces acting on the neighbors of the vacancy at the origin. The force field is fairly short-ranged: the forces acting on more distant neighbors than those shown are very small and unimportant. To give an idea of the scale, we note that the magnitude of the force acting on atom 200 in Mo (the second neighbor) is 1.14eV/a.

calculated. The aim here is to do recursion on a cluster of no more than one or two hundred atoms per bond to a sufficient number of levels for the convergence of the forces in the bond. Preliminary calculations for the canonical d-band model have suggested that this will require three shells of neighbors around a bond, and that convergence of the relative energies of different structures can be helped by a finite temperature population of the discrete cluster levels (Davies, private communication). One should soon be able to simulate localized and extended defects in transition metals, including impurities, and explore the factors affecting for example grain boundary embrittlement. However, at the same time the conceptual basis of the model is still being studied, for example the non-orthogonality of orbitals is an important feature which simulations have neglected hitherto, and it is not clear in general for s-p-bonded materials that it can be satisfactorily accounted for within the pairwise part of the energy, although this may sometimes be the case, e.g., the analysis of Skinner and Pettifor (to be published) of the energetics of metallic hydrogen. Another route to the justification and quantification of tight-binding models is via the transformation of the LMTO equations of Andersen and coworkers (Andersen and Jepsen 1984). While work in this direction is proceeding, it must be recognized that simple tight-binding models will never be as rigorous as density functional theory, with which atomistic simulation of simple defects is now feasible and of increasing importance; see for example the work of Payne, Bristowe and Joannopoulos (1986). However, tight-binding models offer a semi-quantitative way to understand and possible predict trends, as well as to handle much larger systems than a self-consistent density functional calculation.

I have noted that a feature of the TBB model is the condition of charge neutrality on each atom, without which the force and energy expressions are inconsistent. However, the neutral atom model is not a suitable one for all materials modelling. A current topic of research is aimed at the simulation of metal-insulator or metal-semiconductor interfaces, at which large transfer must be taken into account. For example, the energy of interaction of an alkali halide with a metal is to a first approximation described by the classical image interaction, which depends on the charge redistribution induced at a metal surface by the potential of the ions. However, one needs a discrete atom version of this electrostatic interaction in order to proceed with atomistic simulation, and furthermore one needs to ensure the quantum mechanical aspects of the response of the metal are not ignored by a classical or semi-classical treatment. A discretized classical model of the image interaction gives surprisingly similar results to a simple quantum mechanical description, which suggests it could be incorporated into an empirical model of metal non-metal interfaces. We have learned this by studying a refinement of the TB model to include on-site and intersite Coulomb energies. Consider the Hamiltonian

$$\sum_{i \neq j} \beta |i\rangle\langle j| + \sum_{i} U \cdot (\rho_i - q_i)|i\rangle\langle i| + \sum_{i \neq j} C_{ij} (\rho_j - q_j)|i\rangle\langle i| \qquad (17)$$

and suppose it is solved self-consistently for the electronic charges ρ_i, where C_{ij} is the Coulomb interaction and q_i is the fixed positive charge on i - th ion. The self-energy parameter U determines the degree of charge imbalance of an atom. For example, a neutral atom solution corresponds to the limit as U tends to infinity.

I have solved this model for the simplest case of just one nearest neighbor hopping matrix element β linking s-orbitals in a simple cubic cluster of 3 x 3 x 3 atoms. A positive charge was brought up towards one face of the cluster. The solution for the energy of the system was calculated as a function of distance. An analogous classical solution was obtained, which was essentially a discretized version of an image interaction calculation, in which the classic potential at each site of the 'metal' was constrained to be constant. This is a much simpler calculation than the self-consistent solution of the above Hamiltonian. It turned out that the TB model response was approximately linear and the interaction energy was surprisingly similar to the classical value (see Fig. 2).

This calculation suggests that a reasonable approach to the problem of modelling the charge transfer for the purpose of simulating the energetics of interfaces would be a discrete classical model. Such simulations have not yet been attempted.

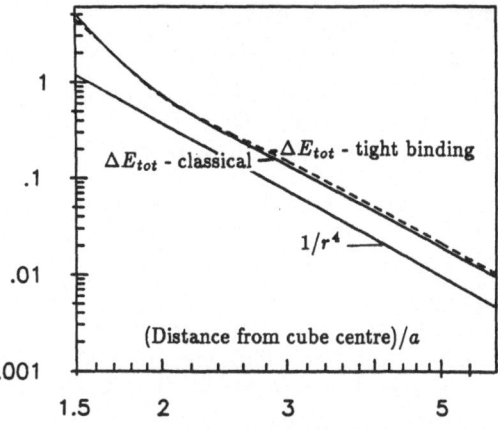

Figure 2. The binding energy versus distance of a positive charge above a 27 atom cube of lattice parameter a. Classical and quantum cases are shown and the $1/r^4$ dependence from the center of the cube expected at large separations is shown for comparison. The energy is in units of the hopping parameter β. $U = 6\beta$. 20 electrons.

REFERENCES

Ackland, G.J. and Finnis, M.W., 1986, Phil. Mag. A, 54:301.
Ackland, G.J., Finnis, M.W. and Vitek, V., 1988, J. Phys. F ,18:L153.
Ackland, G.J. and Thetford, R., 1987, Phil. Mag. A, 56:15.
Allan G., 1970, Ann. Phys., 5:169.
Allan, G and Lannoo, J., 1976, J. Phys. Chem. Solids 37:699.
Allan, G and Lannoo, J., 1976, Phys. Stat. Solids B, 74:409.
Andersen, O.K. and Jepsen, O., 1984, Phys. Rev. Lett., 53:2571.
Carlsson, A.E. and Ashcroft, N.W., 1983, Phys. Rev. B, 27:2101.
Chadi, D.J., 1979a, J. Vac. Sci. Technol., 16:1290.
Chadi, D.J., 1979b, Phys. Rev. B, 19:2074.
Chadi, D.J., 1984, Phys. Rev. B, 29:785.
Coulson, C.A., 1939, Proc. R. Soc. Lond. A, 169:413.
Daw, M.S. and Baskes, M.I., 1983, Phys. Rev. Lett. 50:1285.
Ercolessi, F., Tosatti, E. and Parinello, M., 1986, Phys. Rev. Lett., 57:719.
Finnis, M.W. and Sinclair, J.E., 1984, Phil. Mag. A, 50:45, ibid., 53:161 (Erratum).
Foulkes, M., 1987, PhD Thesis, University of Cambridge.
Glanville, S., Paxton, A.T. and Finnis, M.W., 1988, J. Phys. F, 18:693.
Harder, J.M. and Bacon, D.J., 1988, Phil. Mag. A, 58:165.
Harris, J., 1985, Phys. Rev. B, 31:1770.
Haydock, R., Heine, V. and Kelly, M.J., 1972, J. Phys. C, 5:2845.
Heine, V., Haydock, R., Bullett, D.W. and Kelly, M.J., 1980, in: "Solid State Physics",
        Vol. 35, H. Ehrenreich, F. Seitz and D. Turnbull , eds., Academic, New York.
Inoue, J. and Ohta, Y., 1987, J. Phys. C 20:1947.
Jones, R. and Lewis, M.W., 1984, Phil. Mag. B, 49:95.
Kohyama, M. Yamamoto, R., Ebata, Y. and Kinoshita, B., 1988, J. Phys. C, 21:3205.
Legrand, B., 1985, Phil. Mag.A 52:83.
Marchese, M., Jacucci, G. and Flynn, C.P., 1988, Phil. Mag. Lett. 57:25.
Masudo-Jindo, K., 1982, J. Physique, 43:921.
Masuda-Jindo, K., Hamada, N. and Terakura, K., 1984, J. Phys. C, 17:1271.
Masuda, K. and Sato, A., 1981, Phil. Mag. A, 44:799.
Masuda, K., Yamamoto, R. and Doyama, M., 1983, J. Phys. F, 13:1407.
Matthai, C.C. and Bacon, D.J., 1985, Phil. Mag. A, 52:1.

Maysenhölder, W., 1986, Phil. Mag. A, 53:783.

Moraitis, G. and Gautier, F., 1979, J. Phys. F, 9:2025.

Nex, C.M.M., 1978, J. Phys. A, 11:653.

Nex, C.M.M., Paper to be presented at MSI workshop on Practical Iterative Methods for Large-Scale Computations, October 1988.

Nørskov, J.K. and Lang, N.D., 1980, Phys. Rev. B, 21:2131.

Nørskov, J.K., 1982, Phys. Rev. B, 26:2875.

Ohta, Y., Finnis, M.W., Pettifor, D.G. and Sutton, A.P., 1987, J. Phys. C, 17:L273.

Payne, M.C., Bristowe, P.D. and Joannopoulos, J.D., 1986, Phys. Rev. Lett., 58:1348.

Paxton, A.T. and Sutton, A.P., 1988, J. Phys. C, 21:L481.

Pettifor, D.G., in "Solid State Physics" Vol. 40, 1987, H. Ehrenreich and D. Turnbull, eds., Academic, New York, pp. 43-92.

Pettifor, D.G. and Podloucky, R., 1986, J. Phys. C, 19:315.

Rebonato, R. and Broughton, J.Q., 1987, Phil. Mag. A, 55:225.

Rebonato, R., Welch, D.O., Hatcher, R.D. and Bilello, J.C., 1987, Phil. Mag. A, 55:655.

Sayers, C.M., 1984, Phil. Mag. B, 50:635.

Sutton, A.P., Finnis, M.W., Pettifor, D.G. and Ohta, Y., 1988, J. Phys. C, 21:35.

Terakura, I., Terakura, K. and Hamada, N., 1981, Surf. Sci., 111:479.

Tréglia, G., Ducastelle, F. and Spanjaard, D., 1980, J. Phys. Paris, 41:281.

Tréglia, G., Desjonquères, M.C. and Spanjaard, D., 1983, J. Phys. C, 16:2407.

INTERATOMIC FORCES FROM THE RECURSION METHOD WITH GAUSSIAN PSEUDOPOTENTIALS

Roger Haydock

Department of Physics and Materials Science Institute
University of Oregon
Eugene, OR 97403

INTRODUCTION - ELECTRONIC STRUCTURE FROM THE RECURSION METHOD

Many properties of materials can be understood from the fundamental viewpoint of their electronic structure which in turn can be determined by the solution of independent particle equations in many cases. Interatomic forces result from the ground state electronic structure and are thus obtainable from density functional theory[1] whose applications are well established. The success of this theory in predicting the structure and elastic properties of metals suggest that it is also capable of describing the complicated motion of atoms during plastic deformation and fracture processes for which there is no fundamental understanding at present.

The purpose of this paper is to propose an accurate and efficient computational scheme for the application of density functional theory to the mechanical properties of metals and alloys. Reliable theoretical exploration of these properties for new materials would lead to great savings in their evaluation for technological applications. Thus the problem addressed here is one of computational physics which one hopes will lead to new understanding of plasticity in metals.

The recursion method[2] is a computationally efficient way of calculating the properties of quantum mechanical systems. The method consists of two parts: The first is the transformation (which can always be done) of an arbitrary quantum system to a chain model - a semi-infinite sequence of states $\{u_0, u_1, ...\}$ in terms of which the Hamiltonian H can be expressed as a three term recurrence,

$$H u_n = a_n u_n + b_{n+1} u_{n+1} + b_n u_{n-1}, \tag{1}$$

where u_{-1} is zero and the parameters of the model, $\{a_n, b_n\}$ completely describe the quantum mechanical motion of the system. The second step is the representation of the system by orthogonal polynomials of energy, $p_{-1}(E)$ is zero, $P_0(E)$ is unity, and

$$E P_n(E) = a_n P_n(E) + b_{n+1} P_{n+1}(E) + b_n P_{n-1}(E). \tag{2}$$

As suggested by the similarity of the above equations, functions of energy correspond to states of the system. The properties of the system can be expressed simply in terms of these polynomials. The power of this method is that regardless of symmetry any system can be efficiently transformed to a chain model, and the sequence of state $\{u_0, u_1, ...\}$ is optimal for the representaton of the stationary states of H with non-zero components in u_0.

The first part of the method proceeds by projecting Hu_0 back on u_0 to determine a_0, and normalizing $Hu_0-a_0u_0$ to get b_1 and u_1. In the general step u_n, b_n, and u_{n-1} are known; Hu_n is projected on u_n to give a_n, then $Hu_n-a_nu_n-b_nu_{n-1}$ is normalized to give b_{n+1} and u_{n+1}. Similarly $P_{n+1}(E)$ is calculated from $P_n(E)$, $P_{n-1}(E)$, b_{n+1}, a_n, and b_{n-1} by solving Eq. (2) for $P_{n+1}(E)$. These are all processes well suited to digital computers once H has been specified in some discrete form such as a matrix.

The matrix form of H need not be kept in random access memory because its elements are always used in the same order to calculate Hu_n. Furthermore, if H is sparse - many zero elements - only the non-zero elements are needed. Because H is not modified during the process, it stays sparse in contrast to methods which rotate H and fill in its zeros. Other advantages of the method are that the mathematics of orthogonal polynomials is well developed and contains such useful theorems as a more powerful version of the variational principle which bounds internal band limits as well as the extremal ones.

It is only within the last few years that the numerical stability of procedures like the recursion method has been understood, see Ref. 3 for example. The conclusion is that diagonalizaton of the three term recurrence in Eq. (1) yields the eigenvalues of H to the full precision, and the eigenstates to the root of the precision, in which the states $\{u_n\}$ and the parameters $\{a_n, b_n\}$ were calculated. Despite its efficiency, there is no additional rounding error introduced by the method. In most applications the speed is greatly increased at the cost of some accuracy by only calculating a few of the basis states and parameters, although when used at full accuracy, the method is still faster than any other for sparse matrices.

The remainder of this paper is organized in four sections, the first of which contains a general discussion of sparse hamiltonian representations. The succeeding section describes the gaussian pseudopotential scheme, the third outlines the calculation of forces, and the last contains comments on the accuracy of the scheme.

SPARSE REPRESENTATION OF THE HAMILTONIAN

The recursion method is most efficient for representations in which matrix elements of powers of the hamiltonian may be simply and rapidly calculated. This amounts to finding a set of normalizable functions $\{\phi_\alpha\}$ whose first property is that the hamiltonian is conveniently represented in the form,

$$H = -\nabla^2 + \sum v_\alpha \, \phi_\alpha \tag{3}$$

where v_α are the coefficients on the expansion. Matrix elements of powers of H involve integrals over products of the $\{\phi_\alpha\}$ and their derivatives, so the second property of the representation is that these expressions be simple to evaluate.

One way to achieve this is by requiring the representation to be sparse:

$$\phi_\alpha \, \phi_\beta = \sum c_{\alpha\beta\gamma} \phi_\gamma, \tag{4}$$

$$\nabla^2 \, \phi_\alpha = \sum d_{\alpha\beta} \, \phi_\beta, \tag{5}$$

where the sums over β and γ are both finite for any choice of α or α and β. This means that if the potential is a finite sum of the $\{\phi_\alpha\}$, then so is any power of H. Viewing H as a matrix in the basis $\{\phi_\alpha\}$, then sparseness means having only a finite number of non-zero elements in each row or column.

Matrix elements are calculated from integrals, so in addition to sparseness, the integrals

$$\Phi_\alpha = \int \phi_\alpha(r) \, dr^3 \tag{6}$$

must be known. Thus, if states u and v have finite expansions in the $\{\phi_\alpha\}$, then uH^nv has a finite expansion, and the matrix element,

$$\int u \, H^n \, v \, dr^3 = \sum h_{n,\alpha} \, \Phi_\alpha \tag{7}$$

can be obtained from a finite computation. The shorter the expansions in Eqs. (4) and (5), the more sparse the representation and the faster the calculation. Indeed, the tridiagonal form is the most sparse short of diagonalization. Many different sets of functions lead to sparse representations. For example, plane waves in a finite or periodic system satisfy the sparseness conditions, but for an infinite system, normalization is not possible. More generally, polynomials times exponentials of polynomials also give sparse representations provided they are normalizable over the system. Still more generally, the functions must be infinitely differentiable where the potential is infinitely differentiable, and singular where the potential is singular.

There are further advantages in choosing the functions to be stationary states of some hamiltonian. This makes the functions orthogonal which simplifies expansions and integrations. Plane waves, the stationary states of the free praticle, and gaussians, the stationary states of the isotropic oscillator are immediate examples although there may be others. Because plane waves are non-normalizable in infinite, non-periodic systems, the remainder of this paper will be devoted to the use of gaussians as a sparse representation.

GAUSSIAN PSEUDOPOTENTIALS

The idea of a pseudopotential[4] is to replace the deep atomic potentials which produce valence wavefunctions having many nodes in the core region by shallow pseudopotentials whose wavefunctions are smooth and nodeless inside the core region, but the same as the atomic wavefunctions outside the core region. For the purposes of this work, the concept of the pseudopotential is modified to that of replacing the atomic potential by the minimum number of guassians which still have the atomic wavefunctions outside the range of the gaussians. More precisely, the neutral atomic potential is to be replaced by a minimal sum of gaussians which have a bound state of the same energy, angular momentum, and charge density as the atom outside the gaussians. By varying the strength and range of a single gaussian, it should be possible to fit the charge density at a single energy and angular momentum. The idea is easily generalized to non-local pseudopotentials, and further gaussians can be added to increase the energy range over which the pseudopotential is accurate.

Following Foulkes and Haydock[5], the band structure energy for a given charge density can be calculated by solving the Schroedinger equation for non-interacting, density functional electrons in the potential for that charge density. For what follows, it is assumed that this potential is well approximated by a product of a low order polynomial and a gaussian for each atomic site. If the charge density is a superposition of neutral atoms, then this potential will be a sum of neutral atom gaussian pseudopotentials with additional terms to correct the exchange-correlation potential where atomic charge densities overlap.

Let the hamiltonian for the density functional electrons be,

$$H = -\nabla^2 + \sum P_\alpha (x,y,z) \exp \{-\lambda_\alpha (r - r_\alpha)^2\}, \tag{8}$$

where the sum is over a site index α, the $P_\alpha(x,y,z)$ are polynomials in $x, y,$ and z, and λ_α is the range parameter for the gaussian centered at r_α. Supposing a recursion is started with a single polynomial times a gaussian,

$$u_0 = u(x,y,z) \exp \{-\lambda r^2\} \tag{9}$$

then in order to estimate the accuracy of the results, it is necessary to estimate how many new basis states are introduced with each successive multiplication by H.

It can be seen that the isotropic oscillator wavefunctions tridiagonalize the free particle hamiltonian by noting that the oscillator wavefunctions satisfy a three-term recurrence in r^2 which is just the difference between the two hamiltonians. As a result, the effect of the kinetic energy operator on a product of a polynomial and a gaussian generates the products of two new polynomials and the same gaussian.

Multiplying u_0 by the potential in H produces terms of the form,

$$P_\alpha (x,y,z) \, u_0 \, (x,y,z) \, \exp \{ - \lambda \, r^2 - \lambda_\alpha \, (r - r_\alpha)^2 \}, \qquad (10)$$

which is the product of a new polynomial and a new gaussian centered at $\lambda_\alpha (\lambda_\alpha + \lambda)^{-1} r_\alpha$. If the centers of the two gaussians in the product are far apart, the product is small and, if less than the error in the original potential, it may be neglected. Products of gaussians and their nearest neighbors must be significant, so the total number M of new basis functions generated by the operation of the hamiltonian on u_0 is two for the kinetic energy plus the coordination number for the potential energy. This varies from six for a four-coordinated structure to sixteen for a fourteen-coordinated structure, and so on.

The above arguement leads to an estimate of the number of exact recursion levels which can be calculated with H starting from u_0. Each multiplication by H introduces M new components for each old component giving of order M^L components for L levels of recursion. If M is ten, then six exact recursion levels requires store for only about a million components.

This can be compared with current tight-binding calculations[6] where the number of exact recursion levels is again limited by the number of new components produced by each multiplication by H. In tightbinding, this is just the coordination number, two less than for the gaussian pseudopotential. However, the tightbinding makes two approximations beyond the gaussian pseudopotential: They are that tightbinding also approximates the kinetic energy operator by projecting it onto the chosen set of orbitals, and the three-center integrals are usually neglected in tightbinding. Thus the gaussian pseudopotential can produce about the same number of exact levels, but with a much more general hamiltonian.

CALCULATION OF FORCES

Forces are defined as the negative derivatives of the total energy of a system with respect to the coordinates. Following Foulkes and Haydock[5], the total energy may be written as the sum of a bandstructure term and other terms due to electrostatics, exchange, and correlation. The bandstructure term is the one to which the recursion method can be applied, and to which this Section will be devoted. The bandstructure force is then the negative derivative of the bandstructure energy with respect to the coordinates.

The bandstructure energy is the sum of the energies of the occupied single particle states, and its negative derivative is the sum of the negative derivatives of the occupied single particle energies,

$$f = - \sum \partial E_\alpha / \partial \delta, \qquad (11)$$

where δ is a displacement of one coordinate. This may be rewritten in terms of a ratio of the derivative of a determinant and the determinant,

$$f = \sum \{ \partial \det [E_\alpha - H] \, /\partial E \}^{-1} \partial \det [E_\alpha - H] \, / \, \partial \delta. \tag{12}$$

This can finally be approximated as in Foulkes and Haydock[7],

$$f_N = \sum [\partial P_N (E_{N,\alpha}) \, / \, \partial E]^{-1} [\partial P_N (E_{N,\alpha}) \, / \, \partial \delta] \tag{13}$$

where $\{P_n(E)\}$ are the orthogonal polynomials obtained from a starting state u_0, $E_{N,\alpha}$ are the zeros of $P_N(E)$, and u_0 spans the occupied levels whose energies depend on δ. This way of calculating forces is numerically stable and efficient, provided that care is taken in the choice of u_0. Only those stationary states of H with non-zero projections on u_0 contribute to f_N, so if there is degeneracy in H, then

$$f_N = \sum g_\alpha [\partial P_N (E_{N,\alpha}) \, / \partial E]^{-1} \partial P_N (E_{N,\alpha}) \, / \, \partial \delta, \tag{14}$$

where g_α is the degeneracy of $E_{N,\alpha}$. For efficiency, u_0 should be chosen so that f_N converges rapidly with N. This is the case when $\{u_0,...,u_N\}$ spans the states on which $\partial H / \partial \delta$ is non-zero. If H contains a gaussian pseudopotential of the form in Eq. (8) and δ is the change in an atomic coordinate r_0, then the choice

$$u_0 = \partial H / \partial \delta = P(x - x_0, y - y_0, z - z_0) \exp \{ - \lambda (r - r_0)^2 \}, \tag{15}$$

ensures that f_N converges rapidly and includes all the states which contribute to this component of force. The scheme can be improved futher by the use of terminators[8], sets of states $\{u_{N+1}, u_{N+2},...\}$ and parameters $\{a_{N+1}, b_{N+2},...\}$ which incorporate the long range properties of the hamiltonian into the chain model.

ACCURACY

The accuracy of force and energy calculations is usually assessed from the accuracy of the hamiltonian and the sensitivity of the results to the basis set used. Similar criteria apply to the scheme presented here. The accuracy of the hamiltonian depends on the local density approximation together with the quality of the gaussian fit, and so is comparable with other methods.

In most methods the basis is determined before the calculation begins and the variational principle ensures that the results improve with the inclusion of new degrees of freedom. The gaussian scheme is very different in that the basis set is generated by the hamiltonian and optimized with respect to that hamiltonian and the stationary states in u_0. Generally speaking, such an optimized basis should be more efficient provided that the overheads of optimization are not large. This remains to be tested, but there are reasons for optimism.

ACKNOWLEDGEMENTS

This work was supported by the National Science Foundation under grant DMR-8712346. The Author wishes to thank the Cavendish Laboratory and Pembroke College, Cambridge for their hospitality, the Royal Society for travel support, Matthew Foulkes and Richard Needs for useful discussions.

REFERENCES

1. C.J. Callaway and N.H. March, Solid State Physics 38, ed. Ehrenreich and Turnbull, Academic Press, New York, 1984.
2. R. Haydock, V. Heine, and M.J. Kelly, Solid State Physics 35 ed. Ehrenreich and Turnbull, Academic Press, New York, 1980.
3. B.N. Partlett, "The Symmetric Eigenvalue Problem", Prentice Hall, Englewood Cliffs, New Jersey, 1980.
4. G.P. Kerker, J. Phys. C, 13, L189-94 (1980).
5. W.M.C. Foulkes and R. Haydock, "Tight Binding Models and Density Functional Theory", submitted for publication.
6. D. Pettifor and D. Weaire, "The Recursion Method and Its Applications", Springer Press, Berlin, 1984.
7. W.M.C. Foulkes and R. Haydock, J. Phys.C, 19, 6573-87 (1986).
8. R. Haydock and C.M.M. Nex, J. Phys. C, 18, 2235-48 (1985).

APPLICATION OF TIGHT-BINDING RECURSION METHOD

TO LATTICE DEFECTS IN METALS AND ALLOYS

K. Masuda-Jindo

K. Kimura and S. Takeuchi

Department of Materials Science
and Engineering
Tokyo Institute of Technology
Nagatsuta, Midori-ku
Yokohama 227, Japan

Institute for Solid State Physics
University of Tokyo
Roppongi, Minato-ku
Tokyo 106, Japan

INTRODUCTION

It is well-known that the macroscopic mechanical properties of metals and alloys are closely related to the microscopic dislocation motion and grain boundary properties. There have been a large amount of experimental and theoretical works on the dislocations and grain boundary properties in metals and alloys. However, fundamental understandings, from the atomistic point of view, have still been quite limited. Our purpose is, therefore, to investigate the fundamental properties of dislocations and grain boundaries in metals and alloys on the basis of a microscopic electronic theory.

In the present paper, we consider two typical problems on dislocations and grain boundaries: Firstly, we calculate the relative stability of the cores of 1/2 <111> screw dislocation in bcc transition metal Mo by using the TB recursion method. This calculation is performed from the following reasons. The prediction of the stable core structure of the screw dislocation is important for the understanding of the low temperature plasticity of bcc metals. There have been two different interpretations as to the origin of the large resistance to the glide of screw dislocations: sessile-dissociation interpretation, first suggested by Hirsch, due to a three-fold-symmetrically extended strain-field of a polarized core (see, a review article by Vitek[1]) and the ordinary Peierls potential interpretation[2]. Furthermore, the core structure can be considered to be related to the characteristic slip geometry of bcc metals[3].

We also discuss the grain boundary properties of Ll$_2$ type intermetallic compounds, in relation to the grain boundary segregation and embrittlement. We focus our attention on an intermetallic compound Ni$_3$Al. For this intermetallic compound, there have been a considerable amount of experimental works on the grain boundary segregation and embrittlement, because of technological importance. In particular, it has been shown that the addition of a small amount of solute atoms like B, Fe and Mn, can drastically improve the ductility of the polycrystalline Ni3Al.

PRINCIPLE OF CALCULATION

The total energy of the system is assumed to be given by the sum of the band structure energy E_b and the pairwise repulsive potential energy E_r[4]. The band structure energy E_b is calculated by using the conventional scalar recursion method by Haydock et al.[5] We have chosen a different termination method and a recursion level n, depending on the lattice defect problems.

Bcc Transition Metals

For the screw dislocation in bcc transition metals, we use the standard formula of the band structure energy

$$\Delta E_b = \sum \int^{E_F} (E - E_F) \{ \rho_i (E) - \rho_o (E) \} \, dE, \tag{1}$$

where $\rho_i (E)$ ($\rho_o(E)$) represents the local density of states (DOS) at site i (perfect lattice site), and E_F the Fermi energy. The first order expansion approximation[6] of eq. (1) may be justifiied for this problem, since the perturbation introduced by a screw dislocation in bcc metals is not strong. The character of the weak perturbation can actually be seen in the local DOS on the atomic sites around the center of the screw dislocation, as will be shown below. A similar scheme has already been applied to the calculation of dislocation core and slip geometry of hcp transition metals.[7]

The local DOS $\rho_i (E)$ on the atomic site i for a given atomic configuration can be calculated by the recursion method by using the (d-electron) site energy and (d-d) two center hopping integrals. Once the recursion coefficients (a_i, b_i ; $i \leq n$) are known, the (d-electron) Green's function $G_{ii} (E)$ is simply wirtten as

$$G_{ii} (E) = 1/\left[E - a_1 - b_1 / (E - a_2 - ... b_n / (E - a_{n+1} - b_{n+1} / E - ... \right]. \tag{2}$$

The DOS $\rho_i (E)$ on atomic site i can then be calculated from the Green's function $G_{ii} (E)$ as

$$\rho_i (E) = - (1/\pi) \lim_{s \to 0} G_{ii} (E + is). \tag{3}$$

The two center hopping integrals of the d-bands, $dd\sigma = -0.08594$, $dd\pi = 0.06444$ and $dd\delta = -0.02402$ in Ryd unit, are taken from Ref. 8, and are assumed to vary as R_{ij}^{-Q} (Q = 3.57 for Mo crystal), where R_{ij} denotes the interatomic distance between atomic sites i and j. These d-band parameters are determined so as to fit the unhybridized TB energy levels to the eigenvalues (high symmetry special points) obtained by the APW band structure calculations.

For the pairwise repulsive potential between the adjacent atomic sites i and j, we take the Born-Mayer potential of $C_0 \exp(-pR_{ij})$. The parameter values C_0 and p can be determined so as to satisfy the equilbrium condition (Born crystal stability) and to reproduce the experimental bulk modulus B. C_0 and p values for Mo are 1678.0 Ryd and 1.99 a_H^{-1}, respectively, a_H being the Bohr radius.

In the present study, the band structure energy E_b is calculated for each atomic row by using the 15-level recursion coefficients and atomic clusters composed of approximately 2000 atoms. The termination of the recursion coefficients and the resulting band energy calculation are performed by the method proposed by Nex[9]. For the calculation of the repulsive potential energy, the contribution of 14 neighboring atoms are taken into account both for the perfect and dislocated crystals, and no particular truncation scheme with a certain cut-off distance for neighboring atoms is used.

Ll2 Compound

For impurity segregations or grain boundary fracture problem in Ll2 type intermetallic compounds, we use the termination method by Beer and Pettifor,[10] and a recursion level of n = 3 (6-th order moment approximation). We think that this approximation scheme of the recursion method is sufficient for the impurity segregation in fracture problems, since the changes in the lowest order moments are large and play a dominant role in the band structure energy calculations. Within this approximation scheme the local DOS ρ_i (E) on site i can be written as

$$\rho_i^\lambda(E) = (1/\pi) \left\{ b_1 b_2 b_3 / D^\lambda (E) \right\} \sqrt{4b_\infty - (E - a_\infty)^2} / 2b_\infty , \qquad (4)$$

with

$$D^\lambda(E) = A_1^\lambda E^6 + A_2^\lambda E^5 + ... + A_7^\lambda , \qquad (5)$$

where all the recursion coefficients (a_m, b_m) after 3-level are assumed to be equal to their infinite values a_∞ and b_∞. λ specifies nine valence orbitals s, p_x, p_y, p_z, xy, yz, zx, x^2-y^2 and $3z^2-r^2$.

In eqs. (4) and (5), the expansion coefficients A_i^λ for $D^\lambda(E)$ are given by simple functions of the recursion coefficients,[11] and asymptotic recursion coefficients a_∞ and b_∞ are related to the band width and the center of the band, and can be determined by the method by Beer and Pettifor[10]. Using eqs. (4) and (5), the local density of states $\rho_i^\lambda(E)$ are now given by the form

$$\rho_i^\lambda (E) = \sum_{j=1}^{6} \{ C_j / (E - E_j) \} \sqrt{4b_\infty - (E - a_\infty)^2} / 2b_\infty , \qquad (6)$$

where E_j denotes the j-th (complex) root of the equation $D^\lambda(E) = 0$, and C_j the weight of the partial fractions $1/(E - E_j)$. Then, the integration of the DOS can be evaluated analytically in terms of the elementary integrals,[11] and results for E_b are much more accurate compared to those obtained by using a direct numerical method.

For the repulsive potential between atomic sites i and j, we take three kinds of power law dependence:

$$\Phi_{AA}(R_{ij}) = C_{AA} \bullet (R_o/R_{ij})^{n_{AA}} , \tag{7}$$

$$\Phi_{BB}(R_{ij}) = C_{BB} \bullet (R_o/R_{ij})^{n_{BB}} \tag{8}$$

$$\Phi_{AB}(R_{ij}) = C_{AB} \bullet (R_o/R_{ij})^{n_{AB}} , \tag{9}$$

where Φ_{AA}, Φ_{AB}, and Φ_{BB} denote the repulsive potential for transition-metal atom pair (Ni-Ni pair), transition-metal sp-valence metal atom pair (Ni-Al) and sp-valence atom pair (Al-Al), respectively. R_0 is the nearest-neighbor interatomic distance of the perfect crystal. We have chosen this type of power law dependence, since within the TB formalism a repulsive contribution falls off as the square of the TB hopping integrals[12]. It is known that such repulsive potentials are quite successful for the studies of the structural trends of p-d bonded AB-compounds[13].

In view of the success in Refs, 12 and 13, the spatial dependences of the repulsive potentials are chosen such that $n_{AA} = 10$, $n_{AB} = 7$ and $n_{BB} = 4$. C_{AA}, C_{BB} and C_{AB} are determined so as to satisfy the equilibrium condition and to reproduce the experimental values of elastic constants, using the condition of $\Phi_{AB} = \sqrt{\Phi_{AA} \bullet \Phi_{BB}}$. The repulsive potential parameter between an impurity and host atoms is determined by taking into account the atomic volumes of the impurity and host atoms. For this determination, we use the formula of volume change ΔV due to the solute atom[14].

$$\Delta V = \sum_i (\vec{R}_i \bullet \vec{F}_i) / 3K \tag{10}$$

where $\vec{R}_i$ and $\vec{F}_i$ denote the position and force vector, respectively, for host atomic site i around a solute atom, and K the bulk modulus of the crystal.

RESULTS AND DISCUSSIONS

Screen Dislocation in Mo Crystal

The atomic relaxation calculations of the 1/2<111> screw dislocation core are performed for the non-degenerate, unpolarized core and doubly-degenerate, polarized core. The unpolarized core of the screw dislocation is introduced in the crystallite by using the solution of the atomic displacements due to the isotropic linear elasticity theory as an initial condition. The polarized cores are introduced by superposing particular displacements on the atomic displacements of the elasticity solution. A typical polarized core can be constructed by atomic row displacement Δz_1 (parallel to the dislocation axis) of the three central atomic rows and by Δz_2 displacement in -z direction of the next three atomic rows.

Firstly, we calculate the equilibrium configuration of the nondegenerate, unpolarized core. By the symmetry requirement for the nondegenerate core, atomic rows 1, 2 and 5 in Fig. 1 (located in a symmetry axis) should be unchanged from the elastic solution, while those of 3, 4 and 6 may be displaced to equilibrium positions in the opposite direction from those of 3', 4', and 6', respectively. The relaxation of these atomic rows are performed by using a direct total energy minimization procedure,[15] without allowing the atomic displacements perpendicular to the dislocation line. In Table 1, we present the atomic displacements parallel to the dislocation line and the energy reduction accompanying the displacements for the atomic rows near the center of the screw dislocation. One can see in Table 1 that both the atomic displacements and associated energy reduction are quite small. In order to understand the effect of the atomic row displacements perpendicular to the dislocation line (core dilatation), we have also performed the three dimensional relaxation calculations for the atomic rows near the center of the screw dislocation. To see this effect in a simple and efficient way, we have performed the structural relaxation by taking into account only the repulsive energy contributions. In this procedure, we obtained the minimum repulsive energy configurations for a fixed boundary condition, keeping the nondegeneracy of the core structure. The calculated atomic row displacements in the radial direction, Δr_i, and the associated energy changes by this procedure are presented in Table 2.

We have performed the similar calculations for the doubly degenerate cores. At first, we choose specifically the displacements $\Delta z_2 = -(1/2)\Delta z_1$, and investigate the stability of the polarized core. This choice of the atomic row displacements gives rise to the typical polarized core[2,16]:

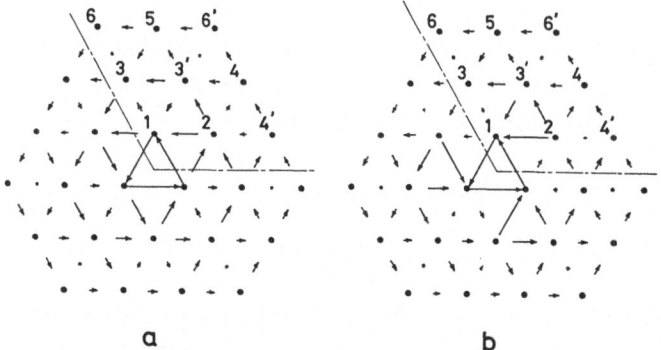

a   b

Fig.1. Differential displacement map for typical 1/2<111> screw dislocation with nondegenerate (a) and doubly degenerate (b) cores. Arrows indicate magnitude of relative displacement between adjacent atomic rows in the direction of the dislocation axis.

For instance, the polarized core with the atomic displacements $\Delta z_1 = 0.08b$ and $\Delta z_2 = -0.04b$ are shown in Fig. 1(b). The calculated results showed that the total energy increases monotonically in a quadratic way with increasing Δz_1 value. In the next step, we have calculated the minimum repulsive energy configurations for the parized core both without and with dilatational relaxation, and then calculated the total energy for these configurations. The results of the displacements of atomic rows and the accompanying

Table 1. Atomic row displacements parallel to the dislocation axis for nondegenerate core, and the local energy change for each atomic row with length b (Burgers vector).

site i	1	2	3	4	5	6
$\Delta z_i/b$	0.0	0.0	3.0×10^{-3}	6.0×10^{-4}	0.0	0.0
$\Delta E_i(eV/b)$	0.0	0.0	-2.4×10^{-4}	-1.1×10^{-5}	0.0	0.0

energy changes are given in Table 2. Summarizing the results, we come to the conclusions: (1) The nondegenerate, unpolarized core is stable compared to the doubly degenerate, polarized core in bcc Mo, irrespective of the inclusion of the perpendicular atomic row displacements. (2) The effect of perpendicular, or dilatational atomic row displacements is not crucial in determining the relative stability of the screw dislocation core structures. This is because the band structure and repulsive energy contributions are largely cancelled with each other. The present result on the core structure indicates that the sessile dissociation interpretation for the low temperature plasticity is not valid at least for Mo. The unpolarized core structure in Mo is consistent with the slip geometry of Mo at low temperature; one can predict {110} type slip for this core[16] in agreement with experiments[17].

As a final remark of this subsection, we briefly discuss the d-band contributions to the stability of the core structure. For this purpose, we calculate the local DOS on each atomic site around the center of the screw dislocation: Figure 2 shows the d-band DOS on five atomic sites for the unpolarized core (solid curve), the polarized core (dotted curve) and the difference between the two, magnified by four times (dashed curve). One can see that the d-band deformation (an important feature is the increase in the DOS near the dip position) is larger for the polarized core. The increase in the DOS at the dip position indicates the increase of the non-bonding states and consequently leads to the increase of the band structure energies of d-band metals with nearly half-filled d-band such as Mo. We can, therefore, conclude that the stability of the unpolarized core is brought about by the attractive band structure energy which is related to the characteristic atomic configuration around the center of the screw dislocation.

Table 2. Displacements of atomic rows and the energies of the unpolarized and polarized cores with minimum repulsive energy configurations.

	$\Delta z_1/b$	$\Delta z_2/b$	$\Delta r_1/b$ $(\Delta r_1/r_1)$	$\Delta r_2/b$ $(\Delta r_2/r_2)$	$\Delta r_3/b$ $(\Delta r_3/r_3)$	$\Delta r_3'/b$ $(\Delta r_3'/r_3')$	$\Delta E_r(eV/b)$	$\Delta E_b(eV/b)$	$\Delta E_{tot}(eV/b)$
Unpolarized core without dilatational relaxation	0	0	0	0	0	0	0.8006	0.2402	1.0407
Unpolarized core with dilatational relaxation	0	0	0.0029 (0.0053)	0.0234 (0.0215)	-0.0031 (-0.0021)	-0.0031 (-0.0021)	0.7222	0.2838	1.0060 (-0.0347)
Polarized core without dilatational relaxation	0.1002	-0.0451	0	0	0	0	0.7958	0.4410	1.2368
Polarized core with dilatational relaxation	0.1002	-0.0451	0.0027 (0.0050)	0.0197 (0.0181)	-0.0083 (-0.0058)	0.0072 (0.0050)	0.7262	0.4781	1.2044 (-0.0324)

Grain Boundary Segregation and Embrittlement in Ll2 Compound

The TB recursion formalism can also be applied to lattice defect problems in Ll2 type intermetallic compounds such as Ni_3Al, Ni_3Si and Ni_3Ga. In this subsection, we discuss the grain boundary properties of Ni_3Al compounds in relation to the grain boundary segregation and embrittlement. The present numerical calculations on the grain boundaries in the intermetallic compound Ni_3Al are performed using the two center hopping integrals proposed by Harrison[18]. The atomic energy levels E_s, E_p and E_d for s, p and d-basis orbitals are taken from atomic structure calculations by Herman and Skillman[19]. The charge neutrality condition is imposed for each atom.

The cleavage strength of the grain boundary is examined as follows: the two halves of the ordered alloy lattice with surface index (l,m,n) are separated in the <l,m,n> direction with no surface relaxation and no reconstruction. This is based on the fact that for transition metal systems the energy reduction due to the surface relaxation or reconstruction is generally small (less than a few percent) compared to the surface energy, and not of primary importance for the present problem. Actual numerical calculations are performed for the symmetric tilt [001] grain boundaries with $\Sigma = 5$ (210) and $\Sigma = 13$ (320) structures in the Ni_3Al crystal. The stable atomic structure of the grain boundary is calculated by using the quenched molecular dynamics method[20].

In Fig. 3, we present the calculated atomic configurations of the tilt grain boundaries with $\Sigma = 5$ (210) (a) and $\Sigma = 13$ (320) (b) structures in Ni_3Al crystal. Symbols $\bigcirc$ and $\square$ represent Ni and Al atoms, respectively. One can notice in Fig. 3 that the interplanar distances between (210) atomic planes or between (320) atomic plances are changed substantially near the grain boundaries. We show in Fig. 4 the calculated cleavage force f as a function of the separation R_{sep} between a pair of the crystallites for $\Sigma = 13$ (320) tilt grain boundary in Ni_3Al crystal. The cleavage force f and the separation distance R_{sep} are given in units of 10^{10} Pa and 2.524Å (nearest-neighbor distance of Ni_3Al), respectively. The maximum value f_{max} of the cleavage force characterizes the fracture strength of the crystal.

We have also calculated the cleavage forces of the impurity segregated grain boundaries and Ni- or Al-enriched grain boundaries. In the present calculation, boron and iron atoms are chosen as the segregation elements at the grain boundaries. The repulsive potential parameter C between the impurity and host atoms are determined so as to reproduce experimental values of the lattice parameters of the Ni_3Al-based ternary solid solutions[21]; $C_{Fe-Ni} = 1.104C_{Ni-Ni}$ and $C_{B-Ni} = 0.95C_{Al-Ni}$. In the Fe (or B) segregated grain boundaries, all of the Al sites at the boundary Ni/Al plane are substituted by Fe (or B) atoms, because iron solute atoms can occupy the Al-sublattice sites in the Ni_3Al crystal[21,22], and the bonding nature of boron is similar to that of Al in the crystal. The enrichment of constituent Al or Ni atoms is also considered at the boundary Ni/Al planes. All of the Al (Ni) sites at the boundary plane are substituted by Ni (Al) atoms in order to simulate the Ni (Al) enrichment (or the deviation from stoichiometry).

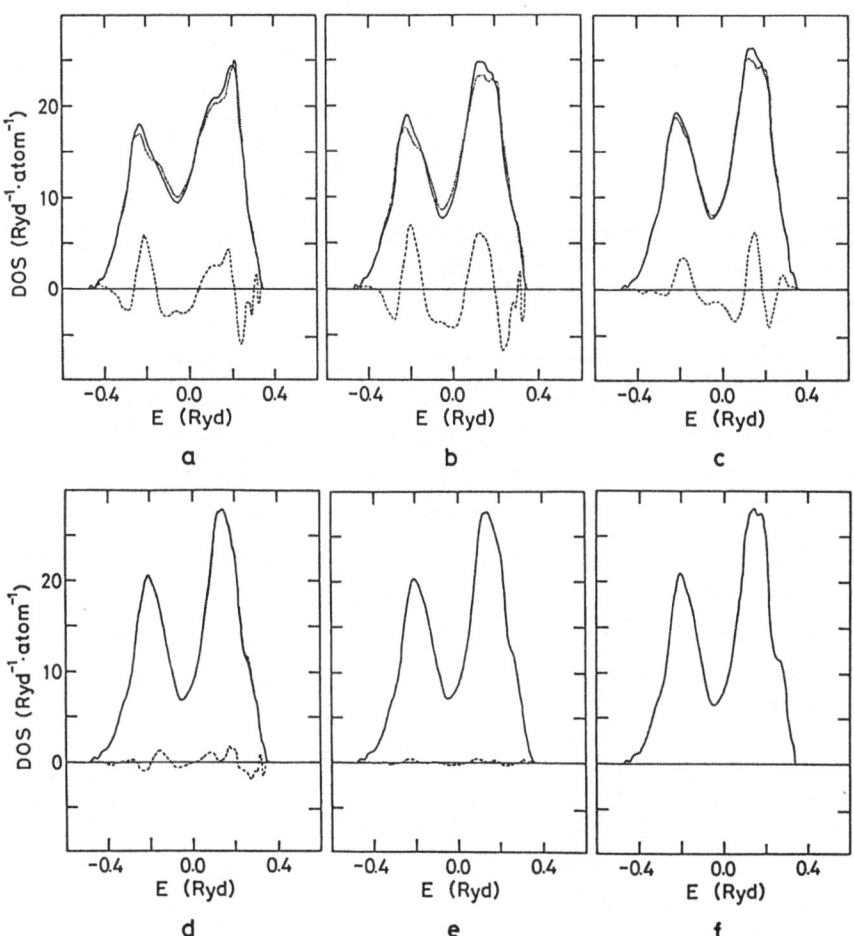

Fig. 2. Local d-band DOS (see text). (a), (b), (c), (d) and (e) are for site 1. 2. 3. 4 . and 5 in Fig. 1, respectively. (f) is for the perfect lattice.

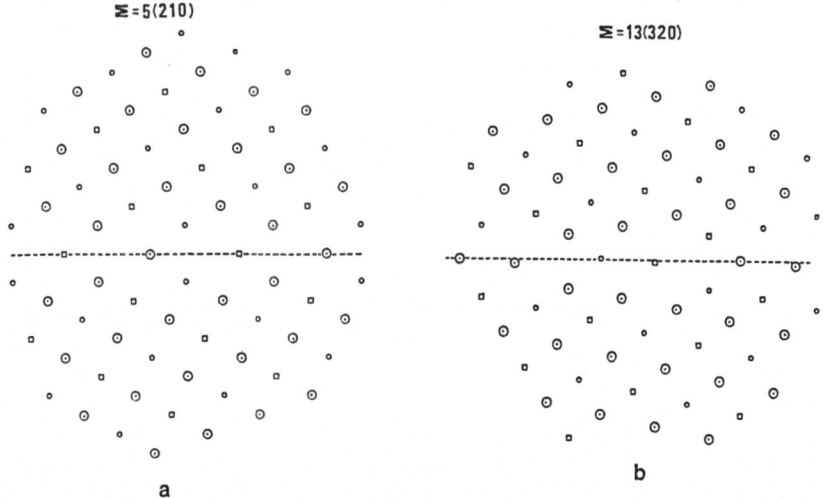

Fig. 3 Calculated atomic configurations around $\Sigma = 5$ (210) (a) and $\Sigma = 13$ (320) (b) tilt grain boundaries in Ni₃Al crystal.

We have also obtained quite similar results for $\Sigma = 5$ (210) tilt grain boundary in Ni₃Al crystal. From these cleavage force calculation we can discuss the intrinsic strength of the grain boundary as well as the effects of impurity segregation on the strength of the grain boundaries. The interesting features of our cleavage force calculations are summarized as follows: As shown in Fig. 4, the cleavage strength of the tilt grain boundaries in Ni₃Al reduces significantly (about 80%) compared to that for the perfect single crystal. One can see in Fig. 4 that the effects of impurity segregation (or enrichment of the constituent Ni or Al atoms) on the strength of the grain boundary are fairly important. The segregated boron and iron atoms at the grain boundary can increase considerably the cleavage strength of the grain boundary, but the strengthening rate of the

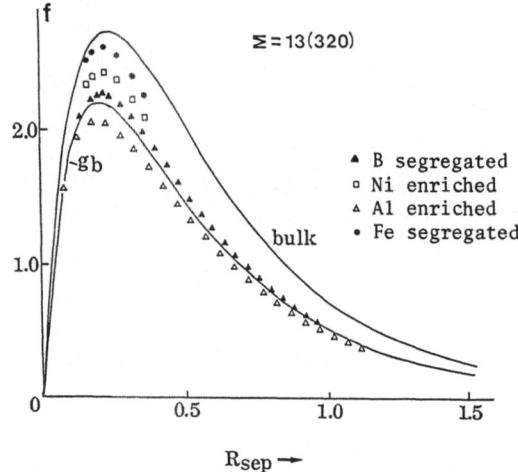

Fig. 4 Calculated cleavage force f as a function of the separation R_{sep} between a pair of the crystallites for $\Sigma = 13$ (320) tilt grain boundary in Ni₃Al crystal.

iron segregation, per single atom, is much larger than that of boron segregation. On the other hand, it is remarkable that the enrichment of Ni(Al) atoms leads to the significant increase (decrease) of the grain boundary strength of Ni_3Al crystal. This indicates that the increase of d-electron bonding, due to the enriched Ni or segregated Fe atoms, strongly enhances the grain boundary cleavage strength of Ni_3Al compound. Finally, we would like to emphasize that the theoretical findings mentioned above are in good agreement with the experimental results on the grain boundary strength of Ni_3Al compounds[23-27].

CONCLUSIONS

We have used the tight binding recursion method to investigate the fundamental properties of lattice defects in metals and alloys. In the present paper, particular attention has been focused on the dislocation in Mo and grain boundaries in an Ll_2 type intermetallic compound Ni_3Al. It has been shown that the present TB recursion method is quite useful for the interpretation of various experimental results on the mechanical properties of the materials.

REFERENCES

1. V. Vitek, Cryst. Lattice Defects, 5:1 (1974).
2. H. Suzuki, in: "Dislocation Dynamics", A.R. Rosenfield, G.T. Hahan, A.L. Bement, Jr. and R.I. Jaffee, ed., McGraw-Hill, New York (1968) p.679; S. Takeuchi, in: "Interatomic Potentials and Crystalline Defects", J.K. Lee, ed., The Met. Soc. AIME, Warrendale (1981) p. 201.
3. L.P. Kubin, Rev. Deform. Behav. Mater. I:244 (1976); IV:181 (1982).
4. F. Ducastelle, J. Phys. (Paris) 31:1055 (1970).
5. R. Haydock, V. Heine and M.J. Kelly, J. Phys. C 8:2591 (1975).
6. T.L. Einstein and J. R. Schrieffer, Phys. Rev. B 7:3629 (1973).
7. B. Legrand, Phil. Mag.B, 49:171 (1984); Phil. Mag. A, 52:83 (1985).
8. I. Terakura, K. Terakura and N. Hamada, Surf. Sci. 103:103 (1981); Surf. Sci. 111:479 (1981).
9. C.M.M. Nex, J. Phys.A 11:653 (1978).
10. N. Beer and D.G. Pettifor, in: "Electronic Structure of Complex Systems", P. Phariscau and W.M. Temmerman, ed., Plenum Press, New York (1984), p. 769.
11. K. Masuda-Jindo, J. Phys. (Paris) 47:2087 (1986).
12. J.M. Wills and W.A. Harrison, Phys. Rev. B 28:4363 (1984).
13. D.G. Pettifor and R. Podloucky, J. Phys. C, 19:1389 (1986).
14. V.K. Tewary, Adv. Phys. 22:757 (1973).
15. K. Masuda and A. Sato, J. Phys. Soc. Jpn. 50:569 (1981).
16. S. Takeuchi, Phil. Mag. A 39:661 (1979).
17. T. Aono, E. Kuramoto and K. Kitajima, in: "Strength of Metals and Alloys", R.C. Gifkins, ed., Pergamon Press, Oxfore (1982) Vol. 1, p. 9.
18. W.A. Harrison, "Electronic Structure and Properties of Solids", Freeman, San Francisco (1980).
19. F. Herman and S. Skillman, "Atomic Structure Calculations", Prentice-Hall, New Jersey (1963).
20. B. Legrand, G. Treglia, M.C. Desjouguerès and D. Spanjaard, J. Phys. C, 19:4463 (1986).
21. S. Ochiai, Y. Oya and T. Suzuki, Acta Met. 32:289 (1984).
22. Y. Mishima, S. Ochiai and T. Suzuki, Acta Met. 33:1161 (1985).
23. K. Aoki and O. Izumi, J. Jpn. Inst. Met. 43:1190 (1979).

24. C.L. White and D.F. Stein, <u>Metall. Trans.A</u>, 9:13 (1978).
25. C.T. Lui, C.L. White and J.A. Horton, <u>Acta Met</u> 33:213 (1985).
26. A.I. Taub, S.C. Huang and K.M. Chang, <u>Metall. Trans. A</u>, 42:399 (1984).
27. T. Takasugi, O. Izumi and N. Masahashi, <u>Acta Met</u>. 33:213 (1985).

ATOMISTIC SIMULATION OF SUPERDISLOCATION DISSOCIATION IN Ni3Al

M. H. Yoo

Metals and Ceramics Division
Oak Ridge National Laboratory
Oak Ridge, Tennessee 37831-6115

M.S. Daw and M.I. Baskes

Theoretical Division
Sandia National Laboratories
Livermore, CA 94550

INTRODUCTION

The source of the anomalous temperature dependence (positive) of yield and flow · strength in Ni3Al is believed to be the intrinsic lattice resistance to the motion of screw superdislocations. According to the cross-slip pinning (CSP) model,[1,2] active dislocations of the $(111)[\bar{1}01]$ primary slip system would acquire cross-slipped segments in the form of a sessile configuration on the (010) plane.[3] These segments are statistically distributed along the leading superpartial dislocation and act as the pinning points on it. In a thermally activated process of double-kink formation, the number density of the pinning points is found to increase with increasing temperature because of two anisotropy factors, the anisotropy of antiphase boundary (APB) energy[4] and the elastic shear anisotropy.[5]

The glissile-sessile transformations of <101> superdislocations were predicted on the basis of the atomistic calculations by Yamaguchi, et al.[3] who used a central force model. The results of these calculations explain why <101> dislocations become immobile when dissociated on {010} planes and thus justify the assumption made earlier by Kear and Wilsdorf.[1] Recently, Farkas and Savino[6] carried out a computer simulation study of <101> dislocation core structure using the embedded atom method (EAM).[7,8] They[6] concluded that in all cases the core structure is nonplanar, regardless of the starting configuration, extended on the $(1\bar{1}1)$ plane and not on the (111) APB plane. This result implies a low mobility of the primary slip dislocations, $(111)[\bar{1}01]$, in Ni3Al.

The purpose of this paper is to determine the core structure of <101> screw superdislocations in Ni3Al using the EAM.[9] First, the bulk properties and fault energies of Ni3Al pertinent to superdislocations are discussed. Second, the initial and boundary conditions for the atomic cell are specified. Third, the calculated results of both spontaneous dissociations and stable equilibrium configurations are presented. Finally, discussion is given on the comparison of the present results with the two previous results[3,6] including the effect of applied stress on dislocation mobility.

BULK PROPERTIES AND FAULT ENERGIES

The embedding functions and electrostatic interactions for nickel and aluminum used in the EAM were determined empirically by Foiles and Daw.[9] A reasonable overall agreement was found between the theoretical calculations and the available experimental data on phase stability, lattice vibration frequencies, point defect properties, and interfacial energies. The elastic constants calculated by the EAM are listed in Table 1 together with

Table 1. Bulk Properties of Ni_3Al

Method		a_0 (nm)	E_c (eV)	C_{ij} (10^{11} N/m^2)			A	References
				C_{11}	C_{12}	C_{44}		
Exp.		0.357	4.62	2.30	1.49	1.32	3.25	[10-12]
Central Force	(a)	0.361		1.92	0.99	0.99	2.13	[13]
	(b)				1.30	1.30	4.19	
EAM		0.357	4.63	2.52	1.37	1.26	2.20	[9]
		0.357	4.59	2.46	1.37	1.23	2.26	[7]
FLAPW		0.349		2.35	1.45	1.32	2.93	[14]

those by other calculations and the experimental measurements. The experimental values at the ground state (T = 0 K), which give the shear anisotropy factor of A = $2C_{44}/(C_{11} - C_{12})$ = 3.25, were obtained by an extrapolation.[10] The extrapolation was accomplished by scaling the temperature dependent elastic constants measured by Ono and Stern[11] in reference to the room temperature values determined by Kayser and Stassis.[12]

The central force model used by Yamaguchi, et al.[13] was based on the two sets of Ni-Al interatomic potentials, (a) and (b), such that a stable $L1_2$ structure was obtained with the lattice parameter of a_0 = 0.361 nm, C_{11} = 1.92 x 10^{11} N/m^2, and (a) C_{12} = C_{44} = 0.99 x 10^{11} N/m^2; (b) C_{12} = C_{44} = 1.30 x 10^{11} N/m^2. These are the lower and upper bound Cauchy relationships, which are consistent with the extrapolated elastic constants at 1150 K.[10]

The equilibrium lattice constant, a_0, and the cohesive energy, E_c, by the EAM listed in Table 1 are the empirically determined values, self-consistent with the experimental data. The elastic constants calculated by the EAM[9] give a relatively weaker anisotropy factor,

A = 2.20, as compared to the experimentally measured value. Also listed in Table 1 are the *ab initio* equilibrium lattice constant and elastic constants at 0 K determined recently by Fu and Yoo[14] using the first-principles full-potential linearized augmented plane-wave (FLAPW) method[15,16] based on the local-density-functional (LDF) theory.[17,18] The predicted anisotropy factor is A = 2.93.

Two types of dissociation of a superdislocation in Ni_3Al are considered, viz., (a) APB type and (b) superlattice intrinsic stacking fault (SISF) type, according to the following respective dislocation reactions:

$$[\bar{1}01] \rightarrow \tfrac{1}{2}[\bar{1}01] + \tfrac{1}{2}[\bar{1}01] \ , \tag{1}$$

$$[\bar{1}01] \rightarrow \tfrac{1}{3}[\bar{2}11] + \tfrac{1}{3}[\bar{1}\bar{1}2] \ . \tag{2}$$

The spacing between two superpartials, d, can be determined by use of the weak-beam method of transmission electron microscopy (TEM). Some of the recently determined fault energies of Ni_3Al[19,20] are listed in Table 2. In Ni_3Al, the APB-type dissociation is observed to be more prevalent than the SISF type. Each superpartial may split into two (or three) 1/6<112> type Shockley partial dislocations bounding the complex stacking fault (CSF) on a {111} plane. The CSF energy has been calculated using the EAM[9] to be $\gamma_c = 259$ mJ/m², which is considerably larger than the APB energy, $\gamma = 156$ mJ/m². This indicates that the equilibrium spacing between Shockley partials is of the order of 0.1 nm, which is the resolution limit of the weak-beam TEM.

Table 2. Fault Energies of Ni_3Al (mJ/m²)

Method	APB		(111)		Remarks	References
	(111)	(100)	CSF	SISF		
Weak-beam	110	90		~10	~Ni_3Al	[19]
TEM	190	185	210		Ni_3(Al, 1 at. % Ta)	[20]
	156	28	259	96		[9]*
EAM					Ni_3Al at 0 K	
	142	83		13		[21]

*Including further calculations in this work.

ATOMISTIC SIMULATION

The atomistic calculations were performed on two slabs with different crystalline orientations. Both slabs had a z-axis along the core and Burgers vector of the screw superdislocation, taken to be $B = a_0[\bar{1}01]$. The orientation of the first slab was with x ∥

[1$\bar{2}$1] and y ‖ [111]. The second slab was rotated around the z-axis, so that x ‖ [101] and y ‖ [010]. The atomic positions were made periodic along the z-axis, with a period of $\sqrt{2}$ a_0. In the x-y plane, a finite, rectangular region was constructed. The atoms in the region were displaced from perfect lattice sites by an anisotropic elastic displacement solution, which included some number of dislocations. After the displacement, atoms within 10 Å of the boundary in the x-y plane were held fixed, and the atomistic potential energy was minimized by relaxing the rest of the atoms.

The first set of calculations involved putting a total screw superdislocation in a slab which was nearly square in the x-y plane. The displacement field was $U = U \overset{a}{z}$ (x,y) = $(B/2\pi)$ atan2$(q\Delta y, \Delta x + r\Delta y)$, where $\Delta x = (x - x_0)$, $\Delta y = (y - y_0)$, (x_0, y_0) is the position of the dislocation core, and r and q are the real and imaginary parts of the roots, $p = r \pm iq$, of a quadratic characteristic equation. For the first orientation $r = \sqrt{A}(A - 1)/(A + 2)$ and q = $3\sqrt{A}/(A + 2)$. For the second, r = 0 and q = $1/\sqrt{A}$. Several positions relative to the atomic lattice were tried for the dislocation core. The goal of this calculation was to observe the spontaneous splitting of the superscrew. It was found that the final geometry was sensitive to the initial starting position. In all cases, the superdislocation split into two screw superpartials, defined by Eq. (1). The superpartials always split along {111} planes, though sometimes they dissociated on two different {111} planes, creating a "V" shape. The most stable configuration was for the case where the superscrew split into two superpartials on the same {111} plane, separated by a (111) APB.

Superpartials Dissociated on the (111) Plane

The second set of calculations started with two screw superpartials separated on the (111) plane. The displacement field in that case was the same as before, but with two contributions, of Burgers vector b = B/2, at some initial separation. In this way, the boundary atoms were brought into closer consistency with the structure of the separated superpartials. The boundary atoms were again held fixed and the energy of the atomistic region was relaxed. In this case, it was observed that the superpartials were apparently "stuck"; that is, the relaxation process could not move the superpartials. Evidently, there is a barrier to moving the superpartials on the (111) plane. When the superpartials are not very close together, the elastic force pushing them apart is too weak to overcome this barrier. Part of this barrier may be due to the boundary conditions, where the boundary atoms are fixed at certain positions.

In order to find the correct separation between the superpartials, we calculated the total energy at different separations. The total energy has two contributions: the energy of the atomistic region and the energy of the strained medium outside. The strain energy outside was calculated for an infinite medium *excluding* the rectangular region covered by the atomistic calculation. The atomistic region was made as large as possible (over 30,000 atoms) and square. When these two contributions were added, the minimum energy was found to be at a separation of 60 Å. This separation is in close agreement with what one would expect from elasticity, based on the APB energies given by the EAM.

Figure 1 shows the stable equilibrium configuration of the left superpartial dislocation core. The elastic center (marked with a "+") was set at $(x_O, y_O) = (0.73, 1.03)$ Å with respect to the Al atom at the origin. The figure displays the net atomic displacements both in plane (top panel) and axial (bottom panel). The displacements shown are the difference between the final atomic displacement and the initial anisotropic elastic displacement, and are therefore due entirely to the atomistic relaxation. The size of circles indicates the magnitude of the axial displacement in the +z direction, and similarly, the cross in the -z direction. The arrows indicate the displacement in the plane of drawing, $(\bar{1}01)$. In all cases, Figs. 1-4, these symbols were magnified by a factor of ten. The displacements demonstrate that the superpartial is itself dissociating into Shockley partials also separated (but by only a few Å) on the (111) plane.

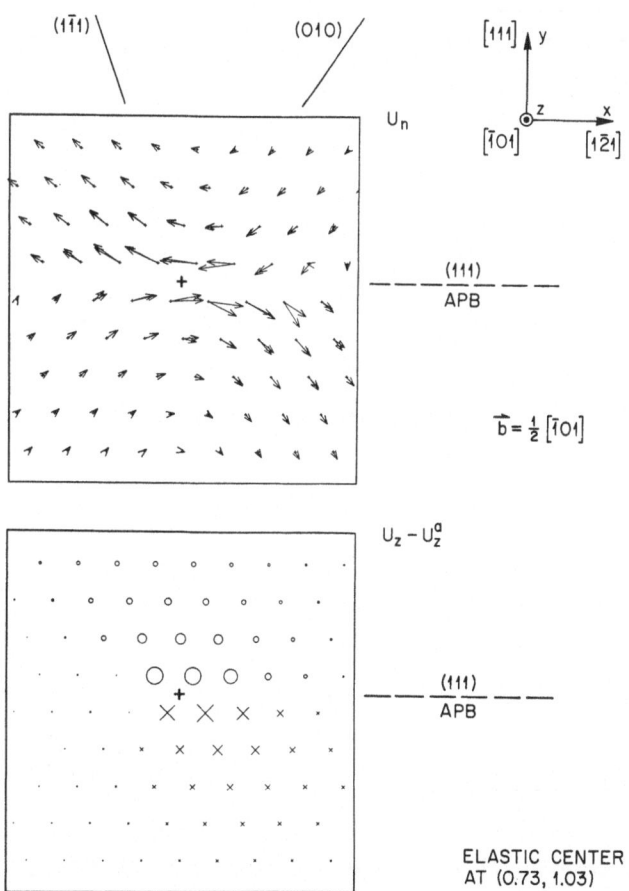

Figure 1.   Stable equilibrium configuration of planar (111) dissociation of the left superpartial, 1/2 [$\bar{1}$01] on the (1$\bar{1}$1) plane. Magnification factor = 10.

Figure 2 shows the same calculation, but where the elastic center was set at (x_O, y_O) = (0.0146, 0.0206) Å with respect to the Al atom. The result here shows the superpartial dissociating onto the (1$\bar{1}$1) plane. This nonplanar configuration is, however, at metastable equilibrium: its total energy is higher than that of Fig. 1 by $\Delta E \sim 0.15$ eV/B. The nonplanar configuration was achieved by choice of the initial starting position and demonstrates that one must be careful in exploring the effects of initial and boundary conditions. We have also repeated these calculations with the EAM functions of ref. 21 with the same results: the planar dissociation on the (111) is more stable than the nonplanar dissociation.

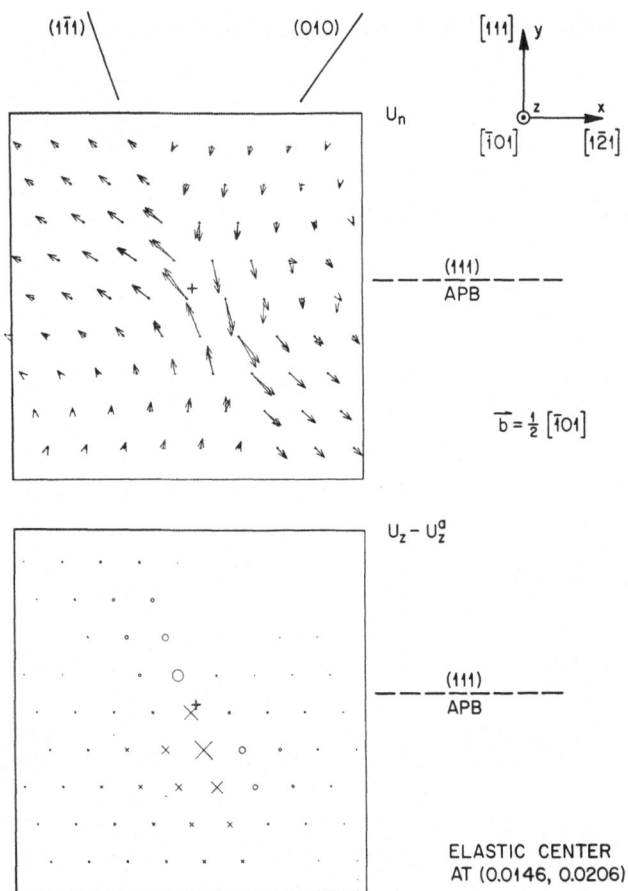

Figure 2. Metastable equilibrium configuration of nonplanar (111) dissociation of the left superpartial, 1/2 [$\bar{1}$01], on the (1$\bar{1}$1) plane. Magnification factor = 10.

A nonplanar dissociation on the (1$\bar{1}$1) plane was reported by Farkas and Savino,[6] using the EAM functions of ref. 21. Our calculations with the same functions seem to indicate that the configuration reported by Farkas and Savino[6] was in fact a metastable one, and that their calculations were biased by their choice of initial position.

Superpartials Dissociated on the (010) Plane

The next set of calculations explores the dissociation on the (010) plane. This dissociation can be induced by starting with the elasticity solutions corresponding to two superpartials separated along the (010) plane. As before, the details of the superpartial core depend on the initial position of the elastic center relative to the atomistic lattice. Figures 3 and 4 show the two stable cases of the dissociation of the superpartial. The dissociation occurs on either the (111) or ($1\bar{1}1$) plane, with a difference in energy less than 0.01 eV/B. These configurations are in excellent agreement with ones reported in recent high resolution electron microscopy work by M. J. Mills.[20]

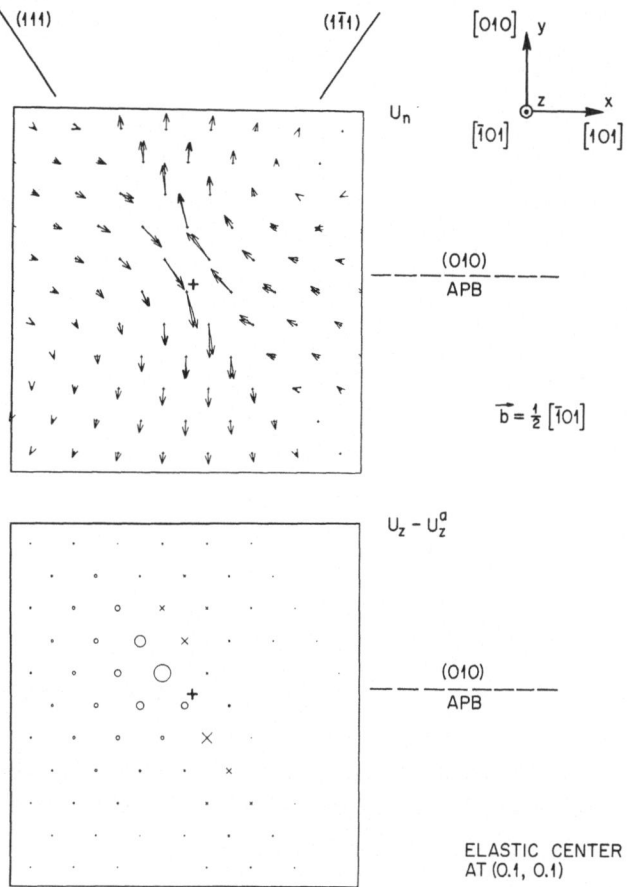

Figure 3.   Stable equilibrium configuration on non-planar (010) dissociation of the left superpartial, 1/2 [$\bar{1}$01], on the (111) plane. Magnification factor = 10

DISCUSSION

Depending on the initial position of the elastic center, a variety of dissociation configurations are possible. For the (111) slip plane, the stable configuration is a planar dissociation, i.e., spreading of the superpartial core region on the (111) plane. A

407

nonplanar dissociation for the (111) case can be produced by a different initial starting configuration, but this is metastable. For the (010) slip plane, two stable configurations of similar energy were obtained, both corresponding to a spreading on a {111} plane, producing a nonplanar structure. The details of the superpartial core appear quite similar in all cases, indicating that the effect of the APB is secondary.

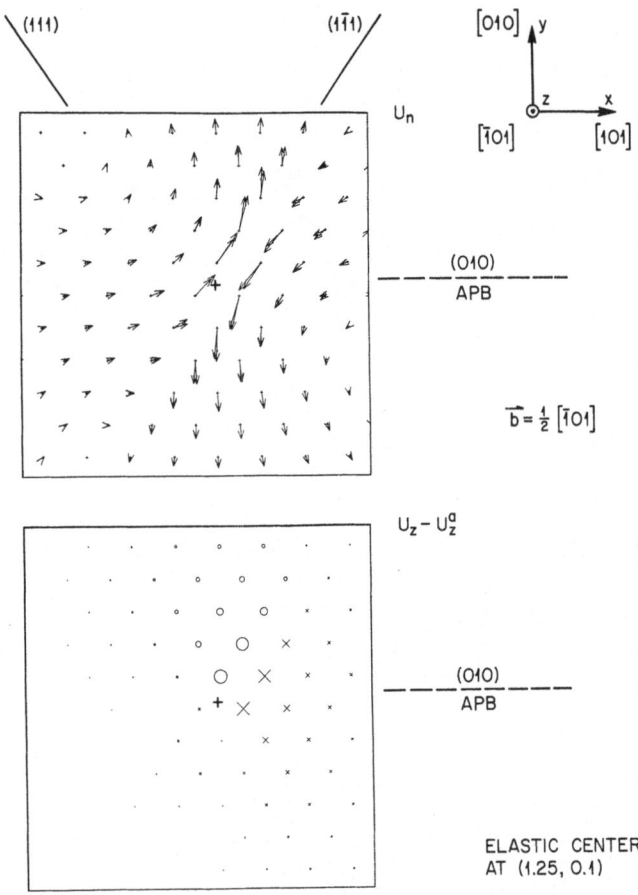

Figure 4.   Stable equilibrium configuration of nonplanar (010) dissociation of the left superpartial, 1/2 [$\bar{1}$01], on the (1$\bar{1}$1) plane. Magnification factor $\doteq$ 10.

The salient features of the calculated core structures are in good agreement with those obtained by Yamaguchi, et al.[3] However, there appears to be a large difference in the description of the atomic relaxation at an APB interface. Beauchamp, et al.[23] used the interatomic potentials constructed by Yamaguchi, et al.[13] to determine the atomic equilibrium positions and energy minima for a wide range of APB plane orientations. Beauchamp, et al.[23] report a large relaxation of the (111) APB plane, viz., $\Delta = (\gamma^* - \gamma)/\gamma^*$ = 0.62, where γ^* and γ are the unrelaxed and relaxed APB energies on the (111) plane, respectively. Our calculations based on the EAM give considerably lower values of $\Delta_{111} =$

0.22 and $\Delta_{010} = 0.15$. The central force model used for APB energy determination clearly overestimates the contribution of atomic relaxation to the (111) APB energy.

It is conceivable that under an applied stress any superpartial core at the stable equilibrium (Fig. 1) can be activated into the metastable higher energy configuration, Fig. 2. On one hand, if such a glissile-sessile transformation occurred on the leading superpartial, this would constitute the first stage of (111)-($1\bar{1}1$) cross-slip process. On the other hand, if such a transformation occurred only on the trailing superpartial, but not on the leading one, then one expects that the width of an APB ribbon would appear much larger than the equilibrium width, d = 0.6 nm. The experimental evidence obtained by an *in situ* straining TEM[24] may be a case in point. This provides an additional source of glide resistance via the fault dragging mechanism, which is consistent with the force couplet model (FCM) for anomalous yield behavior.[25] Our preliminary calculations to study the effect of applied stress show that the mobility of a superpartial is much higher on the (111) plane than on the (010) plane, which is consistent with the work by Paidar, et al.[26] A full account of the dislocation mobility in Ni_3Al will be presented elsewhere.

SUMMARY

1. A screw superdislocation, [$\bar{1}01$], in Ni_3Al dissociates spontaneously into two superpartials bounding an APB ribbon.
2. Depending on the orientation and the initial position of the elastic center, a variety of dissociation configurations are possible.
3. The stable equilibrium configurations are found to be a (111) planar dissociation on the (111) APB plane, which gives d = 0.6 nm and $\gamma_{111} = 143$ mJ/m^2.
4. On the (010) slip plane a nonplanar dissociation occurs on either the (111) or the ($1\bar{1}1$) plane, which is at a stable equilibrium.

ACKNOWLEDGMENTS

The authors would like to thank M. J. Mills and C. L. Fu for helpful discussions and Connie Dowker for preparation of the manuscript. This research was sponsored by the Office of Basic Energy Sciences, Division of Materials Sciences, U.S. Department of Energy, under contract DE-AC05-84OR21400 with Martin Marietta Energy Systems, Inc., and under contract DE-AC05-76DP00789 with Sandia National Laboratories.

REFERENCES

1. B. H. Kear and H.G.F. Wilsdorf, *Trans. TMS-AIME* **224**, 382 (1962).
2. S. Takeuchi and E. Kuramoto, *Acta Metall.* **21**, 45 (1973).
3. M. Yamaguchi, V. Paidar, D. P. Pope, and V. Vitek, *Philos. Mag. A* **45**, 867 (1982).
4. V. Paidar, D. P. Pope, and V. Vitek, *Acta Metall.* **32**, 435 (1984).
5. M. H. Yoo, *Scr. Metall.* **20**, 915 (1986).
6. D. Farkas and E. J. Savino, *Scr. Metall.* **22**, 557 (1988).

7. A. F. Voter, D. Srolovitz, and S. P. Chen, *MRS Symposium I Proceedings* (1986).
8. M. S. Daw and M. I. Baskes, *Phys. Rev. B* **29,** 6443 (1984).
9. S. M. Foiles and M. S. Daw, *J. Mater. Res.* **2,** 5 (1987).
10. M. H. Yoo, *Acta Metall.* **35,** 1559 (1987).
11. K. Ono and R. Stern, *Trans. TMS-AIME* **245,** 171 (1969).
12. F. X. Kayser and C. Stassis, *Phys. Status Solidi* **A64,** 335 (1981).
13. M. Yamaguchi, V. Vitek, and D. P. Pope, *Philos. Mag.* **43,** 1027 (1981).
14. C. L. Fu and M. H. Yoo, *Philos. Mag. Lett.* **58(4),** 199-204 (1988).
15. E. Wimmer, H. Krakauer, M. Weinert, and A. J. Freeman, *Phys. Rev. B* **24,** 864 (1981)
16. C. L. Fu, M. Weinert, and A. J. Freeman, *Phys. Rev.* (1988).
17. P. Hohenberg and W. Kohn, *Phys. Rev. B* **136,** 864 (1964).
18. W. Kohn and L. J. Sham, *Phys. Rev. A* **140,** 1133 (1965).
19. J. Douin, P. Veyssiere, and P. Beauchamp, *Philos. Mag. A* **54,** 375 (1986).
20. M. J. Mills, Sandia National Laboratories (Livermore), private communication (1988).
21. S. P. Chen, A. F. Voter, and D. J. Srolovitz, *Scr. Metall.* **20,** 1389 (1986).
22. M. H. Yoo and B.T.M. Loh, *J. Appl. Phys.* **43,** 1373 (1972).
23. P. Beauchamp, J. Douin, and P. Veyssiere, *Philos. Mag. A* **55,** 565 (1987).
24. I. Baker, J. A. Horton, and E. M. Schulson, *Philos. Mag. Lett.* **55,** 3 (1987).
25. M. H. Yoo, J. A. Horton, and C. T. Liu, *Acta Metall.* **36,** 2935 (1988).
26. V. Paidar, M. Yamaguchi, D. P. Pope, and V. Vitek, *Philos. Mag. A* **45,** 883 (1982).

A NEW METHOD FOR COUPLED ELASTIC-ATOMISTIC
MODELLING

Stephan Kohlhoff and Siegfried Schmauder

Max-Planck-Institut für Metallforschung
Institut für Werkstoffwissenschaft
Seestr. 92, D-7000 Stuttgart
Federal Republic of Germany

INTRODUCTION

Molecular Dynamics and Molecular Statics have become important tools for model investigations of crystal defects. In spite of the ever increasing computer power the size of the models which can be treated by these methods is very limited. Therefore, in order to avoid surface effects it is common practice to employ one of the following techniques:

- periodic boundary conditions or
- semidiscrete methods.

The latter combines lattice and continuum theory. Various methods of this type have been used successfully to study crystal defects [1,2,3,4,5]. Since an analytical description of the continuum is used simplifying assumptions have to be made, which may contain isotropic or/and linear material behaviour. Finite strains, i.e., non-linear elasticity theory is very seldom considered. Furthermore, modelling capabilities are quite limited because of complicated expressions arising from boundary conditions when arbitrary geometries are considered.

A relatively new approach in the context of semidiscrete methods is the approximation of the continuum by the finite element method (FEM) [6,7]. Beside the model used by Mullins no further developments based on this idea have been published. This model, however, does not properly describe the transition between the non-local lattice and the local continuum.

In this paper a new, improved semidiscrete method based on FEM is presented which makes some contributions to the advancement of elastic-atomistic calculations. As an application of this method the analysis of a crack in α-iron is presented.

THE METHOD

In semidiscrete methods the model is divided into continuum, lattice, and transition

(Fig. 1). The FEM which is employed for the continuum is desribed in detail in the literature (see for example [8]). It can cope with geometrical as well as physical non-linearities, i.e., full non-linear elasticity theory may be applied. The only factors which impose restrictions on three-dimensional calculations are computer time and storage. The FEM as will be shown later is also particularly suitable because it requires a discretization which can be matched to the lattice geometry. The atomistic part may be modelled by pair-potentials or other, more advanced techniques, e.g. embedded atom method (EAM) [9]. The transition where continuum and lattice overlap deserves special attention.

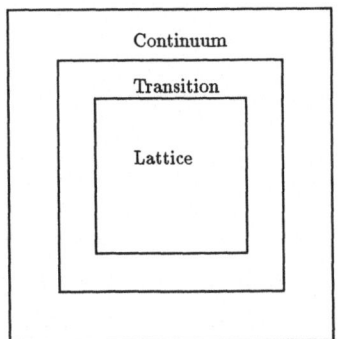

Fig. 1.   Scheme of a semidiscrete model

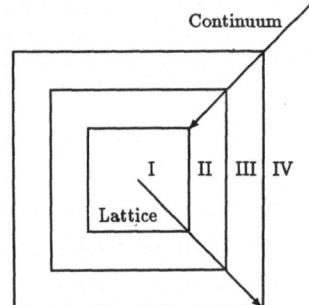

Fig.2. Overlapping lattice and continuum.

The Transition

An inherent difficulty in semidiscrete methods is the proper description of the transition from the lattice to the continuum. Cauchy's stress principle [10] on which classical continuum mechanics is based defines internal forces as equivalent to surface tractions, i.e., it neglects the finiteness of interatomic forces. This clearly contradicts lattice theory. On the other hand, since continuum mechanics is used very successfully and makes reasonable predictions about material behaviour there must be domains within the model where both approximations are valid.

In the so called non-local elasticity theory which can be derived from lattic theory [11] and which provides the framework for the model described below the stress is not a function but a functional of the coordinate vector **x** and may be written [12] as:

$$\sigma_{ij}(x) = \int_{V'} f(x,x',\beta)\ \sigma_{ij}^*(x')dV' \tag{1}$$

where σ^* is the conventional stress tensor as defined in local elasticity theory and f is a two point vector function with the following properties:

- the maximum is located at $\mathbf{x}' = \mathbf{x}$
- $\int_{V'} f(x, x', \beta)\ dV' = 1$.

The parameter β determines the degree of non-locality and is a material constant proportional to l/L where l is an internal characteristic length associated with the range of the interatomic force law and L is an external characteristic length and may, in the present case, be associated with the dimension of a finite element at the transition. As β approaches zero the function f reduces to a delta function $\delta(|x' - x|)$ and the theory of non-local elasticity renders local response. The total stress may, therefore, be split into a local and a non-local part [13] $\sigma = \sigma_l + \sigma_n$ with

$$\sigma_{l_{ij}} = \sigma_{ij}^*, \tag{2}$$

$$\sigma_{n_{ij}} = \int_{V'} [f(x,x',\beta) - \delta(|x' - x|)] \ \sigma_{ij}^* (x') \ dV'. \tag{3}$$

Substituting σ^* by the constitutive equation $\sigma_{ij}^* = c_{ijkl}\varepsilon_{kl}$ the last expression may be rewritten as:

$$\sigma_{n_{ij}} = c_{ijkl} \int_{V'} [f(x,x',\beta) - \delta(|x' - x|)] \ \varepsilon_{kl} (x') \ dV'. \tag{4}$$

By virtue of the properties of f it can directly be concluded that the non-local stresses will vanish in the case of homogeneous deformation, i.e. when the strain ε is constant within the interaction volume V'. However, if the strain gradient is finite the non-local stress is only present in the lattice and remains, therefore, unbalanced at the transition. For this reason the stress fields in continuum and lattice are incompatible. In the present model we circumvent this difficulty by mutually applying displacement, i.e. Dirichlet boundary conditions to the continuum and the lattice.

The Model

The model is set up by defining four regions (Fig. 2). Region I contains the lattice, IV the continuum, and II and III the transition where both parts are superimposed. Region II defines the boundary for the continuum and actually collapses to a line or surface, respectively. Region III contains the boundary atoms of the lattice and must, therefore, have a finite thickness equal to the range of the interatomic potential. The atoms in III are displaced according to the corresponding nodal points in the finite element mesh and the nodes on line II are displaced according to the corresponding atoms. The present model shows the following properties.

1. Both parts of the model must sustain the same displacements within the transition. Therefore, we can conclude that the strains at the transition from region II to III are also equal.

2. Since continuum and lattice are only coupled via displacements equilibrium is not explicitly assumed between the two domains. This guarantees that the incompatible stress fields will not affect each other.

The following conclusions can be drawn from these statements. If the continuum is loaded with constant stress the corresponding strain will be introduced into the lattice, too, independent of the force law. The complete model can reproduce constant strain and should, therefore, be able to reproduce constant stress, too. This has not always been possible ;with previous methods ([5,7]).

Since the second statement would violate basic mechanical principles we will introduce an additional requirement equivalent to the equilibrium condition which will be derived as follows. If the elastic energy W is a continuous differentiable function of strain ε we may expand it in a series:

$$W(\varepsilon_{ij}) = W(0) + \frac{\partial W}{\partial \varepsilon_{ij}}\bigg|\varepsilon_{ij} + \frac{1}{2}\frac{\partial^2 W}{\partial \varepsilon_{ij}\varepsilon_{kl}}\bigg|_0 \varepsilon_{kl} + ... \tag{5}$$

If we drop the irrelevant term $W(0)$ we may conclude that the strain energy and all its derivatives in lattice and continuum are equal if the strain and all constants of this series have the same value for both. In practice, however, we cut off this series after the second or third derivative. These constants are identified as the elastic constants of first, second, etc. ... order.

$$1.\quad \text{order:}\quad b_{ij} = \frac{\partial W}{\partial \varepsilon_{ij}}\bigg|_0 \qquad 2.\ \text{order:}\quad c_{ijkl} = \frac{\partial^2 W}{\partial \varepsilon_{ij}\,\partial \varepsilon_{kl}}\bigg|_0 \tag{6}$$

In conclusion, equilibrium is guaranteed if the interatomic potential reproduces the elastic constants, or vice versa the elastic constants used for the continuum are set to those derived from the potential. In particular, the potential must yield zero stress for the perfect lattice. In addition, it should be noted that in equation (5) only local quantites are used, i.e., continuum and lattice are indirectly coupled via local stresses when the elastic constants are derived from the potential as, for example, shown by Johnson [14]. This equilibrium condition is equivalent to the principle of virtual work. A last important remark on this method is that phase transitions must not coincide with the coupling because this would yield non-equilibrium. This new coupling method has been implemented in the non-linear finite element program system **LARSTRAN**[1].

ANALYSIS OF A CRACK IN α–IRON UNDER MODE-I LOADING

For illustration of this new coupling method the analysis of an atomistically sharp crack is presented. Three different orientations are considered: (010)[001], (010)[101], and (10$\bar{1}$)[101]. The cleavage plane is denoted by the Miller indices in parentheses and the crack front direction by those in brackets. Fig. 3 depicts the two-dimensional finite element mesh surrounding the lattice of the (010)[001] orientation consisting of 1800 atoms. The

---

[1]**LARSTRAN** is a trademark of Lasso Ingenieurgesellschaft Diez, Hindenlang, Kurz; Nobelstrasse 15; D-7000 Stuttgart 80.

complete model has a width of 200 lattice constants. The crack front is initially always located in the center of the model.

The structures are loaded by displacement boundary conditions according to the anisotropic linear elastic solution [15]. The loading is, therefore, completely determined by the K_I-value. For the atomistic part the Finnis-Sinclair [16] potential is used and for the continuum anisotropic non-linear plane strain elasticity with the corresponding elastic constants is employed. All models were analyzed dynamically with a constant loading rate of 0.5 K_G/ps. K_G is the Griffith value of the stress intensity factor calculated from the potential and the anisotropic elastic constants. As initial conditions the models were subcritically loaded to prevent receding of the crack and relaxed so that no unbalanced stresses were present.

From subsequent views of the models it can be seen that the finite element part of the transition has been matched to lattice geometry, i.e., nodal points are situated at the same positions as the corresponding atoms. In the Mullins model many atoms had been pinned to one element which must yield a too stiff continuum which in turn is unable to react to local stress concentrations at the transition.

RESULTS

Fig. 4 shows the (010)[001] crack after 6 ps. The crack has proceeded on a (110) plane in a purely brittle fashion. This analysis was repeated with a lower loading rate of 0.1 K_G/ps but no significant difference was observed except that the crack inclination turned upwards on the ($\bar{1}$10) plane. The same result was also obtained by Gehlen et al. [17]. Markworth et al. [18] and Mullins [7], on the other hand, observed cleavage on the (100) plane. In all these calculation the Johnson pair-potential [19] was employed.

At about 7 ps the cleavage plane changes again to (100) (Fig. 5). This is probably caused by the external loading which favors (100) cleavage and the redistribution of the stresses due to the change in geometry. This kind of behaviour has also already been observed by deCelis et al. [20]. These authors associated the two different failure modes with the type of boundary condition, i.e. being derived from the isotropic or anisotropic solution. This, however, does not give a physical interpretation of the phenomenon. It is interesting to note that these results may be obtained by simple pair-potentials as well as by multi-body potentials. Another important point is that all previous analyses were made by applying the displacement field of the continuum solution for the corresponding load to the whole model and subsequently relaxing the unbalanced forces. Since the behavior of the model may be loading path dependent this seems to be a somewhat unrealistic procedure. These observations lead to the conclusion that the (010)[001] crack favours (110) cleavage no matter what type of potential used. However, a too small model and inappropriate boundary conditions may enforce (100) cleavage.

The (010)[101] crack (Fig. 6) fails in a perfect brittle manner. The critical stress level relative to K_G is also lower than before which has been observed by Mullins, too. These results strongly suggest that the crack front in a three-dimensional (010)[001] model

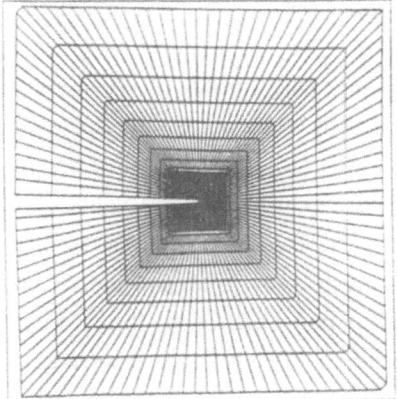

Fig. 3   Model overview.

Fig. 4   Inclined cleavage in the (010)
[00$\bar{1}$] crack.

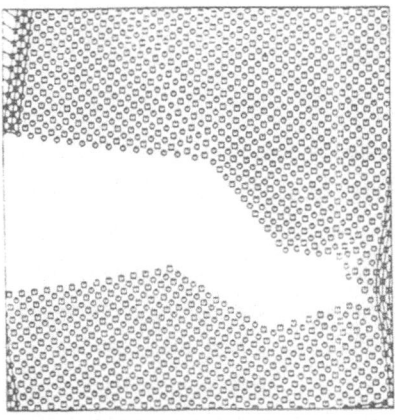

Fig. 5   Change of cleavage in the
(0$\bar{1}$0)[001] crack.

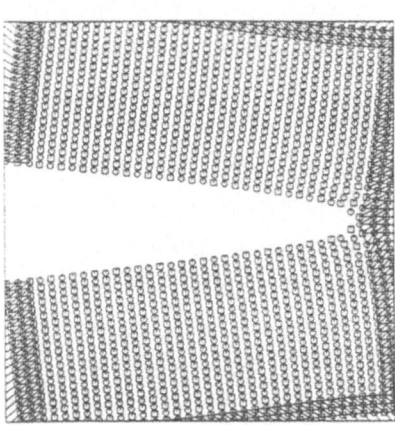

Fig. 6   Propagated (010)[101] crack.

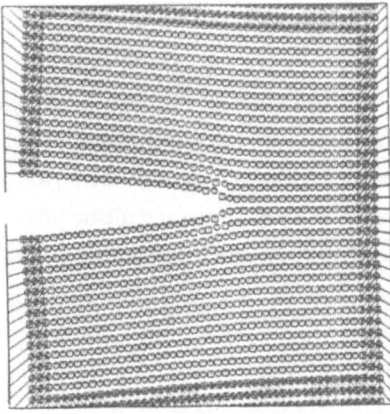

Fig. 7   Propagated (10$\bar{1}$)[101] crack.

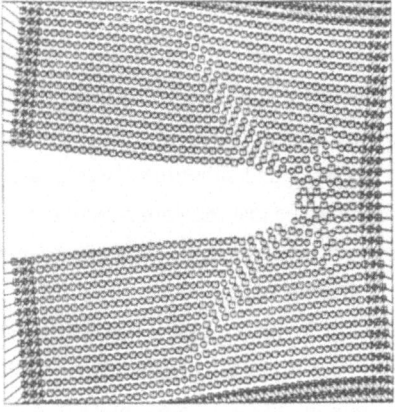

Fig. 8   Dislocations emitted from the
(10$\bar{1}$)[101] crack.

will be faceted to allow easy crack growth within the (010) crack plane.

The behaviour of the $(10\bar{1})[101]$ crack differs significantly from the previous ones. While, at first, brittle crack growth is to be observed (Fig. 7) dislocations were emitted after 5 ps (Fig. 8). Because the angle of maximum shear stress $\tau_{r\phi}$ - 53° in the isotropic case - coincides with the inclination of the $(12\bar{1})$ plane (55°) very well this plane could have been activated as a glide plane. In addition, $(12\bar{1})$ twins have developed backwards from the crack tip due to the high shear strains.

CONCLUSIONS

A new method for elastic atomistic analysis has been presented. It combines a correct description of the transition with a versatile, easy-to-use continuum method. The major advantages are summarized as follows: The present method leaves less uncertainties on the coupling between continuum and lattice. All standard algorithms to solve the dynamic or static problem may be used. The application of boundary conditions is easy and is achieved in a most realistic way. Anisotropic, non-linear matrial behaviour and arbitrary geometries may also be considered.

The present analyses support previous results obtained by other researchers using simple pair-potentials. The superposition of the results for the (010) cleavage leads to a prediction for the shape of the crack front which needs experimental confirmation. The $(10\bar{1})[101]$ crack which apparently has not previously been analyzed yields dislocation emission. Twinning was also observed which seems reasonable because of the shock-like loading at low temperature. This analysis has shown that if the geometrical requirements are optimally fulfilled plastic crack growth is possible in α-iron even at very low temperatures.

ACKNOWLEDGMENTS

The authors gratefully acknowledge careful proofreading of the manuscript and helpful discussion with A.T. Paxton. We also would like to thank P. Blöchl for assistance concerning the implementation of the Finnis-Sinclair potential. The support by **Lasso** Ingenieurgesellschaft is greatly appreciated.

REFERENCES

1.   J.E. Sinclair, J. Appl. Phys. 42:5321 (1971).
2.   P.C. Gehlen, J.P. Hirth, R.G. Hoagland, M.F. Kanninen, J. Appl. Phys. 43:3921 (1972).
3.   R.G. Hoagland, J .P. Hirth, P.C. Gehlen, Phil. Mag., 34:413 (1976).
4.   J.E. Sinclair, P.C. Gehlen, R.G. Hoagland, J.P. Hirth, J. Appl. Phys. 49:3890 (1978).
5.   C. Teodosiu, in: "Continuum Models of Discrete Systems 4", O. Brulin, R.K.T. Hsieh, eds. (North Holland, 1981).
6.   M.I. Baskes, C.F. Melius, W.D. Wilson in: "Interatomic Potentials and Crystalline Defects", October 1980.
7.   M. Mullins, M.A. Dokainish, Phil. Mag. 46:771 (1982).

8. O.C. Zienkiewicz, "The Finite Element Method" (McGraw-Hill, London, 1977).
9. M.S. Daw, M.I. Baskes, Phys. Rev. Lett. 50:1285 (1983).
10. C. Teodosiu, "Elastic Models of Crystal Defects" (Springer, 1982).
11. E. Kröner,, B.K. Datta, Z.f. Physik 196:203 (1966).
12. A.C. Eringen, J. Appl. Phys. 54:4703 (1983).
13. D. Kessel, E. Kröner, Z. Naturforsch 25a:1046 (1970).
14. R.A. Johnson, Phys. Rev. 134:2094 (1972).
15. G.C. Sih, H. Liebowitz, Mathematical theories of brittle fracture, in: "Fracture", Vol. 2, (Academic Press, 1968).
16. M.W. Finnis, J.E. Sinclair, Phil. Mag., 50:45 (1984). Erratum *ibid* 53:161 (1986).
17. P.C. Gehlen, G.T. Hahn, M.F. Kannen, Scripta Metall. 6:1087 (1972).
18. A.J. Markworth, L.R. Kahn, P.C. Gehlen, G.T. Hahn, Res Mechanica 2:141 (1981).
19. R.A. Johnson, Phys. Rev. 134:1329 (1964).
20. B. deCelis, A.S. Argon, S. Yip, J. Appl. Phys. 54:4864 (1983).

SELF-DIFFUSION AND IMPURITY DIFFUSION OF FCC METALS USING THE EMBEDDED ATOM METHOD*

J.B. Adams, S.M. Foiles, and W.G. Wolfer

Theoretical Division
Sandia National Laboratories
Livermore, CA 94550

INTRODUCTION

Diffusion in FCC metals at medium and high temperatures occurs primarily by a vacancy mechanism. Diffusion is dominated by the contribution of mono-vacancies, but the contribution of di-vacancies is significant at high temperatures. Thus, the total diffusion rate D is given by:

$$D = D_{1v} + D_{2v} = D_{iv}^{o} \exp(-Q_1/kT) + D_{2v}^{o} \exp(-Q_{2v}/kT) \tag{1}$$

where D_{1v} is the contribution of mono-vacancies, D_{2v} is the contribution of di-vacancies, D_{1v}^{o} and D_{2v}^{o} are constants, Q_{1v} and Q_{2v} are the activation energies for the mono-vacancy and di-vacancy mechanism, respectively, k is Boltzmann's constant, and T is the temperature.

For the case of <u>self-diffusion</u>, Q_{1v} and Q_{2v} are the sum of the formation and migration energies of mono-vacancies and di-vacancies, respectively. For <u>impurity diffusion</u>, the mono-vacancy mechanism is well-described by the classic "five-frequency formula"[1]. The formula allows the calculation of the diffusion rate of impurities, D_2, relative to self-diffusion, D_0, according to the formula:

$$\frac{D_2}{D_o} = \frac{f_2}{f_o} \frac{w_2}{w_1} \frac{w_4}{w_0} \frac{w_1}{w_3} \tag{2}$$

Here the f's are correlation factors; $f_0 = 0.7815$ (for fcc metals) and f_2 is a function of the w's[1]. The w's are jump frequencies for the following atomic motions: w_0 - vacancy

* Work supported by Department of Energy, Office of Basic Energy Sciences.

419

motion in the bulk; w_1 - vacancy motion from a site adjacent to an impurity to a similar adjacent site; w_2 - motion of an impurity atom into an adjacent vacancy; w_3 - vacancy motion from a site adjacent to an impurity to a non-adjacent site (2nd, 3rd, or 4th nearest neighbor); w_4 - the reverse of w_3.

In a previous paper[2] we demonstrated how the Embedded Atom Method (EAM)[3] could be used to calculate the diffusion rate of He in Ni by four different diffusion mechanisms. In the present paper we use the EAM to calculate the activation energies for self-diffusion of FCC metals by both a mono-vacancy and di-vacancy mechanism. We also calculate the w's in Eq. (2) to determine the activation energies for impurity diffusion. The EAM is a semi-empirical method based on density functional theory. It is analogous to pair potentials, but it involves both pairwise and many-body interactions. In this paper we use the EAM functions known as the AFW functions[4], which are quite similar to the previous FBD functions[5] but based on new estimates of vacancy formation energies. These functions were previously shown to well-describe many physical properties of metals, including diffusion[4].

SELF-DIFFUSION DUE TO MONO-VACANCY AND DI-VACANCY MECHANISMS

Using the EAM, it is possible to calculate the formation and migration energies of vacancies, the sum of which is the activation energy for self-diffusion by a mono-vacancy mechanism. In Table I we compare the EAM results with experimental measurements, as analyzed by Neumann and Tolle[6], Peterson[7], and Schule[8]. The analysis by Neumann and Tolle is perhaps the most accurate, since it resulted in the smallest error limits on the divacancy parameters for Cu, Ag, and Au. The experimentally measured diffusion coefficient for Pd has not been divided into the contributions due to vacancies and di-vacancies, so the actual value for Q_{1v} should be slightly lower than the experimentally measured value. This is especially likely since the measurements by Peterson[9] were conducted at temperatures up to the melting point, where the effect of di-vacancies becomes significant.

The EAM was also used to calculate the formation and migration energies of divacancies, the sum of which is the activation energy for self-diffusion by a di-vacancy mechanism. In Table II we compare the calculations with the three experimental analyses. In general, the AFW functions yield reasonable agreement with experiment. It should be noted that the experimental uncertainty is much larger for di-vacancies, since they are significant only at high temperatures.

IMPURITY DIFFUSION BY A MONO-VACANCY MECHANISM

The ratio of impurity diffusion to self-diffusion is given by the five-frequency formula of Eq. (2). The purpose of this section is to present the calculations of the jump frequencies which in Eq. (2) determine the ratio between impurity diffusion and self-diffusion. The experimental values of self-diffusion may then be used with our results to determine the rate of impurity diffusion.

420

TABLE I Comparison of the activation energies (in eV) for self-diffusion by a mono-vacancy mechanism. Each of the experimental analyses were based on several sets of experimental data.

Metal	EAM (FBD)	EAM (AFW)	Analyses of Experiments [6]	[7]	[8]	[9]
Cu	1.95	2.02	2.05±0.02	2.07	2.09	
Ag	1.77	1.74	1.77±0.02	1.76	1.82	
Au	1.69	1.69	1.70±0.02	1.76	1.75	
Ni	2.70	2.81	2.87±0.01	2.88	2.88	
Pd	2.20	2.41				2.76*
Pt	2.50	2.63	2.64±0.04		2.67	

*Effective activation energy due to effect of both vacancies and di-vacancies, so actual Q_{1v} should be lower.

Table II Comparison of the activation energies (in eV) for self-diffusion by a di-vacancy mechanism. Each of the experimental analyses were based on several sets of experimental data.

Metal	EAM (FBD)	EAM (AFW)	Analysis of Experiments [6]	[7]	[8]
Cu	2.78	2.88	2.46±0.12	2.59	2.50
Ag	2.37	2.36	2.35±0.14	2.19	2.95
Au	2.47	2.49	2.20±0.14	2.37	2.30
Ni	3.69	3.86	4.15±0.69	3.70	3.58
Pd	3.22	3.51			
Pt	3.78	3.87	4.05±1.11		3.39

The jump frequencies are assumed to be given by an Arrhenius expression:

$$w_i = v_i \exp(S_i/k) \exp(-E_i/kT) \tag{3}$$

where v_i is a vibrational frequency, S_i is the activation entropy and E_i is the activation energy. We assume that host atoms have the same jump frequency ($v_0 = v_1 = v_3 = v_4$), and that the vibrational frequency of the impurity atom may be approximated by[10]:

$$\frac{v_2}{v_o} = \sqrt{\frac{m_o T_{m2}}{m_2 T_{mo}}} \tag{4}$$

where m_0 and m_2 are the mass of the host and the impurity atom, respectively, and T_{mo} and T_{m2} are the melting point of the host and impurity. Similarly, the activation entropies of the host atoms are assumed to be equal to the value in the pure host metal ($S_0 = S_1 = S_3 = S_4$). Following Neumann's example[11], we use the Wert-Zener relation to determine S_2.

Given the above vibrational frequencies and entropies, only the activation energies need to be evaluated. The values for the activation energies for the five jump frequencies of Figure I were calculated using the AFW functions[4]. Inserting those values in Eq. (2) and using Neumann and Tolle's estimates D_{1v} for self-diffusion yielded the impurity diffusion rate. An Arrhenius expression fit the diffusion rate to within 2% accuracy over the temperature range .5 T_{mp} to .75 T_{mp}. The Arrhenius expression yielded effective values of Q_{1v} and D_{1v}^{o}, which are listed in Table III.

In order to compare the experimental data with our calculations for a mono-vacancy mechanism, we assumed that the di-vacancy contribution to impurity diffusion was identical to that for self-diffusion. This assumption is best for medium temperatures and small ΔQ_{1v}, the difference between Q_{1v} for self diffusion and impurity diffusion. Using Neumann and Tolle's estimates of the di-vacancy contribution, we used Eqs. (1) and (2) to determine D for the relevant temperature range. An Arrhenius expression fitted the diffusion rate to within 5% over each experimental temperature range analyzed. The Arrhenius expression yielded effective D^o and Q, which are listed in Table III and compared with the most reliable experimental data, as determined by Neumann[12].

Comparison of Impurity Diffusion Experiments and Calculations

The low-temperature data is generally in good agreement with the EAM values of

TABLE III   Comparison of theoretical and experimental values of Q and Q_{1v}. The experimental data is that which Neumann[12] determined to be reliable. The theoretical Q's are determined from calculations of ΔQ_{1v} and the assumption that Q_{2v} for impurities is the same as for self-diffusion.

System	Exp. Ref.	Temp. Range (T/T_{mp})	Q Exp	Q EAM	Q_{1v} EAM	D_{1v}^{o} EAM
Ag in Cu	15	.78-.99	2.02	2.05	1.93	.10
Au in Cu	16	.49-.72	1.98	1.88	1.88	.15
Ni in Cu	17,18	.75-.99	2.46	2.37	2.22	.26
Pd in Cu	19	.79-.98	2.36	2.06	1.94	.11
Pt in Cu	20	.85-.99	2.42	2.20	2.00	.084
Cu in Ag	21	.79-.99	2.00	1.92	1.73	.053
Au in Ag	12	.75-.99	2.06	2.04	1.77	.042
Pd in Ag	19	.82-.98	2.46	2.19	1.89	.087
Pt in Ag	20	.88-.89	2.44	2.33	2.07	.13
Ag in Au	22	.75-.99	1.75	1.80	1.72	.036
Cu in Au	23#	.77-.95	1.76	1.82	1.75	.051
Ni in Au	24	.73-.99	1.95	1.90	1.83	.090
Pd in Au	25	.73-.95	2.02	1.93	1.83	.050
Pt in Au	25	.73-.95	2.09	2.07	1.95	.070
Ag in Ni	26	.75-.98	2.89	2.68	2.66	.96
Cu in Ni	27	.62-.93	2.64	2.76	2.74	.86
Au in Pt	28	.55-.75	2.61	2.49	2.48	.030
Ag in Pt	29#	.67-.92	2.68	2.71	2.59	.036

#Not included in Neumann's choice of the best experimental data[12].

Q_{1v}, since the di-vacancies are not important. In general, our calculations show that Do for impurity diffusion by mono-vacancies is within a factor of three of D_0 for self-diffusion. Similarly, Q for impurity diffusion is typically within a few tenths of an eV for Q self-diffusion.

The most reliable experimental data (as determined by Neumann[12]) generally covers only elevated temperatures ($\sim.85\ T_{mp}$), where di-vacancies are important. Thus, the EAM calculations of ΔQ_{1v} are combined with Neumann and Tolles' estimates of D_{1v} and D_{2v} for self-diffusion[60] to determine the effective D_0 and Q for a particular temperature range, as described above. These effective values of D_0 and Q were compared with the experimental data, and the best agreement occurred for <u>noble metal impurities diffusing in noble metals</u>. For impurity diffusion of Ni, Pd, and Pt in the noble metals, the EAM tends to yield values of Q that are too low by .1-.2eV. For impurity diffusion of Au and Ag in Ni or Pt, the EAM values of Q are also too low by .1-.2 eV. However, for impurity diffusion of Cu in Ni or Pt, the EAM yields values of Q that are about .1 eV too high.

Part of the discrepancy between theory and experiment is due to the assumption that D_{2v} for self-diffusion is identical to D_{2v} for impurity diffusion. However, part of the discrepancy is due to small inaccuracies of the AFW functions. This is consistent with the fact that the AFW functions for the noble metals were better fitted to their elastic constants than were those for Ni, Pd, or Pt[4]. The functions for the noble metals are also known to more accurately predict many physical properties, such as phonons[13], thermal expansion[14], and Gibbs free energies[14].

CONCLUSION

In this paper we used the EAM with the AFW functions to determine activation energies for self-diffusion by mono-vacancies and di-vacancies, and these results agreed well with experimental analyses.[6-8] By approximating the divacancy contribution to impurity diffusion as that for self-diffusion, EAM calculations for impurity diffusion by a mono-vacancy mechanism could also be compared with experimental data at high temperatures, where di-vacancies are important. The values for Q were typically within a couple tenths of an eV of the experimental values, and the best agreement was for combinations of the noble metals. Thus, the EAM is the first model capable of accurately calculating activation energies for <u>both</u> self-diffusion and impurity diffusion.

It should be noted that our estimates of D_{1v}^0 for impurity diffusion are consistently within a factor of 3 of D_{1v}^0 for self-diffusion. Similarly, Q_{1v} for impurity diffusion is typically within 0.4 eV of Q_{1v} for self-diffusion. This result appears to be consistent with the low temperature data.

Due to the high accuracy of much of the diffusion data, diffusion studies are seen to be exacting tests of the accuracy of the interatomic functions of both pure metals and their alloys. The alloys best described by the AFW functions are those composed of noble metals.

ACKNOWLEDGMENTS

We thank Denise Vickers for her invaluable assistance with the computer programming tasks. We would like to thank Drs. Murray Daw and Mike Baskes of this laboratory for many useful comments and suggestions. We also thank Prof. Dr. G. Neumann for carefully reviewing an early draft of this paper and making several useful suggestions.

REFERENCES

1.   A.D. Le Claire, J. Nuc. Mater. 69/70: 70 (1978).
2.   J.B. Adams and W.G. Wolfer, J. Nuc. Mater., to be published.
3.   M.S. Daw and M.I. Baskes, Phys. Rev. Lett. 50:1285 (1983).
4.   J.B. Adams, S.M. Foiles, W.G. Wolfer, in preparation.
5.   S.M. Foiles, M.I. Baskes, M.S. Daw, Phys. Rev. B.33:7983 (1986); Errata: to be published.
6.   G. Neumann and V. Tolle, Phil. Mag. A, V54(5):619 (1986).
7.   N.L. Peterson, J. Nuc. Mater., 69/70:3 (1978).
8.   W. Schule, "Point Defects and Defect Interactions in Metals", Ed., J.I. Takamura, M. Doyama, and M. Kiritani, Un. of Tokyo Press (1982), p. 551.
9.   N.L. Peterson, Phys. Rev.136, A568 (1964).
10.   G. Neumann and W. Hirschwald, Z. Phys. Chem. Neue Folge. Bd, 89:309 (1974). G. Neumann and W. Hirschwald, Phys. Stat. Sol. (b), 55:99 (1973).
11.   G. Neumann, Materials Science Forum, V 15-18:413 (1987). G. Neumann, Phys. Stat. Sol. (b), 144:329 (1987).
12.   A.S. LeClaire, G. Neumann, Chapter 3 in "Diffusion in Metals and Alloys", Landolt-Bornstein, New Series, Ed. O. Madelung, Vol. Ed. H. Mehrer, Springer, Heidelberg, in the press.
13.   J.S. Nelson, E. Sowa, and M.S. Daw, in preparation.
14.   S.M. Foiles and J.B. Adams, in preparation.
15.   N.H. Nachtrieb, C.T. Tomizuika, L.G. Schulz: Report AFOSR-TR-60-23, the University of Chicago (1960).
16.   S. Fujikawa, M. Werner, H. Mehrer and A. Seeger, Mat. Sci. Forum, V. 15-18: 431 (1987).
17.   M.P. Macht, V. Naundorf, and R. Dohl, Proceedings of Int. Conf. on Diffusion in Metals and Alloys at Tihany, Hungary, Eds., F.J. Kedves and D.L. Beke, Diffusion and Defect Monograph Series No. 7 (1983), Trans. Tech. Pub., Switzerland.
18.   G. Neumann and V. Tolle, Phil. Mag. A57(4): 621 (1988).
19.   N.L. Peterson, Phys. Rev. 132:2471 (1963).
20.   G. Neumann, M. Pfundstein, and P. Reimers, Phil. Mag. A45:499 (1982).
21.   P. Dorner, W. Gust, M.B. Hintz, A. Lodding, H. Odelius, and B. Predel, Acta Met. 28:291 (1980).
22.   C. Herzig and D. Wolter, Z. Metallkunde 4:273 (1974).
23.   A. Vignes and J.P. Haeussler, Mem. Sci. Rev. Metall. 63:1091 (1966), (translation available from NTIS at TT 70-57660).
24.   R.L. Fogel'son, Y.A.A. Ugay, and I.A. Akimova, Fiz. Met. Metalloved 41(3): 653 (1976).
25.   R.L. Fogel'son, I.M. Voronina and T.I. Somova, Phys. Met. Metall. 46(1): 163 (1979).
26.   A.B. Vladmirov, V.N. Kaigorodov, S.M. Klotsman, and I. Sh. Trachtenberg, Proceedings of Int. Conf. on Diffusion in Metals and Alloys at Tihany, Hungary, Eds., F.J. Kedves and D.L. Beke, Diffusion and Defect Monograph Series No. 7 (1983), Trans. Tech. Pub., Switzerland.
27.   O. Taguchi, Y. Iijima, and K. Hirano, J. Jap. Inst. Met. 48(1):20 (1984).
28.   G. Rein, H. Mehrer, and K. Maier, Phys. Stat. Sol. (a) 45:253 (1978).
29.   D. Bergner and K. Schwarz, Neue Hutte, 23(6) 210 (1978).

SIMULATION OF ATOMIC AND MOLECULAR PROCESSES AT SOLID SURFACES

Madhu Menon and Roland E. Allen

Center for Theoretical Physics
Department of Physics
Texas A&M University
College Station, Texas 77843, U. S. A.

INTRODUCTION

As the papers of this conference make clear, there are many approaches to the calculation of total energies and atomic forces. During the past four years, we have developed a technique that is particularly useful for computer simulations of atomic motion and chemical reactions at solid surfaces [1-18]. The long-range goal of such simulations is microscopic understanding of complex processes like interfacial growth and catalysis. A feasible short-range goal is to study the more specific atomic processes that occur when an atom, molecule, or cluster interacts with a solid surface.

In a molecular dynamics simulation [19-33], one numerically solves Newton's equations of motion $\vec{F} = m\vec{a}$ for a system of interacting atoms. The hard part of the calculation, of course, is determining the forces $\vec{F}$. For systems with van der Waals or ionic bonding, it is a good approximation to compute $\vec{F}$ from simple pair potentials--e.g., Lennard-Jones or Coulomb plus Born-Mayer. For systems with covalent or metallic bonding, however, pair potentials are unrealistic. For example, a tetrahedral or even body-centered cubic structure will typically collapse to a close-packed structure if the atoms are taken to interact through a simple pair potential.

Because of the importance of the various problems involving atomic motion and atomic geometries in semiconductors, many investigators have attempted to construct 3-body potentials or other schemes that would mimic the effects of covalent bonding [34-41]. Some of these attempts--like the potential of Stillinger and Weber [24]--have been rather successful. Such phenomenological models will undoubtedly have some range of validity. However, one would also like to have a method which determines the atomic forces directly from the electronic structure of the system, as nature itself does. Covalent and metallic bonding are inherently quantum-mechanical, and there are limits to the applicability of classical models.

The general technique that we have introduced is exact in the limit $(T/t_0) \to \infty$ and $M \to \infty$, where T is the available computer time, t_0 is the time for an arithmetic operation, and M is the available memory. In principle, one can use a self-consistent Hamiltonian and even include many-body effects, since the technique is based on Green's functions. The use of Green's functions also reduces the electronic problem for a semi-infinite solid to a finite problem, involving the perturbed part of the surface where a molecule or cluster is interacting with the solid. The atomic forces are computed from the electronic structure via the Hellmann-Feynman theorem, novel Green's function methods, and integration in the complex energy plane. All the mathematical details are given in a series of three papers [3,11,18].

In the initial simulations reviewed here, we have used a simple physical model which is not as accurate as self-consistent calculations in the local-density approximation [42], but which provides a reasonably good description of the basic physics and chemistry. The sum of the one-electron energies, $\Sigma_i \varepsilon_i$, is treated with a tight-binding Hamiltonian. A repulsive potential is used to represent the remaining contributions to the total energy, $(V_{ions} - V_{ee} + V_{xc})$. Here V_{ions} is the ion-ion repulsion, V_{ee} is the doubly-counted Coulomb interaction between electrons, and V_{xc} is the exchange and correlation energy. All the details of this simplified physical model are given in Ref. 11.

RESULTS

Let us now review some of the results obtained in simulations using the general mathematical technique and specific physical model mentioned above, beginning with the chemisorption of individual atoms. Fig. 1 shows a top view of the GaAs (110) surface, with the origin of coordinates taken to be the unrelaxed position of a Ga surface atom. This figure also shows schematically the positions of 6 chemisorption sites that have been observed in our simulations. These same sites have been observed as energy minima along the surface in calculations of static chemisorption [43,44].

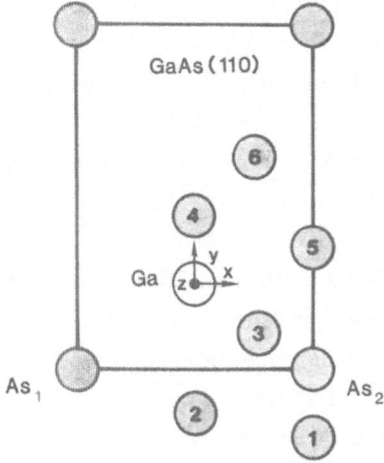

Figure 1. Top view of GaAs (110) surface, with choice of coordinates, labeling of coordinates, labeling of atoms, and labeling of six chemisorption sites that have been observed in the present simulations.

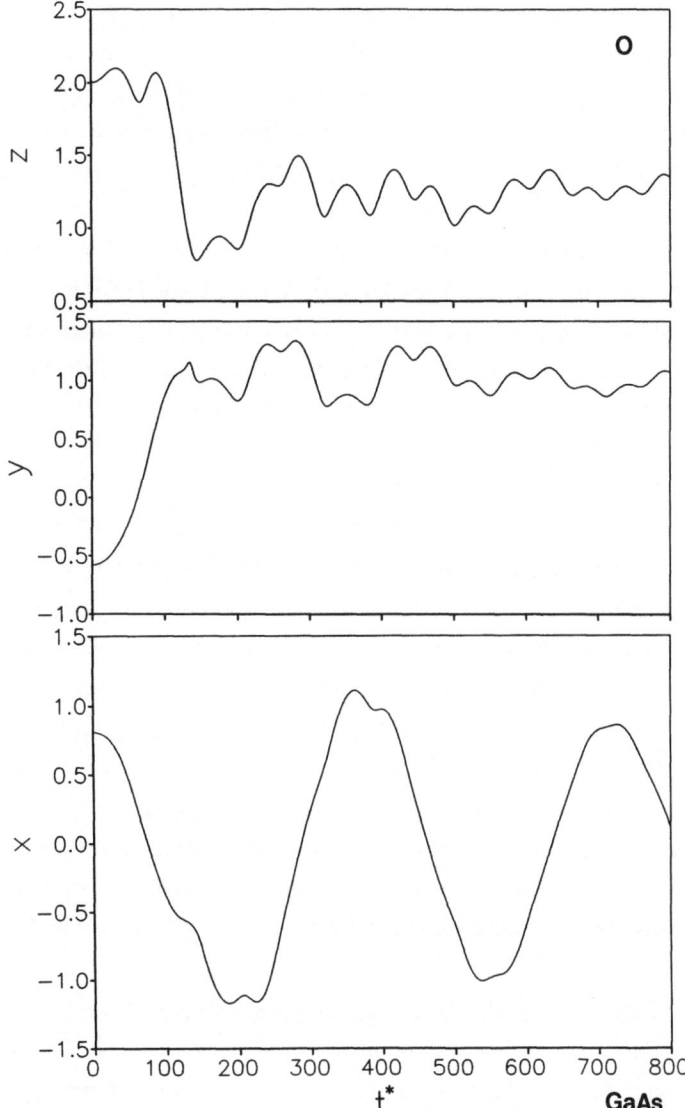

Figure 2. Motion of oxygen atom deposited on GaAs (110) surface.

Fig. 2 shows a simulation for monatomic oxygen on GaAs, with the O atom released with zero initial velocity just above the surface at a point between Ga and As surface atoms. The unit of length is $r_0/2$, where r_0 is the bulk bond length (2.45Å for GaAs and 2.54Å for InP). The unit of time is 1.04 femtosecond for GaAs and 0.91 femtosecond for InP. As shown in Fig. 1, the z axis points out from the surface, and the y axis points away from the midpoint of a line between the two As atoms to which the Ga atom is bonded. Notice that the O atom moves away from the nearest As atom (y direction) as it falls to a position above the surface (z direction) where it is bonded to the nearest Ga atom at a symmetric site (mean value of $x=0$). since O is relatively light, it vibrates with relatively high frequency in the y and z directions. However, the period of vibration in the x direction is long, and the amplitude of vibration large, because the restoring force is small.

We have performed simulations for other group VI, VII, and V elements on GaAs and InP (110). For the same initial conditions as in Fig. 2, each of these electronegative atoms tends to bind to the cation, as one might naively expect, but there are considerable variations in both the time-dependent behavior and the final chemisorption site. For example, for these initial conditions Se bonds above the surface at a "Ga-Ga bridge site," midway between two Ga atoms ($x=1.6$), rather than to a single Ga atom ($x=0$), although it belongs to the same chemical group as O.

Fig. 3 shows the motion of a chemisorbing Al atom, which bonds at a "Ga-As bridge site," above the surface and approximately midway between Ga and As surface atoms. Initially motionless, the Ga atom to which this Al bonds is found to vibrate with a characteristic angular frequency ω of about 25 THz (y and z motion), in satisfactory agreement with estimates based on the measured bulk phonon frequencies [45].

In Fig. 4, we show a simulation for Si on GaAs. Starting with the same initial conditions as for the O and Al atoms of Figs. 2 and 3, the Si atom moves to an "As-As bridge site" above the surface and midway between two As atoms.

We have performed simulations for various group III, II, and I elements on GaAs and InP, and find that these electropositive atoms usually prefer to bond to As for these initial conditions. On the other hand, as seen above, Al bonded at the Ga-As bridge site 3 of Fig. 1.

Let us now consider what happens if we change the initial conditions. Since there are various rather shallow energy minima along the (110) surfaces of these zincblende-structure semiconductors [44], we might expect to observe more than one initial chemisorption site. In the following three simulations for Al, representative of many that were performed for various chemical species, we will see that this is indeed the case. In fact, for Al alone we have observed initial chemisorption (in the first picosecond) for all six of the sites of Fig. 1.

Fig. 5 shows a simulation in which the Al atom is attracted to site 1, where it bonds to a single As. This is a site that one might naturally expect, since it is where a Ga would bond if an epitaxial layer of GaAs were deposited on the surface. Notice that the Al

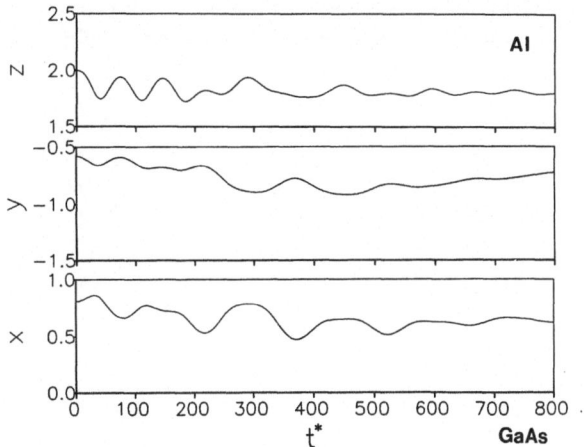

Figure 3. Motion of A*l* atom on GaAs (110).

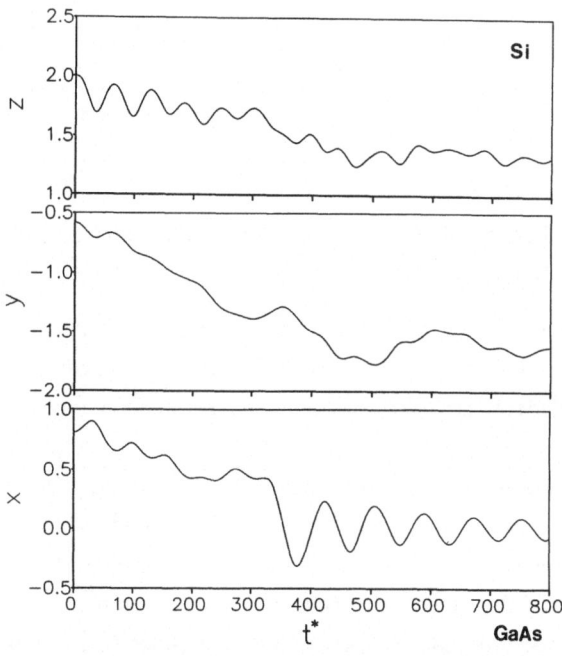

Figure 4. Motion of Si atom on GaAs (110).

vibrates with a large amplitude, low frequency "wagging" mode since there is a weak restoring force perpendicular to the Al-As bond.

In Fig. 6, with different initial conditions, the Al chooses to bond to a pair of As atoms at the As-As bridge site 2. The more complicated vibrational motion reflects the fact that the chemisorbed Al and the surface atoms (As_1 and As_2 of Fig. 1) vibrate as a coupled system.

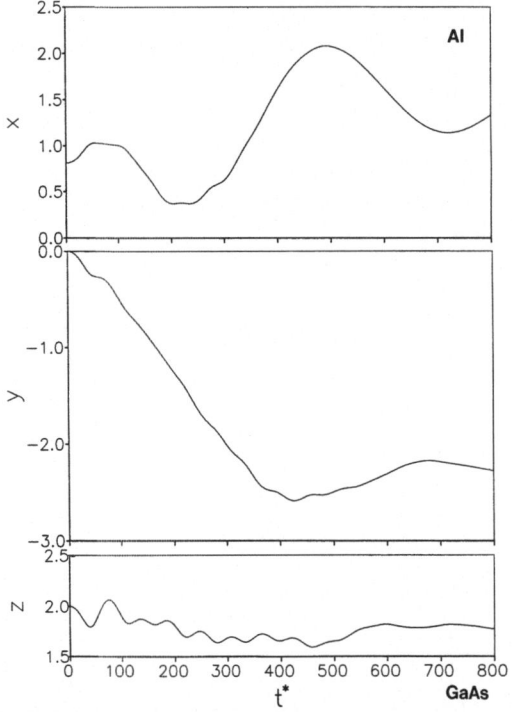

Figure 5. Simulation of A*l* chemisorbing on GaAs (110).

In Fig. 7, with still another set of initial conditions, the Al atom bonds to a pair of Ga atoms, at the Ga-Ga bridge site 5. The high vibrational frequency in the $\underline{x}$ direction results from the relatively large restoring force in that direction, together with the relatively small mass of Al.

For small atoms, we have observed another interesting phenomenon--indiffusion. First consider Fig. 8, which represents a simulation of C on GaAs (110). For these initial conditions (position and velocity of the incident atom), C bonds to a single Ga atom at a site that is essentially the same as that occupied by an As atom in bulk GaAs. Notice that there are short-period vibrations--which we interpret as bond-stretching vibrations parallel to the Ga-C bond--superimposed on much longer period and larger amplitude vibrations-- which we interpret as angle-bending vibrations perpendicular to the Ga-C bond, with a very small restoring force. These results indicate that C can exhibit normal chemisorption above the surface.

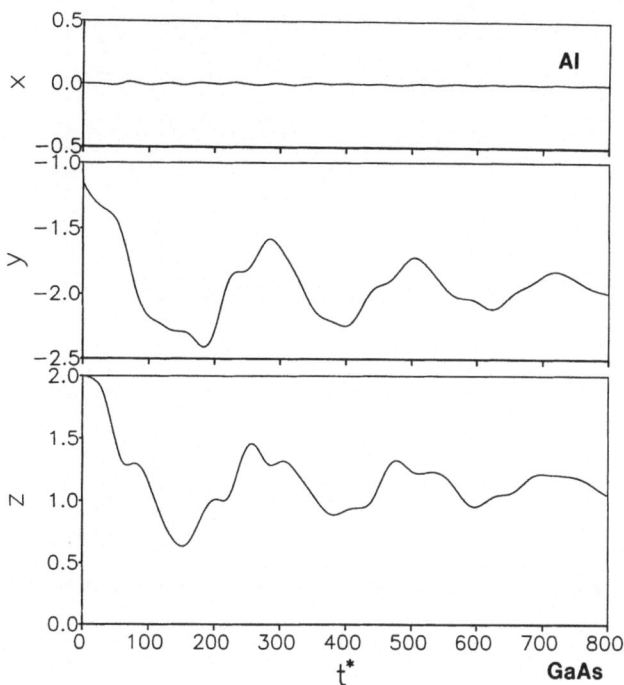

Figure 6. A*l* on GaAs (110) with a different set of initial conditions.

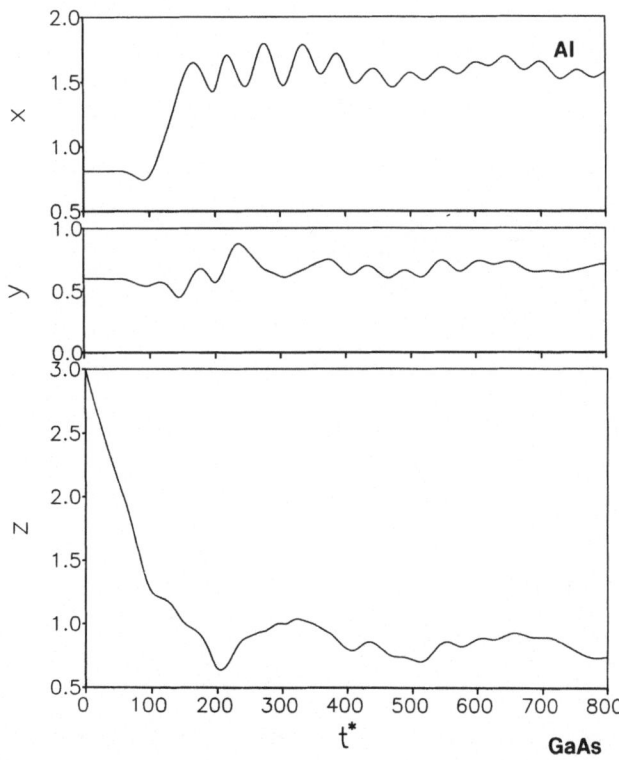

Figure 7. A*l* on GaAs (110) with still another set of initial conditions.

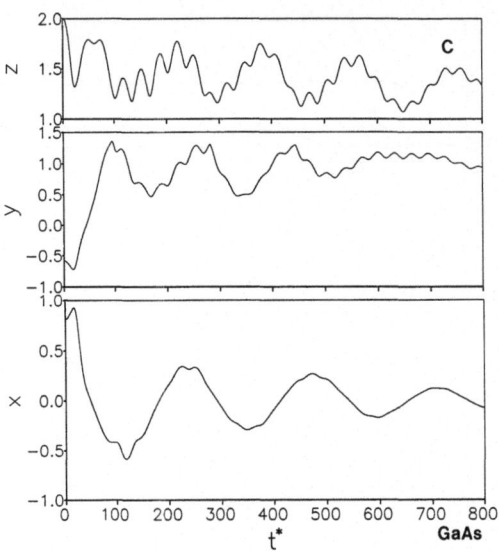

Figure 8. C chemisorbing on GaAs, bonding to a single Ga atom.

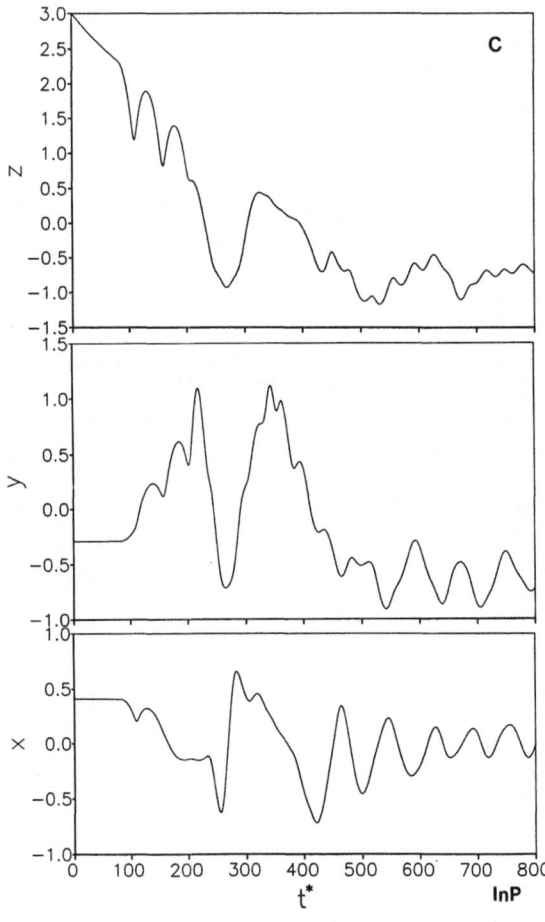

Figure 9. C indiffusing into InP.

In Fig. 9, however, we show a simulation for C on InP (110), with different initial conditions. Notice that the C almost immediately indiffuses to a site beneath the surface P atom. Fig. 10 shows the response of the P. Notice that it is only slightly displaced from the original, relaxed position that it normally occupies. We interpret this as a consequence of the relatively small size of P, and the fact that its original relaxation is outward.

Fig. 11 shows a simulation for B on InP. B also indiffuses, with its behavior very similar to that of C. In a simulation not shown here, N is again found to indiffuse with the same general behavior. Our prediction that each of these small atoms indiffuses to a position beneath the surface should be experimentally testable. For example, the vibrational frequencies should be measurable with electron energy-loss spectroscopy.

Fig. 12 represents a simulation of C on GaAs (110). Again, indiffusion occurs almost immediately for these initial conditions. In this case the final position of the C is beneath the cation Ga (as opposed to the anion P). However, B and (perhaps surprisingly) N were both found to indiffuse to positions beneath the As_1 atom of Fig. 1.

Notice that the Ga atom responds by moving to a position well above the surface, in Fig. 13. This may in part be associated with the fact that Ga is somewhat larger than P. In the simulations for B and N, the As_1 atom was also found to protrude above the surface once the indiffused atom was in place beneath it.

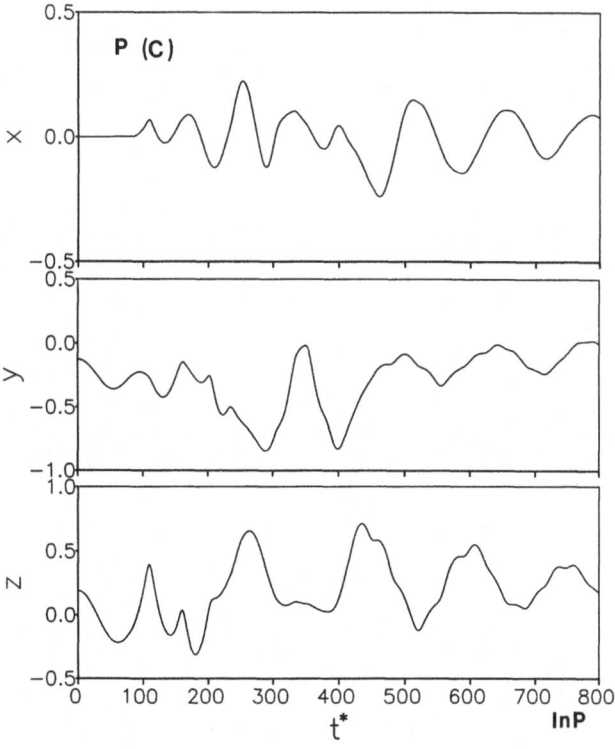

Figure 10. Response of P in simulation of Fig. 9.

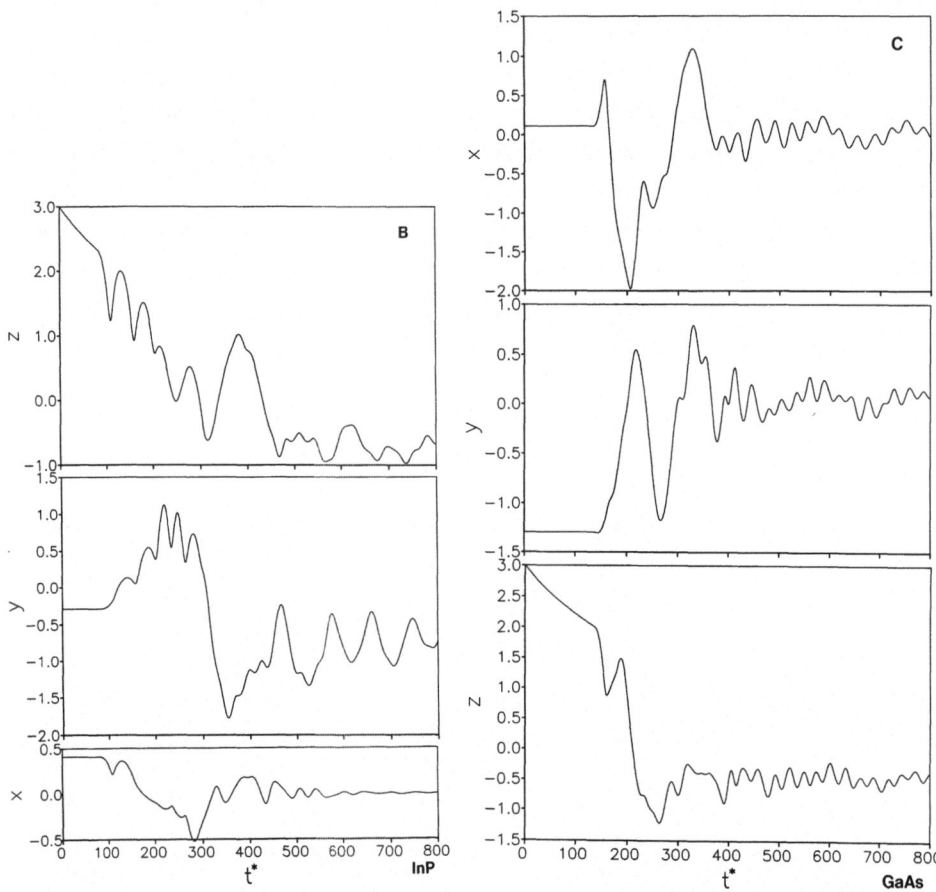

Figure 11. C indiffusing into GaAs. Figure 12. C indiffusing into GaAs.

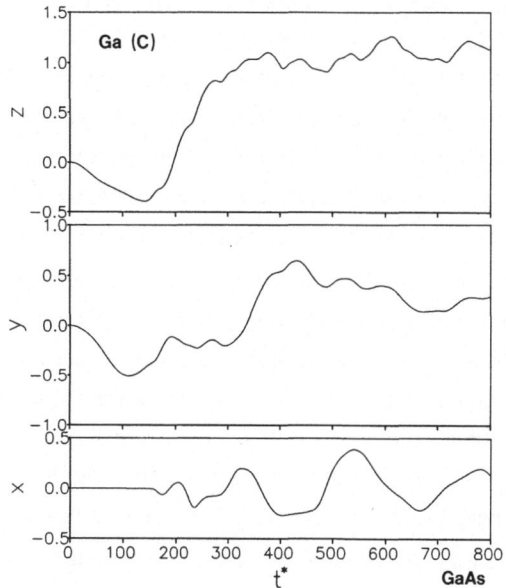

Figure 13. Response of Ga in simulation of Fig. 12.

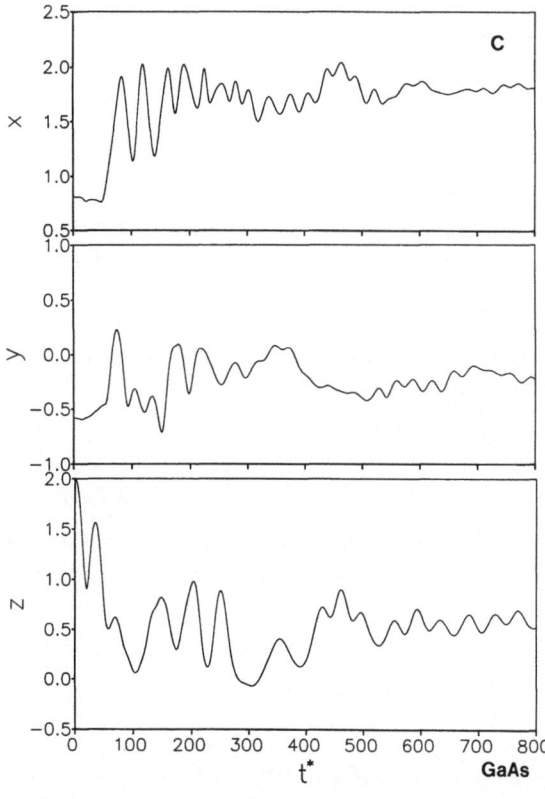

Figure 14. C chemisorbing on GaAs, bonding at a Ga-Ga bridge site.

Finally, in Fig. 14 we show a simulation, with different initial conditions, in which the C atom chooses to chemisorb above the surface--and in a different chemisorption site than that of Fig. 8. In this case it bonds at a Ga-Ga bridge site. We thus observe that there are various possibilities for what the small atoms B, C, and N will do when deposited on a semiconductor surface. Unlike all the larger atoms we have studied, they may almost immediately indiffuse to a position beneath the surface. Alternatively, they may choose to chemisorb above the surface--and again in more than one initial chemisorption site.

As mentioned above, we have found that Al chemisorbs at all six of the sites of Fig. 1 during the first picosecond, with a rather sensitive dependence on the initial position and velocity. Presumably the Al atom finds its way to the most stable site on a longer time scale. Using our method, it is a simple matter to monitor the total energy, and in Table I we give the energies of the various sites of Fig. 1 relative to the energy of the isolated atom plus unrelaxed surface. The most stable position is found to be the long Ga-As bridge site 6, in agreement with the self-consistent calculations of Ihm and Joannopoulos [44]. The energy of -2.15 eV is also in good agreement with their value of -2.0 eV. For the other sites, there are differences in both the ordering of the sites and the absolute energies, but this is neither surprising nor significant, since the energy differences are relatively small and neither theoretical approach can be trusted completely at the 0.1 eV level.

Al Sites and Chemisorption Energies

Site	Energy (eV)
1	-1.907
2	-2.095
3	-0.71
4	-1.37
5	-1.76
6	-2.15

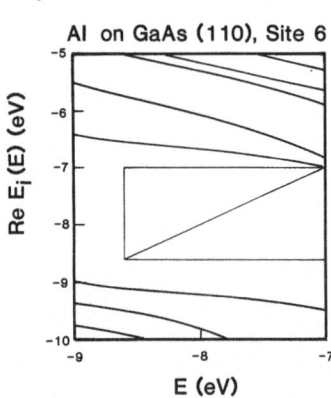

Al on GaAs (110), Site 6

Figure 15. $E_i(E)$ versus E for Al at the most stable site 6. No states are found in the GaAs band gap, between -8.6 and -7.0 eV.

With the present method, one can also calculate the bound states and resonances associated with an adsorbed atom. Once the atom has settled into its final position, with a consequent relaxation of the substrate atoms, we obtain the subspace Hamiltonian [3] and its eigenvalues $E_i(E)$ for a set of real energies E within and near the band gap. The criterion for a bound state is that

$$E_i(E) = E \qquad (1)$$

for some energy E within the band gap. We find that this criterion is only sometimes satisfied for Al chemisorbing on GaAs (110). To be more precise, we find no bound states

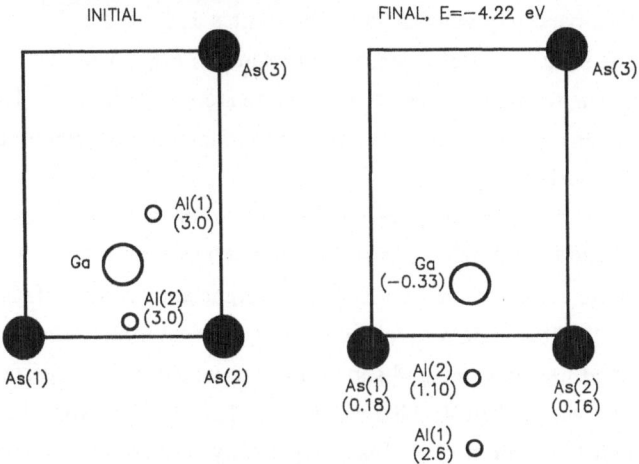

Figure 16. Initial and final positions of incoming Al$_2$ dimer and GaAs surface atoms.

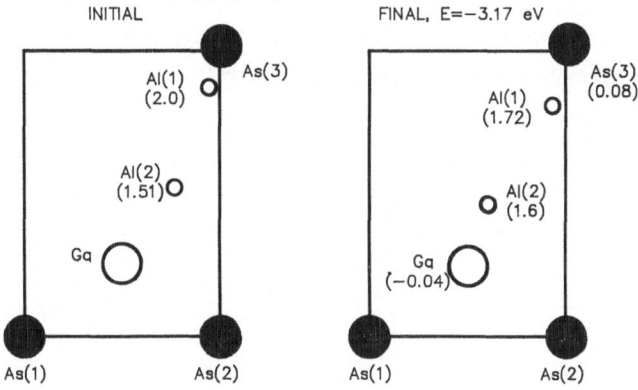

Figure 17. Positions for a different set of initial conditions.

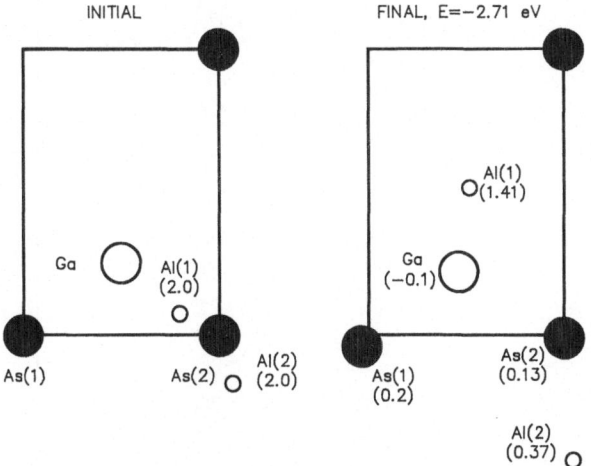

Figure 18. Another set of initial conditions. In Figs. 15 and 16, the Al$_2$ stayed together. Here it dissociates, but the energy is found to be substantially higher.

within the GaAs band gap when Al adsorbs at sites 1, 3, 4, and 6 of Fig. 1; there are, however, bound states for sites 2 and 5. For example, Fig. 15 shows $E_i(E)$ versus E for the most stable site 6, and it can be seen that there are no solutions to (1) in the indicated band gap--between -8.6 and -7.0 eV, relative to the vacuum level, within the tight-binding model that we are using [46].

Now let us consider Al_2 dimers impinging on the surface. Figs. 16 and 17 show the initial (t=0) and final (t=800Δt, Δt=1.04 femtosecond) positions of the Al atoms and of a Ga and two As atoms at the surface. The height above the surface is indicated in parenthesis. Notice that in both simulations, with different initial conditions, the Al atoms moved but stayed together. Fig. 18 shows a run in which the Al_2 dissociated; however, the final energy was -2.71 eV, whereas in Figs. 16 and 17 it was -4.22 and -3.17 eV. This means that the 2-atom cluster is energetically preferable to the dissociated state (although the energy of the cluster depends on its location). We conclude that Al atoms atop a GaAs (110) surface tend to form clusters. This result is also in agreement with previous theoretical and experimental work [47].

As one might expect, O_2 behaves differently from Al_2. In Figs. 19 and 20 we show the detailed motion of the two incoming oxygen atoms in a representative simulation.

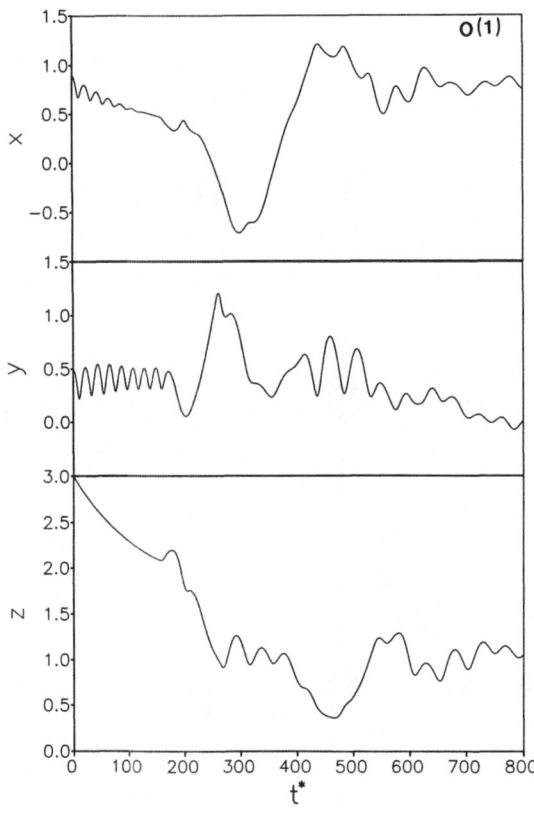

Figure 19. Detailed motion of one O atoms, O(1), in a simulation for an O_2 molecule.

The two atoms move apart, with one going to a position beneath the surface plane. Fig. 21 shows the final configuration with the O atoms clearly separated. This configuration suggests the beginning of oxide formation.

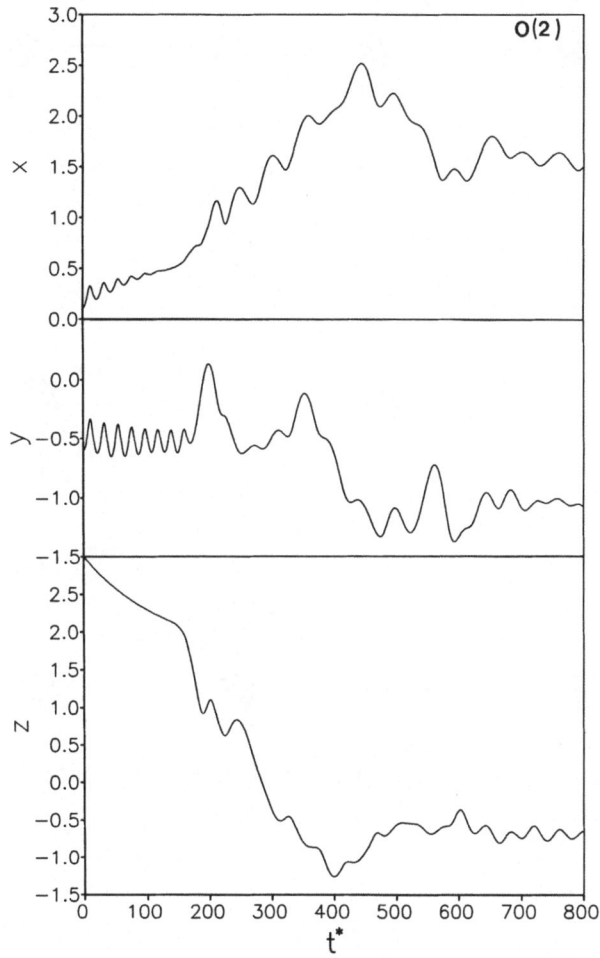

Figure 20. Detailed motion of other O atom, O(2), in simulation of Fig. 19.

In Fig. 22, we show the final positions after a run in which the O_2 stayed together, exhibiting molecular chemisorption. However, the final energy was -1.8 eV, larger than the energy of Fig. 21 for dissociated atomic chemisorption. We conclude that molecular chemisorption of O_2 is possible, but is metastable with respect to dissociation.

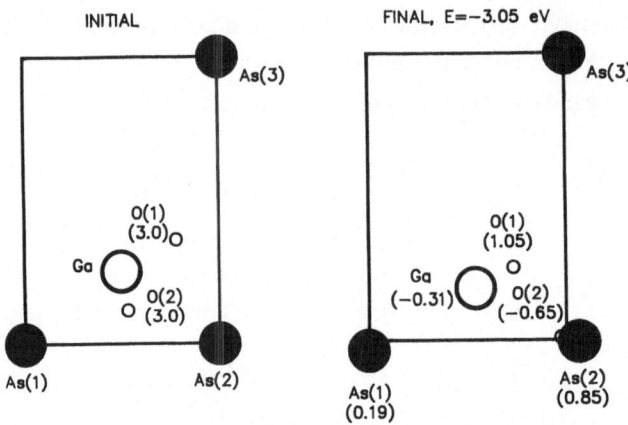

Figure 21. Initial and final positions of atoms in simulation of Figs. 19 and 20.

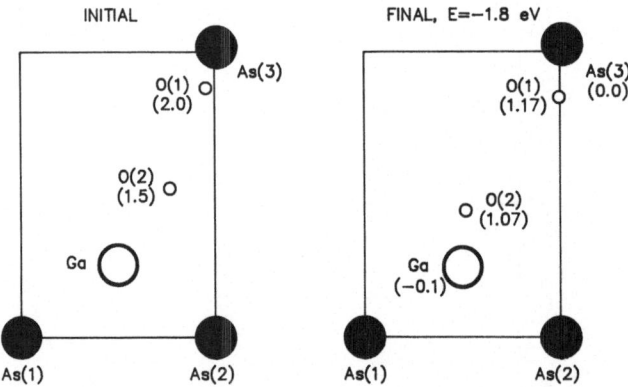

Figure 22. Positions for another O_2 simulation.

ACKNOWLEDGEMENT

This work was supported by the Office of Naval Research (N00014-82-K-0447) and by the Robert A. Welch Foundation. Additional support was provided by the Texas A&M Board of Regents.

REFERENCES

1.   M. Menon and R. E. Allen, Bull. Am. Phys. Soc. 30:362 (1985).

2.	M. Menon and R. E. Allen, in: "Proceedings of the 1985 Computer Simulation Conference", edited by the Society for Computer Simulation (North Holland, Amserdam, 1985).
3.	R. E. Allen and M. Menon, Phys. Rev. B 33:5611 (1986).
4.	M. Menon and R. E. Allen, Phys. Rev. B 33:7099 (1986).
5.	O. F. Sankey and R. E. Allen, Phys. Rev. B 33:7164 (1986).
6.	M. Menon and C. W. Myles, J. Phys. Chem. Solids 48:621 (1987).
7.	M. Menon and R. E. Allen, Superlattices and Microstructures 3:295 (1987).
8.	M. Menon and R. E. Allen, Solid State Commun. 64:353 (1987).
9.	M. Menon and R. E. Allen, Progress in Surface Science 25:317 (1987).
10.	M. Menon and R. E. Allen, Surf. Sci. 197:L237 (1988).
11.	M. Menon and R. E. Allen, Phys. Rev. B. 38: 6196(1988).
12.	M. Menon and R. E. Allen, in: "The Structure of Surfaces II", edited by J. F. van der Veen and M. A. Van Hove (Springer-Verlag, Berlin, 1988), p. 399.
13.	M. Menon and R. E. Allen, J. Vac. Sci. Technol. A 6:1491 (1988).
14.	M. Menon and R. E. Allen, J. Vac. Sci. Technol. B 6:1302 (1988).
15.	M. Menon and R. E. Allen, Superlattices and Microstructures, 4:629 (1988).
16.	M. Menon and R. E. Allen, Proceedings of the 19th International Conference on the Physics of Semiconductors (Warsaw, August, 1988).
17.	M. Menon and R. E. Allen, submitted for the Proceedings of the 35th National Symposium of the American Vacuum society (Atlanta, October, 1988).
18.	M. Menon and R. E. Allen, to be published.
19.	A. Rahman, Phys. Rev. 136:A405 (1964).
20.	R. E. Allen, F. W. deWette, and A. Rahman, Phys. Rev. 179:887 (1969).
21.	F. W. deWette, R. E. Allen, D. S. Hughes, and A. Rahman, Phys. Lett. 29A:548 (1969).
22.	S. Shumway and R. E. Allen, Surf. Sci. 177:L999 (1986).
23.	F. H. Stillinger and A. Rahman, J. Chem. Phys. 60:1545 (1974).
24.	F. H. Stillinger and T. A. Weber, Phys. Rev. B 31:5262 (1985).
25.	M. Parrinello and A. Rahman, Phys. Rev. Lett. 45:1196 (1980).
26.	M. Parrinello, A. Rahman, and P. Vashishta, Phys. Rev. Lett. 50:1073 (1983).
27.	R. Car and M. Parrinello, Phys. Rev. Lett. 55:2471 (1985); Solid State Commun. 62:403 (1987).
28.	H. C. Andersen, J. Chem. Phys. 72:2384 (1980).
29.	J. C. Tully, Surf. Sci. 125:282 (1983).
30.	F. F. Abraham, J. Vac. Sci. Technol. 2:534 (1984) and references therein.
31.	U. Landman, R. N. Barnett, C. L. Cleveland, and R. H. Rast, J. Vac. Sci. Technol. A 3:1574 (1985) and references therein.
32.	D. W. Brenner and B. J. Garrison, Phys. Rev. B 34:1304 (1986).
33.	S. M. Paik, A. Kobayashi, S. Das Sarma, and K. E. Knor in: "The Physics of Semiconductors", edited by O. Engstrom (World Scientific, Singapore, 1987), p. 1189.
34.	N. Keating, Phys. Rev. 145:637 (1966).
35.	E. Pearson, T. Takai, H. Halicioglu, and W. A. Tiller, J. Cryst. Growth 70:33 (1984).
36.	R. Biswas and D. R. Hamann, Phys. Rev. Lett. 55:2001 (1985); Phys. Rev. B 34:895 (1986); and in: "The Physics of Semiconductors", edited by O. Engstrom (World Scientific, Singapore, 1987), p. 1173.
37.	J. Tersoff, Phys. Rev. Lett. 56:632 (1986).
38.	B. W. Dodson, Phys. Rev. B 33:7361 (1986); Phys. Rev. B 35:2795 (1987).
39.	K. Ding and H. C. Andersen, Phys. Rev. B 34:6987 (1986).
40.	M. I. Baskes, Phys. Rev. Lett. 59:2666 (1987).
41.	See also the various approaches discussed at this conference.
42.	M. Needels, M. C. Payne, and J. D. Joannopoulos, Phys. Rev. Lett. 58:1765 (1987).

43. D. J. Chadi and R. Z. Bachrach, <u>J. Vac. Sci. Technol</u>. 15:1159 (1979).

44. J. Ihm and J. D. Joannopoulos, <u>Phys. Rev. B</u> 26:4429 (1982) and references therein.

45. G. Dolling and J. L. T. Waugh, in: "Lattice Dynamics", edited by R. F. Wallis Pergamon, Oxford, (1965).

46. W. A. Harrison, "Electronic Structure and the Properties of Solids" (W. H. Freeman, San Francisco, 1980).

47. See e.g., R. Cao, K. Miyano, T. Kendelewicz, K.K. Chin, I. Lindau, and W. E. Spicer, <u>J. Vac. Sci. Technol</u>. B5:998 (1987), and references therein.

MOLECULAR DYNAMICS SIMULATIONS OF MATERIALS:

BEYOND PAIR INTERACTIONS

Uzi Landman and W.D. Luedtke

School of Physics
Georgia Institute of Technology
Atlanta, GA 30332

INTRODUCTION

Basic understanding of the structure and dynamics of materials and their properties often requires knowledge on a microscopic level of the underlying energetics and interaction mechanisms, whose consequences we observe and measure. Answers to material science problems are in principle possible, embodied in solutions to the Schrodinger equation subject to the appropriate boundary conditions. However, the full implementation of such a program is impossible for most (one may dare say all) materials science and condensed matter systems and we must resort to various approximations and simplification. The degree of microscopic detail with which we probe physical phenomena is determined mainly by the resolution of our experimental tools, by the ability to found the theoretical analysis on microscopic principles and by the complexity, hence solubility, of the model. In many situations the level of complexity of the model, which is necessary in order to describe faithfully the physical phenomena, is such that analytical approaches fail to provide a solution. In these situations, which include the majority of material systems and phenomena, the use of computer-based methods [1-7] is essential.

Computer simulations [1-7] where the evolution of a physical system is simulated, with refined temporal and spatial resolution, via a direct numerical solution to the model equations are in a sense computer experiments which open new avenues in investigations of the microscopic origins of material phenomena. These methods alleviate certain of the major difficulties which hamper other theoretical approaches, particularly for complex systems such as those characterized by a large number of degrees of freedom, lack of symmetry, non-linearities and complicated interactions. In addition to comparison with experimental data, computer simulations can be used as a source of physical information which is not accessible to laboratory experiments, and in some instances the computer experiment itself serves as a testing ground for theoretical concepts.

In this short review we focus on computer simulations which employ realistic non pair-wise interaction potentials. Our aim is to demonstrate the level of reality which can be achieved via computer simulations of materials systems and the wealth of information which they provide about the fundamental mechanisms and processes which govern the properties and response of material systems.

MOLECULAR DYNAMICS SIMULATIONS AND INTERACTION POTENTIALS

The classical molecular-dynamics (MD) method [1-7] consists of a numerical generation of the phase-space trajectories for a system of N particles, (and, for extended systems, employing periodic boundary conditions), interacting via a potential function $V(\vec{r}_1,...,\vec{r}_N)$, where $\vec{r}_i$ is the coordinate of particle i. Formulations have been developed which allow simulations of various equilibrium ensembles, such as constant volume microcanonical (E,Ω,N) constant pressure (H,P,N), or constant temperature macrocanonical (T,Ω,N) ensembles.

The starting point of a MD simulation is a well-defined microscopic description of the physical system, in terms of a Hamiltonian or a Lagrangian from which the equations of motions are derived and the phase-space trajectories which are generated can be used in studies of equilibrium and nonequilibrium phenomena.

The correspondence between the simulation model and the physical system depends to a large extent on the nature of the potential energy employed in the simulations. In general the potential energy of a system of N atoms is a function of the atomic coordinates and may also contain a contribution from a term which does not depend explicitly on the atomic coordinates but on the density (such as in metallic systems). Thus in general the potential energy may be expanded in a series of n-body interatomic potentials

$$V = \phi_o + \Sigma\phi_1(i) + \Sigma\phi_2(ij) + \Sigma\phi_3(ijk) + ... + \phi_N(ijkl,...), \tag{1}$$

where ϕ_n is the n-body interaction potential which is a function of the positions of n atoms ijk.... The sums in Eq. (1) are over all combinations (excluding redundant contributions) of n atoms in the system.

The potentials, ϕ_n, and the number of terms which are retained in a practical application depend on the nature of the system under investigation. Thus for example rare gases and ionic systems can be adequately described in terms of pair potentials [8], which depend only upon the distance between interacting pairs, (such potentials include the 6-12 Lennard Jones potentials for rare gases and potentials which include electrostatic interactions and Born-Meyer repulsive terms for ionic systems [9]).

For systems characterized by covalent bonding (such as semiconductors) one must go beyond the pair interactions due to the directionality of the localization of charge in the bonding regions. Potentials which include 3-body terms have been used for sometime [10] for the description of equilibrium structures of crystalline semiconductors and recently generalizations for arbitrary atomic locations have been developed [11-13]. In our work on silicon we have employed the potential developed by Stillinger and Weber (SW) [11] which has the form

$$\phi_2(r_{ij}) = A[Br_{ij}^P - 1]\, g_\beta(r_{ij}),\qquad(2a)$$

$$\phi_3(r_i, r_j, r_k) = U_{jik} + U_{ijk} + U_{ikj},\qquad(2b)$$

$$U_{jik} \equiv \lambda\, g_\gamma(r_{ij})\, [\cos\theta_{jik} + 1/3]^2,\qquad(2c)$$

$$g_\gamma(r) = \exp[\gamma/(r - a)].\qquad(2d)$$

The parameters of the potential have been determined [11] via empirical fits to properties of silicon in the crystalline and liquid phases.

For metallic systems it is well known [14] that the cohesive energy contains density (volume) dependent contributions. Consequently, the cohesive energy of a simple metal can be written as

$$V = E_\Omega + \frac{1}{2}\sum_{i,j}\phi_2(r_{ij}; r_s);\qquad(3a)$$

$$E_\Omega = N\,E_{el}(r_s) + N\phi_1(r_s)\qquad(3b)$$

where Ω is the volume of the system, $r_s = (3/4\,\pi\rho)^{1/3}$, E_Ω is the volume energy expressed in terms of the energy of the uniform electron gas, $E_{el}(r_s)$, and the single particle contribution $\phi_1(r_s)$, and $\phi_2(r_{ij}, r_s)$ is a density dependent effective pair-potential [14]. ϕ_1 and ϕ_2 are derived from pseudo-potential theory [14] and their specific form depends upon the choice of ionic pseudopotentials (model, local or non-local). Clearly, due to the explicit and-implicit dependence of V on the density of the system, simulations of metal system phenomena which may involve changes in volume or structural transformations require a method in which the volume of the system varies dynamically [15,16].

Alternatively, based on the philosophy of density-functional theory, a description of metallic systems which is amenable to molecular dynamics simulations is provided by effective medium theory (EMT) [17] or the related embedded atom method (EAM) [18]. The basic feature of the EMT is that the effect of the surroundings on each atom in the system can be described in terms of the average electron density which other atoms in the system provide around the atom in question. The electronic structure problem is then converted to that of the embeddment of an atom in a homogeneous electron gas, which can be described in terms of a universal density dependent energy function. Thus, the density dependent term gives rise to many-body interactions. Invoking the adiabatic (Born-Oppenheimer) approximation we have used the EMT method in MD simulations of premelting phenomena in aluminum described below [19].

SAMPLE RESULTS
Amorphous Silicon

The difficulty in studying matter in the amorphous state originates from the lack of periodicity which hinders the use of conventional condensed-matter structural determination and modeling methods leading to the development of various models [20-24] in which disordered structures are hand built or computer generated according to well-defined algorithms. Although the above structural models aid the elucidation and interpretation of data, alternative descriptions in terms of interatomic potentials are required in order to investigate the response of materials to changes in external physical conditions and the dynamics and mechanisms of phase transformations (such as melting, crystallization, and amorphization).

Recently, several MD simulations of the preparation of amorphous silicon [25-28] (and germanium [29]) via cooling of a melt have been performed. While the SW potential which was parameterized [11] on the basis of solid and liquid phase data for Si yields an adequate description of these aggregation states (bulk [11] and surface phenomena [30], point defects [31], and studies of equilibrium [32,33] and nonequilibrium [34] solid-melt interfaces of silicon), it appears [25-28] that upon quenching, this potential (as well as other proposed potentials [27]) produces a range of degenerate disordered structures, and it was suggested that the difference of the configurational entropies of the low-temperature liquid and amorphous phases is in error resulting in a too small value of the heat of fusion for the liquid to amorphous transition.

In an attempt to alleviate these difficulties methods for preparing amorphous silicon, using MD simulations, have been developed [25-28]. In one of the preparation procedures which we have explored [28] the system is first prepared at high temperature and then cooled using an increased three-body potential term (see Eq. 2c), thus enhancing the tendency toward tetrahedral coordination. Once the path which the system traverses in its configurational space has been redirected in this manner, restoring the value of the coefficient (λ) of the three-body potential term to its original SW value ($\lambda = 21$) results in an amorphous material which exhibits structural, dynamical, and thermodynamical properties in agreement with the experimental data. Furthermore, upon fast heating, the so-prepared amorphous silicon melts via a first-order phase transition at a temperature which is 230 K below the crystalline melting temperature in agreement with experimental measurements. When the behavior of more fully relaxed configurations of the amorphous material at several temperatures is analyzed we find that the nature of the transition and the latent heat associated with it maintains, but melting starts at a lower temperature [28].

While the amorphous sample prepared in the above indirect manner can be utilized for studies of the properties of the amorphous phase under various conditions, and in investigations of transformations of the amorphous phase, a principal question remains: do the above documented failure to achieve the amorphous state via <u>direct</u> quenching of the melt [25-28] reflect an inherent defect of the interaction potentials which were employed, or is the difficulty of a kinetic origin, i.e., related to the quench rates used. To resolve this question (which is often raised, in various variants, in the context of MD simulations, and in particular those involving nonequilibrium phenomena) we have performed lengthy simulations in which a SW silicon bulk melt (588 particles) was cooled at a rate smaller than has been attempted in all previous MD studies. These simulations employed a periodically replicated calculational cell possessing full dynamical freedom to change shape and volume [35] under zero external pressure, and the equations of motion were integrated using Gear's 5th order predictor-corrector algorithm with a time-step $\Delta t = 1.5 \times 10^{-3}$ psec. Cooling of the system was achieved by reducing the temperature at selected stages of the simulation over a 2000 Δt interval by 0.005ε ($\varepsilon = 50$ kcal/mole) followed by prolonged relaxations at constant energy.

The time evolution of characteristic quantities of the system are shown in Fig. 1 (the overall time span is 4.5×10^4 tu $= 3.447 \times 10^{-9}$ sec). As seen from Fig. 1 longer periods of relaxation were required as the temperature of the system decreased (to convert to T in K multiply the reduced temperature by 2.5173×10^4). To demonstrate the slow rate by which the system explores the accessible phase-space we show in Fig. 2 the temporal relaxation of system properties for the time span $4.75 \times 10^3 \leq t \leq 2.25 \times 10^4$ tu. A comparison of the total energy and density of the system in these simulations with our previous results obtained via heating at two rates of the amorphous sample prepared via the indirect route is given in Fig. 5. The quality of the amorphous material obtained in these simulations can be assessed via inspection of the pair distributions and density of states (obtained via Fourier transformation of the velocity autocorrelation function) given in Figs. 4 and 5, respectively. Other structural characteristics obtained in these simulations compare favorably with our previous results [28] and are in accord with experimental data. The average coordination at 300 K and 800 K are 4.21 and 4.31, respectively, compared to 4.14 and 4.26 for the a' (indirect preparation [28]) sample; the average bonding angles and their standard deviations for the two temperatures are 107.8 ± 16.0 ° and $107.1 \pm 18.3°$, as compared to $108.4 \pm 14.2°$ and $107.3 \pm 18.0°$ for the a' sample [37]; the average bonding angles corresponding to 4-fold coordinated sites at the two temperatures are $109.0 \pm 10.3°$ and $108.9 \pm 11.1°$ and $108.8 \pm 12.0°$ for the a' sample, and those for 5-fold coordinated sites are 105.2 ± 23.8 and $104.8 \pm 24.7°$ for the a' system; the density of the system is ~3% higher than that of the a' sample. In addition, analysis of the correlation between the dihedral angles of neighboring bonds indicates the lack of medium-range order in the material.

These results demonstrate that the origin of previous failures to obtain amorphous silicon via direct cooling of the melt, using the SW potentials, is of kinetic origin. This conclusion may be extended as a general cautionary remark concerning investigations in the area of nonequilibrium phenomena involving phase transformations and "ill condensed phases" (i.e., amorphous and glassy states).

Solid-Melt Interfaces of Silicon and Liquid-Phase Epitaxial (LPE) Growth

Understanding the equilibrium properties of interfaces and of the non-equilibrium mechanisms, kinetics, and dynamics of crystal-growth processes and identification of the microscopic material properties and of the macroscopic control parameters related to the method of growth, which govern the growth processes and quality of the grown crystals, are of fundamental importance from both basic and applied perspectives.

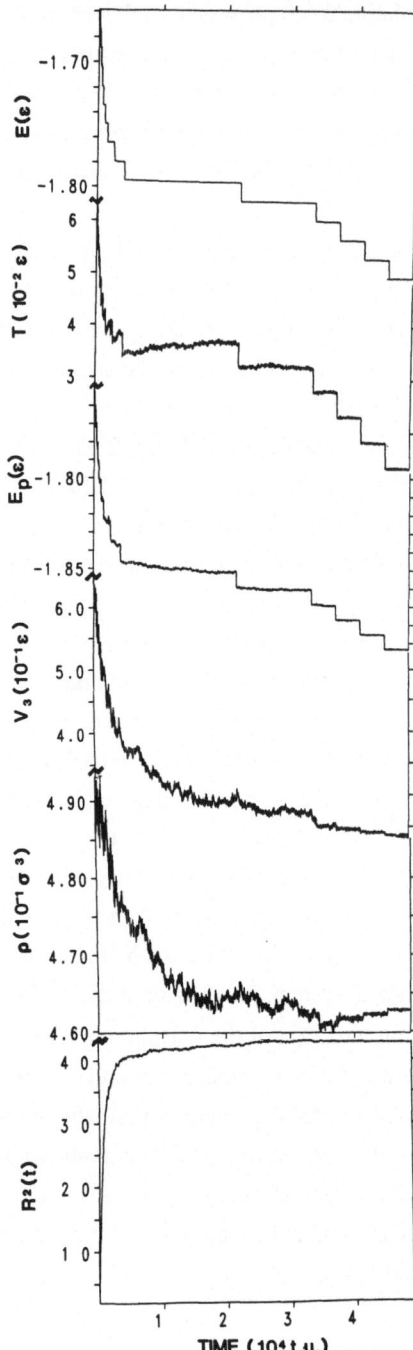

Figure 1. Total energy (E), temperature (T), potential energy (E_p), 3-body potential contribution (V_3), density (ρ) and measure of particle mobilities, $R^2(t) = N^{-1} \sum_{i=1}^{N} [\vec{r}_i(t) - \vec{r}_i(o)]^2$. E = in units of 50 kcal/mole, density in units of σ^{-3} ($\sigma = 2.0951$Å R^2 in units of of t.u. = 7.66 x 10^{-14} sec.

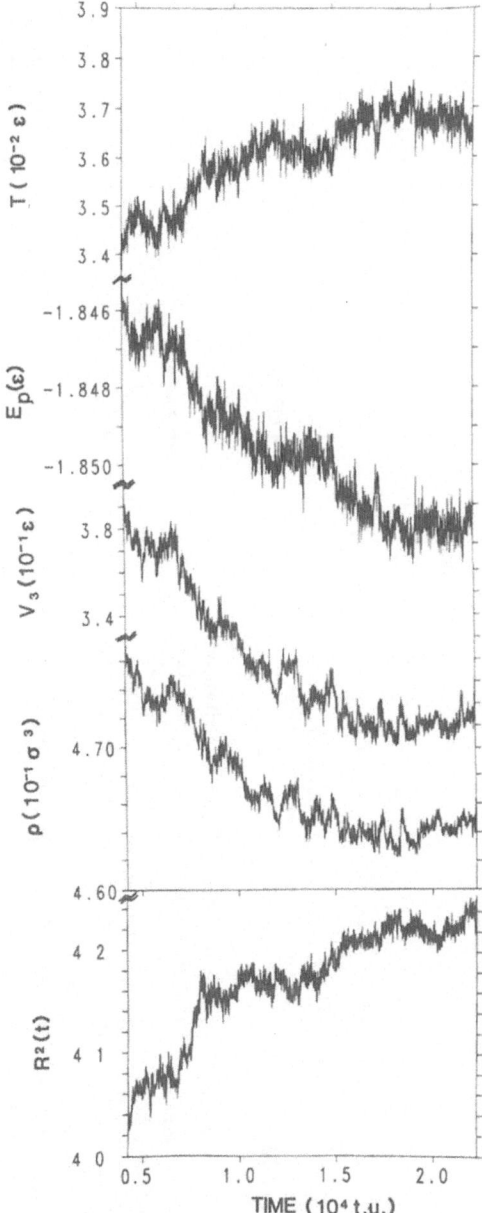

Figure 2. Same as Fig. 1 for the time interval
$4.75 \times 10^3 \leq t \leq 2.25 \ 10^4$ t.u., demonstrating
the slow relaxation kinetics.

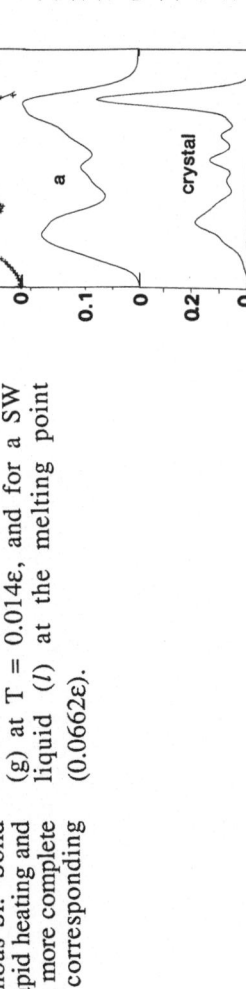

Figure 5. Density of states, D(ω), for room temperature SW crystal, indirect amorphous (a'), glass at T ≃ 0.014ε and SW liquid at the melting point. Frequency in units of inverse SW time unit (t.u.). D(ω) is normalized such that the integral over ω is unity. The experimental D(ω) [36] is also shown.

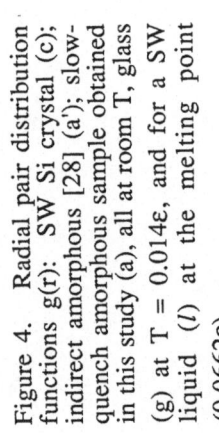

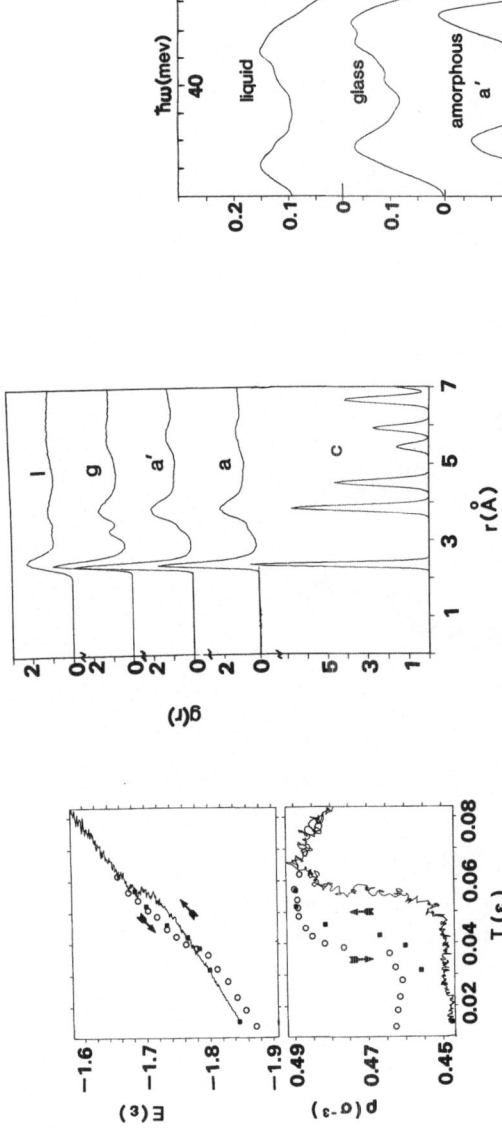

Figure 4. Radial pair distribution functions g(r): SW Si crystal (c); indirect amorphous [28] (a'); slow-quench amorphous sample obtained in this study (a), all at room T, glass (g) at T = 0.014ε, and for a SW liquid (l) at the melting point (0.0662ε).

Figure 3. Total energy per particle, E, and density, ρ, of the slowly cooled melt (empty circles) and those obtained by heating of the indirectly prepared [28] amorphous Si. Solid lines correspond to rapid heating and the solid squares to a more complete relaxation at the corresponding temperatures.

Ample evidence exists that surfaces of crystalline silicon become faceted upon melting and the solid-melt interface of growing Si establishes itself on the (111) crystal plane [38]. Motivated by these observations comprehensive MD studies of the solid-melt interfaces of Si [32,33] and simulations of liquid-phase epitaxial growth of Si [34] have been performed, employing the SW potentials. It was discovered that the nature and morphology of the interphase-interface exhibits crystalline face anisotropy [32,33]. Thus while the Si (111) equilibrium solid-melt interface was found to be atomically smooth, the Si (100) interface is structured exhibiting facets established along (111) directions [32], in qualitative agree with experiment [32]. Furthermore, it was found that faceting initiates upon melting and further refines upon achieving equilibrium at a temperature of 1665K (compared to the experimental value of 1683K).

Liquid-phase epitaxial growth [34] was initiated by driving the crystal-melt systems out of equilibrium via allowing heat conduction to the underlying substrate. It was found that for both interfaces the system initially supercools by about 150K (in agreement with some experimental estimates), followed by ordering and crystallization processes. On the (100) surface the growth processes yielded a perfect crystal at a growth velocity of ~18m/sec while growth on the (111) surface, driven by the same rate of heat removal, resulted in an imperfect crystal containing an assortment of defects (stacking faults and disordered region), at a growth velocity of ~14m/sec. A slower growth rate of the Si (111) system results in a grown crystal possessing a markedly higher degree or crystalline perfection (stacking faults and a very narrow region of disorder limited to the crystal-vacuum interface). The crystallization rates and their orientational dependences and the crystal face dependence of the maximum growth velocities for the formation of defect-free crystals found in the simulations are in agreement with those deduced from experimental measurements [34]. Furthermore, the growth mode found in the simulations emphasizes the dynamical nature of the two interacting phases (solid and melt), the kinetics of ordering processes in the melt in the vicinity of the moving interface [34,38] and the interface morphology. While the (111) microfaceting remains during growth on Si (100), which proceeds in an almost continuous, monotonic manner, growth on Si (111) proceeds in a layer-wise manner, with periods of crystallization interrupted by stages of no apparent movement of the interface. Close inspection reveals that during these stages dynamical self-annealing of defects occurs. As the defective region achieves a high level of structural perfection crystallization resumes [34].

Amorphous Film Growth Via Molecular Beam Epitaxy (MBE)

Thin film fabrication via direct condensation of vapor atoms onto a substrate at temperatures below the melting point has been used for over 100 years [40]. More recently, low-energy beam deposition (generally below keV) methods have emerged as important techniques for growth of electronic materials of specified properties [41]. Full atomistic simulations of these processes [43-45] are difficult due to system size and time-span considerations.

Illustrations of results obtained via MD simulations [45] of the growth of amorphous thin films of silicon, using the SW potentials, are shown in Figs. 6 and 7. In previous studies the structural, dynamical and thermodynamical properties of amorphous silicon prepared via cooling of bulk molten silicon have been investigated (see beginning of this section) yielding results in agreement with experimental data. In the current study we focus on growth of amorphous silicon via low energy beam deposition. In Fig. 6 results for deposition of silicon atoms onto a cold (room temperature) substrate are shown. In these simulations the (111) crystalline substrate contained several layers with 49 dynamic atoms per layer (with periodic boundary conditions in the XY directions) and was cooled at the bottom. The directions in the plane of the calculational cell are $[1\bar{1}0]$ and $[10\bar{1}]$ and the outward normal is in the [111] direction (distance is in units of $\sigma = 2.0951$ Å). The atomic beam is normal to the surface and atoms are released at random positions from a planar source above the substrate at a rate of one atom per $500\Delta t$ ($\Delta t = 1.15 \times 10^{-3}$ psec) with translational kinetic energy corresponding to $3k_B T_M/2$, where T_M is the melting point of silicon. Real space trajectories (viewed along the $[1\bar{1}0]$ direction) at a late stage of the deposition process are shown in Fig. 6a. In Fig. 6b the temperature gradient in the system at this stage (plotted versus distance along the (111) direction) is shown. The density and 4-fold coordination, C_4 and 3-body potential energy, V_3, profiles shown in Figs. 6c-6e, respectively, illustrate that in the interface between the substrate crystal (the first 3 layers from the left) and the grown film, the amorphous material is layered and possesses properties intermediate between the crystal and amorphous phase. Similar results were obtained when the angles of the deposited particles are randomly distributed.

The effect of the angle of deposition on the characteristics of the grown film are shown in Fig. 7 where results are displayed for deposition at the same rate as that shown in Fig. 6 but with an angle of incidence of 60° from the normal (with the plane of incidence defined by the [111] and $[1\bar{2}1]$ directions). In these simulations 96 dynamic atoms per substrate layer were used. A view of the system from the top (along the [111] direction) shown in Fig. 7b reveals regions of smaller density. The columnar microstructure of the grown film is evident from Fig. 7c where a slice through the system containing the upper part of the plane (Fig. 7b) is shown. (Note that the calculational cell was doubled for visual impression in the $[1\bar{2}1]$ direction. The periodicity of the columns is simply a consequence of the periodic boundary conditions). Analysis shows that the orientation of the columns and that of the incident beam obey approximately the tangent rule [41] (i.e., 2 $\tan\theta_{column} = \tan\theta_{beam}$). The relevance and importance of these studies relates the remarkable consequences of the microstructure on the film chemical and physical characteristics such as oxygen uptake, reactivity, optical, electrical and mechanical properties [41].

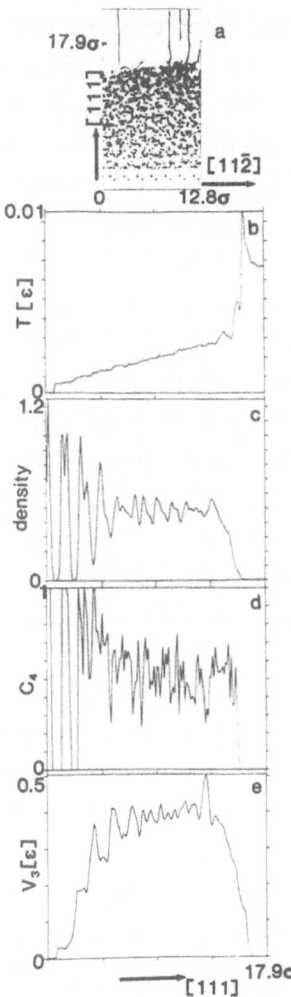

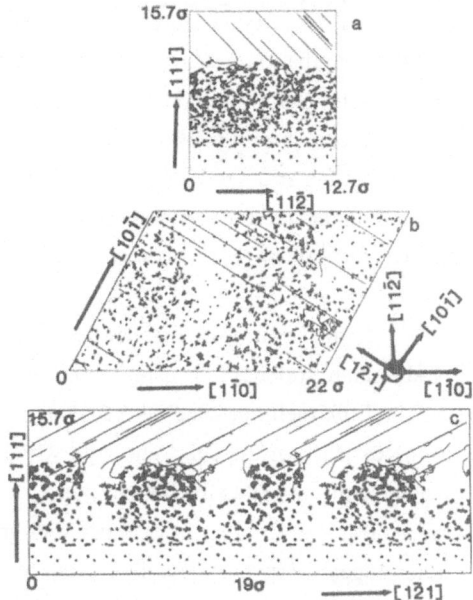

Figure 7. Molecular beam epitaxial growth of Si at 60° incidence; (a)real space trajectories for the whole system. (b) A view from above along the [111] direction. (c) A slice through the system including particles in the upper half of the plane shown in (b) and going through the system in the [111] direction, demonstrating the columnar structure of the film (the calculational cell was doubled in the [1̄21] for visual impression.

Figure 6. Molecular beam epitaxial growth of Si at normal incidence (a) real-space particle trajectories (the bottom-most 3 layers belong to the crystalline (111) Si substrate). (b) Temperature profile of the system, (c-e) density, 4-fold coordination (C_4) and 3-body potential energy, V_3, profiles exhibiting characteristics of amorphous Si, and a layered region at the substrate-amorphous-film interface.

Metallic Systems

Simulations of metallic systems require special consideration of the nature of interactions and cohesion in these systems. These considerations led to the development of a new method of simulations [15] which allows for volume and shape variations and incorporates explicitly the dependence on the density of the "volume energy" and the effective pair potentials (see Eq. 3). Application of this method to studies of a liquid metal (Mg) yielded good agreement with experiments for several properties, including internal energy, density, structure factors, the adiabatic bulk modulus and trends in the diffusion, electrical resistivity and thermopower at several temperatures and pressures [15]. In addition, simulations of the formation and properties of a metallic glass ($Ca_{67}Mg_{33}$) [16], i.e., the solid phase formed by ultra rapid cooling of the liquid alloy, yielded results [16] in good agreement with neutron scattering data for the static and dynamic structure factors, thus providing structural information and revealing the nature of elementary excitations in disordered media. Furthermore, we found that the accessible local minima involve a large displacement of single atoms accompanied by local structural relaxation of the environment [16,1].

In this section we discuss the dynamics and energetics of disordering and melting of aluminum surfaces [19] using MD simulations in conjunction with the EMT [17].

Bulk melting, which is a first-order transition, involves growth of the liquid phase at the expense of the solid one with which it co-exists. Among the possible nucleation mechanisms and sites for the melting process, which include crystalline defects (such as vacancies, interstitials, dislocations and grain boundaries), impurities, and surfaces, the latter appear to be the most likely, due to the weakening of the binding of surface atoms caused by missing neighbors [46]. (In this context we caution against the often occurring confusion between "surface melting" and "surface roughening"). Experimentally, it has been found that disordering of the outermost layers of (110) surfaces of Pb [47] and Al [48] begins at temperatures as low as 150K below the bulk melting temperature, T_M, with a gradual thickening of the melted region as T_M is approached. In contrast, melting of the close packed (111) surfaces shows only weak disordering and that at a temperature close to T_M.

In Fig. 8, we exhibit for Al (110) the calculated layer by layer absolute square of the structure factor, $S_l(\vec{k},T)$, where l denotes the layer between z_{l-1} and z_l averaged over particle positions generated in a constant temperature simulation. The wave vector $\vec{k}$ was chosen as the first allowed Bragg peak $\vec{k} = (2,0,0)2\pi/d$, where d is the nearest neighbor Al - Al distance. Following a slow, monotonic decrease with temperature due to incoherence caused by lattice vibrations, sharp drops, indicating complete loss of order, occur. The incipient loss of order at T ~ T_M - 150K (where the bulk melting temperature T_M is determined to be at T_M ~ 800K) is associated with the formation of a liquid (or quasi-liquid) region of a thickness of ~3 atomic layers, whose thickness grows gradually as the temperature is increased. Furthermore, the calculated total scattering intensity of low-energy electrons (LEED) as a function of temperature is found to be in good agreement with measured LEED intensities [48]. We have also found that while the (100) surface shows pronounced disordering, the periodic crystalline layer structure on the (111) face is preserved up to temperatures close to the bulk melting point (see Fig. 9). Coupled withdetailed analysis of the atomic trajectories we find that the surface premelting process is triggered by the formation of vacancies at the surface. This provides a clue for understanding the mechanism of the initiation of melting and the observed crystal face anisotropy. Indeed, calculations using the EMT yield that the energy required for the formation of a Frenkel pair on the (110) surface is only 0.3 eV while on the (111) surface the energy cost is 1.3 eV and in the bulk it is even larger (4.2 eV).

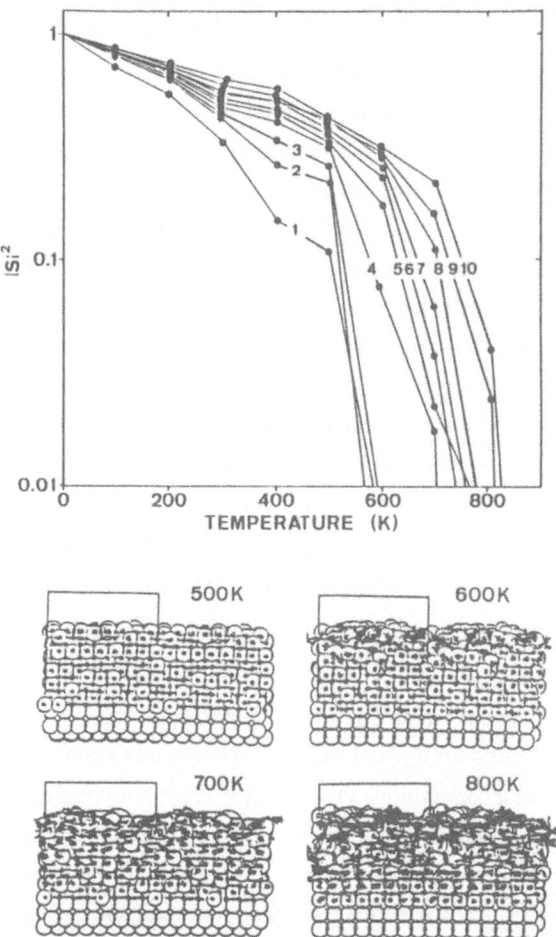

Figure 8.   Structure factors for layers 1-10 (1 being the topmost surface layer) of A*l* (110) versus temperature. The sudden decrease between 500 and 800K is an indication of premelting. Bottom: real space particle trajectories at the indicated temperatures.

Tip-Substrate Interaction in Atomic Force Microscopy

The developments of scanning tunneling microscopy [49] (STM) and of the related atomic force microscopy (AFM) [50] revolutionized our perspectives and abilities to probe the morphological and electronic structure and the nature of interatomic forces in materials, as well as opened new avenues [51] for microscopic investigations and manipulations of technological systems and phenomena such as tribology [52], lithography and in biochemical applications. In both techniques a sharp tip is brought close to the surface and either the tunneling current (STM) or deflection of the cantilever holding the tip (AFM) are monitored as the surface is scanned. Of particular interest for the development of these techniques are questions related to the dynamical response of the substrate and tip which due to their proximity may result in temporary or permanent modifications of the local properties, and be reflected in the recorded data.

To investigate the consequences of the dynamical interactions between the tip and the substrate we have performed MD simulations, employing the SW potentials for silicon [11]. Since both the substrate and tip consist of the same material, these simulations correspond to the case of a reactive tip-substrate system. Our simulations [53], in both the constant-tip-height and constant-force scan modes, reveal that the <u>local</u> structure of the surface can be stressed and modified as a consequence of the tip-substrate dynamical interaction, even at tip-substrate separations which correspond to weak interaction. For large separations these perturbations anneal upon advancement of the tip while permanent damage can occur for smaller separations. For the material that we simulated (Si), we do not find long-range elastic deformations, which may occur in other circumstances [54] depending upon the elastic properties of the material and the nature of interactions.

We illustrate the method via results for constant-force simulations employing a tip consisting of 102 atoms, arranged in 4 layers and exposing a 16 atom (111) facet. The dynamic tip is supported by a static Si holder consisting of two Si (111) layers and scans in the [1$\bar{2}$1] direction over a 6 layer dynamical Si (111) substrate with 100 atoms/layer. In the constant-force simulations in addition to the particle equations of motion the center of mass of the tip-holder assembly, Z, is required to obey $M\ddot{Z} = (\vec{F}(t) - \vec{F}_{ext}) \cdot \hat{Z} - \gamma\dot{Z}$ where $\vec{F}$ is the total force exerted by the tip atoms on the static holder at time t, which corresponds to the force acting on the tip atoms due to their interaction with the substrate, $\vec{F}_{ext}$ is the desired (prescribed) force for a given scan, γ is a damping factor and M is the mass of the tip and holder. In these simulations the system is brought to equilibrium for a prescribed for a value of F_{ext}, and lateral scans proceed in a manner which maintains fully relaxed tip and substrate configurations throughout the scan, while the height of the tip-holder assembly adjusts dynamically according to the above feed-back mechanism.

In Fig. 10 we show results for a scan, for a constant force value $F_{z,ext} = -13.0$ (i.e., 2.15×10^{-8} N). Side views of the system trajectories at the beginning and end stages of the scan are shown in Fig. (10a,b) and (10c), respectively. As seen the tip-substrate interactions induce local modifications of the substrate and tip structure, which are transient (compare the surface structure under the tip at the beginning of the scan (Fig. 10a), exhibiting outward atomic displacements of the top layer atoms, to that at the end of the scan (Fig. 10c), where that region relaxed to the unperturbed configuration). The recorded force on the tip-holder along the scan direction (X) is shown in Fig. (10d), exhibiting a periodic modulation, portraying the periodicity of the substrate. Most significant is the stick-slip behavior signified by the asymmetry in F_x (observed also in the real-space atomic trajectories in Figs. 10(a,b)). Here, the tip atoms closest to the substrate attempt to remain in a favorable bonding environment as the tip-holder assembly proceeds to scan. When the forces on these atoms due to the other tip-atoms exceed the forces from the substrate, they move rapidly by breaking their current bonds to the surface and forming new bonds in a

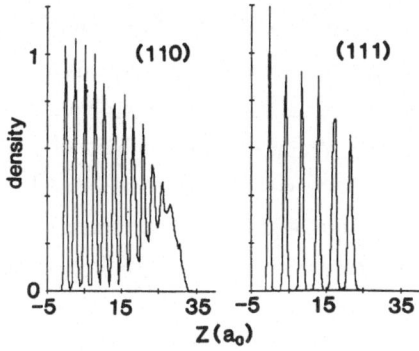

Figure 9. Comparison of the atomic density profiles along the normal to the surface at 600 K for Al (110), left, and Al (111), right, showing disordering at the surface for the former while the latter remains ordered.

region translated by one unit cell along the scan direction. We note that our constant-force simulation method corresponds to the experiments in Ref. 52 in the limit of a stiff wire (lever) and that the stick-slip phenomena which we observed are a direct consequence of the interplay between the surface forces and interactions among the tip atoms. The F_x force which we record corresponds to the frictional force. From the extrema in F_x (Fig. 10d)

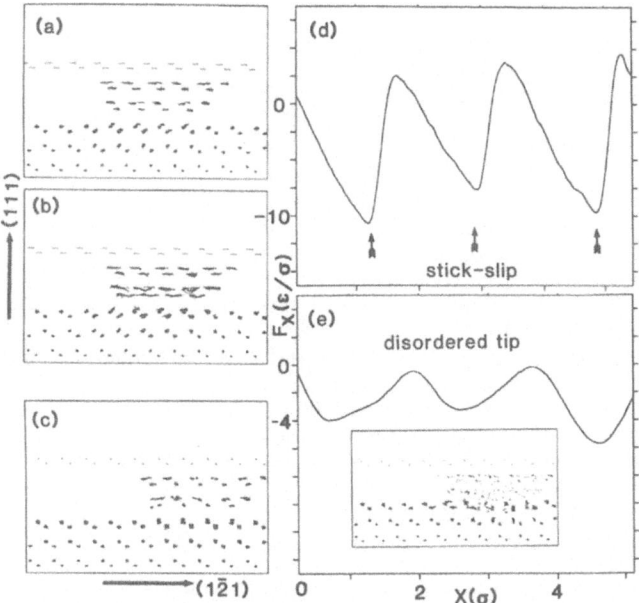

Figure 10. (a-c) Particle trajectories in a constant-force simulation viewed along the [1$\bar{0}$1] direction just before (a) and after (b) a stick-slip event and towards the end of the scan (c), for a large, initially ordered, dynamic tip. (d) The recorded Fx, exhibiting stick-slip behavior. (e) The F_x force in a constant-force scan ($F_{z,ext} = 1.0$) employing a glassy static tip, exhibiting the periodicity of the substrate. Shown in the inset are the real-space trajectories towards the end of the scan, demonstrating tip-induced substrate local modifications.

and the load ($F_{z,ext}$) used, we obtain a coefficient of friction $\mu = |F_x| / |F_{z,ext}| = 0.77,$ in the range of typical values obtained from tribological measurements in vacuum.

Finally, we show in Fig. 10e the frictional force obtained in simulations employing a disordered static 102-atom tip, prepared by quenching of a molten droplet, scanning under a load $F_{z,ext} = 1.0$. The significance of this result lies in the periodic variation of the force reflecting the atomic structure of the substrate. This demonstrates that microscopic investigations of structural characteristics and tribological properties of crystalline substrates are not limited to ordered tips.

Acknowledgments

Fruitful collaboration with R.N. Barnett, C.L. Cleveland, W.L. Ribarsky, J.K. Norskov and P. Stoltze is gratefully acknowledged. These studies were supported by the U.S. DOE grant number FG05-86ER-45234 and the work on AFM in part by the Hughes Aircraft Company, DARPA-Hughes Tribological Fundamentals Program and the NSF Tribology Program. The simulations were performed on the CRAY-XMP at NMFECC, Livermore, California. The assistance of A. Ralston and V. Mallette in the preparation of the manuscript is gratefully acknowledged.

References

1. U. Landman et al., Mat. Res. Soc. Symp. Proc. 63:273 (1985).
2. F.F. Abraham, Adv. Phys. 35:1 (1986); J. Vac. Sci. Technol. B2:534 (1984).
3. MRS Bull. Volume XIII (2), February 1988, p. 14-39.
4. U. Landman, R.N. Barnett, C.L. Cleveland, J. Luo, D. Scharf and J. Jortner, in: "Few-Body Systems and Multiparticle Dynamics", ed., D.A. Micha, AIP Conf. Proc. 162 (AIP, New York, 1987), p. 200.
5. D.W. Heerman, in: "Computer Simulation Methods" (Springer, Berlin, 1986).
6. "Computer Simulations of Solids", eds. C.R.A. Catlow and W.C. Machord (Springer, Berlin, 1982).
7. U. Landman, in: "Computer Simulation Studies in Condensed Matter Physics: Recent Developments", eds., D.P. Landau, K.K. Mon and H.-B. Schuttler (Springer, Berlin, 1988), p.108.
8. G.C. Maitland, M. Rigby, E.B. Smith and W.A. Wakeham, in: "Intermolecular Forces" (Clarendon, Oxford, 1981).
9. M.J. Sangster and M. Dixson, Adv. Phys. 25: 247 (1976).
10. P.N. Keating, Phys. Rev. 145:637 (1966).
11. F.H. Stillinger and T.A. Weber, Phys. Rev. B31:5262 (1985).
12. R. Biswas and D.R. Hamann, Phys. Rev. Lett. 55:2001 (1985).
13. T. Tersoff, Phys. Rev. B37:6991 (1988).
14. See, e.g., W.A. Harrison, Pseudopotentials in the Theory of Metals (Benajamin, Reading, Mass., 1966).
15. R.N. Barnett, C.L. Cleveland and U. Landman, Phys. Rev. Lett. 54:1679 (1985).
16. R.N. Barnett, C.L. Cleveland and U. Landman, Phys. Rev. Lett. 55:2035 (1985).
17. K.W. Jacobsen, J.K. Norskov and M.J. Puska, Phys. Rev. B 35:7423 (1987).
18. See M. Basklas, M. Daw, B. Dodson and S. Foils in ref. 3, p. 28.
19. P. Stoltze, J.K. Norskov and U. Landman, Phys. Rev. Lett. 61:440 (1988).
20. D.E. Polk, J. Non-Cryst. Solids 5:365 (1971).
21. D. Henderson and F. Herman, J. Non-Cryst. Solids 8-10:359 (1972).
22. D. Henderson, J. Non-Cryst. Solids 16:317 (1974).
23. L. Guttman, Phys. Rev. B 23:1866 (1981).

24. F. Wooten, K. Winer, and D. Weaire, Phys. Rev. Lett. 54:1392 (1985).
25. J.Q. Broughton and X.P. Li, Phys. Rev. B 35:9120 (1987).
26. M.D. Kluge, J.R. Ray, and A. Rahman, Phys. Rev. B 36:4234 (1987).
27. R. Biswas, G.S. Grest and C.M. Soukoulis, Phys. Rev. B 36:7437 (1987).
28. W.D. Luedtke and U. Landman, Phys. Rev. B 37:4656 (1988).
29. K. Ding and H.C. Andersen, Phys. Rev. B 34:6987 (1986).
30. F.F. Abraham and I.P. Batra, Surf. Sci. 163:L752 (1985).
31. I.P. Batra, F.F. Abraham and S. Ciraci, Phys. Rev. B 35:9552 (1987).
32. U. Landman, W.D. Luedtke, R.N. Barnett, C.L. Cleveland, M.W. Ribarsky,
    E. Arnold, S. Ramesh, H. Baumgart, A. Martinez, and B. Khan,
    Phys. Rev. Lett. 56:155 (1986).
33. F.F. Abraham and J.Q. Broughton, Phys. Rev. Lett. 56:734 (1986).
34. U. Landman, W.D. Luedtke, M.W. Ribarsky, R.N. Barnett, and C.L. Cleveland
    Phys. Rev. B 37:4637,4647 (1988).
35. M. Parrinello and A. Rahman, J. Appl. Phys. 52:7182 (1981); see also
    M.W. Ribarsky and U. Landman, Phys. Rev. B 38:9522 (1988).
36. W.A. Kamitakahara, C.M. Soukoulis, H.R. Shanks, U. Buchenau and
    G.S. Grest, Phys. Rev. B 36:6539 (1987).
37. In calculating the coordination numbers and average angles, cutoff distances
    (first minimum in g(r), see Fig. 4) of 2.87Å at room temperature and
    2.93 Å at the higher temperature were used. The a' system [28] is
    better relaxed at the higher temperature.
38. See citations in reference 32.
39. See also earlier MD studies of liquid-phase epitaxy: U. Landman, R.N. Barnett,
    C.L. Cleveland and R.H. Rast, J. Vac. Sci. Technol. A3:1574 (1985);
    U. Landman, C.L. Cleveland and C.S. Brown, Phys. Rev. Lett. 45:2032
    (1980) and in: "Nonlinear Phenomena of Phase Transitions and In-
    stabilities", ed. T. Riste (Plenum, NY, 1982), p. 379.
40. M. Faraday, Phil. Trans. 147:145 (1857).
41. For a recent review see H.J. Leamy, G.H. Gilmer, and A.G. Dirks, in "Current
    Topics in Materials Science", ed. E. Kaldis (North-Holland, Amsterdam,
    1980), Vol. 6, Chap. 4.
42. B.R. Appelton, R.A. Zuhr, T.S. Noggle, N. Herbots and S.J. Pennycook in
    "Beam-Solid Interactions and Transient Processes", eds., M.O. Thompson,
    S.T. Picraux and J.S. Williams (MRS. Symp. Proc. 74, Pittsburgh, PA,
    1987), p. 45.
43. M. Schneider, I .K. Schuller and A. Rahman, Phys. Rev. B36:1340 (198.
44. P.A. Taylor and B.W. Dodson, Phys. Rev. B36:1355 (1987); B.W. Dodson,
    Phys. Rev. B36:1068 (1987).
45. W.D. Luedtke and U. Landman, Phys. Rev. B (to be published, 1988).
46. See review by J.F. Van der Veen, B. pluis and A.W. Denier van der Gon in
    "Chemistry and Physics of Solids", Vol. VII (Springer, Berlin, 1988).
47. J.W.M. Frenken and J.F. van der Veen, Phys. Rev. Lett. 54:134 (1985);
    B. Pluris, A.W. van der Gon, J.W.M. Frenken and J.F. van der Veen,
    Phys. Rev. Lett. (1987).
48. P. von Blackenhagen, W. Schommer and V. Voegel, J. Vac. Sci. Technol. A5:
    649 (1987).
49. G. Binnig and H. Rohrer, IBM J. Res. Develop. 30:355 (1986).
50. G. Binnig, C.F. Quate and Ch. Gerber, Phys. Rev. Lett. 56:930 (1986).
51. P.H. Hansma and J. Tersoff, J. Appl. Phys. 61:R1 (1986).
52. C.M. Mate, G.M. McClelland, R. Erlandsson and S. Chiang, Phys. Rev.
    Lett. 59:1942 (1987).
53. U. Landman, W.D. Luedtke and A. Nitzan (Surface Sci., in press); see also
    ref. 7 and U. Landman, W.D. Luedtke and M.W. Ribarsky,
    J Vac. Sci. Technol., (to be published).
54. J.M. Soler, A.M. Baro, N. Garcia and H. Rohrer, Phys. Rev. Lett. 57:444
    (1986); see comment by J.B. Pethica, ibid 57:3235 (1986).

ON SOME SPECTACULAR SURFACE SEGREGATION BEHAVIORS IN CuNi AND PtNi ALLOYS ANALYZED WITHIN THE TIGHT-BINDING ISING MODEL

B. Legrand

S.R.M.P.
C.E.N. Saclay
91191 Gif sur Yvette Cedex
France

G. Tréglia

Laboratoire de Physique
des Solides
Bât. 510
91405 Orsay Cedex, France

F. Ducastelle

ONERA
B.P. 72
92322, Châtillon Cedex, France

INTRODUCTION

In an alloy, the concentration near the surface can be different from the one in the bulk. This is the so-called surface segregation phenomenon [1], which is usually modelled on the basis of a description of the total energy as a sum of pair interactions. Unfortunately, it is well known that the cohesive energy of transition metals and alloys does not reduce to such a sum [2]. We will show here than, when interested in surface segregation processes, one can derive from the electronic structure of the disordered alloy an effective Ising hamiltonian containing a linear term (which proves to be very close to the difference in surface tensions between the constituents) and a quadratic one (which involves <u>effective</u> pair interactions). This is the recently developed Tight-Binding Ising Model (TBIM) [3].

THE T.B.I.M.

Let us start from the grand-potential $\Omega(\zeta)$ of a semi-infinite alloy A_cB_{1-c} for a given configuration ζ, i.e. a set of occupation numbers p_n^i ($p_n^i = 1$ (0) if the site n is (not) occupied by an atom of type i):

$$\Omega(\zeta) = \int^{E_F(\zeta)} E\, n\,(\zeta,E)\, dE - E_F\,(\zeta)\, N\,(\zeta,E_F) = - \int^{E_F(\zeta)} N\,(\zeta,E)\, dE \qquad (1)$$

where $E_F(\zeta)$ is the Fermi level, $n(\zeta,E)$ the electronic density of states and $N(\zeta,E) = \int^E n(\zeta,E')\,dE'$ is the integrated density of states which can be rewritten in terms of the usual Green functions $G(\zeta,E) = (E - H(\zeta))^{-1}$ as:

$$N(\zeta,E) = \frac{Im}{\pi}\,Tr\,Log\,G\,(\zeta,E) \tag{2}$$

Tr denotes the trace over all the atomic states |n> involved in the tight-binding hamiltonian:

$$H(\zeta) = \sum_{n,i} p_n^i\,\varepsilon^i\,|n><n| + \sum_{n,m}{}' \beta_{nm}\,|n><m| \tag{3}$$

where ε^i is the d level for i atoms, β_{nm} the hopping integrals and prime means $m \neq n$ (for the sake of clarity we will write here the equations for non degenerate bands, the calculations being performed however for a realistic tenfold degenerate d band). The main idea of TBIM is to separate this hamiltonian into two contributions: $H(\zeta) = \bar{H} + H_d(\zeta)$, the first term $\bar{H}$ being defined in the disordered state (treated in the CPA [4] and characterized by the self-energy σ_n on each site n) and therefore configuration independent, and the second one $H_d(\zeta)$ diagonal and configuration dependent [5]:

$$\bar{H} = \sum_n \sigma_n\,|n><n| + \sum_{n,m}{}' \beta_{nm}\,|n><m| \text{ and } H_d(\zeta) = \sum_{n,i} p_n^i\,(\varepsilon^i - \sigma_n)\,|n><n| \tag{4}$$

Using this decomposition, $\Omega(\zeta)$ can be rewritten [3]:

$$\Omega(\zeta) = \bar{\Omega} + \sum_{n,i} p_n^i\,h_n^i + \frac{1}{2} \sum_{n,m,i,j}{}' p_n^i\,p_m^j\,V_{nm}^{ij} + ... \tag{5}$$

with

$$h_n^i = \frac{Im}{\pi} \int^{E_F} Log\,[1 - (\varepsilon^i - \sigma_n)\,\bar{G}_{nn}(E)]\,dE$$

and

$$V_{nm}^{ij} = -\frac{Im}{\pi} \int^{E_F} t_n^i(E)\,t_m^j(E)\,\bar{G}_{nm}^2(E)\,dE$$

and $\Omega\,(= -\int^{E_F} \bar{N}(E)\,dE)$ does not depend on the configuration. $\bar{G}$ is the Green function of the disordered state, $\bar{N}$ is obtained from $\bar{G}$ using (2), and t_n^i is the t-matrix:

$$t_n^i(E) = \frac{\varepsilon^i - \sigma_n(E)}{1 - (\varepsilon^i - \sigma_n(E))\,\bar{G}_{nn}(E)} \tag{6}$$

It has been checked [3] that the expansion of $\Omega(\zeta)$ is rapidly convergent and can be limited to the linear and quadratic terms in (5). It is then justified to describe surface segregation in terms of an effective Ising hamiltonian which, for a binary alloy ($p_n^A = 1 - p_n^B = p_n$), is written up to a constant:

$$H^{eff} = \sum_n p_n \{ h_{eff}^n - {\sum_m}' V_{nm} \} + {\sum_{n,m}}' p_n p_m V_{nm} \tag{7}$$

with

$$h_n^{eff} = h_n^A - h_n^B + \frac{1}{2} {\sum_m}' \left(V_{nm}^{AA} - V_{nm}^{BB} \right) \text{ and } V_{nm} = \frac{1}{2} \left(V_{nm}^{AA} + V_{nm}^{BB} - 2V_{nm}^{AB} \right)$$

By analogy with the commonly used phenomenological models, $\Delta h_p^{eff} = (h_n^{eff}{}_{\in bulk} - h_n^{eff}{}_{\in p\text{-plane}})$ plays a role equivalent to the difference in p-layer tensions between pure constituents $\Delta \tau_p = \tau_p^B - \tau_p^A$, the p-layer tension τ_p^i being defined as the difference in energy between an atom in the p-plane parallel to the surface and a bulk atom in the pure metal i. This is fully confirmed by Figure 1 in which we compare Δh_p^{eff} to $\Delta \tau_p$ for p = 0 (surface plane) and p = 1 (first underlayer) for the (110) face.

The calculation has been performed for a homogeneous fcc alloy as a function of the concentration and of the average d-band filling $\bar{N}_d$ for a diagonal disorder parameter $\delta = \frac{(\epsilon^B - \epsilon^A)}{W} = 2$ (W : d bandwidth) using canonical ratios for the hopping integrals [6]; the respective d band fillings of the pure metals have been taken equal to the partial d band fillings in the disordered alloy. The agreement between $\Delta h_0^{eff}(1)$ and $\Delta \tau_{0(1)}$ is almost perfect! In particular, the results are almost concentration independent which completely justifies the success of the empirical criteria based on the surface tension arguments [1]. A similar agreement has been obtained for the other low index faces [3].

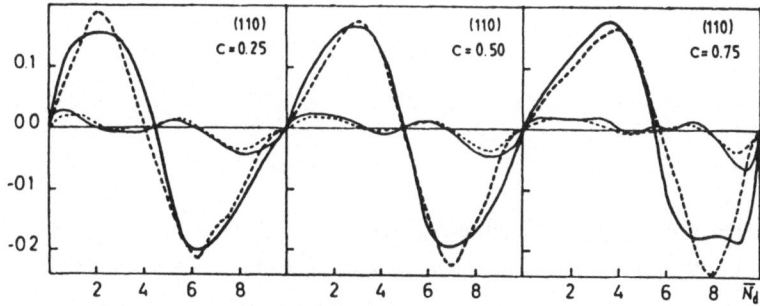

Figure 1.   Variation of Δh_p^{eff} (full lines) and $\Delta \tau_p$ (dotted lines) as a function of the average d band filling $\bar{N}_d$ for p = 0 (surface : thick line) and p = 1 (first underlayer : thin line) in units of W.

Figure 2 illustrates the variation of the effective pair interaction V_{nm} at the (110) surface. One sees that their trends are similar to those encountered in the bulk [6] but that their amplitude are much larger. More precisely one finds for a wide range of d band fillings:

$$V_{ss} \cong V_{sb} \cong 2\, V_{bb} \text{ for the (110) face}$$

$$V_{ss} \cong V_{sb} \cong 1.5\, V_{bb} \text{ for the (111) and (100) faces} \tag{8}$$

This variation may be important in some peculiar cases, as will be illustrated in the following, even though the linear term appears in general as the leading term (compare fig. 1 and fig. 2).

MEAN FIELD APPROXIMATION

Let us now define p-layer concentrations c_p so that $\forall\, n \in$ p-plane, $c_n = <p_n> = c_p$. The experimental temperature for the Pt_cNi_{1-c} being above the critical temperature ($T_c = 910$ K for $c = 0.5$), it is justified to use a mean field approach ($<p_np_m> = <p_n><p_m>$). Then the free energy, F, is obtained by averaging the TBIM hamiltonian (7) and the entropy over all configurations. Its minimization ($\frac{\partial F}{\partial c_p} = 0, \forall\, c_p$) leads to the following system of coupled equations:

$$\frac{c_o}{1-c_o} = \frac{c}{1-c} \exp -\beta\ [\ -\Delta h_o^{eff} + (Z+Z'+Z'')(V-V_o) + (Z'+Z'')V + 2\ \{Z(c_oV_o-cV)$$

$$+ Z'(c_1V_o-2cV) + Z''\ (c_2V_o - 2cV)\} + \Delta H_o^{s.e.}]$$

$$\frac{c_1}{1-c_1} = \frac{c}{1-c} \exp -\beta\ [\ -\Delta h_1^{eff} + (Z'+Z'')(V-V_o) + Z''V + 2\ \{Z(c_1-c)V$$

$$\tag{9}$$

$$+ Z'(c_2V + c_oV_o - 2cV) + Z''\ (c_3 - 2c)V\} + \Delta H_1^{s.e.}\ ]$$

$$\frac{c_p}{1-c_p} = \frac{c}{1-c}\ \exp - 2\beta\ [Z\ (c_p - c) + Z'\ (c_{p+1} + c_{p-1} - 2c) + Z''\ (c_{p+2} + c_{p-2} - 2c)]\ V$$

where V is the effective pair interaction V_{nm} between two first neighbors for sites n and m in the bulk and V_o the corresponding quantity when n or m is at the surface (see (8)). Z, Z' and Z" are the numbers of first neighbors of a site in a given bulk layer parallel to the surface respectively in the same plane, in the first and second plane below or above; $\beta = \frac{1}{kT}$. Finally it is worth noticing that eq. (7) does not account for a possible size effect (s.e.). This effect has been introduced in eqs. (9) through the layer contributions $\Delta H_p^{s.e.}$.

calculated using simultaneously a tight binding atomistic model and a relaxation process to minimize the strain energy [7].

Let us now illustrate the ability of TBIM in the spectacular case of PtNi. Actually this system has given rise to a great deal of debates due to its peculiar behavior that the usual phenomenological theories failed to explain.

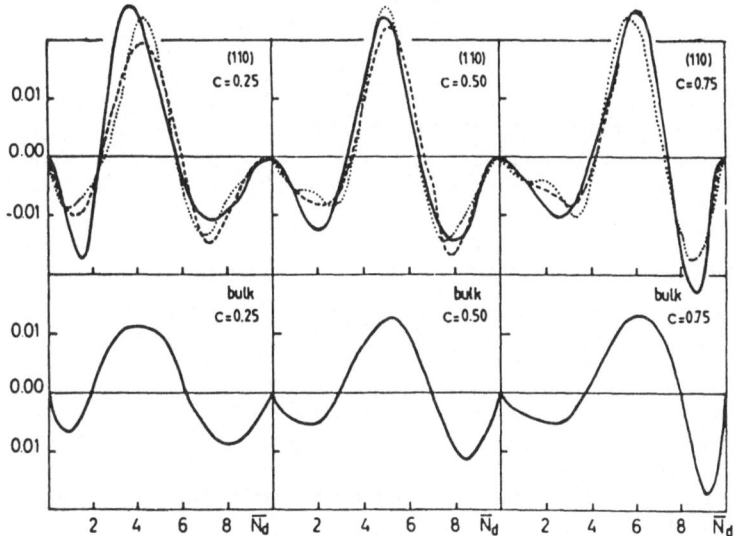

Figure 2. Variation of the first neighbor effective pair interaction V_{nm} as a function of the average d band filling N_d in units of W for n <u>and</u> m in the bulk: V_{bb} (—); n <u>and</u> m in the (110) surface : V_{ss} (—) ; n at the surface and m in the first (----) or second (·····) underlayer: V_{sb}.

APPLICATION TO PtNi

LEED studies of the (111) surface of Pt_cNi_{1-c} at T = 1120 K have exhibited a strong composition oscillation across the outermost layers with Pt on top which has been called surface sandwich segregation [8]. Recently, the same authors observed for $Pt_{0.5}Ni_{0.5}$ (110) a reversal of this concentration profile, namely they found a sandwich with Ni on top [9]. In Fig. 3 we show the variation of the first three layer concentrations as a function of the bulk one calculated from (9) at T = 1120 K for the two crystallographic orientations. The corresponding values of the parameters can be found in [10]. It is important to note that in this case $V_0 = V$ for the (111) face and $V_0 = 2V$ for the (110) one.

The main tendencies which emerge from this figure is the existence of a two layer sandwich with Pt on top for the (111) face on the whole range of concentration and a reversed three layer sandwich with Ni on top for the (110) face for c > 0.5. These unexpected results are in good qualitative agreement with the experimental observations confirming the unique capabilities of TBIM to study surface segregation.

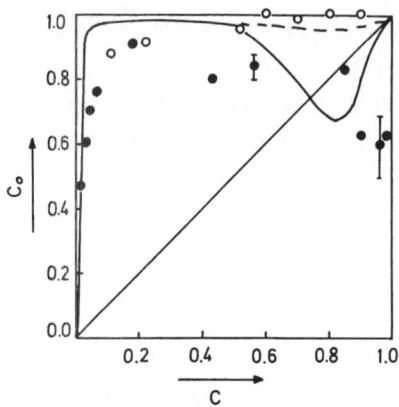

Figure 3. Variation of the Pt concentration at the surface (c_0 : ———), on the first (c_1 : ----) and second (c_2 : ····) underlayers as a function of the Pt bulk concentration : c, calculated at T = 1120 K for the (111) and (110) faces within the TBIM.

REFERENCES

1.   M.J. Kelley and V. Ponec, Progr. in Surf. Sci. 11: 139 (1981).
2.   J. Friedel, "The Physics of Metals", J.M. Ziman ed., (Cambridge University Press) 340, 1969.
3.   G. Tréglia, B. Legrand and F. Ducastelle, Europhys. Lett., 7:575 (1988).
4.   B. Velicky, F. Kirkpatrick and H. Ehrenreich, Phys. Rev. 175:747 (1968).
5.   F. Ducastelle, "Advanced Study Institute on Alloy Phase Stability", Maleme (Grèce), 1987.
6.   A. Bieber, F. Gautier, G. Tréglia and F. Ducastelle, Solid State Comm., 39:149 (1981).
7.   G. Tréglia and B. Legrand, Phys. Rev. B, 35:4338 (1987).
8.   R. Baudoing, Y. Gauthier, M. Lundberg and J. Rundgren, J. Phys. C, 19:2825, (1986).
9.   Y. Gauthier, R. Baudoing, M. Lundberg and J. Rundgren, Phys. Rev. B, 35:7867 (1987).
10.   B. Legrand and G. Tréglia, "The Structure of Surfaces II", J.F. Van der Veen, M.A. Van Hove, eds., (Springer Verlag), 167, 1987.

INDEX